Undergraduate Texts in Mathematics

Undergraduate Texts in Mathematics are generally aimed at third- and fourth-year undergraduate mathematics students at North American universities. These texts strive to provide students and teachers with new perspectives and novel approaches. The books include motivation that guides the reader to an appreciation of interrelations among different aspects of the subject. They feature examples that illustrate key concepts as well as exercises that strengthen understanding.

More information about this series at http://www.springer.com/series/666

Undergraduate Texts in Mathematics

Undergraduate Texts in Mathematics are generally aimed at third- and fourth-year undergraduate mathematics students at North American universities. These texts strive to provide students and teachers with new perspectives and novel approaches. The books include motivation that guides the reader to an appreciation of interrelations among different aspects of the subject. They feature examples that illustrate key concepts as well as exercises that strengthen understanding.

More information about this series at http://www.springer.com/series/666

Jeffrey Hoffstein • Jill Pipher
Joseph H. Silverman

An Introduction to Mathematical Cryptography

Second Edition

Springer

Jeffrey Hoffstein
Department of Mathematics
Brown University
Providence, RI, USA

Jill Pipher
Department of Mathematics
Brown University
Providence, RI, USA

Joseph H. Silverman
Department of Mathematics
Brown University
Providence, RI, USA

ISSN 0172-6056 ISSN 2197-5604 (electronic)
ISBN 978-1-4939-3938-1 ISBN 978-1-4939-1711-2 (eBook)
DOI 10.1007/978-1-4939-1711-2
Springer New York Heidelberg Dordrecht London

Softcover reprint of the hardcover 2nd edition 2014

Printed on acid-free paper

Springer is part of Springer Science+Business Media (www.springer.com)

Preface

The creation of public key cryptography by Diffie and Hellman in 1976 and the subsequent invention of the RSA public key cryptosystem by Rivest, Shamir, and Adleman in 1978 are watershed events in the long history of secret communications. It is hard to overestimate the importance of public key cryptosystems and their associated digital signature schemes in the modern world of computers and the Internet. This book provides an introduction to the theory of public key cryptography and to the mathematical ideas underlying that theory.

Public key cryptography draws on many areas of mathematics, including number theory, abstract algebra, probability, and information theory. Each of these topics is introduced and developed in sufficient detail so that this book provides a self-contained course for the beginning student. The only prerequisite is a first course in linear algebra. On the other hand, students with stronger mathematical backgrounds can move directly to cryptographic applications and still have time for advanced topics such as elliptic curve pairings and lattice-reduction algorithms.

Among the many facets of modern cryptography, this book chooses to concentrate primarily on public key cryptosystems and digital signature schemes. This allows for an in-depth development of the necessary mathematics required for both the construction of these schemes and an analysis of their security. The reader who masters the material in this book will not only be well prepared for further study in cryptography, but will have acquired a real understanding of the underlying mathematical principles on which modern cryptography is based.

Topics covered in this book include Diffie–Hellman key exchange, discrete logarithm based cryptosystems, the RSA cryptosystem, primality testing, factorization algorithms, digital signatures, probability theory, information theory, collision algorithms, elliptic curves, elliptic curve cryptography, pairing-based cryptography, lattices, lattice-based cryptography, and the NTRU cryptosystem. A final chapter very briefly describes some of the many other aspects of modern cryptography (hash functions, pseudorandom number generators, zero-knowledge proofs, digital cash, AES, etc.) and serves to point the reader toward areas for further study.

Electronic Resources: The interested reader will find additional material and a list of errata on the Mathematical Cryptography home page:

`www.math.brown.edu/~jhs/MathCryptoHome.html`

This web page includes many of the numerical exercises in the book, allowing the reader to cut and paste them into other programs, rather than having to retype them.

No book is ever free from error or incapable of being improved. We would be delighted to receive comments, good or bad, and corrections from our readers. You can send mail to us at

`mathcrypto@math.brown.edu`

Acknowledgments: We, the authors, would like the thank the following individuals for test-driving this book and for the many corrections and helpful suggestions that they and their students provided: Liat Berdugo, Alexander Collins, Samuel Dickman, Michael Gartner, Nicholas Howgrave-Graham, Su-Ion Ih, Saeja Kim, Yuji Kosugi, Yesem Kurt, Michelle Manes, Victor Miller, David Singer, William Whyte. In addition, we would like to thank the many students at Brown University who took Math 158 and helped us improve the exposition of this book.

Acknowledgments for the Second Edition: We would like to thank the following individuals for corrections and suggestions that have been incorporated into the second edition: Stefanos Aivazidis, Nicole Andre, John B. Baena, Carlo Beenakker, Robert Bond, Reinier Broker, Campbell Hewett, Rebecca Constantine, Stephen Constantine, Christopher Davis, Maria Fox, Steven Galbraith, Motahhareh Gharahi, David Hartz, Jeremy Huddleston, Calvin Jongsma, Maya Kaczorowski, Yamamoto Kato, Jonathan Katz, Chan-Ho Kim, Ariella Kirsch, Martin M. Lauridsen, Kelly McNeilly, Ryo Masuda, Shahab Mirzadeh, Kenneth Ribet, Jeremy Roach, Hemlal Sahum, Ghassan Sarkis, Frederick Schmitt, Christine Schwartz, Wei Shen, David Singer, Michael Soltys, David Spies, Bruce Stephens, Paulo Tanimoto, Patrick Vogt, Ralph Wernsdorf, Sebastian Welsch, Ralph Wernsdorf, Edward White, Pomona College Math 113 (Spring 2009), University of California at Berkeley Math 116 (Spring 2009, 2010).

Providence, USA

Jeffrey Hoffstein
Jill Pipher
Joseph H. Silverman

Contents

Introduction

Principal Goals of (Public Key) Cryptography
• Allow two people to exchange confidential information, even if they have never met and can communicate only via a channel that is being monitored by an adversary.
• Allow a person to attach a digital signature to a document, so that any other person can verify the validity of the signature, but no one can forge a signature on any other document.

The security of communications and commerce in a digital age relies on the modern incarnation of the ancient art of codes and ciphers. Underlying the birth of modern cryptography is a great deal of fascinating mathematics, some of which has been developed for cryptographic applications, but much of which is taken from the classical mathematical canon. The principal goal of this book is to introduce the reader to a variety of mathematical topics while simultaneously integrating the mathematics into a description of modern public key cryptography.

For thousands of years, all codes and ciphers relied on the assumption that the people attempting to communicate, call them Bob and Alice, share a *secret key* that their adversary, call her Eve, does not possess. Bob uses the secret key to encrypt his message, Alice uses the same secret key to decrypt the message, and poor Eve, not knowing the secret key, is unable to perform the decryption. A disadvantage of these *private key cryptosystems* is that Bob and Alice need to exchange the secret key before they can get started.

During the 1970s, the astounding idea of *public key cryptography* burst upon the scene.[1] In a public key cryptosystem, Alice has two keys, a public encryption key K^{Pub} and a private (secret) decryption key K^{Pri}. Alice publishes her public key K^{Pub}, and then Adam and Bob and Carl and everyone else can use K^{Pub} to encrypt messages and send them to Alice. The idea underlying public key cryptography is that although everyone in the world knows K^{Pub} and can use it to encrypt messages, only Alice, who knows the private key K^{Pri}, is able to decrypt messages.

[1] A brief history of cryptography is given is Sects. 1.6, 2.1, 6.5, and 7.7.

The advantages of a public key cryptosystem are manifold. For example, Bob can send Alice an encrypted message even if they have never previously been in direct contact. But although public key cryptography is a fascinating theoretical concept, it is not at all clear how one might create a public key cryptosystem. It turns out that public key cryptosystems can be based on hard mathematical problems. More precisely, one looks for a mathematical problem that is initially hard to solve, but that becomes easy to solve if one knows some extra piece of information.

Of course, private key cryptosystems have not disappeared. Indeed, they are more important than ever, since they tend to be significantly more efficient than public key cryptosystems. Thus in practice, if Bob wants to send Alice a long message, he first uses a public key cryptosystem to send Alice the key for a private key cryptosystem, and then he uses the private key cryptosystem to encrypt his message. The most efficient modern private key cryptosystems, such as DES and AES, rely for their security on repeated application of various mixing operations that are hard to unmix without the private key. Thus although the subject of private key cryptography is of both theoretical and practical importance, the connection with fundamental underlying mathematical ideas is much less pronounced than it is with public key cryptosystems. For that reason, this book concentrates almost exclusively on public key cryptography, especially public key cryptosystems and digital signatures.

Modern mathematical cryptography draws on many areas of mathematics, including especially number theory, abstract algebra (groups, rings, fields), probability, statistics, and information theory, so the prerequisites for studying the subject can seem formidable. By way of contrast, the prerequisites for reading this book are minimal, because we take the time to introduce each required mathematical topic in sufficient depth as it is needed. Thus this book provides a self-contained treatment of mathematical cryptography for the reader with limited mathematical background. And for those readers who have taken a course in, say, number theory or abstract algebra or probability, we suggest briefly reviewing the relevant sections as they are reached and then moving on directly to the cryptographic applications.

This book is not meant to be a comprehensive source for all things cryptographic. In the first place, as already noted, we concentrate on public key cryptography. But even within this domain, we have chosen to pursue a small selection of topics to a reasonable mathematical depth, rather than providing a more superficial description of a wider range of subjects. We feel that any reader who has mastered the material in this book will not only be well prepared for further study in cryptography, but will have acquired a real understanding of the underlying mathematical principles on which modern cryptography is based.

However, this does not mean that the omitted topics are unimportant. It simply means that there is a limit to the amount of material that can be included in a book (or course) of reasonable length. As in any text, the choice of particular topics reflects the authors' tastes and interests. For the convenience of the reader, the final chapter contains a brief survey of areas for further study.

A Guide to Mathematical Topics: This book includes a significant amount of mathematical material on a variety of topics that are useful in cryptography. The following list is designed to help coordinate the mathematical topics that we cover with subjects that the class or reader may have already studied.

Congruences, primes, and finite fields	— Sects. 1.2, 1.3, 1.4, 1.5, 2.10.4
The Chinese remainder theorem	— Sect. 2.8
Euler's formula	— Sect. 3.1
Primality testing	— Sect. 3.4
Quadratic reciprocity	— Sect. 3.9
Factorization methods	— Sects. 3.5, 3.6, 3.7, 6.6
Discrete logarithms	— Sects. 2.2, 3.8, 5.4, 5.5, 6.3
Group theory	— Sect. 2.5
Rings, polynomials, and quotient rings	— Sects. 2.10 and 7.9
Combinatorics and probability	— Sects. 5.1 and 5.3
Information and complexity theory	— Sects. 5.6 and 5.7
Elliptic curves	— Sects. 6.1, 6.2, 6.7, 6.8
Linear algebra	— Sects. 7.3
Lattices	— Sects. 7.4, 7.5, 7.6, 7.13

Intended Audience and Prerequisites: This book provides a self-contained introduction to public key cryptography and to the underlying mathematics that is required for the subject. It is suitable as a text for advanced undergraduates and beginning graduate students. We provide enough background material so that the book can be used in courses for students with no previous exposure to abstract algebra or number theory. For classes in which the students have a stronger background, the basic mathematical material may be omitted, leaving time for some of the more advanced topics.

The formal prerequisites for this book are few, beyond a facility with high school algebra and, in Chap. 6, analytic geometry. Elementary calculus is used here and there in a minor way, but is not essential, and linear algebra is used in a small way in Chap. 3 and more extensively in Chap. 7. No previous knowledge is assumed for mathematical topics such as number theory, abstract algebra, and probability theory that play a fundamental role in modern cryptography. They are covered in detail as needed.

However, it must be emphasized that this is a mathematics book with its share of formal definitions and theorems and proofs. Thus it is expected that the reader has a certain level of mathematical sophistication. In particular, students who have previously taken a proof-based mathematics course will find the material easier than those without such background. On the other hand, the subject of cryptography is so appealing that this book makes a good text for an introduction-to-proofs course, with the understanding that the instructor will need to cover the material more slowly to allow the students time to become comfortable with proof-based mathematics.

Suggested Syllabus: This book contains considerably more material than can be comfortably covered by beginning students in a one semester course. However, for more advanced students who have already taken courses in number theory and abstract algebra, it should be possible to do most of the remaining material. We suggest covering the majority of the topics in Chaps. 1–4, possibly omitting some of the more technical topics, the optional material on the Vigènere cipher, and the section on ring theory, which is not used until much later in the book. The next three chapters on information theory (Chap. 5), elliptic curves (Chap. 6), and lattices (Chap. 7) are mostly independent of one another, so the instructor has the choice of covering one or two of them in detail or all of them in less depth. We offer the following syllabus as an example of one of the many possibilities. We have indicated that some sections are optional. Covering the optional material leaves less time for the later chapters at the end of the course.

Chapter 1. An Introduction to Cryptography.
Cover all sections.

Chapter 2. Discrete Logarithms and Diffie–Hellman.
Cover Sects. 2.1–2.7. Optionally cover the more mathematically sophisticated Sects. 2.8–2.9 on the Pohlig–Hellman algorithm. Omit Sect. 2.10 on first reading.

Chapter 3. Integer Factorization and RSA.
Cover Sects. 3.1–3.5 and 3.9–3.10. Optionally, cover the more mathematically sophisticated Sects. 3.6–3.8, dealing with smooth numbers, sieves, and the index calculus.

Chapter 4. Digital Signatures.
Cover all sections.

Chapter 5. Probability Theory and Information Theory.
Cover Sects. 5.1, 5.3, and 5.4. Optionally cover the more mathematically sophisticated sections on Pollard's ρ method (Sect. 5.5), information theory (Sect. 5.6), and complexity theory (Sect. 5.7). The material on the Vigenère cipher in Sect. 5.2 nicely illustrates the use of statistics in cryptanalysis, but is somewhat off the main path.

Chapter 6. Elliptic Curves.
Cover Sects. 6.1–6.4. Cover other sections as time permits, but note that Sects. 6.7–6.10 on pairings require finite fields of prime power order, which are described in Sect. 2.10.4.

Chapter 7. Lattices and Cryptography.
Cover Sects. 7.1–7.8. (If time is short, one may omit either or both of Sects. 7.1 and 7.2.) Cover either Sects. 7.13–7.14 on the LLL lattice reduction algorithm or Sects. 7.9–7.11 on the NTRU cryptosystem, or

both, as time permits. (The NTRU sections require the material on polynomial rings and quotient rings covered in Sect. 2.10.)

Chapter 8. Additional Topics in Cryptography.
The material in this chapter points the reader toward other important areas of cryptography. It provides a good list of topics and references for student term papers and presentations.

Further Notes for the Instructor: Depending on how much of the harder mathematical material in Chaps. 2–5 is covered, there may not be time to delve into both Chaps. 6 and 7, so the instructor may need to omit either elliptic curves or lattices in order to fit the other material into one semester.

We feel that it is helpful for students to gain an appreciation of the origins of their subject, so we have scattered a handful of sections throughout the book containing some brief comments on the history of cryptography. Instructors who want to spend more time on mathematics may omit these sections without affecting the mathematical narrative.

Changes in the Second Edition:

- The chapter on digital signatures has been moved, since we felt that this important topic should be covered earlier in the course. More precisely, RSA, Elgamal, and DSA signatures are now described in the short Chap. 4, while the material on elliptic curve signatures is covered in the brief Sect. 6.4.3. The two sections on lattice-based signatures from the first edition have been extensively rewritten and now appear as Sect. 7.12.
- Numerous new exercises have been included.
- Numerous typographical and minor mathematical errors have been corrected, and notation has been made more consistent from chapter to chapter.
- Various explanations have been rewritten or expanded for clarity, especially in Chaps. 5–7.
- New sections on digital cash and on homomorphic encryption have been added to the additional topics in Chap. 8; see Sects. 8.8 and 8.9.

Chapter 1

An Introduction to Cryptography

1.1 Simple Substitution Ciphers

As Julius Caesar surveys the unfolding battle from his hilltop outpost, an exhausted and disheveled courier bursts into his presence and hands him a sheet of parchment containing gibberish:

```
jsjrdkfqqnslgfhpgwjfpymwtzlmnrrnsjsyqzhnzx
```

Within moments, Julius sends an order for a reserve unit of charioteers to speed around the left flank and exploit a momentary gap in the opponent's formation.

How did this string of seemingly random letters convey such important information? The trick is easy, once it is explained. Simply take each letter in the message and shift it five letters up the alphabet. Thus j in the *ciphertext* becomes e in the *plaintext*,[1] because e is followed in the alphabet by f,g,h,i,j. Applying this procedure to the entire ciphertext yields

```
j s j r d k f q q n s l g f h p g w j f p y m w t z l m n r r n s j s y q z h n z x
e n e m y f a l l i n g b a c k b r e a k t h r o u g h i m m i n e n t l u c i u s
```

The second line is the decrypted plaintext, and breaking it into words and supplying the appropriate punctuation, Julius reads the message

```
Enemy falling back. Breakthrough imminent. Lucius.
```

[1]The *plaintext* is the original message in readable form and the *ciphertext* is the encrypted message.

J. Hoffstein et al., *An Introduction to Mathematical Cryptography*,
Undergraduate Texts in Mathematics, DOI 10.1007/978-1-4939-1711-2_1

There remains one minor quirk that must be addressed. What happens when Julius finds a letter such as `d`? There is no letter appearing five letters before `d` in the alphabet. The answer is that he must wrap around to the end of the alphabet. Thus `d` is replaced by `y`, since `y` is followed by `z`,`a`,`b`,`c`,`d`.

This wrap-around effect may be conveniently visualized by placing the alphabet `abcd...xyz` around a circle, rather than in a line. If a second alphabet circle is then placed within the first circle and the inner circle is rotated five letters, as illustrated in Fig. 1.1, the resulting arrangement can be used to easily encrypt and decrypt Caesar's messages. To decrypt a letter, simply find it on the inner wheel and read the corresponding plaintext letter from the outer wheel. To encrypt, reverse this process: find the plaintext letter on the outer wheel and read off the ciphertext letter from the inner wheel. And note that if you build a cipherwheel whose inner wheel spins, then you are no longer restricted to always shifting by exactly five letters. Cipher wheels of this sort have been used for centuries.[2]

Although the details of the preceding scene are entirely fictional, and in any case it is unlikely that a message to a Roman general would have been written in modern English(!), there is evidence that Caesar employed this early method of cryptography, which is sometimes called the *Caesar cipher* in his honor. It is also sometimes referred to as a *shift cipher*, since each letter in the alphabet is shifted up or down. *Cryptography*, the methodology of concealing the content of messages, comes from the Greek root words kryptos, meaning hidden,[3] and graphikos, meaning writing. The modern scientific study of cryptography is sometimes referred to as *cryptology*.

In the Caesar cipher, each letter is replaced by one specific substitute letter. However, if Bob encrypts a message for Alice[4] using a Caesar cipher and allows the encrypted message to fall into Eve's hands, it will take Eve very little time to decrypt it. All she needs to do is try each of the 26 possible shifts.

Bob can make his message harder to attack by using a more complicated replacement scheme. For example, he could replace every occurrence of `a` by `z` and every occurrence of `z` by `a`, every occurrence of `b` by `y` and every occurrence of `y` by `b`, and so on, exchanging each pair of letters `c` $\leftrightarrow$ `x`,..., `m` $\leftrightarrow$ `n`.

This is an example of a *simple substitution cipher*, that is, a cipher in which each letter is replaced by another letter (or some other type of symbol). The

[2] A cipher wheel with mixed up alphabets and with encryption performed using different offsets for different parts of the message is featured in a fifteenth century monograph by Leon Batista Alberti [63].

[3] The word cryptic, meaning hidden or occult, appears in 1638, while crypto- as a prefix for concealed or secret makes its appearance in 1760. The term cryptogram appears much later, first occurring in 1880.

[4] In cryptography, it is traditional for Bob and Alice to exchange confidential messages and for their adversary Eve, the eavesdropper, to intercept and attempt to read their messages. This makes the field of cryptography much more personal than other areas of mathematics and computer science, whose denizens are often X and Y!

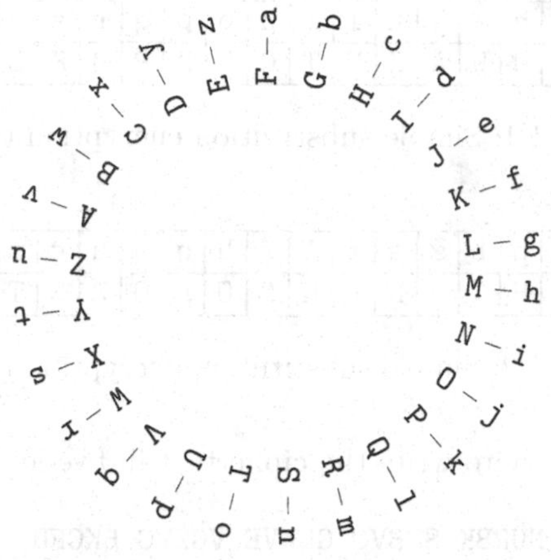

Figure 1.1: A cipher wheel with an offset of five letters

Caesar cipher is an example of a simple substitution cipher, but there are many simple substitution ciphers other than the Caesar cipher. In fact, a simple substitution cipher may be viewed as a rule or function

$$\{\mathtt{a,b,c,d,e,\ldots,x,y,z}\} \longrightarrow \{\mathtt{A,B,C,D,E,\ldots,X,Y,Z}\}$$

assigning each plaintext letter in the domain a different ciphertext letter in the range. (To make it easier to distinguish the plaintext from the ciphertext, we write the plaintext using lowercase letters and the ciphertext using uppercase letters.) Note that in order for decryption to work, the encryption function must have the property that no two plaintext letters go to the same ciphertext letter. A function with this property is said to be *one-to-one* or *injective.*

A convenient way to describe the encryption function is to create a table by writing the plaintext alphabet in the top row and putting each ciphertext letter below the corresponding plaintext letter.

Example 1.1. A simple substitution encryption table is given in Table 1.1. The ciphertext alphabet (the uppercase letters in the bottom row) is a randomly chosen permutation of the 26 letters in the alphabet. In order to encrypt the plaintext message

`Four score and seven years ago,`

we run the words together, look up each plaintext letter in the encryption table, and write the corresponding ciphertext letter below.

f	o	u	r	s	c	o	r	e	a	n	d	s	e	v	e	n	y	e	a	r	s	a	g	o
N	U	R	B	K	S	U	B	V	C	G	Q	K	V	E	V	G	Z	V	C	B	K	C	F	U

a	b	c	d	e	f	g	h	i	j	k	l	m	n	o	p	q	r	s	t	u	v	w	x	y	z
C	I	S	Q	V	N	F	O	W	A	X	M	T	G	U	H	P	B	K	L	R	E	Y	D	Z	J

Table 1.1: Simple substitution encryption table

j	r	a	x	v	g	n	p	b	z	s	t	l	f	h	q	d	u	c	m	o	e	i	k	w	y
A	B	C	D	E	F	G	H	I	J	K	L	M	N	O	P	Q	R	S	T	U	V	W	X	Y	Z

Table 1.2: Simple substitution decryption table

It is then customary to write the ciphertext in five-letter blocks:

`NURBK SUBVC GQKVE VGZVC BKCFU`

Decryption is a similar process. Suppose that we receive the message

`GVVQG VYKCM CQQBV KKWGF SCVKC B`

and that we know that it was encrypted using Table 1.1. We can reverse the encryption process by finding each ciphertext letter in the second row of Table 1.1 and writing down the corresponding letter from the top row. However, since the letters in the second row of Table 1.1 are all mixed up, this is a somewhat inefficient process. It is better to make a decryption table in which the ciphertext letters in the lower row are listed in alphabetical order and the corresponding plaintext letters in the upper row are mixed up. We have done this in Table 1.2. Using this table, we easily decrypt the message.

G	V	V	Q	G	V	Y	K	C	M	C	Q	Q	B	V	K	K	W	G	F	S	C	V	K	C	B
n	e	e	d	n	e	w	s	a	l	a	d	d	r	e	s	s	i	n	g	c	a	e	s	a	r

Putting in the appropriate word breaks and some punctuation reveals an urgent request!

`Need new salad dressing. -Caesar`

1.1.1 Cryptanalysis of Simple Substitution Ciphers

How many different simple substitution ciphers exist? We can count them by enumerating the possible ciphertext values for each plaintext letter. First we assign the plaintext letter a to one of the 26 possible ciphertext letters A–Z. So there are 26 possibilities for a. Next, since we are not allowed to assign b to the same letter as a, we may assign b to any one of the remaining 25 ciphertext letters. So there are $26 \cdot 25 = 650$ possible ways to assign a and b. We have now used up two of the ciphertext letters, so we may assign c to any one of the remaining 24 ciphertext letters. And so on.... Thus the total number of ways to assign the 26 plaintext letters to the 26 ciphertext letters, using each ciphertext letter only once, is

$$26 \cdot 25 \cdot 24 \cdots 4 \cdot 3 \cdot 2 \cdot 1 = 26! = 403291461126605635584000000.$$

There are thus more than 10^{26} different simple substitution ciphers. Each associated encryption table is known as a *key*.

Suppose that Eve intercepts one of Bob's messages and that she attempts to decrypt it by trying every possible simple substitution cipher. The process of decrypting a message without knowing the underlying key is called *cryptanalysis*. If Eve (or her computer) is able to check one million cipher alphabets per second, it would still take her more than 10^{13} years to try them all.[5] But the age of the universe is estimated to be on the order of 10^{10} years. Thus Eve has almost no chance of decrypting Bob's message, which means that Bob's message is secure and he has nothing to worry about![6] Or does he?

It is time for an important lesson in the practical side of the science of cryptography:

> Your opponent always uses her best strategy to defeat you, not the strategy that you want her to use. Thus the security of an encryption system depends on the best known method to break it. As new and improved methods are developed, the level of security can only get worse, never better.

Despite the large number of possible simple substitution ciphers, they are actually quite easy to break, and indeed many newspapers and magazines feature them as a companion to the daily crossword puzzle. The reason that Eve can easily cryptanalyze a simple substitution cipher is that the letters in the English language (or any other human language) are not random. To take an extreme example, the letter `q` in English is virtually always followed by the letter `u`. More useful is the fact that certain letters such as `e` and `t` appear far more frequently than other letters such as `f` and `c`. Table 1.3 lists the letters with their typical frequencies in English text. As you can see, the most frequent letter is `e`, followed by `t`, `a`, `o`, and `n`.

Thus if Eve counts the letters in Bob's encrypted message and makes a frequency table, it is likely that the most frequent letter will represent `e`, and that `t`, `a`, `o`, and `n` will appear among the next most frequent letters. In this way, Eve can try various possibilities and, after a certain amount of trial and error, decrypt Bob's message.

[5] Do you see how we got 10^{13} years? There are $60 \cdot 60 \cdot 24 \cdot 365$ s in a year, and 26! divided by $10^6 \cdot 60 \cdot 60 \cdot 24 \cdot 365$ is approximately $10^{13.107}$.

[6] The assertion that a large number of possible keys, in and of itself, makes a cryptosystem secure, has appeared many times in history and has equally often been shown to be fallacious.

By decreasing frequency			
E	13.11 %	M	2.54 %
T	10.47 %	U	2.46 %
A	8.15 %	G	1.99 %
O	8.00 %	Y	1.98 %
N	7.10 %	P	1.98 %
R	6.83 %	W	1.54 %
I	6.35 %	B	1.44 %
S	6.10 %	V	0.92 %
H	5.26 %	K	0.42 %
D	3.79 %	X	0.17 %
L	3.39 %	J	0.13 %
F	2.92 %	Q	0.12 %
C	2.76 %	Z	0.08 %

In alphabetical order			
A	8.15 %	N	7.10 %
B	1.44 %	O	8.00 %
C	2.76 %	P	1.98 %
D	3.79 %	Q	0.12 %
E	13.11 %	R	6.83 %
F	2.92 %	S	6.10 %
G	1.99 %	T	10.47 %
H	5.26 %	U	2.46 %
I	6.35 %	V	0.92 %
J	0.13 %	W	1.54 %
K	0.42 %	X	0.17 %
L	3.39 %	Y	1.98 %
M	2.54 %	Z	0.08 %

Table 1.3: Frequency of letters in English text

```
LOJUM YLJME PDYVJ QXTDV SVJNL DMTJZ WMJGG YSNDL UYLEO SKDVC
GEPJS MDIPD NEJSK DNJTJ LSKDL OSVDV DNGYN VSGLL OSCIO LGOYG
ESNEP CGYSN GUJMJ DGYNK DPPYX PJDGG SVDNT WMSWS GYLYS NGSKJ
CEPYQ GSGLD MLPYN IUSCP QOYGM JGCPL GDWWJ DMLSL OJCNY NYLYD
LJQLO DLCNL YPLOJ TPJDM NJQLO JWMSE JGGJG XTUOY EOOJO DQDMM
YBJQD LLOJV LOJTV YIOLU JPPES NGYQJ MOYVD GDNJE MSVDN EJM
```

Table 1.4: A simple substitution cipher to cryptanalyze

In the remainder of this section we illustrate how to cryptanalyze a simple substitution cipher by decrypting the message given in Table 1.4. Of course the end result of defeating a simple substitution cipher is not our main goal here. Our key point is to introduce the idea of statistical analysis, which will prove to have many applications throughout cryptography. Although for completeness we provide full details, the reader may wish to skim this material.

There are 298 letters in the ciphertext. The first step is to make a frequency table listing how often each ciphertext letter appears (Table 1.5).

	J	L	D	G	Y	S	O	N	M	P	E	V	Q	C	T	W	U	K	I	X	Z	B	A	F	R	H
Freq	32	28	27	24	23	22	19	18	17	15	12	12	8	8	7	6	6	5	4	3	1	1	0	0	0	0
%	11	9	9	8	8	7	6	6	6	5	4	4	3	3	2	2	2	2	1	1	0	0	0	0	0	0

Table 1.5: Frequency table for Table 1.4—Ciphertext length: 298

The ciphertext letter J appears most frequently, so we make the provisional guess that it corresponds to the plaintext letter e. The next most frequent ciphertext letters are L (28 times) and D (27 times), so we might guess from Table 1.3 that they represent t and a. However, the letter frequencies in a

th	he	an	re	er	in	on	at	nd	st	es	en	of	te	ed
168	132	92	91	88	86	71	68	61	53	52	51	49	46	46

(a) Most common English bigrams (frequency per 1000 words)

LO	OJ	GY	DN	VD	YL	DL	DM	SN	KD	LY	NG	OY	JD	SK	EP	JG	SV	JM	JQ
9	7	6 each		5 each					4 each										

(b) Most common bigrams appearing in the ciphertext in Table 1.4

Table 1.6: Bigram frequencies

short message are unlikely to exactly match the percentages in Table 1.3. All that we can say is that among the ciphertext letters L, D, G, Y, and S are likely to appear several of the plaintext letters t, a, o, n, and r.

There are several ways to proceed. One method is to look at *bigrams*, which are pairs of consecutive letters. Table 1.6a lists the bigrams that most frequently appear in English, and Table 1.6b lists the ciphertext bigrams that appear most frequently in our message. The ciphertext bigrams LO and OJ appear frequently. We have already guessed that J = e, and based on its frequency we suspect that L is likely to represent one of the letters t, a, o, n, or r. Since the two most frequent English bigrams are th and he, we make the tentative identifications

$$\mathtt{LO} = \mathtt{th} \qquad \text{and} \qquad \mathtt{OJ} = \mathtt{he}.$$

We substitute the guesses J = e, L = t, and O = h, into the ciphertext, writing the putative plaintext letter below the corresponding ciphertext letter.

```
LOJUM YLJME PDYVJ QXTDV SVJNL DMTJZ WMJGG YSNDL UYLEO SKDVC
the-- -te-- ----e ----- --e-t ---e- --e-- ----t --t-h -----
-----------------------------------------------------------
GEPJS MDIPD NEJSK DNJTJ LSKDL OSVDV DNGYN VSGLL OSCIO LGOYG
---e- ----- --e-- --e-e t---t h---- ----- ---tt h---h t-h--
-----------------------------------------------------------
ESNEP CGYSN GUJMJ DGYNK DPPYX PJDGG SVDNT WMSWS GYLYS NGSKJ
----- ----- --e-e ----- ----- -e--- ----- ----- --t-- ----e
-----------------------------------------------------------
CEPYQ GSGLD MLPYN IUSCP QOYGM JGCPL GDWWJ DMLSL OJCNY NYLYD
----- ---t- -t--- ----- -h--- e---t ----e --t-t he--- --t--
-----------------------------------------------------------
LJQLO DLCNL YPLOJ TPJDM NJQLO JWMSE JGGJG XTUOY EOOJO DQDMM
te-th -t--t --the --e-- -e-th e---- e--e- ---h- -hheh -----
-----------------------------------------------------------
YBJQD LLOJV LOJTV YIOLU JPPES NGYQJ MOYVD GDNJE MSVDN EJM
--e-- tthe- the-- --ht- e---- ----e -h--- ---e- ----- -e-
```

At this point, we can look at the fragments of plaintext and attempt to guess some common English words. For example, in the second line we see the three blocks

```
VSGLL OSCIO LGOYG,
---tt h---h t-h--.
```

Looking at the fragment th---ht, we might guess that this is the word thought, which gives three more equivalences,

$$S = o, \qquad C = u, \qquad I = g.$$

This yields

```
LOJUM YLJME PDYVJ QXTDV SVJNL DMTJZ WMJGG YSNDL UYLEO SKDVC
the-- -te-- ----e ----- o-e-t ---e- --e-- -o--t --t-h o---u
GEPJS MDIPD NEJSK DNJTJ LSKDL OSVDV DNGYN VSGLL OSCIO LGOYG
---eo --g-- --eo- --e-e to--t ho--- ----- -o-tt hough t-h--
ESNEP CGYSN GUJMJ DGYNK DPPYX PJDGG SVDNT WMSWS GYLYS NGSKJ
-o--- u--o- --e-e ----- ----- -e--- o---- --o-o --t-o --o-e
CEPYQ GSGLD MLPYN IUSCP QOYGM JGCPL GDWWJ DMLSL OJCNY NYLYD
u---- -o-t- -t--- g-ou- -h--- e-u-t ----e --tot heu-- --t--
LJQLO DLCNL YPLOJ TPJDM NJQLO JWMSE JGGJG XTUOY EOOJO DQDMM
te-th -tu-t --the --e-- -e-th e--o- e--e- ---h- -hheh -----
YBJQD LLOJV LOJTV YIOLU JPPES NGYQJ MOYVD GDNJE MSVDN EJM
--e-- tthe- the-- -ght- e---o ----e -h--- ---e- -o--- -e-
```

Now look at the three letters ght in the last line. They must be preceded by a vowel, and the only vowels left are a and i, so we guess that Y = i. Then we find the letters itio in the third line, and we guess that they are followed by an n, which gives N = n. (There is no reason that a letter cannot represent itself, although this is often forbidden in the puzzle ciphers that appear in newspapers.) We now have

```
LOJUM YLJME PDYVJ QXTDV SVJNL DMTJZ WMJGG YSNDL UYLEO SKDVC
the-- ite-- --i-e ----- o-ent ---e- --e-- ion-t -it-h o---u
GEPJS MDIPD NEJSK DNJTJ LSKDL OSVDV DNGYN VSGLL OSCIO LGOYG
---eo --g-- n-eo- -ne-e to--t ho--- -n-in -o-tt hough t-hi-
ESNEP CGYSN GUJMJ DGYNK DPPYX PJDGG SVDNT WMSWS GYLYS NGSKJ
-on-- u-ion --e-e --in- ---i- -e--- o--n- --o-o -itio n-o-e
CEPYQ GSGLD MLPYN IUSCP QOYGM JGCPL GDWWJ DMLSL OJCNY NYLYD
u--i- -o-t- -t-in g-ou- -hi-- e-u-t ----e --tot heuni niti-
LJQLO DLCNL YPLOJ TPJDM NJQLO JWMSE JGGJG XTUOY EOOJO DQDMM
te-th -tunt i-the --e-- ne-th e--o- e--e- ---hi -hheh -----
YBJQD LLOJV LOJTV YIOLU JPPES NGYQJ MOYVD GDNJE MSVDN EJM
i-e-- tthe- the-- ight- e---o n-i-e -hi-- --ne- -o--n -e-
```

So far, we have reconstructed the following plaintext/ciphertext pairs:

	J	L	D	G	Y	S	O	N	M	P	E	V	Q	C	T	W	U	K	I	X	Z	B	A	F	R	H
	e	t	-	-	i	o	h	n	-	-	-	-	-	u	-	-	-	-	g	-	-	-	-	-	-	-
Freq	32	28	27	24	23	22	19	18	17	15	12	12	8	8	7	6	6	5	4	3	1	1	0	0	0	0

Recall that the most common letters in English (Table 1.3) are, in order of decreasing frequency,

e, t, a, o, n, r, i, s, h.

We have already assigned ciphertext values to e, t, o, n, i, h, so we guess that D and G represent two of the three letters a, r, s. In the third line we notice that GYLYSN gives -ition, so clearly G must be s. Similarly, on the fifth line we have LJQLO DLCNL equal to te-th -tunt, so D must be a, not r. Substituting these new pairs G = s and D = a gives

```
LOJUM YLJME PDYVJ QXTDV SVJNL DMTJZ WMJGG YSNDL UYLEO SKDVC
the-- ite-- -ai-e ---a- o-ent a--e- --ess ionat -it-h o-a-u

GEPJS MDIPD NEJSK DNJTJ LSKDL OSVDV DNGYN VSGLL OSCIO LGOYG
s--eo -ag-a n-eo- ane-e to-at ho-a- ansin -ostt hough tshis

ESNEP CGYSN GUJMJ DGYNK DPPYX PJDGG SVDNT WMSWS GYLYS NGSKJ
-on-- usion s-e-e asin- a--i- -eass o-an- --o-o sitio nso-e

CEPYQ GSGLD MLPYN IUSCP QOYGM JGCPL GDWWJ DMLSL OJCNY NYLYD
u--i- sosta -t-in g-ou- -his- esu-t sa--e a-tot heuni nitia

LJQLO DLCNL YPLOJ TPJDM NJQLO JWMSE JGGJG XTUOY EOOJO DQDMM
te-th atunt i-the --ea- ne-th e--o- esses ---hi -hheh a-a--

YBJQD LLOJV LOJTV YIOLU JPPES NGYQJ MOYVD GDNJE MSVDN EJM
i-e-a tthe- the-- ight- e---o nsi-e -hi-a sane- -o-an -e-
```

It is now easy to fill in additional pairs by inspection. For example, the missing letter in the fragment atunt i-the on the fifth line must be l, which gives P = l, and the missing letter in the fragment -osition on the third line must be p, which gives W = p. Substituting these in, we find the fragment e-p-ession on the first line, which gives Z = x and M = r, and the fragment -on-lusion on the third line, which gives E = c. Then consi-er on the last line gives Q = d and the initial words the-riterclai-e- must be the phrase "the writer claimed," yielding U = w and V = m. This gives

```
LOJUM YLJME PDYVJ QXTDV SVJNL DMTJZ WMJGG YSNDL UYLEO SKDVC
thewr iterc laime d--am oment ar-ex press ionat witch o-amu

GEPJS MDIPD NEJSK DNJTJ LSKDL OSVDV DNGYN VSGLL OSCIO LGOYG
scleo ragla nceo- ane-e to-at homam ansin mostt hough tshis

ESNEP CGYSN GUJMJ DGYNK DPPYX PJDGG SVDNT WMSWS GYLYS NGSKJ
concl usion swere asin- alli- leass oman- propo sitio nso-e

CEPYQ GSGLD MLPYN IUSCP QOYGM JGCPL GDWWJ DMLSL OJCNY NYLYD
uclid sosta rtlin gwoul dhisr esult sappe artot heuni nitia

LJQLO DLCNL YPLOJ TPJDM NJQLO JWMSE JGGJG XTUOY EOOJO DQDMM
tedth atunt ilthe -lear nedth eproc esses --whi chheh adarr

YBJQD LLOJV LOJTV YIOLU JPPES NGYQJ MOYVD GDNJE MSVDN EJM
i-eda tthem the-m ightw ellco nside rhima sanec roman cer
```

It is now a simple matter to fill in the few remaining letters and put in the appropriate word breaks, capitalization, and punctuation to recover the plaintext:

> The writer claimed by a momentary expression, a twitch of a muscle or a glance of an eye, to fathom a man's inmost thoughts. His conclusions were as infallible as so many propositions of Euclid. So startling would his results appear to the uninitiated that until they learned the processes by which he had arrived at them they might well consider him as a necromancer.[7]

1.2 Divisibility and Greatest Common Divisors

Much of modern cryptography is built on the foundations of algebra and number theory. So before we explore the subject of cryptography, we need to develop some important tools. In the next four sections we begin this development by describing and proving fundamental results in these areas. If you have already studied number theory in another course, a brief review of this material will suffice. But if this material is new to you, then it is vital to study it closely and to work out the exercises provided at the end of the chapter.

At the most basic level, *Number Theory* is the study of the natural numbers

$$1, 2, 3, 4, 5, 6, \ldots,$$

or slightly more generally, the study of the integers

$$\ldots, -5, -4, -3, -2, -1, 0, 1, 2, 3, 4, 5, \ldots.$$

The set of integers is denoted by the symbol $\mathbb{Z}$. Integers can be added, subtracted, and multiplied in the usual way, and they satisfy all the usual rules of arithmetic (commutative law, associative law, distributive law, etc.). The set of integers with their addition and multiplication rules are an example of a *ring*. See Sect. 2.10.1 for more about the theory of rings.

If a and b are integers, then we can add them, $a + b$, subtract them, $a - b$, and multiply them, $a \cdot b$. In each case, we get an integer as the result. This property of staying inside of our original set after applying operations to a pair of elements is characteristic of a ring.

But if we want to stay within the integers, then we are not always able to divide one integer by another. For example, we cannot divide 3 by 2, since there is no integer that is equal to $\frac{3}{2}$. This leads to the fundamental concept of divisibility.

Definition. Let a and b be integers with $b \neq 0$. We say that b *divides* a, or that a *is divisible by* b, if there is an integer c such that

$$a = bc.$$

We write $b \mid a$ to indicate that b divides a. If b does not divide a, then we write $b \nmid a$.

[7] *A Study in Scarlet* (Chap. 2), Sir Arthur Conan Doyle.

Example 1.2. We have $847 \mid 485331$, since $485331 = 847 \cdot 573$. On the other hand, $355 \nmid 259943$, since when we try to divide 259943 by 355, we get a remainder of 83. More precisely, $259943 = 355 \cdot 732 + 83$, so 259943 is not an exact multiple of 355.

Remark 1.3. Notice that every integer is divisible by 1. The integers that are divisible by 2 are the *even integers*, and the integers that are not divisible by 2 are the *odd integers*.

There are a number of elementary divisibility properties, some of which we list in the following proposition.

Proposition 1.4. *Let $a, b, c \in \mathbb{Z}$ be integers.*
(a) *If $a \mid b$ and $b \mid c$, then $a \mid c$.*
(b) *If $a \mid b$ and $b \mid a$, then $a = \pm b$.*
(c) *If $a \mid b$ and $a \mid c$, then $a \mid (b + c)$ and $a \mid (b - c)$.*

Proof. We leave the proof as an exercise for the reader; see Exercise 1.6. □

Definition. A *common divisor* of two integers a and b is a positive integer d that divides both of them. The *greatest common divisor* of a and b is, as its name suggests, the largest positive integer d such that $d \mid a$ and $d \mid b$. The greatest common divisor of a and b is denoted $\gcd(a, b)$. If there is no possibility of confusion, it is also sometimes denoted by (a, b). (If a and b are both 0, then $\gcd(a, b)$ is not defined.)

It is a curious fact that a concept as simple as the greatest common divisor has many applications. We'll soon see that there is a fast and efficient method to compute the greatest common divisor of any two integers, a fact that has powerful and far-reaching consequences.

Example 1.5. The greatest common divisor of 12 and 18 is 6, since $6 \mid 12$ and $6 \mid 18$ and there is no larger number with this property. Similarly,

$$\gcd(748, 2024) = 44.$$

One way to check that this is correct is to make lists of all of the positive divisors of 748 and of 2024.

$$\begin{aligned} \text{Divisors of } 748 &= \{1, 2, 4, 11, 17, 22, 34, 44, 68, 187, 374, 748\}, \\ \text{Divisors of } 2024 &= \{1, 2, 4, 8, 11, 22, 23, 44, 46, 88, 92, 184, 253, \\ &\qquad 506, 1012, 2024\}. \end{aligned}$$

Examining the two lists, we see that the largest common entry is 44. Even from this small example, it is clear that this is not a very efficient method. If we ever need to compute greatest common divisors of large numbers, we will have to find a more efficient approach.

The key to an efficient algorithm for computing greatest common divisors is *division with remainder*, which is simply the method of "long division" that you learned in elementary school. Thus if a and b are positive integers and if you attempt to divide a by b, you will get a quotient q and a remainder r, where the remainder r is smaller than b. For example,

```
        13 R 9
 17 ) 230
      17
       60
       51
        9
```

so 230 divided by 17 gives a quotient of 13 with a remainder of 9. What does this last statement really mean? It means that 230 can be written as

$$230 = 17 \cdot 13 + 9,$$

where the remainder 9 is strictly smaller than the divisor 17.

Definition. (Division With Remainder) Let a and b be positive integers. Then we say that a *divided by* b *has quotient* q *and remainder* r if

$$a = b \cdot q + r \qquad \text{with } 0 \le r < b.$$

The values of q and r are uniquely determined by a and b; see Exercise 1.14.

Suppose now that we want to find the greatest common divisor of a and b. We first divide a by b to get

$$a = b \cdot q + r \qquad \text{with } 0 \le r < b. \tag{1.1}$$

If d is any common divisor of a and b, then it is clear from Eq. (1.1) that d is also a divisor of r. (See Proposition 1.4(c).) Similarly, if e is a common divisor of b and r, then (1.1) shows that e is a divisor of a. In other words, the common divisors of a and b are the same as the common divisors of b and r; hence

$$\gcd(a, b) = \gcd(b, r).$$

We repeat the process, dividing b by r to get another quotient and remainder, say

$$b = r \cdot q' + r' \qquad \text{with } 0 \le r' < r.$$

Then the same reasoning shows that

$$\gcd(b, r) = \gcd(r, r').$$

Continuing this process, the remainders become smaller and smaller, until eventually we get a remainder of 0, at which point the final value $\gcd(s, 0) = s$ is equal to the gcd of a and b.

We illustrate with an example and then describe the general method, which goes by the name *Euclidean algorithm*.

Example 1.6. We compute gcd(2024, 748) using the Euclidean algorithm, which is nothing more than repeated division with remainder. Notice how the b and r values on each line become the new a and b values on the subsequent line:

$$\begin{aligned} 2024 &= 748 \cdot 2 + 528 \\ 748 &= 528 \cdot 1 + 220 \\ 528 &= 220 \cdot 2 + 88 \\ 220 &= 88 \cdot 2 + 44 \quad \leftarrow \boxed{\gcd = 44} \\ 88 &= 44 \cdot 2 + 0 \end{aligned}$$

Theorem 1.7 (The Euclidean Algorithm). *Let a and b be positive integers with $a \ge b$. The following algorithm computes $\gcd(a, b)$ in a finite number of steps.*

(1) *Let $r_0 = a$ and $r_1 = b$.*

(2) *Set $i = 1$.*

(3) *Divide r_{i-1} by r_i to get a quotient q_i and remainder r_{i+1},*

$$r_{i-1} = r_i \cdot q_i + r_{i+1} \qquad \textit{with} \quad 0 \le r_{i+1} < r_i.$$

(4) *If the remainder $r_{i+1} = 0$, then $r_i = \gcd(a, b)$ and the algorithm terminates.*

(5) *Otherwise, $r_{i+1} > 0$, so set $i = i + 1$ and go to Step 3.*

The division step (Step 3) is executed at most

$$2 \log_2(b) + 2 \quad \textit{times.}$$

Proof. The Euclidean algorithm consists of a sequence of divisions with remainder as illustrated in Fig. 1.2 (remember that we set $r_0 = a$ and $r_1 = b$).

$$\begin{array}{rcll} a &=& b \cdot q_1 + r_2 & \text{with } 0 \le r_2 < b, \\ b &=& r_2 \cdot q_2 + r_3 & \text{with } 0 \le r_3 < r_2, \\ r_2 &=& r_3 \cdot q_3 + r_4 & \text{with } 0 \le r_4 < r_3, \\ r_3 &=& r_4 \cdot q_4 + r_5 & \text{with } 0 \le r_5 < r_4, \\ \vdots && \vdots & \vdots \\ r_{t-2} &=& r_{t-1} \cdot q_{t-1} + r_t & \text{with } 0 \le r_t < r_{t-1}, \\ r_{t-1} &=& r_t \cdot q_t & \end{array}$$

$$\text{Then } r_t = \gcd(a, b).$$

Figure 1.2: The Euclidean algorithm step by step

The r_i values are strictly decreasing, and as soon as they reach zero the algorithm terminates, which proves that the algorithm does finish in a finite number of steps. Further, at each iteration of Step 3 we have an equation of the form

$$r_{i-1} = r_i \cdot q_i + r_{i+1}.$$

This equation implies that any common divisor of r_{i-1} and r_i is also a divisor of r_{i+1}, and similarly it implies that any common divisor of r_i and r_{i+1} is also a divisor of r_{i-1}. Hence

$$\gcd(r_{i-1}, r_i) = \gcd(r_i, r_{i+1}) \qquad \text{for all } i = 1, 2, 3, \ldots. \tag{1.2}$$

However, as noted earlier, we eventually get to an r_i that is zero, say $r_{t+1} = 0$. Then $r_{t-1} = r_t \cdot q_t$, so

$$\gcd(r_{t-1}, r_t) = \gcd(r_t \cdot q_t, r_t) = r_t.$$

But Eq. (1.2) says that this is equal to $\gcd(r_0, r_1)$, i.e., to $\gcd(a, b)$, which completes the proof that the last nonzero remainder in the Euclidean algorithm is equal to the greatest common divisor of a and b.

It remains to estimate the efficiency of the algorithm. We noted above that since the r_i values are strictly decreasing, the algorithm terminates, and indeed since $r_1 = b$, it certainly terminates in at most b steps. However, this upper bound is far from the truth. We claim that after every two iterations of Step 3, the value of r_i is at least cut in half. In other words:

$$\textbf{Claim:} \quad r_{i+2} < \tfrac{1}{2} r_i \quad \text{for all } i = 0, 1, 2, \ldots.$$

We prove the claim by considering two cases.

Case I: $r_{i+1} \le \frac{1}{2} r_i$

We know that the r_i values are strictly decreasing, so

$$r_{i+2} < r_{i+1} \le \tfrac{1}{2} r_i.$$

Case II: $r_{i+1} > \frac{1}{2} r_i$

Consider what happens when we divide r_i by r_{i+1}. The value of r_{i+1} is so large that we get

$$r_i = r_{i+1} \cdot 1 + r_{i+2} \quad \text{with} \quad r_{i+2} = r_i - r_{i+1} < r_i - \tfrac{1}{2} r_i = \tfrac{1}{2} r_i.$$

We have now proven our claim that $r_{i+2} < \frac{1}{2} r_i$ for all i. Using this inequality repeatedly, we find that

$$r_{2k+1} < \frac{1}{2} r_{2k-1} < \frac{1}{4} r_{2k-3} < \frac{1}{8} r_{2k-5} < \frac{1}{16} r_{2k-7} < \cdots < \frac{1}{2^k} r_1 = \frac{1}{2^k} b.$$

Hence if $2^k \ge b$, then $r_{2k+1} < 1$, which forces r_{2k+1} to equal 0 and the algorithm to terminate. In terms of Fig. 1.2, the value of r_{t+1} is 0, so we have

$t+1 \le 2k+1$, and thus $t \le 2k$. Further, there are exactly t divisions performed in Fig. 1.2, so the Euclidean algorithm terminates in at most $2k$ iterations. Choose the smallest such k, so $2^k \ge b > 2^{k-1}$. Then

$$\#\text{ of iterations} \le 2k = 2(k-1)+2 < 2\log_2(b)+2,$$

which completes the proof of Theorem 1.7. □

Remark 1.8. We proved that the Euclidean algorithm applied to a and b with $a \ge b$ requires no more than $2\log_2(b)+2$ iterations to compute $\gcd(a,b)$. This estimate can be somewhat improved. It has been proven that the Euclidean algorithm takes no more than $1.45\log_2(b)+1.68$ iterations, and that the average number of iterations for randomly chosen a and b is approximately $0.85\log_2(b)+0.14$; see [66].

Remark 1.9. One way to compute quotients and remainders is by long division, as we did on page 12. You can speed up the process using a simple calculator. The first step is to divide a by b on your calculator, which will give a real number. Throw away the part after the decimal point to get the quotient q. Then the remainder r can be computed as

$$r = a - b \cdot q.$$

For example, let $a = 2387187$ and $b = 27573$. Then $a/b \approx 86.57697748$, so $q = 86$ and

$$r = a - b \cdot q = 2387187 - 27573 \cdot 86 = 15909.$$

If you need just the remainder, you can instead take the decimal part (also sometimes called the *fractional part*) of a/b and multiply it by b. Continuing with our example, the decimal part of $a/b \approx 86.57697748$ is 0.57697748, and multiplying by $b = 27573$ gives

$$27573 \cdot 0.57697748 = 15909.00005604.$$

Rounding this off gives $r = 15909$.

After performing the Euclidean algorithm on two numbers, we can work our way back up the process to obtain an extremely interesting formula. Before giving the general result, we illustrate with an example.

Example 1.10. Recall that in Example 1.6 we used the Euclidean algorithm to compute $\gcd(2024, 748)$ as follows:

$$\begin{aligned} 2024 &= 748 \cdot 2 + 528 \\ 748 &= 528 \cdot 1 + 220 \\ 528 &= 220 \cdot 2 + 88 \\ 220 &= 88 \cdot 2 + 44 \quad \leftarrow \boxed{\gcd = 44} \\ 88 &= 44 \cdot 2 + 0 \end{aligned}$$

We let $a = 2024$ and $b = 748$, so the first line says that

$$528 = a - 2b.$$

We substitute this into the second line to get

$$b = (a - 2b) \cdot 1 + 220, \qquad \text{so} \qquad 220 = -a + 3b.$$

We next substitute the expressions $528 = a - 2b$ and $220 = -a + 3b$ into the third line to get

$$a - 2b = (-a + 3b) \cdot 2 + 88, \qquad \text{so} \qquad 88 = 3a - 8b.$$

Finally, we substitute the expressions $220 = -a + 3b$ and $88 = 3a - 8b$ into the penultimate line to get

$$-a + 3b = (3a - 8b) \cdot 2 + 44, \qquad \text{so} \qquad 44 = -7a + 19b.$$

In other words,

$$-7 \cdot 2024 + 19 \cdot 748 = 44 = \gcd(2024, 748),$$

so we have found a way to write $\gcd(a, b)$ as a linear combination of a and b using integer coefficients.

In general, it is always possible to write $\gcd(a, b)$ as an integer linear combination of a and b, a simple sounding result with many important consequences.

Theorem 1.11 (Extended Euclidean Algorithm). *Let a and b be positive integers. Then the equation*

$$au + bv = \gcd(a, b)$$

always has a solution in integers u and v. (See Exercise 1.12 *for an efficient algorithm to find a solution.)*

If (u_0, v_0) is any one solution, then every solution has the form

$$u = u_0 + \frac{b \cdot k}{\gcd(a, b)} \quad \text{and} \quad v = v_0 - \frac{a \cdot k}{\gcd(a, b)} \qquad \text{for some } k \in \mathbb{Z}.$$

Proof. Look back at Fig. 1.2, which illustrates the Euclidean algorithm step by step. We can solve the first line for $r_2 = a - b \cdot q_1$ and substitute it into the second line to get

$$b = (a - b \cdot q_1) \cdot q_2 + r_3, \qquad \text{so} \qquad r_3 = -a \cdot q_2 + b \cdot (1 + q_1 q_2).$$

Next substitute the expressions for r_2 and r_3 into the third line to get

$$a - b \cdot q_1 = \bigl(-a \cdot q_2 + b \cdot (1 + q_1 q_2)\bigr) q_3 + r_4.$$

After rearranging the terms, this gives

$$r_4 = a \cdot (1 + q_2 q_3) - b \cdot (q_1 + q_3 + q_1 q_2 q_3).$$

The key point is that $r_4 = a \cdot u + b \cdot v$, where u and v are integers. It does not matter that the expressions for u and v in terms of q_1, q_2, q_3 are rather messy. Continuing in this fashion, at each stage we find that r_i is the sum of an integer multiple of a and an integer multiple of b. Eventually, we get to $r_t = a \cdot u + b \cdot v$ for some integers u and v. But $r_t = \gcd(a, b)$, which completes the proof of the first part of the theorem. We leave the second part as an exercise (Exercise 1.11). □

An especially important case of the extended Euclidean algorithm arises when the greatest common divisor of a and b is 1. In this case we give a and b a special name.

Definition. Let a and b be integers. We say that a and b are *relatively prime* if $\gcd(a, b) = 1$.

More generally, any equation

$$Au + Bv = \gcd(A, B)$$

can be reduced to the case of relatively prime numbers by dividing both sides by $\gcd(A, B)$. Thus

$$\frac{A}{\gcd(A,B)}u + \frac{B}{\gcd(A,B)}v = 1,$$

where $a = A/\gcd(A, B)$ and $b = B/\gcd(A, B)$ are relatively prime and satisfy $au+bv = 1$. For example, we found earlier that 2024 and 748 have greatest common divisor 44 and satisfy

$$-7 \cdot 2024 + 19 \cdot 748 = 44.$$

Dividing both sides by 44, we obtain

$$-7 \cdot 46 + 19 \cdot 17 = 1.$$

Thus $2024/44 = 46$ and $748/44 = 17$ are relatively prime, and $u = -7$ and $v = 19$ are the coefficients of a linear combination of 46 and 17 that equals 1.

In Example 1.10 we explained how to substitute the values from the Euclidean algorithm in order to solve $au + bv = \gcd(a, b)$. Exercise 1.12 describes an efficient computer-oriented algorithm for computing u and v. If a and b are relatively prime, we now describe a more conceptual version of this substitution procedure. We first illustrate with the example $a = 73$ and $b = 25$. The Euclidean algorithm gives

$$73 = 25 \cdot 2 + 23$$

$$\begin{aligned} 25 &= 23 \cdot 1 + \ 2 \\ 23 &= 2 \cdot 11 + \ 1 \\ 2 &= \ \ 1 \cdot 2 + \ 0. \end{aligned}$$

We set up a box, using the sequence of quotients 2, 1, 11, and 2, as follows:

		2	1	11	2
0	1	*	*	*	*
1	0	*	*	*	*

Then the rule to fill in the remaining entries is as follows:

$$\begin{aligned} \text{New Entry} = (\text{Number at Top}) \cdot (\text{Number to the Left}) \\ + (\text{Number Two Spaces to the Left}). \end{aligned}$$

Thus the two leftmost $*$'s are

$$2 \cdot 1 + 0 = 2 \qquad \text{and} \qquad 2 \cdot 0 + 1 = 1,$$

so now our box looks like this:

		2	1	11	2
0	1	2	*	*	*
1	0	1	*	*	*

Then the next two leftmost $*$'s are

$$1 \cdot 2 + 1 = 3 \qquad \text{and} \qquad 1 \cdot 1 + 0 = 1,$$

and then the next two are

$$11 \cdot 3 + 2 = 35 \qquad \text{and} \qquad 11 \cdot 1 + 1 = 12,$$

and the final entries are

$$2 \cdot 35 + 3 = 73 \qquad \text{and} \qquad 2 \cdot 12 + 1 = 25.$$

The completed box is

		2	1	11	2
0	1	2	3	35	73
1	0	1	1	12	25

Notice that the last column repeats a and b. More importantly, the next to last column gives the values of $-v$ and u (in that order). Thus in this example we find that $73 \cdot 12 - 25 \cdot 35 = 1$. The general algorithm is given in Fig. 1.3.

In general, if a and b are relatively prime and if $q_1, q_2, \ldots, q_t$ is the sequence of quotients obtained from applying the Euclidean algorithm to a and b as in Figure 1.2 on page 13, then the box has the form

		q_1	q_2	$\cdots$	q_{t-1}	q_t
0	1	P_1	P_2	$\cdots$	P_{t-1}	a
1	0	Q_1	Q_2	$\cdots$	Q_{t-1}	b

The entries in the box are calculated using the initial values

$$P_1 = q_1, \qquad Q_1 = 1, \qquad P_2 = q_2 \cdot P_1 + 1, \qquad Q_2 = q_2 \cdot Q_1,$$

and then, for $i \geq 3$, using the formulas

$$P_i = q_i \cdot P_{i-1} + P_{i-2} \quad \text{and} \quad Q_i = q_i \cdot Q_{i-1} + Q_{i-2}.$$

The final four entries in the box satisfy

$$a \cdot Q_{t-1} - b \cdot P_{t-1} = (-1)^t.$$

Multiplying both sides by $(-1)^t$ gives the solution $u = (-1)^t Q_{t-1}$ and $v = (-1)^{t+1} P_{t-1}$ to the equation $au + bv = 1$.

Figure 1.3: Solving $au + bv = 1$ using the Euclidean algorithm

1.3 Modular Arithmetic

You may have encountered "clock arithmetic" in grade school, where after you get to 12, the next number is 1. This leads to odd-looking equations such as

$$6 + 9 = 3 \quad \text{and} \quad 2 - 3 = 11.$$

These look strange, but they are true using clock arithmetic, since for example 11 o'clock is 3 h before 2 o'clock. So what we are really doing is first computing $2 - 3 = -1$ and then adding 12 to the answer. Similarly, 9 h after 6 o'clock is 3 o'clock, since $6 + 9 - 12 = 3$.

The theory of *congruences* is a powerful method in number theory that is based on the simple idea of clock arithmetic.

Definition. Let $m \geq 1$ be an integer. We say that the integers a and b are *congruent modulo m* if their difference $a - b$ is divisible by m. We write

$$a \equiv b \pmod{m}$$

to indicate that a and b are congruent modulo m. The number m is called the *modulus*.

Our clock examples may be written as congruences using the modulus $m = 12$:

$$6 + 9 = 15 \equiv 3 \pmod{12} \qquad \text{and} \qquad 2 - 3 = -1 \equiv 11 \pmod{12}.$$

Example 1.12. We have

$$17 \equiv 7 \pmod{5}, \qquad \text{since 5 divides } 10 = 17 - 7.$$

On the other hand,

$$19 \not\equiv 6 \pmod{11}, \qquad \text{since 11 does not divide } 13 = 19 - 6.$$

Notice that the numbers satisfying

$$a \equiv 0 \pmod{m}$$

are the numbers that are divisible by m, i.e., the multiples of m.

The reason that congruence notation is so useful is that congruences behave much like equalities, as the following proposition indicates.

Proposition 1.13. *Let $m \geq 1$ be an integer.*
(a) *If $a_1 \equiv a_2 \pmod{m}$ and $b_1 \equiv b_2 \pmod{m}$, then*

$$a_1 \pm b_1 \equiv a_2 \pm b_2 \pmod{m} \qquad \textit{and} \qquad a_1 \cdot b_1 \equiv a_2 \cdot b_2 \pmod{m}.$$

(b) *Let a be an integer. Then*

$$a \cdot b \equiv 1 \pmod{m} \textit{ for some integer } b \textit{ if and only if } \gcd(a, m) = 1.$$

Further, if $a \cdot b_1 \equiv a \cdot b_2 \equiv 1 \pmod{m}$, then $b_1 \equiv b_2 \pmod{m}$. We call b the (multiplicative) inverse of a modulo m.

Proof. (a) We leave this as an exercise; see Exercise 1.15.
(b) Suppose first that $\gcd(a, m) = 1$. Then Theorem 1.11 tells us that we can find integers u and v satisfying $au + mv = 1$. This means that $au - 1 = -mv$ is divisible by m, so by definition, $au \equiv 1 \pmod{m}$. In other words, we can take $b = u$.

For the other direction, suppose that a has an inverse modulo m, say $a \cdot b \equiv 1 \pmod{m}$. This means that $ab - 1 = cm$ for some integer c. It follows that $\gcd(a, m)$ divides $ab - cm = 1$, so $\gcd(a, m) = 1$. This completes the proof that a has an inverse modulo m if and only if $\gcd(a, m) = 1$. It remains to show that the inverse is unique modulo m.

So suppose that $a \cdot b_1 \equiv a \cdot b_2 \equiv 1 \pmod{m}$. Then

$$b_1 \equiv b_1 \cdot 1 \equiv \beta_1 \cdot (a \cdot b_2) \equiv (b_1 \cdot a) \cdot b_2 \equiv 1 \cdot b_2 \equiv b_2 \pmod{m},$$

which completes the proof of Proposition 1.13. □

Proposition 1.13(b) says that if $\gcd(a,m) = 1$, then there exists an inverse b of a modulo m. This has the curious consequence that the fraction $a^{-1} = 1/a$ has a meaningful interpretation in the world of integers modulo m, namely a^{-1} modulo m is the unique number b modulo m satisfying the congruence $ab \equiv 1 \pmod{m}$.

Example 1.14. We take $m = 5$ and $a = 2$. Clearly $\gcd(2,5) = 1$, so there exists an inverse to 2 modulo 5. The inverse of 2 modulo 5 is 3, since $2 \cdot 3 \equiv 1 \pmod{5}$, so $2^{-1} \equiv 3 \pmod{5}$. Similarly $\gcd(4,15) = 1$ so 4^{-1} exists modulo 15. In fact $4 \cdot 4 \equiv 1 \pmod{15}$ so 4 is its own inverse modulo 15.

We can even work with fractions a/d modulo m as long as the denominator is relatively prime to m. For example, we can compute 5/7 modulo 11 by first observing that $7 \cdot 8 \equiv 1 \pmod{11}$, so $7^{-1} \equiv 8 \pmod{11}$. Then

$$\frac{5}{7} = 5 \cdot 7^{-1} \equiv 5 \cdot 8 \equiv 40 \equiv 7 \pmod{11}.$$

Remark 1.15. In the preceding examples it was easy to find inverses modulo m by trial and error. However, when m is large, it is more challenging to compute a^{-1} modulo m. Note that we showed that inverses exist by using the extended Euclidean algorithm (Theorem 1.11). In order to actually compute the u and v that appear in the equation $au + mv = \gcd(a,m)$, we can apply the Euclidean algorithm directly as we did in Example 1.10, or we can use the somewhat more efficient box method described at the end of the preceding section, or we can use the algorithm given in Exercise 1.12. In any case, since the Euclidean algorithm takes at most $2\log_2(b) + 2$ iterations to compute $\gcd(a,b)$, it takes only a small multiple of $\log_2(m)$ steps to compute a^{-1} modulo m.

We now continue our development of the theory of modular arithmetic. If a divided by m has quotient q and remainder r, it can be written as

$$a = m \cdot q + r \qquad \text{with } 0 \le r < m.$$

This shows that $a \equiv r \pmod{m}$ for some integer r between 0 and $m - 1$, so if we want to work with integers modulo m, it is enough to use the integers $0 \le r < m$. This prompts the following definition.

Definition. We write

$$\mathbb{Z}/m\mathbb{Z} = \{0, 1, 2, \ldots, m-1\}$$

and call $\mathbb{Z}/m\mathbb{Z}$ the *ring of integers modulo m*. We add and multiply elements of $\mathbb{Z}/m\mathbb{Z}$ by adding or multiplying them as integers and then dividing the result by m and taking the remainder in order to obtain an element in $\mathbb{Z}/m\mathbb{Z}$.

Figure 1.4 illustrates the ring $\mathbb{Z}/5\mathbb{Z}$ by giving complete addition and multiplication tables modulo 5.

+	0	1	2	3	4
0	0	1	2	3	4
1	1	2	3	4	0
2	2	3	4	0	1
3	3	4	0	1	2
4	4	0	1	2	3

·	0	1	2	3	4
0	0	0	0	0	0
1	0	1	2	3	4
2	0	2	4	1	3
3	0	3	1	4	2
4	0	4	3	2	1

Figure 1.4: Addition and multiplication tables modulo 5

Remark 1.16. If you have studied ring theory, you will recognize that $\mathbb{Z}/m\mathbb{Z}$ is the quotient ring of $\mathbb{Z}$ by the principal ideal $m\mathbb{Z}$, and that the numbers $0, 1, \ldots, m-1$ are actually coset representatives for the congruence classes that comprise the elements of $\mathbb{Z}/m\mathbb{Z}$. For a discussion of congruence classes and general quotient rings, see Sect. 2.10.2.

Definition. Proposition 1.13(b) tells us that a has an inverse modulo m if and only if $\gcd(a, m) = 1$. Numbers that have inverses are called *units*. We denote the set of all units by

$$\begin{aligned}(\mathbb{Z}/m\mathbb{Z})^* &= \{a \in \mathbb{Z}/m\mathbb{Z} : \gcd(a, m) = 1\} \\ &= \{a \in \mathbb{Z}/m\mathbb{Z} : a \text{ has an inverse modulo } m\}.\end{aligned}$$

The set $(\mathbb{Z}/m\mathbb{Z})^*$ is called the *group of units modulo m*.

Notice that if a_1 and a_2 are units modulo m, then so is $a_1 a_2$. (Do you see why this is true?) So when we multiply two units, we always get a unit. On the other hand, if we add two units, we often do not get a unit.

Example 1.17. The group of units modulo 24 is

$$(\mathbb{Z}/24\mathbb{Z})^* = \{1, 5, 7, 11, 13, 17, 19, 23\}.$$

Similarly, the group of units modulo 7 is

$$(\mathbb{Z}/7\mathbb{Z})^* = \{1, 2, 3, 4, 5, 6\},$$

since every number between 1 and 6 is relatively prime to 7. The multiplication tables for $(\mathbb{Z}/24\mathbb{Z})^*$ and $(\mathbb{Z}/7\mathbb{Z})^*$ are illustrated in Fig. 1.5.

In many of the cryptosystems that we will study, it is important to know how many elements are in the unit group modulo m. This quantity is sufficiently ubiquitous that we give it a name.

Definition. *Euler's phi function* (also sometimes known as *Euler's totient function*) is the function $\phi(m)$ defined by the rule

$$\phi(m) = \#(\mathbb{Z}/m\mathbb{Z})^* = \#\{0 \le a < m : \gcd(a, m) = 1\}.$$

For example, we see from Example 1.17 that $\phi(24) = 8$ and $\phi(7) = 6$.

·	1	5	7	11	13	17	19	23
1	1	5	7	11	13	17	19	23
5	5	1	11	7	17	13	23	19
7	7	11	1	5	19	23	13	17
11	11	7	5	1	23	19	17	13
13	13	17	19	23	1	5	7	11
17	17	13	23	19	5	1	11	7
19	19	23	13	17	7	11	1	5
23	23	19	17	13	11	7	5	1

Unit group modulo 24

·	1	2	3	4	5	6
1	1	2	3	4	5	6
2	2	4	6	1	3	5
3	3	6	2	5	1	4
4	4	1	5	2	6	3
5	5	3	1	6	4	2
6	6	5	4	3	2	1

Unit group modulo 7

Figure 1.5: The unit groups $(\mathbb{Z}/24\mathbb{Z})^*$ and $(\mathbb{Z}/7\mathbb{Z})^*$

1.3.1 Modular Arithmetic and Shift Ciphers

Recall that the Caesar (or shift) cipher studied in Sect. 1.1 works by shifting each letter in the alphabet a fixed number of letters. We can describe a shift cipher mathematically by assigning a number to each letter as in Table 1.7.

a	b	c	d	e	f	g	h	i	j	k	l	m	n	o	p	q	r	s	t	u	v	w	x	y	z
0	1	2	3	4	5	6	7	8	9	10	11	12	13	14	15	16	17	18	19	20	21	22	23	24	25

Table 1.7: Assigning numbers to letters

Then a shift cipher with shift k takes a plaintext letter corresponding to the number p and assigns it to the ciphertext letter corresponding to the number $p + k \bmod 26$. Notice how the use of modular arithmetic, in this case modulo 26, simplifies the description of the shift cipher. The shift amount serves as both the encryption key and the decryption key. Encryption is given by the formula

$$(\text{Ciphertext Letter}) \equiv (\text{Plaintext Letter}) + (\text{Secret Key}) \pmod{26},$$

and decryption works by shifting in the opposite direction,

$$\text{(Plaintext Letter)} \equiv \text{(Ciphertext Letter)} - \text{(Secret Key)} \pmod{26}.$$

More succinctly, if we let

$$p = \text{Plaintext Letter}, \qquad c = \text{Ciphertext Letter}, \qquad k = \text{Secret Key},$$

then

$$\underbrace{c \equiv p + k \pmod{26}}_{\text{Encryption}} \qquad \text{and} \qquad \underbrace{p \equiv c - k \pmod{26}}_{\text{Decryption}}.$$

1.3.2 The Fast Powering Algorithm

In some cryptosystems that we will study, for example the RSA and Diffie–Hellman cryptosystems, Alice and Bob are required to compute large powers of a number g modulo another number N, where N may have hundreds of digits. The naive way to compute g^A is by repeated multiplication by g. Thus

$$g_1 \equiv g \pmod{N}, \quad g_2 \equiv g \cdot g_1 \pmod{N}, \quad g_3 \equiv g \cdot g_2 \pmod{N},$$
$$g_4 \equiv g \cdot g_3 \pmod{N}, \quad g_5 \equiv g \cdot g_4 \pmod{N}, \ldots.$$

It is clear that $g_A \equiv g^A \pmod{N}$, but if A is large, this algorithm is completely impractical. For example, if $A \approx 2^{1000}$, then the naive algorithm would take longer than the estimated age of the universe! Clearly if it is to be useful, we need to find a better way to compute $g^A \pmod{N}$.

The idea is to use the binary expansion of the exponent A to convert the calculation of g^A into a succession of squarings and multiplications. An example will make the idea clear, after which we give a formal description of the method.

Example 1.18. Suppose that we want to compute $3^{218} \pmod{1000}$. The first step is to write 218 as a sum of powers of 2,

$$218 = 2 + 2^3 + 2^4 + 2^6 + 2^7.$$

Then 3^{218} becomes

$$3^{218} = 3^{2+2^3+2^4+2^6+2^7} = 3^2 \cdot 3^{2^3} \cdot 3^{2^4} \cdot 3^{2^6} \cdot 3^{2^7}. \tag{1.3}$$

Notice that it is relatively easy to compute the sequence of values

$$3, \quad 3^2, \quad 3^{2^2}, \quad 3^{2^3}, \quad 3^{2^4}, \ldots,$$

since each number in the sequence is the square of the preceding one. Further, since we only need these values modulo 1000, we never need to store more than three digits. Table 1.8 lists the powers of 3 modulo 1000 up to 3^{2^7}. Creating Table 1.8 requires only 7 multiplications, despite the fact that the number $3^{2^7} = 3^{128}$ has quite a large exponent, because each successive entry in the table is equal to the square of the previous entry.

i	0	1	2	3	4	5	6	7
$3^{2^i} \pmod{1000}$	3	9	81	561	721	841	281	961

Table 1.8: Successive square powers of 3 modulo 1000

We use (1.3) to decide which powers from Table 1.8 are needed to compute 3^{218}. Thus

$$\begin{aligned} 3^{218} &= 3^2 \cdot 3^{2^3} \cdot 3^{2^4} \cdot 3^{2^6} \cdot 3^{2^7} \\ &\equiv 9 \cdot 561 \cdot 721 \cdot 281 \cdot 961 \pmod{1000} \\ &\equiv 489 \pmod{1000}. \end{aligned}$$

We note that in computing the product $9 \cdot 561 \cdot 721 \cdot 281 \cdot 961$, we may reduce modulo 1000 after each multiplication, so we never need to deal with very large numbers. We also observe that it has taken us only 11 multiplications to compute $3^{218} \pmod{1000}$, a huge savings over the naive approach. And for larger exponents we would save even more.

The general approach used in Example 1.18 goes by various names, including the *Fast Powering Algorithm* and the *Square-and-Multiply Algorithm.*[8] We now describe the algorithm more formally.

The Fast Powering Algorithm

Step 1. Compute the binary expansion of A as

$$A = A_0 + A_1 \cdot 2 + A_2 \cdot 2^2 + A_3 \cdot 2^3 + \cdots + A_r \cdot 2^r \quad \text{with } A_0, \ldots, A_r \in \{0, 1\},$$

where we may assume that $A_r = 1$.

Step 2. Compute the powers $g^{2^i} \pmod N$ for $0 \le i \le r$ by successive squaring,

$$\begin{aligned} a_0 &\equiv g && \pmod N \\ a_1 &\equiv a_0^2 \equiv g^2 && \pmod N \\ a_2 &\equiv a_1^2 \equiv g^{2^2} && \pmod N \\ a_3 &\equiv a_2^2 \equiv g^{2^3} && \pmod N \\ &\vdots && \vdots \\ a_r &\equiv a_{r-1}^2 \equiv g^{2^r} && \pmod N. \end{aligned}$$

Each term is the square of the previous one, so this requires r multiplications.

[8]The first known recorded description of the fast powering algorithm appeared in India before 200 BC, while the first reference outside India dates to around 950 AD. See [66, page 441] for a brief discussion and further references.

Step 3. Compute $g^A \pmod N$ using the formula

$$\begin{aligned} g^A &= g^{A_0 + A_1 \cdot 2 + A_2 \cdot 2^2 + A_3 \cdot 2^3 + \cdots + A_r \cdot 2^r} \\ &= g^{A_0} \cdot (g^2)^{A_1} \cdot (g^{2^2})^{A_2} \cdot (g^{2^3})^{A_3} \cdots (g^{2^r})^{A_r} \\ &\equiv a_0^{A_0} \cdot a_1^{A_1} \cdot a_2^{A_2} \cdot a_3^{A_3} \cdots a_r^{A_r} \pmod N. \end{aligned} \tag{1.4}$$

Note that the quantities $a_0, a_1, \ldots, a_r$ were computed in Step 2. Thus the product (1.4) can be computed by looking up the values of the a_i's whose exponent A_i is 1 and then multiplying them together. This requires at most another r multiplications.

Running Time. It takes at most $2r$ multiplications modulo N to compute g^A. Since $A \geq 2^r$, we see that it takes at most $2\log_2(A)$ multiplications[9] modulo N to compute g^A. Thus even if A is very large, say $A \approx 2^{1000}$, it is easy for a computer to do the approximately 2000 multiplications needed to calculate 2^A modulo N.

Efficiency Issues. There are various ways in which the square-and-multiply algorithm can be made somewhat more efficient, in particular regarding eliminating storage requirements; see Exercise 1.25 for an example.

1.4 Prime Numbers, Unique Factorization, and Finite Fields

In Sect. 1.3 we studied modular arithmetic and saw that it makes sense to add, subtract, and multiply integers modulo m. Division, however, can be problematic, since we can divide by a in $\mathbb{Z}/m\mathbb{Z}$ only if $\gcd(a, m) = 1$. But notice that if the integer m is a prime, then we can divide by every nonzero element of $\mathbb{Z}/m\mathbb{Z}$. We start with a brief discussion of prime numbers before returning to the ring $\mathbb{Z}/p\mathbb{Z}$ with p prime.

Definition. An integer p is called a *prime* if $p \geq 2$ and if the only positive integers dividing p are 1 and p.

For example, the first ten primes are $2, 3, 5, 7, 11, 13, 17, 19, 23, 29$, while the hundred thousandth prime is 1299709 and the millionth is 15485863. There are infinitely many primes, a fact that was known in ancient Greece and appears as a theorem in Euclid's *Elements*. (See Exercise 1.28.)

A prime p is defined in terms of the numbers that divide p. So the following proposition, which describes a useful property of numbers that are divisible by p, is not obvious and needs to be carefully proved. Notice that the proposition is false for composite numbers. For example, 6 divides $3 \cdot 10$, but 6 divides neither 3 nor 10.

---

[9] Note that $\log_2(A)$ means the usual logarithm to the base 2, not the so-called discrete logarithm that will be discussed in Chap. 2.

Proposition 1.19. *Let p be a prime number, and suppose that p divides the product ab of two integers a and b. Then p divides at least one of a and b.*

More generally, if p divides a product of integers, say

$$p \mid a_1 a_2 \cdots a_n,$$

then p divides at least one of the individual a_i.

Proof. Let $g = \gcd(a, p)$. Then $g \mid p$, so either $g = 1$ or $g = p$. If $g = p$, then $p \mid a$ (since $g \mid a$), so we are done. Otherwise, $g = 1$ and Theorem 1.11 tells us that we can find integers u and v satisfying $au + pv = 1$. We multiply both sides of the equation by b to get

$$abu + pbv = b. \tag{1.5}$$

By assumption, p divides the product ab, and certainly p divides pbv, so p divides both terms on the left-hand side of (1.5). Hence it divides the right-hand side, which shows that p divides b and completes the proof of Proposition 1.19.

To prove the more general statement, we write the product as $a_1(a_2 \cdots a_n)$ and apply the first statement with $a = a_1$ and $b = a_2 \cdots a_n$. If $p \mid a_1$, we're done. Otherwise, $p \mid a_2 \cdots a_n$, so writing this as $a_2(a_3 \cdots a_n)$, the first statement tells us that either $p \mid a_2$ or $p \mid a_3 \cdots a_n$. Continuing in this fashion, we must eventually find some a_i that is divisible by p. □

As an application of Proposition 1.19, we prove that every positive integer has an essentially unique factorization as a product of primes.

Theorem 1.20 (The Fundamental Theorem of Arithmetic). *Let $a \geq 2$ be an integer. Then a can be factored as a product of prime numbers*

$$a = p_1^{e_1} \cdot p_2^{e_2} \cdot p_3^{e_3} \cdots p_r^{e_r}.$$

Further, other than rearranging the order of the primes, this factorization into prime powers is unique.

Proof. It is not hard to prove that every $a \geq 2$ can be factored into a product of primes. It is tempting to assume that the uniqueness of the factorization is also obvious. However, this is not the case; unique factorization is a somewhat subtle property of the integers. We will prove it using the general form of Proposition 1.19. (For an example of a situation in which unique factorization fails to be true, see the $\mathbb{E}$-zone described in [137, Chapter 7].)

Suppose that a has two factorizations into products of primes,

$$a = p_1 p_2 \cdots p_s = q_1 q_2 \cdots q_t, \tag{1.6}$$

where the p_i and q_j are all primes, not necessarily distinct, and s does not necessarily equal t. Since $p_1 \mid a$, we see that p_1 divides the product $q_1 q_2 q_3 \cdots q_t$. Thus by the general form of Proposition 1.19, we find that p_1 divides one of the q_i. Rearranging the order of the q_i if necessary, we may assume that $p_1 \mid q_1$. But p_1 and q_1 are both primes, so we must have $p_1 = q_1$. This allows us to cancel them from both sides of (1.6), which yields

$$p_2 p_3 \cdots p_s = q_2 q_3 \cdots q_t.$$

Repeating this process s times, we ultimately reach an equation of the form

$$1 = q_{t-s} q_{t-s+1} \cdots q_t.$$

It follows immediately that $t = s$ and that the original factorizations of a were identical up to rearranging the order of the factors. (For a more detailed proof of the fundamental theorem of arithmetic, see any basic number theory textbook, for example [35, 52, 59, 100, 111, 137].) $\square$

Definition. The fundamental theorem of arithmetic (Theorem 1.20) says that in the factorization of a positive integer a into primes, each prime p appears to a particular power. We denote this power by $\operatorname{ord}_p(a)$ and call it the *order* (or *exponent*) *of p in a*. (For convenience, we set $\operatorname{ord}_p(1) = 0$ for all primes.)

For example, the factorization of 1728 is $1728 = 2^6 \cdot 3^3$, so

$$\operatorname{ord}_2(1728) = 6, \quad \operatorname{ord}_3(1728) = 3, \quad \text{and} \quad \operatorname{ord}_p(1728) = 0 \text{ for all primes } p \geq 5.$$

Using the ord_p notation, the factorization of a can be succinctly written as

$$a = \prod_{\text{primes } p} p^{\operatorname{ord}_p(a)}.$$

Note that this product makes sense, since $\operatorname{ord}_p(a)$ is zero for all but finitely many primes.

It is useful to view ord_p as a function

$$\operatorname{ord}_p : \{1, 2, 3, \ldots\} \longrightarrow \{0, 1, 2, 3, \ldots\}. \tag{1.7}$$

This function has a number of interesting properties, some of which are described in Exercise 1.31.

We now observe that if p is a prime, then every nonzero number modulo p has a multiplicative inverse modulo p. This means that when we do arithmetic modulo a prime p, not only can we add, subtract, multiply, but we can also divide by nonzero numbers, just as we can with real numbers. This property of primes is sufficiently important that we formally state it as a proposition.

Proposition 1.21. *Let p be a prime. Then every nonzero element a in $\mathbb{Z}/p\mathbb{Z}$ has a multiplicative inverse, that is, there is a number b satisfying*

$$ab \equiv 1 \pmod{p}.$$

We denote this value of b by $a^{-1} \bmod p$, or if p has already been specified, then simply by a^{-1}.

Proof. This proposition is a special case of Proposition 1.13(b) using the prime modulus p, since if $a \in \mathbb{Z}/p\mathbb{Z}$ is not zero, then $\gcd(a, p) = 1$. □

Remark 1.22. The extended Euclidean algorithm (Theorem 1.11) gives us an efficient computational method for computing $a^{-1} \bmod p$. We simply solve the equation

$$au + pv = 1 \quad \text{in integers } u \text{ and } v,$$

and then $u = a^{-1} \bmod p$. For an alternative method of computing $a^{-1} \bmod p$, see Remark 1.26.

Proposition 1.21 can be restated by saying that if p is prime, then

$$(\mathbb{Z}/p\mathbb{Z})^* = \{1, 2, 3, 4, \ldots, p-1\}.$$

In other words, when the 0 element is removed from $\mathbb{Z}/p\mathbb{Z}$, the remaining elements are units and closed under multiplication.

Definition. If p is prime, then the set $\mathbb{Z}/p\mathbb{Z}$ of integers modulo p with its addition, subtraction, multiplication, and division rules is an example of a *field*. If you have studied abstract algebra (or see Sect. 2.10), you know that a field is the general name for a (commutative) ring in which every nonzero element has a multiplicative inverse. You are already familiar with some other fields, for example the field of real numbers $\mathbb{R}$, the field of rational numbers (fractions) $\mathbb{Q}$, and the field of complex numbers $\mathbb{C}$.

The field $\mathbb{Z}/p\mathbb{Z}$ of integers modulo p has only finitely many elements. It is a *finite field* and is often denoted by $\mathbb{F}_p$. Thus $\mathbb{F}_p$ and $\mathbb{Z}/p\mathbb{Z}$ are really just two different notations for the same object.[10] Similarly, we write $\mathbb{F}_p^*$ interchangeably for the group of units $(\mathbb{Z}/p\mathbb{Z})^*$. Finite fields are of fundamental importance throughout cryptography, and indeed throughout all of mathematics.

Remark 1.23. Although $\mathbb{Z}/p\mathbb{Z}$ and $\mathbb{F}_p$ are used to denote the same concept, equality of elements is expressed somewhat differently in the two settings. For $a, b \in \mathbb{F}_p$, the equality of a and b is denoted by $a = b$, while for $a, b \in \mathbb{Z}/p\mathbb{Z}$, the equality of a and b is denoted by equivalence modulo p, i.e., $a \equiv b \pmod{p}$.

1.5 Powers and Primitive Roots in Finite Fields

The application of finite fields in cryptography often involves raising elements of $\mathbb{F}_p$ to high powers. As a practical matter, we know how to do this efficiently using the powering algorithm described in Sect. 1.3.2. In this section

[10]Finite fields are also sometimes called *Galois fields*, after Évariste Galois, who studied them in the nineteenth century. Yet another notation for $\mathbb{F}_p$ is $\mathrm{GF}(p)$, in honor of Galois. And yet one more notation for $\mathbb{F}_p$ that you may run across is $\mathbb{Z}_p$, although in number theory the notation $\mathbb{Z}_p$ is more commonly reserved for the ring of p-adic integers.

we investigate powers in $\mathbb{F}_p$ from a purely mathematical viewpoint, prove a fundamental result due to Fermat, and state an important property of the group of units $\mathbb{F}_p^*$.

We begin with a simple example. Table 1.9 lists the powers of $1, 2, 3, \ldots, 6$ modulo the prime 7.

$1^1 \equiv 1$	$1^2 \equiv 1$	$1^3 \equiv 1$	$1^4 \equiv 1$	$1^5 \equiv 1$	$1^6 \equiv 1$
$2^1 \equiv 2$	$2^2 \equiv 4$	$2^3 \equiv 1$	$2^4 \equiv 2$	$2^5 \equiv 4$	$2^6 \equiv 1$
$3^1 \equiv 3$	$3^2 \equiv 2$	$3^3 \equiv 6$	$3^4 \equiv 4$	$3^5 \equiv 5$	$3^6 \equiv 1$
$4^1 \equiv 4$	$4^2 \equiv 2$	$4^3 \equiv 1$	$4^4 \equiv 4$	$4^5 \equiv 2$	$4^6 \equiv 1$
$5^1 \equiv 5$	$5^2 \equiv 4$	$5^3 \equiv 6$	$5^4 \equiv 2$	$5^5 \equiv 3$	$5^6 \equiv 1$
$6^1 \equiv 6$	$6^2 \equiv 1$	$6^3 \equiv 6$	$6^4 \equiv 1$	$6^5 \equiv 6$	$6^6 \equiv 1$

Table 1.9: Powers of numbers modulo 7

There are quite a few interesting patterns visible in Table 1.9, including in particular the fact that the right-hand column consists entirely of ones. We can restate this observation by saying that

$$a^6 \equiv 1 \pmod 7 \qquad \text{for every } a = 1, 2, 3, \ldots, 6.$$

Of course, this cannot be true for all values of a, since if a is a multiple of 7, then so are all of its powers, so in that case $a^n \equiv 0 \pmod 7$. On the other hand, if a is not divisible by 7, then a is congruent to one of the values $1, 2, 3, \ldots, 6$ modulo 7. Hence

$$a^6 \equiv \begin{cases} 1 \pmod 7 & \text{if } 7 \nmid a, \\ 0 \pmod 7 & \text{if } 7 \mid a. \end{cases}$$

Further experiments with other primes suggest that this example reflects a general fact.

Theorem 1.24 (Fermat's Little Theorem). *Let p be a prime number and let a be any integer. Then*

$$a^{p-1} \equiv \begin{cases} 1 \pmod p & \text{if } p \nmid a, \\ 0 \pmod p & \text{if } p \mid a. \end{cases}$$

Proof. There are many proofs of Fermat's little theorem. If you have studied group theory, the quickest proof is to observe that the nonzero elements in $\mathbb{F}_p$ form a group $\mathbb{F}_p^*$ of order $p-1$, so by Lagrange's theorem, every element of $\mathbb{F}_p^*$ has order dividing $p-1$. For those who have not yet taken a course in group theory, we provide a direct proof.

If $p \mid a$, then it is clear that every power of a is divisible by p. So we only need to consider the case that $p \nmid a$. We now look at the list of numbers

$$a, \quad 2a, \quad 3a, \quad \ldots, \quad (p-1)a \qquad \text{reduced modulo } p. \tag{1.8}$$

There are $p-1$ numbers in this list, and we claim that they are all different. To see why, take any two of them, say $ja \bmod p$ and $ka \bmod p$, and suppose that they are the same. This means that

$$ja \equiv ka \pmod{p}, \qquad \text{and hence that} \qquad (j-k)a \equiv 0 \pmod{p}.$$

Thus p divides the product $(j-k)a$. Proposition 1.19 tells us that either p divides $j-k$ or p divides a. However, we have assumed that p does not divide a, so we conclude that p divides $j-k$. But both j and k are between 1 and $p-1$, so their difference $j-k$ is between $-(p-2)$ and $p-2$. There is only one number between $-(p-2)$ and $p-2$ that is divisible by p, and that number is zero! This proves that $j-k=0$, which means that $ja=ka$. We have thus shown that the $p-1$ numbers in the list (1.8) are all different. They are also nonzero, since $1, 2, 3, \ldots, p-1$ and a are not divisible by p.

To recapitulate, we have shown that the list of numbers (1.8) consists of $p-1$ *distinct* numbers between 1 and $p-1$. But there are only $p-1$ distinct numbers between 1 and $p-1$, so the list of numbers (1.8) must simply be the list of numbers $1, 2, \ldots, p-1$ in some mixed up order.

Now consider what happens when we multiply together all of the numbers $a, 2a, 3a, \ldots, (p-1)a$ in the list (1.8) and reduce the product modulo p. This is the same as multiplying together all of the numbers $1, 2, 3, \ldots, p-1$ modulo p, so we get a congruence

$$a \cdot 2a \cdot 3a \cdots (p-1)a \equiv 1 \cdot 2 \cdot 3 \cdots (p-1) \pmod{p}.$$

There are $p-1$ copies of a appearing on the left-hand side. We factor these out and use factorial notation $(p-1)! = 1 \cdot 2 \cdots (p-1)$ to obtain

$$a^{p-1} \cdot (p-1)! \equiv (p-1)! \pmod{p}.$$

Finally, we are allowed to cancel $(p-1)!$ from both sides, since it is not divisible by p. (We are using the fact that $\mathbb{F}_p$ is a field, so we are allowed to divide by any nonzero number.) This yields

$$a^{p-1} \equiv 1 \pmod{p},$$

which completes the proof of Fermat's "little" theorem.[11] □

[11] You may wonder why Theorem 1.24 is called a "little" theorem. The reason is to distinguish it from Fermat's "big" theorem, which is the famous assertion that $x^n + y^n = z^n$ has no solutions in positive integers x, y, z if $n \geq 3$. It is unlikely that Fermat himself could prove this big theorem, but in 1996, more than three centuries after Fermat's era, Andrew Wiles finally found a proof.

Example 1.25. The number $p = 15485863$ is prime, so Fermat's little theorem (Theorem 1.24) tells us that

$$2^{15485862} \equiv 1 \pmod{15485863}.$$

Thus without doing any computing, we know that the number $2^{15485862} - 1$, a number having more than two million digits, is a multiple of 15485863.

Remark 1.26. Fermat's little theorem (Theorem 1.24) and the fast powering algorithm (Sect. 1.3.2) provide us with a reasonably efficient method of computing inverses modulo p, namely

$$a^{-1} \equiv a^{p-2} \pmod{p}.$$

This congruence is true because if we multiply a^{p-2} by a, then Fermat's theorem tells us that the product is equal to 1 modulo p. This gives an alternative to the extended Euclidean algorithm method described in Remark 1.22. In practice, the two algorithms tend to take about the same amount of time, although there are variants of the Euclidean algorithm that are somewhat faster in practice; see for example [66, Chapter 4.5.3, Theorem E].

Example 1.27. We compute the inverse of 7814 modulo 17449 in two ways. First,

$$7814^{-1} \equiv 7814^{17447} \equiv 1284 \pmod{17449}.$$

Second, we use the extended Euclidean algorithm to solve

$$7814u + 17449v = 1.$$

The solution is $(u, v) = (1284, -575)$, so $7814^{-1} \equiv 1284 \pmod{17449}$.

Example 1.28. Consider the number $m = 15485207$. Using the powering algorithm, it is not hard to compute (on a computer)

$$2^{m-1} = 2^{15485206} \equiv 4136685 \pmod{15485207}.$$

We did not get the value 1, so it seems that Fermat's little theorem is not true for m. What does that tell us? If m were prime, then Fermat's little theorem says that we would have obtained 1. Hence the fact that we did not get 1 *proves* that the number $m = 15485207$ is not prime.

Think about this for a minute, because it's actually a bit astonishing. By a simple computation, we have conclusively proven that m is not prime, yet we do not know any of its factors![12]

Fermat's little theorem tells us that if a is an integer not divisible by p, then $a^{p-1} \equiv 1 \pmod{p}$. However, for any particular value of a, there may well be smaller powers of a that are congruent to 1. We define the *order of* a *modulo* p to be the smallest exponent $k \geq 1$ such that[13]

[12] The prime factorization of m is $m = 15485207 = 3853 \cdot 4019$.

[13] We earlier defined the *order of* p *in* a to be the exponent of p when a is factored into primes. Thus unfortunately, the word "order" has two different meanings. You will need to judge which one is meant from the context.

$$a^k \equiv 1 \pmod{p}.$$

Proposition 1.29. *Let p be a prime and let a be an integer not divisible by p. Suppose that $a^n \equiv 1 \pmod{p}$. Then the order of a modulo p divides n. In particular, the order of a divides $p-1$.*

Proof. Let k be the order of a modulo p, so by definition $a^k \equiv 1 \pmod{p}$, and k is the smallest positive exponent with this property. We are given that $a^n \equiv 1 \pmod{p}$. We divide n by k to obtain

$$n = kq + r \quad \text{with } 0 \le r < k.$$

Then

$$1 \equiv a^n \equiv a^{kq+r} \equiv (a^k)^q \cdot a^r \equiv 1^q \cdot a^r \equiv a^r \pmod{p}.$$

But $r < k$, so the fact that k is the smallest positive power of a that is congruent to 1 tells us that r must equal 0. Therefore $n = kq$, so k divides n.

Finally, Fermat's little theorem tells us that $a^{p-1} \equiv 1 \pmod{p}$, so k divides $p-1$. □

Fermat's little theorem describes a special property of the units (i.e., the nonzero elements) in a finite field. We conclude this section with a brief discussion of another property that is quite important both theoretically and practically.

Theorem 1.30 (Primitive Root Theorem). *Let p be a prime number. Then there exists an element $g \in \mathbb{F}_p^*$ whose powers give every element of $\mathbb{F}_p^*$, i.e.,*

$$\mathbb{F}_p^* = \{1, g, g^2, g^3, \ldots, g^{p-2}\}.$$

Elements with this property are called primitive roots of $\mathbb{F}_p$ *or* generators of $\mathbb{F}_p^*$. *They are the elements of $\mathbb{F}_p^*$ having order $p-1$.*

Proof. See [137, Chapter 20] or one of the texts [35, 52, 59, 100, 111]. □

Example 1.31. The field $\mathbb{F}_{11}$ has 2 as a primitive root, since in $\mathbb{F}_{11}$,

$$\begin{array}{lllll} 2^0 = 1 & 2^1 = 2 & 2^2 = 4 & 2^3 = 8 & 2^4 = 5 \\ 2^5 = 10 & 2^6 = 9 & 2^7 = 7 & 2^8 = 3 & 2^9 = 6. \end{array}$$

Thus all 10 nonzero elements of $\mathbb{F}_{11}$ have been generated as powers of 2. On the other hand, 2 is not a primitive root for $\mathbb{F}_{17}$, since in $\mathbb{F}_{17}$,

$$\begin{array}{lllll} 2^0 = 1 & 2^1 = 2 & 2^2 = 4 & 2^3 = 8 & 2^4 = 16 \\ 2^5 = 15 & 2^6 = 13 & 2^7 = 9 & 2^8 = 1, & \end{array}$$

so we get back to 1 before obtaining all 16 nonzero values modulo 17. However, it turns out that 3 is a primitive root for 17, since in $\mathbb{F}_{17}$,

$$\begin{array}{llllll} 3^0 = 1 & 3^1 = 3 & 3^2 = 9 & 3^3 = 10 & 3^4 = 13 & 3^5 = 5 \\ 3^6 = 15 & 3^7 = 11 & 3^8 = 16 & 3^9 = 14 & 3^{10} = 8 & 3^{11} = 7 \\ 3^{12} = 4 & 3^{13} = 12 & 3^{14} = 2 & 3^{15} = 6. & & \end{array}$$

Remark 1.32. If p is large, then the finite field $\mathbb{F}_p$ has quite a few primitive roots. The precise formula says that $\mathbb{F}_p$ has exactly $\phi(p-1)$ primitive roots, where ϕ is Euler's phi function (see page 22). For example, you can check that the following is a complete list of the primitive roots for $\mathbb{F}_{29}$:

$$\{2, 3, 8, 10, 11, 14, 15, 18, 19, 21, 26, 27\}.$$

This agrees with the value $\phi(28) = 12$. More generally, if k divides $p-1$, then there are exactly $\phi(k)$ elements of $\mathbb{F}_p^*$ having order k.

1.6 Cryptography Before the Computer Age

We pause for a short foray into the history of pre-computer cryptography. Our hope is that these brief notes will whet your appetite for further reading on this fascinating subject, in which political intrigue, daring adventure, and romantic episodes play an equal role with technical achievements.

The origins of cryptography are lost in the mists of time, but presumably secret writing arose shortly after people started using some form of written communication, since one imagines that the notion of confidential information must date back to the dawn of civilization. There are early recorded descriptions of ciphers being used in Roman times, including Julius Caesar's shift cipher from Sect. 1.1, and certainly from that time onward, many civilizations have used both substitution ciphers, in which each letter is replaced by another letter or symbol, and transposition ciphers, in which the order of the letters is rearranged.

The invention of cryptanalysis, that is, the art of decrypting messages without previous knowledge of the key, is more recent. The oldest surviving texts, which include references to earlier lost volumes, are by Arab scholars from the fourteenth and fifteenth centuries. These books describe not only simple substitution and transposition ciphers, but also the first recorded instance of a homophonic substitution cipher, which is a cipher in which a single plaintext letter may be represented by any one of several possible ciphertext letters. More importantly, they contain the first description of serious methods of cryptanalysis, including the use of letter frequency counts and the likelihood that certain pairs of letters will appear adjacent to one another. Unfortunately, most of this knowledge seems to have disappeared by the seventeenth century.

Meanwhile, as Europe emerged from the Middle Ages, political states in Italy and elsewhere required secure communications, and both cryptography and cryptanalysis began to develop. The earliest known European homophonic substitution cipher dates from 1401. The use of such a cipher suggests contemporary knowledge of cryptanalysis via frequency analysis, since the only reason to use a homophonic system is to make such cryptanalysis more difficult.

In the fifteenth and sixteenth centuries there arose a variety of what are known as polyalphabetic ciphers. (We will see an example of a polyalphabetic cipher, called the Vigenère cipher, in Sect. 5.2.) The basic idea is that each letter of the plaintext is enciphered using a different simple substitution cipher. The name "polyalphabetic" refers to the use of many different cipher alphabets, which were used according to some sort of key. If the key is reasonably long, then it takes a long time for the any given cipher alphabet to be used a second time. It wasn't until the nineteenth century that statistical methods were developed to reliably solve such systems, although there are earlier recorded instances of cryptanalysis via special tricks or lucky guesses of part of the message or the key. Jumping forward several centuries, we note that the machine ciphers that played a large role in World War II were, in essence, extremely complicated polyalphabetic ciphers.

Ciphers and codes[14] for both political and military purposes become increasingly widespread during the eighteenth, nineteenth, and early twentieth centuries, as did cryptanalytic methods, although the level of sophistication varied widely from generation to generation and from country to country. For example, as the United States prepared to enter World War I in 1917, the U.S. Army was using ciphers, inferior to those invented in Italy in the 1600s, that any trained cryptanalyst of the time would have been able to break in a few hours!

The invention and widespread deployment of long-range communication methods, especially the telegraph, opened the need for political, military, and commercial ciphers, and there are many fascinating stories of intercepted and decrypted telegraph messages playing a role in historical events. One example, the infamous Zimmerman telegram, will suffice. With the United States maintaining neutrality in 1917 as Germany battled France and Britain on the Western Front, the Germans decided that their best hope for victory was to tighten their blockade of Britain by commencing unrestricted submarine warfare in the Atlantic. This policy, which meant sinking ships from neutral countries, was likely to bring the United States into the war, so Germany decided to offer an alliance to Mexico. In return for Mexico invading the United States, and thus distracting it from the ground war in Europe, Germany proposed giving Mexico, at the conclusion of the war, much of present-day Texas, New Mexico, and Arizona. The British secret service intercepted this communication, and despite the fact that it was encrypted using one of Germany's

[14] In classical terminology, a code is a system in which each word of the plaintext is replaced with a code word. This requires sender and receiver to share a large dictionary in which plaintext words are paired with their ciphertext equivalents. Ciphers operate on the individual letters of the plaintext, either by substitution, transposition, or some combination. This distinction between the words "code" and "cipher" seems to have been largely abandoned in today's literature.

most secure cryptosystems, they were able to decipher the cable and pass its contents on to the United States, thereby helping to propel the United States into World War I.

The invention and development of radio communications around 1900 caused an even more striking change in the cryptographic landscape, especially in urgent military and political situations. A general could now instantaneously communicate with all of his troops, but unfortunately the enemy could listen in on all of his broadcasts. The need for secure and efficient ciphers became paramount and led to the invention of machine ciphers, such as Germany's Enigma machine. This was a device containing a number of rotors, each of which had many wires running through its center. Before a letter was encrypted, the rotors would spin in a predetermined way, thereby altering the paths of the wires and the resultant output. This created an immensely complicated polyalphabetic cipher in which the number of cipher alphabets was enormous. Further, the rotors could be removed and replaced in a vast number of different starting configurations, so breaking the system involved knowing both the circuits through the rotors and figuring out that day's initial rotor configuration.

Despite these difficulties, during World War II the British managed to decipher a large number of messages encrypted on Enigma machines. They were aided in this endeavor by Polish cryptographers who, just before hostilities commenced, shared with Britain and France the methods that they had developed for attacking Enigma. But determining daily rotor configurations and analyzing rotor replacements was still an immensely difficult task, especially after Germany introduced an improved Enigma machine having an extra rotor. The existence of Britain's ULTRA project to decrypt Enigma remained secret until 1974, but there are now several popular accounts. Military intelligence derived from ULTRA was of vital importance in the Allied war effort.

Another WWII cryptanalytic success was obtained by United States cryptographers against a Japanese cipher machine that they code-named Purple. This machine used switches, rather than rotors, but again the effect was to create an incredibly complicated polyalphabetic cipher. A team of cryptographers, led by William Friedman, managed to reconstruct the design of the Purple machine purely by analyzing intercepted encrypted messages. They then built their own machine and proceeded to decrypt many important diplomatic messages.

In this section we have barely touched the surface of the history of cryptography from antiquity through the middle of the twentieth century. Good starting points for further reading include Simon Singh's light introduction [139] and David Kahn's massive and comprehensive, but fascinating and quite readable, book *The Codebreakers* [63].

1.7 Symmetric and Asymmetric Ciphers

We have now seen several different examples of ciphers, all of which have a number of features in common. Bob wants to send a secret message to Alice. He uses a secret key k to scramble his plaintext message m and turn it into a ciphertext c. Alice, upon receiving c, uses the secret key k to unscramble c and reconstitute m. If this procedure is to work properly, then both Alice and Bob must possess copies of the secret key k, and if the system is to provide security, then their adversary Eve must not know k, must not be able to guess k, and must not be able to recover m from c without knowing k.

In this section we formulate the notion of a cryptosystem in abstract mathematical terms. There are many reasons why this is desirable. In particular, it allows us to highlight similarities and differences between different systems, while also providing a framework within which we can rigorously analyze the security of a cryptosystem against various types of attacks.

1.7.1 Symmetric Ciphers

Returning to Bob and Alice, we observe that they must share knowledge of the secret key k. Using that secret key, they can both encrypt and decrypt messages, so Bob and Alice have equal (or symmetric) knowledge and abilities. For this reason, ciphers of this sort are known as *symmetric ciphers*. Mathematically, a symmetric cipher uses a key k chosen from a space (i.e., a set) of possible keys $\mathcal{K}$ to encrypt a plaintext message m chosen from a space of possible messages $\mathcal{M}$, and the result of the encryption process is a ciphertext c belonging to a space of possible ciphertexts $\mathcal{C}$.

Thus encryption may be viewed as a function

$$e : \mathcal{K} \times \mathcal{M} \to \mathcal{C}$$

whose domain $\mathcal{K} \times \mathcal{M}$ is the set of pairs (k, m) consisting of a key k and a plaintext m and whose range is the space of ciphertexts $\mathcal{C}$. Similarly, decryption is a function

$$d : \mathcal{K} \times \mathcal{C} \to \mathcal{M}.$$

Of course, we want the decryption function to "undo" the results of the encryption function. Mathematically, this is expressed by the formula

$$d\big(k, e(k, m)\big) = m \qquad \text{for all } k \in \mathcal{K} \text{ and all } m \in \mathcal{M}.$$

It is sometimes convenient to write the dependence on k as a subscript. Then for each key k, we get a pair of functions

$$e_k : \mathcal{M} \longrightarrow \mathcal{C} \qquad \text{and} \qquad d_k : \mathcal{C} \longrightarrow \mathcal{M}$$

satisfying the decryption property

$$d_k\big(e_k(m)\big) = m \qquad \text{for all } m \in \mathcal{M}.$$

In other words, for every key k, the function d_k is the inverse function of the function e_k. In particular, this means that e_k must be one-to-one, since if $e_k(m) = e_k(m')$, then

$$m = d_k\big(e_k(m)\big) = d_k\big(e_k(m')\big) = m'.$$

It is safest for Alice and Bob to assume that Eve knows the encryption method that is being employed. In mathematical terms, this means that Eve knows the functions e and d. What Eve does not know is the particular key k that Alice and Bob are using. For example, if Alice and Bob use a simple substitution cipher, they should assume that Eve is aware of this fact. This illustrates a basic premise of modern cryptography called *Kerckhoff's principle*, which says that the security of a cryptosystem should depend only on the secrecy of the key, and not on the secrecy of the encryption algorithm itself.

If $(\mathcal{K}, \mathcal{M}, \mathcal{C}, e, d)$ is to be a successful cipher, it must have the following properties:

1. For any key $k \in \mathcal{K}$ and plaintext $m \in \mathcal{M}$, it must be easy to compute the ciphertext $e_k(m)$.

2. For any key $k \in \mathcal{K}$ and ciphertext $c \in \mathcal{C}$, it must be easy to compute the plaintext $d_k(c)$.

3. Given one or more ciphertexts $c_1, c_2, \ldots, c_n \in \mathcal{C}$ encrypted using the key $k \in \mathcal{K}$, it must be very difficult to compute any of the corresponding plaintexts $d_k(c_1), \ldots, d_k(c_n)$ without knowledge of k.

Here is another property that is desirable, although more difficult to achieve.

4. Given one or more pairs of plaintexts and their corresponding ciphertexts, $(m_1, c_1), (m_2, c_2), \ldots, (m_n, c_n)$, it must be very difficult to decrypt any ciphertext c that is not in the given list without knowing k. This property is called security against a *known plaintext attack*.

Even better is to achieve security while allowing the attacker to choose the known plaintexts.

5. For any list of plaintexts $m_1, \ldots, m_n \in \mathcal{M}$ chosen by the adversary, even with knowledge of the corresponding ciphertexts $e_k(m_1), \ldots, e_k(m_n)$, it is very difficult to decrypt any ciphertext c that is not in the given list without knowing k. This is known as security against a *chosen plaintext attack*. N.B. In this attack, the adversary is allowed to choose $m_1, \ldots, m_n$, as opposed to a known plaintext attack, where the attacker is given a list of plaintext/ciphertext pairs not of his choosing.

Example 1.33. The simple substitution cipher does not have Property 4, since even a single plaintext/ciphertext pair (m, c) reveals most of the encryption table. Similarly, the Vigenère cipher discussed in Sect. 5.2 has the property that a plaintext/ciphertext pair immediately reveals the keyword used for encryption. Thus both simple substitution and Vigenère ciphers are vulnerable to known plaintext attacks. See Exercise 1.43 for a further example.

In our list of desirable properties for a cryptosystem, we have left open the question of what exactly is meant by the words "easy" and "hard." We defer a formal discussion of this profound question to Sect. 5.7; see also Sects. 2.1 and 2.6. For now, we informally take "easy" to mean computable in less than a second on a typical desktop computer and "hard" to mean that all of the computing power in the world would require several years (at least) to perform the computation.

1.7.2 Encoding Schemes

It is convenient to view keys, plaintexts, and ciphertexts as numbers and to write those numbers in binary form. For example, we could take strings of 8 bits,[15] which give numbers from 0 to 255, and use them to represent the letters of the alphabet via

$$\texttt{a} = 00000000, \quad \texttt{b} = 00000001, \quad \texttt{c} = 00000010, \quad \ldots, \quad \texttt{z} = 00011001.$$

To distinguish lowercase from uppercase, we could let A = 00011011, B = 00011100, and so on. This encoding method allows up to 256 distinct symbols to be translated into binary form.

Your computer may use a method of this type, called the ASCII code,[16] to store data, although for historical reasons the alphabetic characters are not assigned the lowest binary values. Part of the ASCII code is listed in Table 1.10. For example, the phrase "Bed bug." (including spacing and punctuation) is encoded in ASCII as

B	e	d		b	u	g	.
66	101	100	32	98	117	103	46
01000010	01100101	01100100	00100000	01100010	01110101	01100111	00101110

Thus where you see the phrase "Bed bug.", your computer sees the list of bits

$$0100001001100101011001000010000001100010011101010110011100101110.$$

Definition. An *encoding scheme* is a method of converting one sort of data into another sort of data, for example, converting text into numbers. The distinction between an encoding scheme and an encryption scheme is one

[15]A *bit* is a 0 or a 1. The word "bit" is an abbreviation for *binary digit*.

[16]ASCII is an acronym for American Standard Code for Information Interchange.

	32	00100000
(	40	00101000
)	41	00101001
,	44	00101100
.	46	00101110

A	65	01000001
B	66	01000010
C	67	01000011
D	68	01000100
⋮	⋮	⋮
X	88	01011000
Y	89	01011001
Z	90	01011010

a	97	01100001
b	98	01100010
c	99	01100011
d	100	01100100
⋮	⋮	⋮
x	120	01111000
y	121	01111001
z	122	01111010

Table 1.10: The ASCII encoding scheme

of intent. An encoding scheme is assumed to be entirely public knowledge and used by everyone for the same purposes. An encryption scheme is designed to hide information from anyone who does not possess the secret key. Thus an encoding scheme, like an encryption scheme, consists of an encoding function and its inverse decoding function, but for an encoding scheme, both functions are public knowledge and should be fast and easy to compute.

With the use of an encoding scheme, a plaintext or ciphertext may be viewed as a sequence of binary blocks, where each block consists of 8 bits, i.e., of a sequence of eight ones and zeros. A block of 8 bits is called a *byte*. For human comprehension, a byte is often written as a decimal number between 0 and 255, or as a two-digit hexadecimal (base 16) number between 00 and FF. Computers often operate on more than 1 byte at a time. For example, a 64-bit processor operates on 8 bytes at a time.

1.7.3 Symmetric Encryption of Encoded Blocks

In using an encoding scheme as described in Sect. 1.7.2, it is convenient to view the elements of the plaintext space $\mathcal{M}$ as consisting of bit strings of a fixed length B, i.e., strings of exactly B ones and zeros. We call B the *blocksize* of the cipher. A general plaintext message then consists of a list of message blocks chosen from $\mathcal{M}$, and the encryption function transforms the message blocks into a list of ciphertext blocks in $\mathcal{C}$, where each block is a sequence of B bits. If the plaintext ends with a block of fewer than B bits, we pad the end of the block with zeros. Keep in mind that this encoding process, which converts the original plaintext message into a sequence of blocks of bits in $\mathcal{M}$, is public knowledge.

Encryption and decryption are done one block at a time, so it suffices to study the process for a single plaintext block, i.e., for a single $m \in \mathcal{M}$. This, of course, is why it is convenient to break a message up into blocks. A message can be of arbitrary length, so it's nice to be able to focus the cryptographic process on a single piece of fixed length. The plaintext block m is a string of B bits, which for concreteness we identify with the corresponding number in binary form. In other words, we identify $\mathcal{M}$ with the set of integers m satisfying $0 \le m < 2^B$ via

$$\overbrace{m_{B-1}m_{B-2}\cdots m_2m_1m_0}^{\text{list of } B \text{ bits of } m} \longleftrightarrow \overbrace{m_{B-1}\cdot 2^{B-1} + \cdots + m_2\cdot 2^2 + m_1\cdot 2 + m_0}^{\text{integer between 0 and } 2^B - 1}.$$

Here $m_0, m_1, \ldots, m_{B-1}$ are each 0 or 1.

Similarly, we identify the key space $\mathcal{K}$ and the ciphertext space $\mathcal{C}$ with sets of integers corresponding to bit strings of a certain blocksize. For notational convenience, we denote the blocksizes for keys, plaintexts, and ciphertexts by B_k, B_m, and B_c. They need not be the same. Thus we have identified $\mathcal{K}$, $\mathcal{M}$, and $\mathcal{C}$ with sets of positive integers

$$\begin{aligned}\mathcal{K} &= \{k \in \mathbb{Z} : 0 \le k < 2^{B_k}\},\\ \mathcal{M} &= \{m \in \mathbb{Z} : 0 \le m < 2^{B_m}\},\\ \mathcal{C} &= \{c \in \mathbb{Z} : 0 \le c < 2^{B_c}\}.\end{aligned}$$

An important question immediately arises: how large should Alice and Bob make the set $\mathcal{K}$, or equivalently, how large should they choose the key blocksize B_k? If B_k is too small, then Eve can check every number from 0 to $2^{B_k} - 1$ until she finds Alice and Bob's key. More precisely, since Eve is assumed to know the decryption algorithm d (Kerckhoff's principle), she takes each $k \in \mathcal{K}$ and uses it to compute $d_k(c)$. Assuming that Eve is able to distinguish between valid and invalid plaintexts, eventually she will recover the message.

This attack is known as an *exhaustive search attack* (also sometimes referred to as a *brute-force attack*), since Eve exhaustively searches through the key space. With current technology, an exhaustive search is considered to be infeasible if the space has at least 2^{80} elements. Thus Bob and Alice should definitely choose $B_k \ge 80$.

For many cryptosystems, especially the public key cryptosystems that form the core of this book, there are refinements on the exhaustive search attack that effectively replace the size of the space with its square root. These methods are based on the principle that it is easier to find matching objects (collisions) in a set than it is to find a particular object in the set. We describe some of these *meet-in-the-middle* or *collision attacks* in Sects. 2.7, 5.4, 5.5, 7.2, and 7.10. If meet-in-the-middle attacks are available, then Alice and Bob should choose $B_k \ge 160$.

1.7.4 Examples of Symmetric Ciphers

Before descending further into a morass of theory and notation, we pause to give a mathematical description of some elementary symmetric ciphers.

Let p be a large prime,[17] say $2^{159} < p < 2^{160}$. Alice and Bob take their key space $\mathcal{K}$, plaintext space $\mathcal{M}$, and ciphertext space $\mathcal{C}$ to be the same set,

$$\mathcal{K} = \mathcal{M} = \mathcal{C} = \{1, 2, 3, \ldots, p-1\}.$$

In fancier terminology, $\mathcal{K} = \mathcal{M} = \mathcal{C} = \mathbb{F}_p^*$ are all taken to be equal to the group of units in the finite field $\mathbb{F}_p$.

Alice and Bob randomly select a key $k \in \mathcal{K}$, i.e., they select an integer k satisfying $1 \le k < p$, and they decide to use the encryption function e_k defined by

$$e_k(m) \equiv k \cdot m \pmod{p}. \tag{1.9}$$

Here we mean that $e_k(m)$ is set equal to the unique positive integer between 1 and p that is congruent to $k \cdot m$ modulo p. The corresponding decryption function d_k is

$$d_k(c) \equiv k' \cdot c \pmod{p},$$

where k' is the inverse of k modulo p. It is important to note that although p is very large, the extended Euclidean algorithm (Remark 1.15) allows us to calculate k' in fewer than $2\log_2 p + 2$ steps. Thus finding k' from k counts as "easy" in the world of cryptography.

It is clear that Eve has a hard time guessing k, since there are approximately 2^{160} possibilities from which to choose. Is it also difficult for Eve to recover k if she knows the ciphertext c? The answer is yes, it is still difficult. Notice that the encryption function

$$e_k : \mathcal{M} \longrightarrow \mathcal{C}$$

is surjective (onto) for any choice of key k. This means that for every $c \in \mathcal{C}$ and any $k \in \mathcal{K}$ there exists an $m \in \mathcal{M}$ such that $e_k(m) = c$. Further, any given ciphertext may represent any plaintext, provided that the plaintext is encrypted by an appropriate key. Mathematically, this may be rephrased by saying that given any ciphertext $c \in \mathcal{C}$ and any plaintext $m \in \mathcal{M}$, there exists a key k such that $e_k(m) = c$. Specifically this is true for the key

$$k \equiv m^{-1} \cdot c \pmod{p}. \tag{1.10}$$

This shows that Alice and Bob's cipher has Properties 1–3 as listed on page 38, since anyone who knows the key k can easily encrypt and decrypt, but it is hard to decrypt if you do not know the value of k. However, this cipher does not have Property 4, since even a single plaintext/ciphertext pair (m, c) allows Eve to recover the private key k using the formula (1.10).

---

[17]There are in fact many primes in the interval $2^{159} < p < 2^{160}$. The prime number theorem implies that almost 1 % of the numbers in this interval are prime. Of course, there is also the question of identifying a number as prime or composite. There are efficient tests that do this, even for very large numbers. See Sect. 3.4.

It is also interesting to observe that if Alice and Bob define their encryption function to be simply multiplication of integers $e_k(m) = k \cdot m$ with no reduction modulo p, then their cipher still has Properties 1 and 2, but Property 3 fails. If Eve tries to decrypt a single ciphertext $c = k \cdot m$, she still faces the (moderately) difficult task of factoring a large number. However, if she manages to acquire several ciphertexts $c_1, c_2, \ldots, c_n$, then there is a good chance that

$$\begin{aligned} \gcd(c_1, c_2, \ldots, c_n) &= \gcd(k \cdot m_1, k \cdot m_2, \ldots, k \cdot m_n) \\ &= k \cdot \gcd(m_1, m_2, \ldots, m_n) \end{aligned}$$

equals k itself or a small multiple of k. Note that it is an easy task to compute the greatest common divisor.

This observation provides our first indication of how reduction modulo p has a wonderful "mixing" effect that destroys properties such as divisibility. However, reduction is not by itself the ultimate solution. Consider the vulnerability of the cipher (1.9) to a known plaintext attack. As noted above, if Eve can get her hands on both a ciphertext c and its corresponding plaintext m, then she easily recovers the key by computing

$$k \equiv m^{-1} \cdot c \pmod{p}.$$

Thus even a single plaintext/ciphertext pair suffices to reveal the key, so the encryption function e_k given by (1.9) does not have Property 4 on page 38.

There are many variants of this "multiplication-modulo-p" cipher. For example, since addition is more efficient than multiplication, there is an "addition-modulo-p" cipher given by

$$e_k(m) \equiv m + k \pmod{p} \qquad \text{and} \qquad d_k(c) \equiv c - k \pmod{p},$$

which is nothing other than the shift or Caesar cipher that we studied in Sect. 1.1. Another variant, called an *affine cipher*, is a combination of the shift cipher and the multiplication cipher. The key for an affine cipher consists of two integers $k = (k_1, k_2)$ and encryption and decryption are defined by

$$\begin{aligned} e_k(m) &\equiv k_1 \cdot m + k_2 \pmod{p}, \\ d_k(c) &\equiv k_1' \cdot (c - k_2) \pmod{p}, \end{aligned} \tag{1.11}$$

where k_1' is the inverse of k_1 modulo p.

The affine cipher has a further generalization called the *Hill cipher*, in which the plaintext m, the ciphertext c, and the second part of the key k_2 are replaced by column vectors consisting of n numbers modulo p. The first part of the key k_1 is taken to be an n-by-n matrix with mod p integer entries. Encryption and decryption are again given by (1.11), but now multiplication $k_1 \cdot m$ is the product of a matrix and a vector, and k_1' is the inverse matrix of k_1 modulo p. Both the affine cipher and the Hill cipher are vulnerable to known plaintext attacks; see Exercises 1.43. and 1.44.

Example 1.34. As noted earlier, addition is generally faster than multiplication, but there is another basic computer operation that is even faster than addition. It is called *exclusive or* and is denoted by XOR or $\oplus$. At the lowest level, XOR takes two individual bits $\beta \in \{0,1\}$ and $\beta' \in \{0,1\}$ and yields

$$\beta \oplus \beta' = \begin{cases} 0 & \text{if } \beta \text{ and } \beta' \text{ are the same,} \\ 1 & \text{if } \beta \text{ and } \beta' \text{ are different.} \end{cases} \tag{1.12}$$

If you think of a bit as a number that is 0 or 1, then XOR is the same as addition modulo 2. More generally, the XOR of 2 bit strings is the result of performing XOR on each corresponding pair of bits. For example,

$$10110 \oplus 11010 = [1 \oplus 1]\,[0 \oplus 1]\,[1 \oplus 0]\,[1 \oplus 1]\,[0 \oplus 0] = 01100.$$

Using this new operation, Alice and Bob have at their disposal yet another basic cipher defined by

$$e_k(m) = k \oplus m \qquad \text{and} \qquad d_k(c) = k \oplus c.$$

Here $\mathcal{K}$, $\mathcal{M}$, and $\mathcal{C}$ are the sets of all binary strings of length B, or equivalently, the set of all numbers between 0 and $2^B - 1$.

This cipher has the advantage of being highly efficient and completely symmetric in the sense that e_k and d_k are the same function. If k is chosen randomly and is used only once, then this cipher is known as *Vernam's one-time pad*. In Sect. 5.57 we show that the one-time pad is provably secure. Unfortunately, it requires a key that is as long as the plaintext, which makes it too cumbersome for most practical applications. And if k is used to encrypt more than one plaintext, then Eve may be able to exploit the fact that

$$c \oplus c' = (k \oplus m) \oplus (k \oplus m') = m \oplus m'$$

to extract information about m or m'. It's not obvious how Eve would proceed to find k, m, or m', but simply the fact that the key k can be removed so easily, revealing the potentially less random quantity $m \oplus m'$, should make a cryptographer nervous. Further, this method is vulnerable in some situations to a known plaintext attack; see Exercise 1.48.

1.7.5 Random Bit Sequences and Symmetric Ciphers

We have arrived, at long last, at the fundamental question regarding the creation of secure and efficient symmetric ciphers. Is it possible to use a single relatively short key k (say consisting of 160 random bits) to securely and efficiently send arbitrarily long messages? Here is one possible construction. Suppose that we could construct a function

$$R : \mathcal{K} \times \mathbb{Z} \longrightarrow \{0,1\}$$

with the following properties:

1. For all $k \in \mathcal{K}$ and all $j \in \mathbb{Z}$, it is easy to compute $R(k, j)$.
2. Given an arbitrarily long sequence of integers $j_1, j_2, \ldots, j_n$ and given all of the values $R(k, j_1), R(k, j_2), \ldots, R(k, j_n)$, it is hard to determine k.
3. Given any list of integers $j_1, j_2, \ldots, j_n$ and given all of the values
$$R(k, j_1), R(k, j_2), \ldots, R(k, j_n),$$
it is hard to guess the value of $R(k, j)$ with better than a 50 % chance of success for any value of j not already in the list.

If we could find a function R with these three properties, then we could use it to turn an initial key k into a sequence of bits

$$R(k, 1), R(k, 2), R(k, 3), R(k, 4), \ldots, \tag{1.13}$$

and then we could use this sequence of bits as the key for a one-time pad as described in Example 1.34.

The fundamental problem with this approach is that the sequence of bits (1.13) is not truly random, since it is generated by the function R. Instead, we say that the sequence of bits (1.13) is a *pseudorandom sequence* and we call R a *pseudorandom number generator*.

Do pseudorandom number generators exist? If so, they would provide examples of the one-way functions defined by Diffie and Hellman in their groundbreaking paper [38], but despite more than a quarter century of work, no one has yet proven the existence of even a single such function. We return to this fascinating subject in Sects. 2.1 and 8.2. For now, we content ourselves with a few brief remarks.

Although no one has yet conclusively proven that pseudorandom number generators exist, many candidates have been suggested, and some of these proposals have withstood the test of time. There are two basic approaches to constructing candidates for R, and these two methods provide a good illustration of the fundamental conflict in cryptography between security and efficiency.

The first approach is to repeatedly apply an ad hoc collection of mixing operations that are well suited to efficient computation and that appear to be very hard to untangle. This method is, disconcertingly, the basis for most practical symmetric ciphers, including the Data Encryption Standard (DES) and the Advanced Encryption Standard (AES), which are the two systems most widely used today. See Sect. 8.12 for a brief description of these modern symmetric ciphers.

The second approach is to construct R using a function whose efficient inversion is a well-known mathematical problem that is believed to be difficult. This approach provides a far more satisfactory theoretical underpinning for a symmetric cipher, but unfortunately, all known constructions of this sort are far less efficient than the ad hoc constructions, and hence are less attractive for real-world applications.

1.7.6 Asymmetric Ciphers Make a First Appearance

If Alice and Bob want to exchange messages using a symmetric cipher, they must first mutually agree on a secret key k. This is fine if they have the opportunity to meet in secret or if they are able to communicate once over a secure channel. But what if they do not have this opportunity and if every communication between them is monitored by their adversary Eve? Is it possible for Alice and Bob to exchange a secret key under these conditions?

Most people's first reaction is that it is not possible, since Eve sees every piece of information that Alice and Bob exchange. It was the brilliant insight of Diffie and Hellman[18] that under certain hypotheses, it is possible. The search for efficient (and provable) solutions to this problem, which is called *public key* (or *asymmetric*) *cryptography*, forms one of the most interesting parts of mathematical cryptography and is the principal focus of this book.

We start by describing a nonmathematical way to visualize public key cryptography. Alice buys a safe with a narrow slot in the top and puts her safe in a public location. Everyone in the world is allowed to examine the safe and see that it is securely made. Bob writes his message to Alice on a piece of paper and slips it through the slot in the top of the safe. Now only a person with the key to the safe, which presumably means only Alice, can retrieve and read Bob's message. In this scenario, Alice's public key is the safe, the encryption algorithm is the process of putting the message in the slot, and the decryption algorithm is the process of opening the safe with the key. Note that this setup is not far-fetched; it is used in the real world. For example, the night deposit slot at a bank has this form, although in practice the "slot" must be well protected to prevent someone from inserting a long thin pair of tongs and extracting other people's deposits!

A useful feature of our "safe-with-a-slot" cryptosystem, which it shares with actual public key cryptosystems, is that Alice needs to put only one safe in a public location, and then everyone in the world can use it repeatedly to send encrypted messages to Alice. There is no need for Alice to provide a separate safe for each of her correspondents. And there is also no need for Alice to open the safe and remove Bob's message before someone else such as Carl or Dave uses it to send Alice a message.

We are now ready to give a mathematical formulation of an asymmetric cipher. As usual, there are spaces of keys $\mathcal{K}$, plaintexts $\mathcal{M}$, and ciphertexts $\mathcal{C}$. However, an element k of the key space is really a pair of keys,

$$k = (k_{\mathsf{priv}}, k_{\mathsf{pub}}),$$

called the *private key* and the *public key*, respectively. For each public key k_{pub} there is a corresponding encryption function

$$e_{k_{\mathsf{pub}}} : \mathcal{M} \longrightarrow \mathcal{C},$$

[18]The history is actually somewhat more complicated than this; see our brief discussion in Sect. 2.1 and the references listed there for further reading.

and for each private key k_{priv} there is a corresponding decryption function

$$d_{k_{\text{priv}}} : \mathcal{C} \longrightarrow \mathcal{M}.$$

These have the property that if the pair $(k_{\text{priv}}, k_{\text{pub}})$ is in the key space $\mathcal{K}$, then

$$d_{k_{\text{priv}}}\bigl(e_{k_{\text{pub}}}(m)\bigr) = m \qquad \text{for all } m \in \mathcal{M}.$$

If an asymmetric cipher is to be secure, it must be difficult for Eve to compute the decryption function $d_{k_{\text{priv}}}(c)$, even if she knows the public key k_{pub}. Notice that under this assumption, Alice can send k_{pub} to Bob using an insecure communication channel, and Bob can send back the ciphertext $e_{k_{\text{pub}}}(m)$, without worrying that Eve will be able to decrypt the message. To easily decrypt, it is necessary to know the private key k_{priv}, and presumably Alice is the only person with that information. The private key is sometimes called Alice's *trapdoor information*, because it provides a trapdoor (i.e., a shortcut) for computing the inverse function of $e_{k_{\text{pub}}}$. The fact that the encryption and decryption keys k_{pub} and k_{priv} are different makes the cipher asymmetric, whence its moniker.

It is quite intriguing that Diffie and Hellman created this concept without finding a candidate for an actual pair of functions, although they did propose a similar method by which Alice and Bob can securely exchange a random piece of data whose value is not known initially to either one. We describe Diffie and Hellman's key exchange method in Sect. 2.3 and then go on to discuss a number of asymmetric ciphers, including Elgamal (Sect. 2.4), RSA (Sect. 3.2), Goldwasser–Micali (Sect. 3.10), ECC (Sect. 6.4), GGH (Sect. 7.8), and NTRU (Sect. 7.10), whose security rely on the presumed difficulty of a variety of different mathematical problems.

Remark 1.35. In practice, asymmetric ciphers tend to be considerably slower than symmetric ciphers such as DES and AES. For that reason, if Bob needs to send Alice a large file, he might first use an asymmetric cipher to send Alice the key to a symmetric cipher, which he would then use to transmit the actual file.

Exercises

Section 1.1. Simple Substitution Ciphers

1.1. Build a cipher wheel as illustrated in Fig. 1.1, but with an inner wheel that rotates, and use it to complete the following tasks. (For your convenience, there is a cipher wheel that you can print and cut out at `www.math.brown.edu/~jhs/MathCrypto/CipherWheel.pdf`.)

(a) Encrypt the following plaintext using a rotation of 11 clockwise.

"A page of history is worth a volume of logic."

(b) Decrypt the following message, which was encrypted with a rotation of 7 clockwise.

AOLYLHYLUVZLJYLAZILAALYAOHUAOLZLJYLAZAOHALCLYFIVKFNBLZZLZ

(c) Decrypt the following message, which was encrypted by rotating 1 clockwise for the first letter, then 2 clockwise for the second letter, etc.

XJHRFTNZHMZGAHIUETXZJNBWNUTRHEPOMDNBJMAUGORFAOIZOCC

a	b	c	d	e	f	g	h	i	j	k	l	m	n	o	p	q	r	s	t	u	v	w	x	y	z
S	C	J	A	X	U	F	B	Q	K	T	P	R	W	E	Z	H	V	L	I	G	Y	D	N	M	O

Table 1.11: Simple substitution encryption table for Exercise 1.3

1.2. Decrypt each of the following Caesar encryptions by trying the various possible shifts until you obtain readable text.

(a) LWKLQNWKDWLVKDOOQHYHUVHHDELOOERDUGORYHOBDVDWUHH

(b) UXENRBWXCUXENFQRLQJUCNABFQNWRCJUCNAJCRXWORWMB

(c) BGUTBMBGZTFHNLXMKTIPBMAVAXXLXTEPTRLEXTOXKHHFYHKMAXFHNLX

1.3. For this exercise, use the simple substitution table given in Table 1.11.

(a) Encrypt the plaintext message

The gold is hidden in the garden.

(b) Make a decryption table, that is, make a table in which the ciphertext alphabet is in order from A to Z and the plaintext alphabet is mixed up.

(c) Use your decryption table from (b) to decrypt the following message.

IBXLX JVXIZ SLLDE VAQLL DEVAU QLB

1.4. Each of the following messages has been encrypted using a simple substitution cipher. Decrypt them. For your convenience, we have given you a frequency table and a list of the most common bigrams that appear in the ciphertext. (If you do not want to recopy the ciphertexts by hand, they can be downloaded or printed from the web site listed in the preface.)

(a) "A Piratical Treasure"

JNRZR BNIGI BJRGZ IZLQR OTDNJ GRIHT USDKR ZZWLG OIBTM NRGJN
IJTZJ LZISJ NRSBL QVRSI ORIQT QDEKJ JNRQW GLOFN IJTZX QLFQL
WBIMJ ITQXT HHTBL KUHQL JZKMM LZRNT OBIMI EURLW BLQZJ GKBJT
QDIQS LWJNR OLGRI EZJGK ZRBGS MJLDG IMNZT OIHRK MOSOT QHIJL
QBRJN IJJNT ZFIZL WIZTO MURZM RBTRZ ZKBNN LFRVR GIZFL KUHIM
MRIGJ LJNRB GKHRT QJRUU RBJLW JNRZI TULGI EZLUK JRUST QZLUK
EURFT JNLKJ JNRXR S

The ciphertext contains 316 letters. Here is a frequency table:

	R	J	I	L	Z	T	N	Q	B	G	K	U	M	O	S	H	W	F	E	D	X	V
Freq	33	30	27	25	24	20	19	16	15	15	13	12	12	10	9	8	7	6	5	5	3	2

The most frequent bigrams are: JN (11 times), NR (8 times), TQ (6 times), and LW, RB, RZ, and JL (5 times each).

(b) "A Botanical Code"

KZRNK GJKIP ZBOOB XLCRG BXFAU GJBNG RIXRU XAFGJ BXRME MNKNG
BURIX KJRXR SBUER ISATB UIBNN RTBUM NBIGK EBIGR OCUBR GLUBN
JBGRL SJGLN GJBOR ISLRS BAFFO AZBUN RFAUS AGGBI NGLXM IAZRX
RMNVL GEANG CJRUE KISRM BOOAZ GLOKW FAUKI NGRIC BEBRI NJAWB
OBNNO ATBZJ KOBRC JKIRR NGBUE BRINK XKBAF QBROA LNMRG MALUF
BBG

The ciphertext contains 253 letters. Here is a frequency table:

	B	R	G	N	A	I	U	K	O	J	L	X	M	F	S	E	Z	C	T	W	P	V	Q
Freq	32	28	22	20	16	16	14	13	12	11	10	10	8	8	7	7	6	5	3	2	1	1	1

The most frequent bigrams are: NG and RI (7 times each), BU (6 times), and BR (5 times).

(c) In order to make this one a bit more challenging, we have removed all occurrences of the word "the" from the plaintext.

"A Brilliant Detective"

GSZES GNUBE SZGUG SNKGX CSUUE QNZOQ EOVJN VXKNG XGAHS AWSZZ
BOVUE SIXCQ NQESX NGEUG AHZQA QHNSP CIPQA OIDLV JXGAK CGJCG
SASUB FVQAV CIAWN VWOVP SNSXV JGPCV NODIX GJQAE VOOXC SXXCG
OGOVA XGNVU BAVKX QZVQD LVJXQ EXCQO VKCQG AMVAX VWXCG OOBOX
VZCSO SPPSN VAXUB DVVAX QJQAJ VSUXC SXXCV OVJCS NSJXV NOJQA
MVBSZ VOOSH VSAWX QHGMV GWVSX CSXXC VBSNV ZVNVN SAWQZ ORVXJ
CVOQE JCGUW NVA

The ciphertext contains 313 letters. Here is a frequency table:

	V	S	X	G	A	O	Q	C	N	J	U	Z	E	W	B	P	I	H	K	D	M	L	R	F
Freq	39	29	29	22	21	21	20	20	19	13	11	11	10	8	8	6	5	5	5	4	3	2	1	1

The most frequent bigrams are: XC (10 times), NV (7 times), and CS, OV, QA, and SX (6 times each).

1.5. Suppose that you have an alphabet of 26 letters.

(a) How many possible simple substitution ciphers are there?

(b) A letter in the alphabet is said to be *fixed* if the encryption of the letter is the letter itself. How many simple substitution ciphers are there that leave:

(i) No letters fixed?
(ii) At least one letter fixed?
(iii) Exactly one letter fixed?
(iv) At least two letters fixed?

(Part (b) is quite challenging! You might try doing the problem first with an alphabet of four or five letters to get an idea of what is going on.)

Section 1.2. Divisibility and Greatest Common Divisors

1.6. Let $a, b, c \in \mathbb{Z}$. Use the definition of divisibility to directly prove the following properties of divisibility. (This is Proposition 1.4.)

(a) If $a \mid b$ and $b \mid c$, then $a \mid c$.

(b) If $a \mid b$ and $b \mid a$, then $a = \pm b$.

(c) If $a \mid b$ and $a \mid c$, then $a \mid (b + c)$ and $a \mid (b - c)$.

1.7. Use a calculator and the method described in Remark 1.9 to compute the following quotients and remainders.

(a) 34787 divided by 353.

(b) 238792 divided by 7843.
(c) 9829387493 divided by 873485.
(d) 1498387487 divided by 76348.

1.8. Use a calculator and the method described in Remark 1.9 to compute the following remainders, without bothering to compute the associated quotients.
(a) The remainder of 78745 divided by 127.
(b) The remainder of 2837647 divided by 4387.
(c) The remainder of 8739287463 divided by 18754.
(d) The remainder of 4536782793 divided by 9784537.

1.9. Use the Euclidean algorithm to compute the following greatest common divisors.
(a) $\gcd(291, 252)$.
(b) $\gcd(16261, 85652)$.
(c) $\gcd(139024789, 93278890)$.
(d) $\gcd(16534528044, 8332745927)$.

1.10. For each of the $\gcd(a, b)$ values in Exercise 1.9, use the extended Euclidean algorithm (Theorem 1.11) to find integers u and v such that $au + bv = \gcd(a, b)$.

1.11. Let a and b be positive integers.
(a) Suppose that there are integers u and v satisfying $au + bv = 1$. Prove that $\gcd(a, b) = 1$.
(b) Suppose that there are integers u and v satisfying $au + bv = 6$. Is it necessarily true that $\gcd(a, b) = 6$? If not, give a specific counterexample, and describe in general all of the possible values of $\gcd(a, b)$?
(c) Suppose that (u_1, v_1) and (u_2, v_2) are two solutions in integers to the equation $au + bv = 1$. Prove that a divides $v_2 - v_1$ and that b divides $u_2 - u_1$.
(d) More generally, let $g = \gcd(a, b)$ and let (u_0, v_0) be a solution in integers to $au + bv = g$. Prove that every other solution has the form $u = u_0 + kb/g$ and $v = v_0 - ka/g$ for some integer k. (This is the second part of Theorem 1.11.)

1.12. The method for solving $au + bv = \gcd(a, b)$ described in Sect. 1.2 is somewhat inefficient. This exercise describes a method to compute u and v that is well suited for computer implementation. In particular, it uses very little storage.
(a) Show that the following algorithm computes the greatest common divisor g of the positive integers a and b, together with a solution (u, v) in integers to the equation $au + bv = \gcd(a, b)$.

1. Set $u = 1$, $g = a$, $x = 0$, and $y = b$
2. If $y = 0$, set $v = (g - au)/b$ and return the values (g, u, v)
3. Divide g by y with remainder, $g = qy + t$, with $0 \le t < y$
4. Set $s = u - qx$
5. Set $u = x$ and $g = y$
6. Set $x = s$ and $y = t$
7. Go To Step (2)

(b) Implement the above algorithm on a computer using the computer language of your choice.

(c) Use your program to compute $g = \gcd(a, b)$ and integer solutions to the equation $au + bv = g$ for the following pairs (a, b).
   (i) $(527, 1258)$
   (ii) $(228, 1056)$
   (iii) $(163961, 167181)$
   (iv) $(3892394, 239847)$

(d) What happens to your program if $b = 0$? Fix the program so that it deals with this case correctly.

(e) It is often useful to have a solution with $u > 0$. Modify your program so that it returns a solution with $u > 0$ and u as small as possible. [*Hint*. If (u, v) is a solution, then so is $(u + b/g, v - a/g)$.] Redo (c) using your modified program.

1.13. Let $a_1, a_2, \ldots, a_k$ be integers with $\gcd(a_1, a_2, \ldots, a_k) = 1$, i.e., the largest positive integer dividing all of $a_1, \ldots, a_k$ is 1. Prove that the equation

$$a_1u_1 + a_2u_2 + \cdots + a_ku_k = 1$$

has a solution in integers $u_1, u_2, \ldots, u_k$. (*Hint*. Repeatedly apply the extended Euclidean algorithm, Theorem 1.11. You may find it easier to prove a more general statement in which $\gcd(a_1, \ldots, a_k)$ is allowed to be larger than 1.)

1.14. Let a and b be integers with $b > 0$. We've been using the "obvious fact" that a divided by b has a unique quotient and remainder. In this exercise you will give a proof.

(a) Prove that the set

$$\{a - bq : q \in \mathbb{Z}\}$$

contains at least one non-negative integer.

(b) Let r be the smallest non-negative integer in the set described in (a). Prove that $0 \le r < b$.

(c) Prove that there are integers q and r satisfying

$$a = bq + r \quad \text{and} \quad 0 \le r < b.$$

(d) Suppose that

$$a = bq_1 + r_1 = bq_2 + r_2 \quad \text{with} \quad 0 \le r_1 < b \quad \text{and} \quad 0 \le r_2 < b.$$

Prove that $q_1 = q_2$ and $r_1 = r_2$.

Section 1.3. Modular Arithmetic

1.15. Let $m \ge 1$ be an integer and suppose that

$$a_1 \equiv a_2 \pmod{m} \qquad \text{and} \qquad b_1 \equiv b_2 \pmod{m}.$$

Prove that

$$a_1 \pm b_1 \equiv a_2 \pm b_2 \pmod{m} \qquad \text{and} \qquad a_1 \cdot b_1 \equiv a_2 \cdot b_2 \pmod{m}.$$

(This is Proposition 1.13(a).)

1.16. Write out the following tables for $\mathbb{Z}/m\mathbb{Z}$ and $(\mathbb{Z}/m\mathbb{Z})^*$, as we did in Figs. 1.4 and 1.5.

(a) Make addition and multiplication tables for $\mathbb{Z}/3\mathbb{Z}$.
(b) Make addition and multiplication tables for $\mathbb{Z}/6\mathbb{Z}$.
(c) Make a multiplication table for the unit group $(\mathbb{Z}/9\mathbb{Z})^*$.
(d) Make a multiplication table for the unit group $(\mathbb{Z}/16\mathbb{Z})^*$.

1.17. Do the following modular computations. In each case, fill in the box with an integer between 0 and $m-1$, where m is the modulus.
(a) $347 + 513 \equiv \boxed{} \pmod{763}$.
(b) $3274 + 1238 + 7231 + 6437 \equiv \boxed{} \pmod{9254}$.
(c) $153 \cdot 287 \equiv \boxed{} \pmod{353}$.
(d) $357 \cdot 862 \cdot 193 \equiv \boxed{} \pmod{943}$.
(e) $5327 \cdot 6135 \cdot 7139 \cdot 2187 \cdot 5219 \cdot 1873 \equiv \boxed{} \pmod{8157}$.
(*Hint.* After each multiplication, reduce modulo 8157 before doing the next multiplication.)
(f) $137^2 \equiv \boxed{} \pmod{327}$.
(g) $373^6 \equiv \boxed{} \pmod{581}$.
(h) $23^3 \cdot 19^5 \cdot 11^4 \equiv \boxed{} \pmod{97}$.

1.18. Find all values of x between 0 and $m-1$ that are solutions of the following congruences. (*Hint.* If you can't figure out a clever way to find the solution(s), you can just substitute each value $x = 1$, $x = 2, \ldots$, $x = m-1$ and see which ones work.)
(a) $x + 17 \equiv 23 \pmod{37}$.
(b) $x + 42 \equiv 19 \pmod{51}$.
(c) $x^2 \equiv 3 \pmod{11}$.
(d) $x^2 \equiv 2 \pmod{13}$.
(e) $x^2 \equiv 1 \pmod{8}$.
(f) $x^3 - x^2 + 2x - 2 \equiv 0 \pmod{11}$.
(g) $x \equiv 1 \pmod{5}$ and also $x \equiv 2 \pmod{7}$. (Find all solutions modulo 35, that is, find the solutions satisfying $0 \le x \le 34$.)

1.19. Suppose that $g^a \equiv 1 \pmod{m}$ and that $g^b \equiv 1 \pmod{m}$. Prove that

$$g^{\gcd(a,b)} \equiv 1 \pmod{m}.$$

1.20. Prove that if a_1 and a_2 are units modulo m, then $a_1 a_2$ is a unit modulo m.

1.21. Prove that m is prime if and only if $\phi(m) = m - 1$, where ϕ is Euler's phi function.

1.22. Let $m \in \mathbb{Z}$.
(a) Suppose that m is odd. What integer between 1 and $m-1$ equals $2^{-1} \bmod m$?
(b) More generally, suppose that $m \equiv 1 \pmod{b}$. What integer between 1 and $m-1$ is equal to $b^{-1} \bmod m$?

1.23. Let m be an odd integer and let a be any integer. Prove that $2m + a^2$ can never be a perfect square. (*Hint.* If a number is a perfect square, what are its possible values modulo 4?)

1.24. (a) Find a single value x that simultaneously solves the two congruences

$$x \equiv 3 \pmod{7} \qquad \text{and} \qquad x \equiv 4 \pmod{9}.$$

(*Hint.* Note that every solution of the first congruence looks like $x = 3 + 7y$ for some y. Substitute this into the second congruence and solve for y; then use that to get x.)

(b) Find a single value x that simultaneously solves the two congruences

$$x \equiv 13 \pmod{71} \qquad \text{and} \qquad x \equiv 41 \pmod{97}.$$

(c) Find a single value x that simultaneously solves the three congruences

$$x \equiv 4 \pmod{7}, \qquad x \equiv 5 \pmod{8}, \qquad \text{and} \qquad x \equiv 11 \pmod{15}.$$

(d) Prove that if $\gcd(m, n) = 1$, then the pair of congruences

$$x \equiv a \pmod{m} \qquad \text{and} \qquad x \equiv b \pmod{n}$$

has a solution for any choice of a and b. Also give an example to show that the condition $\gcd(m, n) = 1$ is necessary.

1.25. Let N, g, and A be positive integers (note that N need not be prime). Prove that the following algorithm, which is a low-storage variant of the square-and-multiply algorithm described in Sect. 1.3.2, returns the value $g^A \pmod{N}$. (In Step 4 we use the notation $\lfloor x \rfloor$ to denote the greatest integer function, i.e., round x down to the nearest integer.)

Input. Positive integers N, g, and A.
1. Set $a = g$ and $b = 1$.
2. Loop while $A > 0$.
  3. If $A \equiv 1 \pmod{2}$, set $b = b \cdot a \pmod{N}$.
  4. Set $a = a^2 \pmod{N}$ and $A = \lfloor A/2 \rfloor$.
  5. If $A > 0$, continue with loop at Step **2**.
6. Return the number b, which equals $g^A \pmod{N}$.

1.26. Use the square-and-multiply algorithm described in Sect. 1.3.2, or the more efficient version in Exercise 1.25, to compute the following powers.
(a) $17^{183} \pmod{256}$.
(b) $2^{477} \pmod{1000}$.
(c) $11^{507} \pmod{1237}$.

1.27. Consider the congruence

$$ax \equiv c \pmod{m}.$$

(a) Prove that there is a solution if and only if $\gcd(a, m)$ divides c.
(b) If there is a solution, prove that there are exactly $\gcd(a, m)$ distinct solutions modulo m.
(*Hint.* Use the extended Euclidean algorithm (Theorem 1.11).)

Section 1.4. Prime Numbers, Unique Factorization, and Finite Fields

1.28. Let $\{p_1, p_2, \ldots, p_r\}$ be a set of prime numbers, and let

$$N = p_1 p_2 \cdots p_r + 1.$$

Prove that N is divisible by some prime not in the original set. Use this fact to deduce that there must be infinitely many prime numbers. (This proof of the infinitude of primes appears in Euclid's *Elements*. Prime numbers have been studied for thousands of years.)

1.29. Without using the fact that every integer has a unique factorization into primes, prove that if $\gcd(a, b) = 1$ and if $a \mid bc$, then $a \mid c$. (*Hint.* Use the fact that it is possible to find a solution to $au + bv = 1$.)

1.30. Compute the following ord_p values:
(a) $\text{ord}_2(2816)$.
(b) $\text{ord}_7(2222574487)$.
(c) $\text{ord}_p(46375)$ for each of $p = 3$, 5, 7, and 11.

1.31. Let p be a prime number. Prove that ord_p has the following properties.
(a) $\text{ord}_p(ab) = \text{ord}_p(a) + \text{ord}_p(b)$. (Thus ord_p resembles the logarithm function, since it converts multiplication into addition!)
(b) $\text{ord}_p(a + b) \ge \min\{\text{ord}_p(a), \text{ord}_p(b)\}$.
(c) If $\text{ord}_p(a) \ne \text{ord}_p(b)$, then $\text{ord}_p(a + b) = \min\{\text{ord}_p(a), \text{ord}_p(b)\}$.
A function satisfying properties (a) and (b) is called a *valuation*.

Section 1.5. Powers and Primitive Roots in Finite Fields

1.32. For each of the following primes p and numbers a, compute $a^{-1} \bmod p$ in two ways: (i) Use the extended Euclidean algorithm. (ii) Use the fast power algorithm and Fermat's little theorem. (See Example 1.27.)
(a) $p = 47$ and $a = 11$.
(b) $p = 587$ and $a = 345$.
(c) $p = 104801$ and $a = 78467$.

1.33. Let p be a prime and let q be a prime that divides $p - 1$.
(a) Let $a \in \mathbb{F}_p^*$ and let $b = a^{(p-1)/q}$. Prove that either $b = 1$ or else b has order q. (Recall that the order of b is the smallest $k \ge 1$ such that $b^k = 1$ in $\mathbb{F}_p^*$. *Hint.* Use Proposition 1.29.)
(b) Suppose that we want to find an element of $\mathbb{F}_p^*$ of order q. Using (a), we can randomly choose a value of $a \in \mathbb{F}_p^*$ and check whether $b = a^{(p-1)/q}$ satisfies $b \ne 1$. How likely are we to succeed? In other words, compute the value of the ratio

$$\frac{\#\{a \in \mathbb{F}_p^* : a^{(p-1)/q} \ne 1\}}{\#\mathbb{F}_p^*}.$$

(*Hint.* Use Theorem 1.30.)

1.34. Recall that g is called a primitive root modulo p if the powers of g give all nonzero elements of $\mathbb{F}_p$.
(a) For which of the following primes is 2 a primitive root modulo p?
(i) $p = 7$ (ii) $p = 13$ (iii) $p = 19$ (iv) $p = 23$

(b) For which of the following primes is 3 a primitive root modulo p?
(i) $p = 5$ (ii) $p = 7$ (iii) $p = 11$ (iv) $p = 17$

(c) Find a primitive root for each of the following primes.
(i) $p = 23$ (ii) $p = 29$ (iii) $p = 41$ (iv) $p = 43$

(d) Find all primitive roots modulo 11. Verify that there are exactly $\phi(10)$ of them, as asserted in Remark 1.32.

(e) Write a computer program to check for primitive roots and use it to find all primitive roots modulo 229. Verify that there are exactly $\phi(228)$ of them.

(f) Use your program from (e) to find all primes less than 100 for which 2 is a primitive root.

(g) Repeat the previous exercise to find all primes less than 100 for which 3 is a primitive root. Ditto to find the primes for which 4 is a primitive root.

1.35. Let p be a prime such that $q = \frac{1}{2}(p-1)$ is also prime. Suppose that g is an integer satisfying

$$g \not\equiv 0 \pmod{p} \quad \text{and} \quad g \not\equiv \pm 1 \pmod{p} \quad \text{and} \quad g^q \not\equiv 1 \pmod{p}.$$

Prove that g is a primitive root modulo p.

1.36. This exercise begins the study of squares and square roots modulo p.

(a) Let p be an odd prime number and let b be an integer with $p \nmid b$. Prove that either b has two square roots modulo p or else b has no square roots modulo p. In other words, prove that the congruence

$$X^2 \equiv b \pmod{p}$$

has either two solutions or no solutions in $\mathbb{Z}/p\mathbb{Z}$. (What happens for $p = 2$? What happens if $p \mid b$?)

(b) For each of the following values of p and b, find all of the square roots of b modulo p.
(i) $(p, b) = (7, 2)$ (ii) $(p, b) = (11, 5)$
(iii) $(p, b) = (11, 7)$ (iv) $(p, b) = (37, 3)$

(c) How many square roots does 29 have modulo 35? Why doesn't this contradict the assertion in (a)?

(d) Let p be an odd prime and let g be a primitive root modulo p. Then any number a is equal to some power of g modulo p, say $a \equiv g^k \pmod{p}$. Prove that a has a square root modulo p if and only if k is even.

1.37. Let $p \geq 3$ be a prime and suppose that the congruence

$$X^2 \equiv b \pmod{p}$$

has a solution.

(a) Prove that for every exponent $e \geq 1$ the congruence

$$X^2 \equiv b \pmod{p^e} \tag{1.14}$$

has a solution. (*Hint.* Use induction on e. Build a solution modulo p^{e+1} by suitably modifying a solution modulo p^e.)

(b) Let $X = \alpha$ be a solution to $X^2 \equiv b \pmod{p}$. Prove that in (a), we can find a solution $X = \beta$ to $X^2 \equiv b \pmod{p^e}$ that also satisfies $\beta \equiv \alpha \pmod{p}$.

(c) Let β and β' be two solutions as in (b). Prove that $\beta \equiv \beta' \pmod{p^e}$.

(d) Use Exercise 1.36 to deduce that the congruence (1.14) has either two solutions or no solutions modulo p^e.

1.38. Compute the value of

$$2^{(p-1)/2} \pmod{p}$$

for every prime $3 \le p < 20$. Make a conjecture as to the possible values of $2^{(p-1)/2} \pmod{p}$ when p is prime and prove that your conjecture is correct.

Section 1.6. Cryptography by Hand

1.39. Write a 2–5 page paper on one of the following topics, including both cryptographic information and placing events in their historical context:

(a) Cryptography in the Arab world to the fifteenth century.

(b) European cryptography in the fifteenth and early sixteenth centuries.

(c) Cryptography and cryptanalysis in Elizabethan England.

(d) Cryptography and cryptanalysis in the nineteenth century.

(e) Cryptography and cryptanalysis during World War I.

(f) Cryptography and cryptanalysis during World War II.

(Most of these topics are too broad for a short term paper, so you should choose a particular aspect on which to concentrate.)

1.40. A *homophonic cipher* is a substitution cipher in which there may be more than one ciphertext symbol for each plaintext letter. Here is an example of a homophonic cipher, where the more common letters have several possible replacements.

a	b	c	d	e	f	g	h	i	j	k	l	m	n	o	p	q	r	s	t	u	v	w	x	y	z
!	4	#	$	1	%	&	*	(	)	3	2	=	+	[	9	]	{	}	:	;	7	<	>	5	?
$\heartsuit$	$\circ$	$\star$	$\aleph$	6	$\nearrow$	$\triangleright$	$\diamondsuit$	$\wedge$			$\searrow$	Δ	∇	8	$\clubsuit$		Ω	$\vee$	$\otimes$	$\spadesuit$				$\flat$	
Θ				∞		$\Uparrow$		$\natural$					$\bullet$	$\odot$				$\triangleleft$	$\oplus$	$\Leftarrow$					
$\swarrow$				$\Downarrow$														$\Rightarrow$	$\nwarrow$						

Decrypt the following message.

$$(\; \% \; \Delta \; \spadesuit \; \Rightarrow \; \natural \; \# \; 4 \; \infty \; : \; \diamondsuit \; 6 \; \nearrow \; \odot \; [\; \aleph \; 8 \; \% \; 2 \; [\; 7 \; \Downarrow \; \clubsuit \; \searrow \; \heartsuit \; 5 \; \odot \; \nabla$$

1.41. A *transposition cipher* is a cipher in which the letters of the plaintext remain the same, but their order is rearranged. Here is a simple example in which the message is encrypted in blocks of 25 letters at a time.[19] Take the given 25 letters and arrange them in a 5-by-5 block by writing the message horizontally on the lines. For example, the first 25 letters of the message

```
Now is the time for all good men to come to the aid...
```

is written as

```
N O W I S
T H E T I
M E F O R
A L L G O
O D M E N
```

[19] If the number of letters in the message is not an even multiple of 25, then extra random letters are appended to the end of the message.

Now the cipehrtext is formed by reading the letters down the columns, which gives the ciphertext

`NTMAO OHELD WEFLM ITOGE SIRON.`

(a) Use this transposition cipher to encrypt the first 25 letters of the message

`Four score and seven years ago our fathers...`

(b) The following message was encrypted using this transposition cipher. Decrypt it.

`WNOOA HTUFN EHRHE NESUV ICEME`

(c) There are many variations on this type of cipher. We can form the letters into a rectangle instead of a square, and we can use various patterns to place the letters into the rectangle and to read them back out. Try to decrypt the following ciphertext, in which the letters were placed horizontally into a rectangle of some size and then read off vertically by columns.

`WHNCE STRHT TEOOH ALBAT DETET SADHE`
`LEELL QSFMU EEEAT VNLRI ATUDR HTEEA`

(For convenience, we've written the ciphertext in 5 letter blocks, but that doesn't necessarily mean that the rectangle has a side of length 5.)

Section 1.7. Symmetric Ciphers and Asymmetric Ciphers

1.42. Encode the following phrase (including capitalization, spacing and punctuation) into a string of bits using the ASCII encoding scheme given in Table 1.10.

`Bad day, Dad.`

1.43. Consider the affine cipher with key $k = (k_1, k_2)$ whose encryption and decryption functions are given by (1.11) on page 43.

(a) Let $p = 541$ and let the key be $k = (34, 71)$. Encrypt the message $m = 204$. Decrypt the ciphertext $c = 431$.

(b) Assuming that p is public knowledge, explain why the affine cipher is vulnerable to a known plaintext attack. (See Property 4 on page 38.) How many plaintext/ciphertext pairs are likely to be needed in order to recover the private key?

(c) Alice and Bob decide to use the prime $p = 601$ for their affine cipher. The value of p is public knowledge, and Eve intercepts the ciphertexts $c_1 = 324$ and $c_2 = 381$ and also manages to find out that the corresponding plaintexts are $m_1 = 387$ and $m_2 = 491$. Determine the private key and then use it to encrypt the message $m_3 = 173$.

(d) Suppose now that p is not public knowledge. Is the affine cipher still vulnerable to a known plaintext attack? If so, how many plaintext/ciphertext pairs are likely to be needed in order to recover the private key?

1.44. Consider the Hill cipher defined by (1.11),

$$e_k(m) \equiv k_1 \cdot m + k_2 \pmod{p} \qquad \text{and} \qquad d_k(c) \equiv k_1^{-1} \cdot (c - k_2) \pmod{p},$$

where m, c, and k_2 are column vectors of dimension n, and k_1 is an n-by-n matrix.

(a) We use the vector Hill cipher with $p = 7$ and the key $k_1 = \left(\begin{smallmatrix}1 & 3\\ 2 & 2\end{smallmatrix}\right)$ and $k_2 = \left(\begin{smallmatrix}5\\ 4\end{smallmatrix}\right)$.
  (i) Encrypt the message $m = \left(\begin{smallmatrix}2\\ 1\end{smallmatrix}\right)$.
  (ii) What is the matrix k_1^{-1} used for decryption?
  (iii) Decrypt the message $c = \left(\begin{smallmatrix}3\\ 5\end{smallmatrix}\right)$.
(b) Explain why the Hill cipher is vulnerable to a known plaintext attack.
(c) The following plaintext/ciphertext pairs were generated using a Hill cipher with the prime $p = 11$. Find the keys k_1 and k_2.

$$m_1 = \left(\begin{smallmatrix}5\\ 4\end{smallmatrix}\right),\quad c_1 = \left(\begin{smallmatrix}1\\ 8\end{smallmatrix}\right),\quad m_2 = \left(\begin{smallmatrix}8\\ 10\end{smallmatrix}\right),\quad c_2 = \left(\begin{smallmatrix}8\\ 5\end{smallmatrix}\right),\quad m_3 = \left(\begin{smallmatrix}7\\ 1\end{smallmatrix}\right),\quad c_3 = \left(\begin{smallmatrix}8\\ 7\end{smallmatrix}\right).$$

(d) Explain how any simple substitution cipher that involves a permutation of the alphabet can be thought of as a special case of a Hill cipher.

1.45. Let N be a large integer and let $\mathcal{K} = \mathcal{M} = \mathcal{C} = \mathbb{Z}/N\mathbb{Z}$. For each of the functions

$$e : \mathcal{K} \times \mathcal{M} \longrightarrow \mathcal{C}$$

listed in (a)–(c), answer the following questions:

- Is e an encryption function?
- If e is an encryption function, what is its associated decryption function d?
- If e is not an encryption function, can you make it into an encryption function by using some smaller, yet reasonably large, set of keys?

(a) $e_k(m) \equiv k - m \pmod{N}$.
(b) $e_k(m) \equiv k \cdot m \pmod{N}$.
(c) $e_k(m) \equiv (k + m)^2 \pmod{N}$.

1.46. (a) Convert the 12 bit binary number 110101100101 into a decimal integer between 0 and $2^{12} - 1$.
(b) Convert the decimal integer $m = 37853$ into a binary number.
(c) Convert the decimal integer $m = 9487428$ into a binary number.
(d) Use exclusive or (XOR) to "add" the bit strings $11001010 \oplus 10011010$.
(e) Convert the decimal numbers 8734 and 5177 into binary numbers, combine them using XOR, and convert the result back into a decimal number.

1.47. Alice and Bob choose a key space $\mathcal{K}$ containing 2^{56} keys. Eve builds a special-purpose computer that can check 10,000,000,000 keys per second.
(a) How many days does it take Eve to check half of the keys in $\mathcal{K}$?
(b) Alice and Bob replace their key space with a larger set containing 2^B different keys. How large should Alice and Bob choose B in order to force Eve's computer to spend 100 years checking half the keys? (Use the approximation that there are 365.25 days in a year.)

For many years the United States government recommended a symmetric cipher called DES that used 56 bit keys. During the 1990s, people built special purpose computers demonstrating that 56 bits provided insufficient security. A new symmetric cipher called AES, with 128 bit keys, was developed to replace DES. See Sect. 8.12 for further information about DES and AES.

1.48. Explain why the cipher

$$e_k(m) = k \oplus m \qquad \text{and} \qquad d_k(c) = k \oplus c$$

defined by XOR of bit strings is not secure against a known plaintext attack. Demonstrate your attack by finding the private key used to encrypt the 16-bit ciphertext $c = 1001010001010111$ if you know that the corresponding plaintext is $m = 0010010000101100$.

1.49. Alice and Bob create a symmetric cipher as follows. Their private key k is a large integer and their messages (plaintexts) are d-digit integers

$$\mathcal{M} = \{m \in \mathbb{Z} : 0 \le m < 10^d\}.$$

To encrypt a message, Alice computes $\sqrt{k}$ to d decimal places, throws away the part to the left of the decimal point, and keeps the remaining d digits. Let α be this d-digit number. (For example, if $k = 87$ and $d = 6$, then $\sqrt{87} = 9.32737905\ldots$ and $\alpha = 327379$.)

Alice encrypts a message m as

$$c \equiv m + \alpha \pmod{10^d}.$$

Since Bob knows k, he can also find α, and then he decrypts c by computing $m \equiv c - \alpha \pmod{10^d}$.

(a) Alice and Bob choose the secret key $k = 11$ and use it to encrypt 6-digit integers (i.e., $d = 6$). Bob wants to send Alice the message $m = 328973$. What is the ciphertext that he sends?

(b) Alice and Bob use the secret key $k = 23$ and use it to encrypt 8-digit integers. Alice receives the ciphertext $c = 78183903$. What is the plaintext m?

(c) Show that the number α used for encryption and decryption is given by the formula

$$\alpha = \left\lfloor 10^d \left(\sqrt{k} - \lfloor \sqrt{k} \rfloor\right) \right\rfloor,$$

where $\lfloor t \rfloor$ denotes the greatest integer that is less than or equal to t.

(d) (Challenge Problem) If Eve steals a plaintext/ciphertext pair (m, c), then it is clear that she can recover the number α, since $\alpha \equiv c - m \pmod{10^d}$. If 10^d is large compared to k, can she also recover the number k? This might be useful, for example, if Alice and Bob use some of the other digits of $\sqrt{k}$ to encrypt subsequent messages.

1.50. Bob and Alice use a cryptosystem in which their private key is a (large) prime k and their plaintexts and ciphertexts are integers. Bob encrypts a message m by computing the product $c = km$. Eve intercepts the following two ciphertexts:

$$c_1 = 12849217045006222, \qquad c_2 = 6485880443666222.$$

Use the gcd method described in Sect. 1.7.4 to find Bob and Alice's private key.

Chapter 2

Discrete Logarithms and Diffie–Hellman

2.1 The Birth of Public Key Cryptography

In 1976, Whitfield Diffie and Martin Hellman published their now famous paper [38] entitled "New Directions in Cryptography." In this paper they formulated the concept of a public key encryption system and made several groundbreaking contributions to this new field. A short time earlier, Ralph Merkle had independently isolated one of the fundamental problems and invented a public key construction for an undergraduate project in a computer science class at Berkeley, but this was little understood at the time. Merkle's work "Secure communication over insecure channels" appeared in 1982 [83].

However, it turns out that the concept of public key encryption was originally discovered by James Ellis while working at the British Government Communications Headquarters (GCHQ). Ellis's discoveries in 1969 were classified as secret material by the British government and were not declassified and released until 1997, after his death. It is now known that two other researchers at GCHQ, Malcolm Williamson and Clifford Cocks, discovered the Diffie–Hellman key exchange algorithm and the RSA public key encryption system, respectively, before their rediscovery and public dissemination by Diffie, Hellman, Rivest, Shamir, and Adleman. To learn more about the fascinating history of public key cryptography, see for example [37, 42, 63, 139].

The Diffie–Hellman publication was an extremely important event—it set forth the basic definitions and goals of a new field of mathematics/computer science, a field whose existence was dependent on the then emerging age of the digital computer. Indeed, their paper begins with a call to arms:

J. Hoffstein et al., *An Introduction to Mathematical Cryptography*, Undergraduate Texts in Mathematics, DOI 10.1007/978-1-4939-1711-2_2

We stand today on the brink of a revolution in cryptography.

An original or breakthrough scientific idea is often called revolutionary, but in this instance, as the authors were fully aware, the term revolutionary was relevant in another sense. Prior to the publication of "New Directions...," encryption research in the United States was the domain of the National Security Agency, and all information in this area was classified. Indeed, until the mid-1990s, the United States government treated cryptographic algorithms as munitions, which meant that their export was prosecutable as a treasonable offense. Eventually, the government realized the futility of trying to prevent free and open discussion about abstract cryptographic algorithms and the dubious legality of restricting domestic use of strong cryptographic methods. However, in order to maintain some control, the government continued to restrict export of high security cryptographic algorithms if they were "machine readable." Their object, to prevent widespread global dissemination of sophisticated cryptography programs to potential enemies of the United States, was laudable,[1] but there were two difficulties that rendered the government's policy unworkable.

First, the existence of optical scanners creates a very blurry line between "machine readable" and "human text." To protest the government's policy, people wrote a three line version of the RSA algorithm in a programming language called perl and printed it on tee shirts and soda cans, thereby making these products into munitions. In principle, wearing an "RSA enabled" tee shirt on a flight from New York to Europe subjected the wearer to a large fine and a 10 year jail term. Even more amusing (or frightening, depending on your viewpoint), tattoos of the RSA perl code made people's bodies into non-exportable munitions!

Second, although these and other more serious protests and legal challenges had some effect, the government's policy was ultimately rendered moot by a simple reality. Public key algorithms are quite simple, and although it requires a certain expertise to implement them in a secure fashion, the world is full of excellent mathematicians and computer scientists and engineers. Thus government restrictions on the export of "strong crypto" simply encouraged the creation of cryptographic industries in other parts of the world. The government was able to slow the adoption of strong crypto for a few years, but it is now possible for anyone to purchase for a nominal sum cryptographic software that allows completely secure communications.[2]

---

[1] It is surely laudable to keep potential weapons out of the hands of one's enemies, but many have argued, with considerable justification, that the government also had the less benign objective of preventing other governments from using communication methods secure from United States prying.

[2] Of course, one never knows what cryptanalytic breakthroughs have been made by the scientists at the National Security Agency, since virtually all of their research is classified. The NSA is reputed to be the world's largest single employer of Ph.D.s in mathematics. However, in contrast to the situation before the 1970s, there are now far more cryptographers employed in academia and in the business world than there are in government agencies.

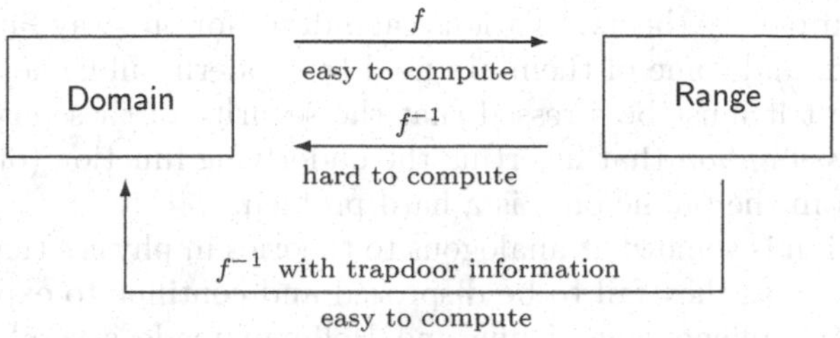

Figure 2.1: Illustration of a one-way trapdoor function

The first important contribution of Diffie and Hellman in [38] was the definition of a *Public Key Cryptosystem* (PKC) and its associated components—one-way functions and trapdoor information. A *one-way function* is an invertible function that is easy to compute, but whose inverse is difficult to compute. What does it mean to be "difficult to compute"? Intuitively, a function is difficult to compute if any algorithm that attempts to compute the inverse in a "reasonable" amount of time, e.g., less than the age of the universe, will almost certainly fail, where the phrase "almost certainly" must be defined probabilistically. (For a more rigorous definition of "hardness," see Sect. 2.6.)

Secure PKCs are built using one-way functions that have a *trapdoor*. The trapdoor is a piece of auxiliary information that allows the inverse to be easily computed. This idea is illustrated in Fig. 2.1, although it must be stressed that there is a vast chasm separating the abstract idea of a one-way trapdoor function and the actual construction of such a function.

As described in Sect. 1.7.6, the key for a public key (or asymmetric) cryptosystem consists of two pieces, a private key k_{priv} and a public key k_{pub}, where in practice k_{pub} is computed by applying some key-creation algorithm to k_{priv}. For each public/private key pair $(k_{\mathsf{priv}}, k_{\mathsf{pub}})$ there is an encryption algorithm $e_{k_{\mathsf{pub}}}$ and a corresponding decryption algorithm $d_{k_{\mathsf{priv}}}$. The encryption algorithm $e_{k_{\mathsf{pub}}}$ corresponding to k_{pub} is public knowledge and easy to compute. Similarly, the decryption algorithm $d_{k_{\mathsf{priv}}}$ must be easily computable by someone who knows the private key k_{priv}, but it should be very difficult to compute for someone who knows only the public key k_{pub}.

One says that the private key k_{priv} is *trapdoor information* for the function $e_{k_{\mathsf{pub}}}$, because without the trapdoor information it is very hard to compute the inverse function to $e_{k_{\mathsf{pub}}}$, but with the trapdoor information it is easy to compute the inverse. Notice that in particular, the function that is used to create k_{pub} from k_{priv} must be difficult to invert, since k_{pub} is public knowledge and k_{priv} allows efficient decryption.

It may come as a surprise to learn that despite years of research, it is still not known whether one-way functions exist. In fact, a proof of the existence of one-way functions would simultaneously solve the famous $\mathcal{P} = \mathcal{NP}$

problem in complexity theory.[3] Various candidates for one-way functions have been proposed, and some of them are used by modern public key encryption algorithms. But it must be stressed that the security of these cryptosystems rests on the *assumption* that inverting the underlying function (or finding the private key from the public one) is a hard problem.

The situation is somewhat analogous to theories in physics that gain credibility over time, as they fail to be disproved and continue to explain or generate interesting phenomena. Diffie and Hellman made several suggestions in [38] for one-way functions, including knapsack problems and exponentiation mod q, but they did not produce an example of a PKC, mainly for lack of finding the right trapdoor information. They did, however, describe a public key method by which certain material could be securely shared over an insecure channel. Their method, which is now called Diffie–Hellman key exchange, is based on the assumption that the discrete logarithm problem (DLP) is difficult to solve. We discuss the DLP in Sect. 2.2, and then describe Diffie–Hellman key exchange in Sect. 2.3. In their paper, Diffie and Hellman also defined a variety of cryptanalytic attacks and introduced the important concepts of digital signatures and one-way authentication, which we discuss in Chap. 4 and Sect. 8.5.

With the publication of [38] in 1976, the race was on to invent a practical public key cryptosystem. Within 2 years, two major papers describing public key cryptosystems were published: the RSA scheme of Rivest, Shamir, and Adleman [110] and the knapsack scheme of Merkle and Hellman [84]. Of these two, only RSA has withstood the test of time, in the sense that its underlying hard problem of integer factorization is still sufficiently computationally difficult to allow RSA to operate efficiently. By way of contrast, the knapsack system of Merkle and Hellman was shown to be insecure at practical computational levels [124]. However, the cryptanalysis of knapsack systems introduces important links to hard computational problems in the theory of integer lattices that we explore in Chap. 7.

2.2 The Discrete Logarithm Problem

The discrete logarithm problem is a mathematical problem that arises in many settings, including the mod p version described in this section and the elliptic curve version that will be studied later, in Chap. 6. The first published public key construction, due to Diffie and Hellman [38], is based on the discrete logarithm problem in a finite field $\mathbb{F}_p$, where recall that $\mathbb{F}_p$ is a field with a prime number of elements. (See Sect. 1.4.) For convenience, we interchangeably use the notations $\mathbb{F}_p$ and $\mathbb{Z}/p\mathbb{Z}$ for this field, and we use equality notation for elements of $\mathbb{F}_p$ and congruence notation for elements of $\mathbb{Z}/p\mathbb{Z}$ (cf. Remark 1.23).

[3]The $\mathcal{P} = \mathcal{NP}$ problem is one of the so-called Millennium Prizes, each of which has a $1,000,000 prize attached. See Sect. 5.7 for more on $\mathcal{P}$ versus $\mathcal{NP}$.

Let p be a (large) prime. Theorem 1.30 tells us that there exists a primitive element g. This means that every nonzero element of $\mathbb{F}_p$ is equal to some power of g. In particular, $g^{p-1} = 1$ by Fermat's little theorem (Theorem 1.24), and no smaller positive power of g is equal to 1. Equivalently, the list of elements

$$1, g, g^2, g^3, \ldots, g^{p-2} \in \mathbb{F}_p^*$$

is a complete list of the elements in $\mathbb{F}_p^*$ in some order.

Definition. Let g be a primitive root for $\mathbb{F}_p$ and let h be a nonzero element of $\mathbb{F}_p$. The *Discrete Logarithm Problem* (DLP) is the problem of finding an exponent x such that

$$g^x \equiv h \pmod{p}.$$

The number x is called the *discrete logarithm of h to the base g* and is denoted by $\log_g(h)$.

Remark 2.1. An older term for the discrete logarithm is the *index*, denoted by $\mathrm{ind}_g(h)$. The index terminology is still commonly used in number theory. It is also convenient if there is a danger of confusion between ordinary logarithms and discrete logarithms, since, for example, the quantity $\log_2$ frequently occurs in both contexts.

Remark 2.2. The discrete logarithm problem is a well-posed problem, namely to find an integer exponent x such that $g^x = h$. However, if there is one solution, then there are infinitely many, because Fermat's little theorem (Theorem 1.24) tells us that $g^{p-1} \equiv 1 \pmod{p}$. Hence if x is a solution to $g^x = h$, then $x + k(p-1)$ is also a solution for every value of k, because

$$g^{x+k(p-1)} = g^x \cdot (g^{p-1})^k \equiv h \cdot 1^k \equiv h \pmod{p}.$$

Thus $\log_g(h)$ is defined only up to adding or subtracting multiples of $p-1$. In other words, $\log_g(h)$ is really defined modulo $p-1$. It is not hard to verify (Exercise 2.3(a)) that $\log_g$ gives a well-defined function[4]

$$\log_g : \mathbb{F}_p^* \longrightarrow \frac{\mathbb{Z}}{(p-1)\mathbb{Z}}. \tag{2.1}$$

Sometimes, for concreteness, we refer to "the" discrete logarithm as the integer x lying between 0 and $p-2$ satisfying the congruence $g^x \equiv h \pmod{p}$.

Remark 2.3. It is not hard to prove (see Exercise 2.3(b)) that

$$\log_g(ab) = \log_g(a) + \log_g(b) \qquad \text{for all } a, b \in \mathbb{F}_p^*.$$

[4]If you have studied complex analysis, you may have noticed an analogy with the complex logarithm, which is not actually well defined on $\mathbb{C}^*$. This is due to the fact that $e^{2\pi i} = 1$, so $\log(z)$ is well defined only up to adding or subtracting multiples of $2\pi i$. The complex logarithm thus defines an isomorphism from $\mathbb{C}^*$ to the quotient group $\mathbb{C}/2\pi i\mathbb{Z}$, analogous to (2.1).

n	$g^n \bmod p$
1	627
2	732
3	697
4	395
5	182
6	253
7	543
8	760
9	374
10	189

n	$g^n \bmod p$
11	878
12	21
13	934
14	316
15	522
16	767
17	58
18	608
19	111
20	904

h	$\log_g(h)$
1	0
2	183
3	469
4	366
5	356
6	652
7	483
8	549
9	938
10	539

h	$\log_g(h)$
11	429
12	835
13	279
14	666
15	825
16	732
17	337
18	181
19	43
20	722

Table 2.1: Powers and discrete logarithms for $g = 627$ modulo $p = 941$

Thus calling $\log_g$ a "logarithm" is reasonable, since it converts multiplication into addition in the same way as the usual logarithm function. In mathematical terminology, the discrete logarithm $\log_g$ is a group isomorphism from $\mathbb{F}_p^*$ to $\mathbb{Z}/(p-1)\mathbb{Z}$.

Example 2.4. The number $p = 56509$ is prime, and one can check that $g = 2$ is a primitive root modulo p. How would we go about calculating the discrete logarithm of $h = 38679$? The only method that is immediately obvious is to compute

$$2^2,\ 2^3,\ 2^4,\ 2^5,\ 2^6,\ 2^7,\dots \pmod{56509}$$

until we find some power that equals 38679. It would be difficult to do this by hand, but using a computer, we find that $\log_2(h) = 11235$. You can verify this by calculating $2^{11235} \bmod 56509$ and checking that it is equal to 38679.

Remark 2.5. It must be emphasized that the discrete logarithm bears little resemblance to the continuous logarithm defined on the real or complex numbers. The terminology is still reasonable, because in both instances the process of exponentiation is inverted—but exponentiation modulo p varies in a very irregular way with the exponent, contrary to the behavior of its continuous counterpart. The random-looking behavior of exponentiation modulo p is apparent from even a cursory glance at a table of values such as those in Table 2.1, where we list the first few powers and the first few discrete logarithms for the prime $p = 941$ and the base $g = 627$. The seeming randomness is also illustrated by the scatter graph of $627^i \bmod 941$ pictured in Fig. 2.2.

Remark 2.6. Our statement of the discrete logarithm problem includes the assumption that the base g is a primitive root modulo p, but this is not strictly necessary. In general, for any $g \in \mathbb{F}_p^*$ and any $h \in \mathbb{F}_p^*$, the discrete logarithm problem is the determination of an exponent x satisfying $g^x \equiv h \pmod{p}$, assuming that such an x exists.

More generally, rather than taking nonzero elements of a finite field $\mathbb{F}_p$ and multiplying them together or raising them to powers, we can take elements of

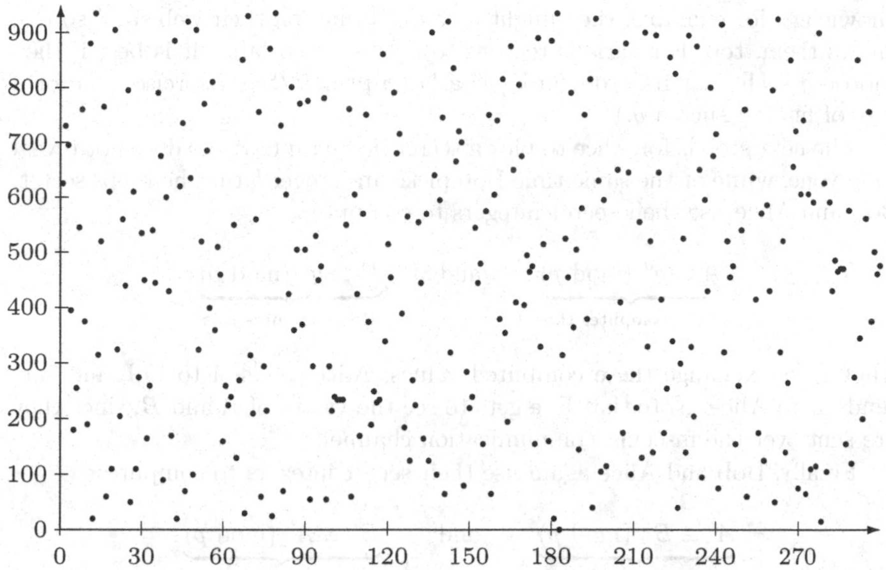

Figure 2.2: Powers 627^i mod 941 for $i = 1, 2, 3, \ldots$

any group and use the group law instead of multiplication. This leads to the most general form of the discrete logarithm problem. (If you are unfamiliar with the theory of groups, we give a brief overview in Sect. 2.5.)

Definition. Let G be a group whose group law we denote by the symbol $\star$. The *Discrete Logarithm Problem* for G is to determine, for any two given elements g and h in G, an integer x satisfying

$$\underbrace{g \star g \star g \star \cdots \star g}_{x \text{ times}} = h.$$

2.3 Diffie–Hellman Key Exchange

The Diffie–Hellman key exchange algorithm solves the following dilemma. Alice and Bob want to share a secret key for use in a symmetric cipher, but their only means of communication is insecure. Every piece of information that they exchange is observed by their adversary Eve. How is it possible for Alice and Bob to share a key without making it available to Eve? At first glance it appears that Alice and Bob face an impossible task. It was a brilliant insight of Diffie and Hellman that the difficulty of the discrete logarithm problem for $\mathbb{F}_p^*$ provides a possible solution.

The first step is for Alice and Bob to agree on a large prime p and a nonzero integer g modulo p. Alice and Bob make the values of p and g public knowledge; for example, they might post the values on their web sites, so Eve knows them, too. For various reasons to be discussed later, it is best if they choose g such that its order in $\mathbb{F}_p^*$ is a large prime. (See Exercise 1.33 for a way of finding such a g.)

The next step is for Alice to pick a secret integer a that she does not reveal to anyone, while at the same time Bob picks an integer b that he keeps secret. Bob and Alice use their secret integers to compute

$$\underbrace{A \equiv g^a \pmod{p}}_{\text{Alice computes this}} \qquad \text{and} \qquad \underbrace{B \equiv g^b \pmod{p}}_{\text{Bob computes this}}.$$

They next exchange these computed values, Alice sends A to Bob and Bob sends B to Alice. Note that Eve gets to see the values of A and B, since they are sent over the insecure communication channel.

Finally, Bob and Alice again use their secret integers to compute

$$\underbrace{A' \equiv B^a \pmod{p}}_{\text{Alice computes this}} \qquad \text{and} \qquad \underbrace{B' \equiv A^b \pmod{p}}_{\text{Bob computes this}}.$$

The values that they compute, A' and B' respectively, are actually the same, since

$$A' \equiv B^a \equiv (g^b)^a \equiv g^{ab} \equiv (g^a)^b \equiv A^b \equiv B' \pmod{p}.$$

This common value is their exchanged key. The Diffie–Hellman key exchange algorithm is summarized in Table 2.2.

Public parameter creation	
A trusted party chooses and publishes a (large) prime p and an integer g having large prime order in $\mathbb{F}_p^*$.	
Private computations	
Alice	**Bob**
Choose a secret integer a. Compute $A \equiv g^a \pmod{p}$.	Choose a secret integer b. Compute $B \equiv g^b \pmod{p}$.
Public exchange of values	
Alice sends A to Bob $\longrightarrow$	A
B	$\longleftarrow$ Bob sends B to Alice
Further private computations	
Alice	**Bob**
Compute the number $B^a \pmod{p}$.	Compute the number $A^b \pmod{p}$.
The shared secret value is $B^a \equiv (g^b)^a \equiv g^{ab} \equiv (g^a)^b \equiv A^b \pmod{p}$.	

Table 2.2: Diffie–Hellman key exchange

Example 2.7. Alice and Bob agree to use the prime $p = 941$ and the primitive root $g = 627$. Alice chooses the secret key $a = 347$ and computes $A = 390 \equiv 627^{347} \pmod{941}$. Similarly, Bob chooses the secret key $b = 781$ and computes $B = 691 \equiv 627^{781} \pmod{941}$. Alice sends Bob the number 390 and Bob sends Alice the number 691. Both of these transmissions are done over an insecure channel, so both $A = 390$ and $B = 691$ should be considered public knowledge. The numbers $a = 347$ and $b = 781$ are not transmitted and remain secret. Then Alice and Bob are both able to compute the number

$$470 \equiv 627^{347 \cdot 781} \equiv A^b \equiv B^a \pmod{941},$$

so 470 is their shared secret.

Suppose that Eve sees this entire exchange. She can reconstitute Alice's and Bob's shared secret if she can solve either of the congruences

$$627^a \equiv 390 \pmod{941} \quad \text{or} \quad 627^b \equiv 691 \pmod{941},$$

since then she will know one of their secret exponents. As far as is known, this is the only way for Eve to find the secret shared value without Alice's or Bob's assistance.

Of course, our example uses numbers that are much too small to afford Alice and Bob any real security, since it takes very little time for Eve's computer to check all possible powers of 627 modulo 941. Current guidelines suggest that Alice and Bob choose a prime p having approximately 1000 bits (i.e., $p \approx 2^{1000}$) and an element g whose order is prime and approximately $p/2$. Then Eve will face a truly difficult task.

In general, Eve's dilemma is this. She knows the values of A and B, so she knows the values of g^a and g^b. She also knows the values of g and p, so if she can solve the DLP, then she can find a and b, after which it is easy for her to compute Alice and Bob's shared secret value g^{ab}. It appears that Alice and Bob are safe provided that Eve is unable to solve the DLP, but this is not quite correct. It is true that one method of finding Alice and Bob's shared value is to solve the DLP, but that is not the precise problem that Eve needs to solve. The security of Alice's and Bob's shared key rests on the difficulty of the following, potentially easier, problem.

Definition. Let p be a prime number and g an integer. The *Diffie–Hellman Problem* (DHP) is the problem of computing the value of $g^{ab} \pmod{p}$ from the known values of $g^a \pmod{p}$ and $g^b \pmod{p}$.

It is clear that the DHP is no harder than the DLP. If Eve can solve the DLP, then she can compute Alice and Bob's secret exponents a and b from the intercepted values $A = g^a$ and $B = g^b$, and then it is easy for her to compute their shared key g^{ab}. (In fact, Eve needs to compute only one of a and b.) But the converse is less clear. Suppose that Eve has an algorithm that efficiently solves the DHP. Can she use it to also efficiently solve the DLP? The answer is not known.

2.4 The Elgamal Public Key Cryptosystem

Although the Diffie–Hellman key exchange algorithm provides a method of publicly sharing a random secret key, it does not achieve the full goal of being a public key cryptosystem, since a cryptosystem permits exchange of specific information, not just a random string of bits. The first public key cryptosystem was the RSA system of Rivest, Shamir, and Adleman [110], which they published in 1978. RSA was, and still is, a fundamentally important discovery, and we discuss it in detail in Chap. 3. However, although RSA was historically first, the most natural development of a public key cryptosystem following the Diffie–Hellman paper [38] is a system described by Taher Elgamal in 1985 [41]. The Elgamal public key encryption algorithm is based on the discrete log problem and is closely related to Diffie–Hellman key exchange from Sect. 2.3. In this section we describe the version of the Elgamal PKC that is based on the discrete logarithm problem for $\mathbb{F}_p^*$, but the construction works quite generally using the DLP in any group. In particular, in Sect. 6.4.2 we discuss a version of the Elgamal PKC based on elliptic curve groups.

The Elgamal PKC is our first example of a public key cryptosystem, so we proceed slowly and provide all of the details. Alice begins by publishing information consisting of a public key and an algorithm. The public key is simply a number, and the algorithm is the method by which Bob encrypts his messages using Alice's public key. Alice does not disclose her private key, which is another number. The private key allows Alice, and only Alice, to decrypt messages that have been encrypted using her public key.

This is all somewhat vague and applies to any public key cryptosystem. For the Elgamal PKC, Alice needs a large prime number p for which the discrete logarithm problem in $\mathbb{F}_p^*$ is difficult, and she needs an element g modulo p of large (prime) order. She may choose p and g herself, or they may have been preselected by some trusted party such as an industry panel or government agency.

Alice chooses a secret number a to act as her private key, and she computes the quantity

$$A \equiv g^a \pmod{p}.$$

Notice the resemblance to Diffie–Hellman key exchange. Alice publishes her public key A and she keeps her private key a secret.

Now suppose that Bob wants to encrypt a message using Alice's public key A. We will assume that Bob's message m is an integer between 2 and p. (Recall that we discussed how to convert messages into numbers in Sect. 1.7.2.) In order to encrypt m, Bob first randomly chooses another number k modulo p.[5] Bob uses k to encrypt one, and only one, message, and then

[5] Most public key cryptosystems require the use of random numbers in order to operate securely. The generation of random or random-looking integers is actually a delicate process. We discuss the problem of generating pseudorandom numbers in Sect. 8.2, but for now we ignore this issue and assume that Bob has no trouble generating random numbers modulo p.

he discards it. The number k is called a *random element*; it exists for the sole purpose of encrypting a single message.

Bob takes his plaintext message m, his random element k, and Alice's public key A and uses them to compute the two quantities

$$c_1 \equiv g^k \pmod{p} \quad \text{and} \quad c_2 \equiv mA^k \pmod{p}.$$

(Remember that g and p are public parameters, so Bob also knows their values.) Bob's ciphertext, i.e., his encryption of m, is the pair of numbers (c_1, c_2), which he sends to Alice.

How does Alice decrypt Bob's ciphertext (c_1, c_2)? Since Alice knows a, she can compute the quantity

$$x \equiv (c_1^a)^{-1} \pmod{p}.$$

She can this by first computing $c_1^1 \pmod{p}$ using the fast power algorithm, and then computing the inverse using the extended Euclidean algorithm. Alternatively, she can just use fast powering to compute $c_1^{p-1-a} \pmod{p}$. Alice next multiplies c_2 by x, and lo and behold, the resulting value is the plaintext m. To see why, we expand the value of $x \cdot c_2$ and find that

$$\begin{aligned}
x \cdot c_2 &\equiv (c_1^a)^{-1} \cdot c_2 && \pmod{p}, && \text{since } x \equiv (c_1^a)^{-1} \pmod{p}, \\
&\equiv (g^{ak})^{-1} \cdot (mA^k) && \pmod{p}, && \text{since } c_1 \equiv g^k,\ c_2 \equiv mA^k \pmod{p}, \\
&\equiv (g^{ak})^{-1} \cdot (m(g^a)^k) && \pmod{p}, && \text{since } A \equiv g^a \pmod{p}, \\
&\equiv m && \pmod{p}, && \text{since the } g^{ak} \text{ terms cancel out.}
\end{aligned}$$

The Elgamal public key cryptosystem is summarized in Table 2.3.

What is Eve's task in trying to decrypt the message? Eve knows the public parameters p and g, and she also knows the value of $A \equiv g^a \pmod{p}$, since Alice's public key A is public knowledge. If Eve can solve the discrete logarithm problem, then she can find a and decrypt the message. More precisely, it's enough for Eve to solve the Diffie–Hellman problem; see Exercise 2.9. Otherwise it appears difficult for Eve to find the plaintext, although there are subtleties, some of which we'll discuss after doing an example with small numbers.

Example 2.8. Alice uses the prime $p = 467$ and the primitive root $g = 2$. She chooses $a = 153$ to be her private key and computes her public key

$$A \equiv g^a \equiv 2^{153} \equiv 224 \pmod{467}.$$

Bob decides to send Alice the message $m = 331$. He chooses a random element, say he chooses $k = 197$, and he computes the two quantities

$$c_1 \equiv 2^{197} \equiv 87 \pmod{467} \quad \text{and} \quad c_2 \equiv 331 \cdot 224^{197} \equiv 57 \pmod{467}.$$

The pair $(c_1, c_2) = (87, 57)$ is the ciphertext that Bob sends to Alice.

Public parameter creation	
A trusted party chooses and publishes a large prime p and an element g modulo p of large (prime) order.	
Alice	**Bob**
Key creation	
Choose private key $1 \le a \le p-1$. Compute $A = g^a \pmod{p}$. Publish the public key A.	
Encryption	
	Choose plaintext m. Choose random element k. Use Alice's public key A to compute $c_1 = g^k \pmod{p}$ and $c_2 = mA^k \pmod{p}$. Send ciphertext (c_1, c_2) to Alice.
Decryption	
Compute $(c_1^a)^{-1} \cdot c_2 \pmod{p}$. This quantity is equal to m.	

Table 2.3: Elgamal key creation, encryption, and decryption

Alice, knowing $a = 153$, first computes

$$x \equiv (c_1^a)^{-1} \equiv c_1^{p-1-a} \equiv 87^{313} \equiv 14 \pmod{467}.$$

Finally, she computes

$$c_2 x \equiv 57 \cdot 14 \equiv 331 \pmod{467}$$

and recovers the plaintext message m.

Remark 2.9. In the Elgamal cryptosystem, the plaintext is an integer m between 2 and $p-1$, while the ciphertext consists of two integers c_1 and c_2 in the same range. Thus in general it takes twice as many bits to write down the ciphertext as it does to write down the plaintext. We say that Elgamal has a 2-*to*-1 *message expansion*.

It's time to raise an important question. Is the Elgamal system as hard for Eve to attack as the Diffie–Hellman problem? Or, by introducing a clever way of encrypting messages, have we unwittingly opened a back door that makes it easy to decrypt messages without solving the Diffie–Hellman problem? One of the goals of modern cryptography is to identify an underlying hard problem like the Diffie–Hellman problem and to *prove* that a given cryptographic construction like Elgamal is at least as hard to attack as the underlying problem.

In this case we would like to prove that anyone who can decrypt arbitrary ciphertexts created by Elgamal encryption, as summarized in Table 2.3, must also be able to solve the Diffie–Hellman problem. Specifically, we would like to prove the following:

Proposition 2.10. *Fix a prime p and base g to use for Elgamal encryption. Suppose that Eve has access to an oracle that decrypts arbitrary Elgamal ciphertexts encrypted using arbitrary Elgamal public keys. Then she can use the oracle to solve the Diffie–Hellman problem described on page 69.*

Conversely, if Eve can solve the Diffie–Hellman problem, then she can break the Elgamal PKC.

Proof. Rather than giving a compact formal proof, we will be more discursive and explain how one might approach the problem of using an Elgamal oracle to solve the Diffie–Hellman problem. Recall that in the Diffie–Hellman problem, Eve is given the two values

$$A \equiv g^a \pmod{p} \qquad \text{and} \qquad B \equiv g^b \pmod{p},$$

and she is required to compute the value of $g^{ab} \pmod{p}$. Keep in mind that she knows both of the values of A and B, but she does not know either of the values a and b.

Now suppose that Eve can consult an Elgamal oracle. This means that Eve can send the oracle a prime p, a base g, a purported public key A, and a purported cipher text (c_1, c_2). Referring to Table 2.3, the oracle returns to Eve the quantity

$$(c_1^a)^{-1} \cdot c_2 \pmod{p}.$$

If Eve wants to solve the Diffie–Hellman problem, what values of c_1 and c_2 should she choose? A little thought shows that $c_1 = B = g^b$ and $c_2 = 1$ are good choices, since with this input, the oracle returns $(g^{ab})^{-1} \pmod{p}$, and then Eve can take the inverse modulo p to obtain $g^{ab} \pmod{p}$, thereby solving the Diffie–Hellman problem.

But maybe the oracle is smart enough to know that it should never decrypt ciphertexts having $c_2 = 1$. Eve can still fool the oracle by sending it random-looking ciphertexts as follows. She chooses an arbitrary value for c_2 and tells the oracle that the public key is A and that the ciphertext is (B, c_2). The oracle returns to her the supposed plaintext m that satisfies

$$m \equiv (c_1^a)^{-1} \cdot c_2 \equiv (B^a)^{-1} \cdot c_2 \equiv (g^{ab})^{-1} \cdot c_2 \pmod{p}.$$

After the oracle tells Eve the value of m, she simply computes

$$m^{-1} \cdot c_2 \equiv g^{ab} \pmod{p}$$

to find the value of $g^{ab} \pmod{p}$. It is worth noting that although, with the oracle's help, Eve has computed $g^{ab} \pmod{p}$, she has done so without knowledge of a or b, so she has solved only the Diffie–Hellman problem, not the discrete logarithm problem.

We leave the proof of the converse, i.e., that a Diffie–Hellman oracle breaks the Elgamal PKC, as an exercise; see Exercise 2.9. □

2.5 An Overview of the Theory of Groups

For readers unfamiliar with the theory of groups, we briefly introduce a few basic concepts that should help to place the study of discrete logarithms, both here and in Chap. 6, into a broader context.

We've just spent some time talking about exponentiation of elements in $\mathbb{F}_p^*$. Since exponentiation is simply repeated multiplication, this seems like a good place to start. What we'd like to do is to underline some important properties of multiplication in $\mathbb{F}_p^*$ and to point out that these attributes appear in many other contexts.

The properties are:

- There is an element $1 \in \mathbb{F}_p^*$ satisfying $1 \cdot a = a$ for every $a \in \mathbb{F}_p^*$.
- Every $a \in \mathbb{F}_p^*$ has an inverse $a^{-1} \in \mathbb{F}_p^*$ satisfying $a \cdot a^{-1} = a^{-1} \cdot a = 1$.
- Multiplication is associative: $a \cdot (b \cdot c) = (a \cdot b) \cdot c$ for all $a, b, c \in \mathbb{F}_p^*$.
- Multiplication is commutative: $a \cdot b = b \cdot a$ for all $a, b \in \mathbb{F}_p^*$.

Suppose that instead of multiplication in $\mathbb{F}_p^*$, we substitute addition in $\mathbb{F}_p$. We also use 0 in place of 1 and $-a$ in place of a^{-1}. Then all four properties are still true:

- $0 + a = a$ for every $a \in \mathbb{F}_p$.
- Every $a \in \mathbb{F}_p$ has an inverse $-a \in \mathbb{F}_p$ with $a + (-a) = (-a) + a = 0$.
- Addition is associative, $a + (b + c) = (a + b) + c$ for all $a, b, c \in \mathbb{F}_p$.
- Addition is commutative, $a + b = b + a$ for all $a, b \in \mathbb{F}_p$.

Sets and operations that behave similarly to multiplication or addition are so widespread that it is advantageous to abstract the general concept and talk about all such systems at once. The leads to the notion of a group.

Definition. A *group* consists of a set G and a rule, which we denote by $\star$, for combining two elements $a, b \in G$ to obtain an element $a \star b \in G$. The composition operation $\star$ is required to have the following three properties:

[Identity Law] There is an $e \in G$ such that

$$e \star a = a \star e = a \quad \text{for every } a \in G.$$

[Inverse Law] For every $a \in G$ there is a (unique) $a^{-1} \in G$ satisfying $a \star a^{-1} = a^{-1} \star a = e$.

[Associative Law] $a \star (b \star c) = (a \star b) \star c \quad$ for all $a, b, c \in G$.

If, in addition, composition satisfies the

[Commutative Law] $a \star b = b \star a \quad$ for all $a, b \in G$,

then the group is called a *commutative group* or an *abelian group*.

If G has finitely many elements, we say that G is a *finite group*. The *order of G* is the number of elements in G; it is denoted by $|G|$ or $\#G$.

Example 2.11. Groups are ubiquitous in mathematics and in the physical sciences. Here are a few examples, the first two repeating those mentioned earlier:

(a) $G = \mathbb{F}_p^*$ and $\star =$ multiplication. The identity element is $e = 1$. Proposition 1.21 tells us that inverses exist. Then G is a finite group of order $p - 1$.

(b) $G = \mathbb{Z}/N\mathbb{Z}$ and $\star =$ addition. The identity element is $e = 0$ and the inverse of a is $-a$. This G is a finite group of order N.

(c) $G = \mathbb{Z}$ and $\star =$ addition. The identity element is $e = 0$ and the inverse of a is $-a$. This group G is an infinite group.

(d) Note that $G = \mathbb{Z}$ and $\star =$ multiplication is not a group, since most elements do not have multiplicative inverses inside $\mathbb{Z}$.

(e) However, $G = \mathbb{R}^*$ and $\star =$ multiplication is a group, since all elements have multiplicative inverses inside $\mathbb{R}^*$.

(f) An example of a noncommutative group is

$$G = \left\{ \begin{pmatrix} a & b \\ c & d \end{pmatrix} : a, b, c, d \in \mathbb{R} \text{ and } ad - bc \neq 0 \right\}$$

with operation $\star =$ matrix multiplication. The identity element is $e = \left(\begin{smallmatrix} 1 & 0 \\ 0 & 1 \end{smallmatrix}\right)$ and the inverse is given by the familiar formula

$$\begin{pmatrix} a & b \\ c & d \end{pmatrix}^{-1} = \begin{pmatrix} \frac{d}{ad-bc} & \frac{-b}{ad-bc} \\ \frac{-c}{ad-bc} & \frac{a}{ad-bc} \end{pmatrix}.$$

Notice that G is noncommutative, since for example, $\left(\begin{smallmatrix} 1 & 1 \\ 0 & 1 \end{smallmatrix}\right)\left(\begin{smallmatrix} 1 & 1 \\ 1 & 0 \end{smallmatrix}\right)$ is not equal to $\left(\begin{smallmatrix} 1 & 1 \\ 1 & 0 \end{smallmatrix}\right)\left(\begin{smallmatrix} 1 & 1 \\ 0 & 1 \end{smallmatrix}\right)$.

(g) More generally, we can use matrices of any size. This gives the *general linear group*

$$\mathrm{GL}_n(\mathbb{R}) = \{n\text{-by-}n \text{ matrices } A \text{ with real coefficients and } \det(A) \neq 0\}$$

and operation $\star =$ matrix multiplication. We can form other groups by replacing $\mathbb{R}$ with some other field, for example, the finite field $\mathbb{F}_p$. (See Exercise 2.15.) The group $\mathrm{GL}_n(\mathbb{F}_p)$ is clearly a finite group, but computing its order is an interesting exercise.

Let g be an element of a group G and let x be a positive integer. Then g^x means that we apply the group operation to x copies of the element g,

$$g^x = \underbrace{g \star g \star g \star \cdots \star g}_{x \text{ repetitions}}.$$

For example, exponentiation g^x in the group $\mathbb{F}_p^*$ has the usual meaning, multiply x copies of g. But "exponentiation" g^x in the group $\mathbb{Z}/N\mathbb{Z}$ means to *add* x copies of g. Admittedly, it is more common to write the quantity "add x copies of g" as $x \cdot g$, but this is just a matter of notation. The key concept underlying exponentiation in a group is repeated application of the group operation to an element of the group.

It is also convenient to give a meaning to g^x when x is not positive. So if x is a negative integer, we define g^x to be $(g^{-1})^{|x|}$. For $x = 0$, we set $g^0 = e$, the identity element of G.

We now introduce a key concept used in the study of groups.

Definition. Let G be a group and let $a \in G$ be an element of the group. Suppose there exists a positive integer d with the property that $a^d = e$. The smallest such d is called the *order* of a. If there is no such d, then a is said to have *infinite order.*

We next prove two propositions describing important properties of the orders of group elements. These are generalizations of Theorem 1.24 (Fermat's little theorem) and Proposition 1.29, which deal with the group $G = \mathbb{F}_p^*$. The proofs are essentially the same.

Proposition 2.12. *Let G be a finite group. Then every element of G has finite order. Further, if $a \in G$ has order d and if $a^k = e$, then $d \mid k$.*

Proof. Since G is finite, the sequence

$$a, a^2, a^3, a^4, \ldots$$

must eventually contain a repetition. That is, there exist positive integers i and j with $j < i$ such that $a^i = a^j$. Multiplying both sides by a^{-j} and applying the group laws leads to $a^{i-j} = e$. Since $i - j > 0$, this proves that some power of a is equal to e. We let d be the smallest positive exponent satisfying $a^d = e$.

Now suppose that $k \geq d$ also satisfies $a^k = e$. We divide k by d to obtain

$$k = dq + r \qquad \text{with } 0 \leq r < d.$$

Using the fact that $a^k = a^d = e$, we find that

$$e = a^k = a^{dq+r} = (a^d)^q \star a^r = e^q \star a^r = a^r.$$

But d is the smallest positive power of a that is equal to e, so we must have $r = 0$. Therefore $k = dq$, so $d \mid k$. □

Proposition 2.13 (Lagrange's Theorem). *Let G be a finite group and let $a \in G$. Then the order of a divides the order G.*

More precisely, let $n = |G|$ be the order of G and let d be the order of a, i.e., a^d is the smallest positive power of a that is equal to e. Then

$$a^n = e \qquad \textit{and} \qquad d \mid n.$$

Proof. We give a simple proof in the case that G is commutative. For a proof in the general case, see any basic algebra textbook, for example [40, §3.2] or [45, §2.3].

Since G is finite, we can list its elements as

$$G = \{g_1, g_2, \ldots, g_n\}.$$

We now multiply each element of G by a to obtain a new set, which we call S_a,

$$S_a = \{a \star g_1, a \star g_2, \ldots, a \star g_n\}.$$

We claim that the elements of S_a are distinct. To see this, suppose that $a \star g_i = a \star g_j$. Multiplying both sides by a^{-1} yields $g_i = g_j$.[6] Thus S_a contains n distinct elements, which is the same as the number of elements of G. Therefore $S_a = G$, so if we multiply together all of the elements of S_a, we get the same answer as multiplying together all of the elements of G. (Note that we are using the assumption that G is commutative.) Thus

$$(a \star g_1) \star (a \star g_2) \star \cdots \star (a \star g_n) = g_1 \star g_2 \star \cdots \star g_n.$$

We can rearrange the order of the product on the left-hand side (again using the commutativity) to obtain

$$a^n \star g_1 \star g_2 \star \cdots \star g_n = g_1 \star g_2 \star \cdots \star g_n.$$

Now multiplying by $(g_1 \star g_2 \star \cdots \star g_n)^{-1}$ yields $a^n = e$, which proves the first statement, and then the divisibility of n by d follows immediately from Proposition 2.12. □

2.6 How Hard Is the Discrete Logarithm Problem?

Given a group G and two elements $g, h \in G$, the discrete logarithm problem asks for an exponent x such that $g^x = h$. What does it mean to talk about the difficulty of this problem? How can we quantify "hard"? A natural measure of hardness is the approximate number of operations necessary for a person or a computer to solve the problem using the most efficient method currently known. For example, we can solve the discrete logarithm problem by computing the list of values $g, g^2, g^3, \ldots$ until we find one that is equal to h. If g has order n, then this algorithm is guaranteed to find the solution

---

[6] We are being somewhat informal here, as is usually done when one is working with groups. Here is a more formal proof. We are given that $a \star g_i = a \star g_j$. We use this assumption and the group law axioms to compute
$g_i = e \star g_i = (a^{-1} \star a) \star g_i = a^{-1} \star (a \star g_i) = a^{-1} \star (a \star g_j) = (a^{-1} \star a) \star g_j = e \star g_j = g_j.$

in at most n multiplications, but if n is large, say $n > 2^{80}$, then it is not a practical algorithm with the computing power available today.

Alternatively, we might try choosing random values of x, compute g^x, and check if $g^x = h$. Using the fast exponentiation method described in Sect. 1.3.2, it takes a small multiple of $\log_2(x)$ modular multiplications to compute g^x. If n and x are k-bit numbers, that is, they are each approximately 2^k, then this trial-and-error approach requires about $k \cdot 2^k$ multiplications. If we are working in the group $\mathbb{F}_p^*$ and if we treat modular addition as our basic operation, then modular multiplication of two k-bit numbers takes (approximately) k^2 basic operations, so solving the DLP by trial-and-error takes a small multiple of $k^2 \cdot 2^k$ basic operations.

We are being somewhat imprecise when we talk about "small multiples" of 2^k or $k \cdot 2^k$ or $k^2 \cdot 2^k$. This is because when we want to know whether a computation is feasible, numbers such as $3 \cdot 2^k$ and $10 \cdot 2^k$ and $100 \cdot 2^k$ mean pretty much the same thing if k is large. The important property is that the constant multiple is fixed as k increases. *Order notation* was invented to make these ideas precise.[7] It is prevalent throughout mathematics and computer science and provides a handy way to get a grip on the magnitude of quantities.

Definition (Order Notation). Let $f(x)$ and $g(x)$ be functions of x taking values that are positive. We say that "f is big-$\mathcal{O}$ of g" and write

$$f(x) = \mathcal{O}\big(g(x)\big)$$

if there are positive constants c and C such that

$$f(x) \le cg(x) \qquad \text{for all } x \ge C.$$

In particular, we write $f(x) = \mathcal{O}(1)$ if $f(x)$ is bounded for all $x \ge C$.

The next proposition gives a method that can sometimes be used to prove that $f(x) = \mathcal{O}\big(g(x)\big)$.

Proposition 2.14. *If the limit*

$$\lim_{x \to \infty} \frac{f(x)}{g(x)}$$

exists (and is finite), then $f(x) = \mathcal{O}\big(g(x)\big)$.

Proof. Let L be the limit. By definition of limit, for any $\epsilon > 0$ there is a constant C_ϵ such that

$$\left| \frac{f(x)}{g(x)} - L \right| < \epsilon \qquad \text{for all } x > C_\epsilon.$$

[7] Although we use the same word for the *order* of a finite group and the *order* of growth of a function, they are two different concepts. Make sure that you don't confuse them.

In particular, taking $\epsilon = 1$, we find that

$$\frac{f(x)}{g(x)} < L + 1 \quad \text{for all } x > C_1.$$

Hence by definition, $f(x) = \mathcal{O}(g(x))$ with $c = L + 1$ and $C = C_1$. □

Example 2.15. We have $2x^3 - 3x^2 + 7 = \mathcal{O}(x^3)$, since

$$\lim_{x\to\infty} \frac{2x^3 - 3x^2 + 7}{x^3} = 2.$$

Similarly, we have $x^2 = \mathcal{O}(2^x)$, since

$$\lim_{x\to\infty} \frac{x^2}{2^x} = 0.$$

(If you don't know the value of this limit, use L'Hôpital's rule twice.)

However, note that we may have $f(x) = \mathcal{O}(g(x))$ even if the limit of $f(x)/g(x)$ does not exist. For example, the limit

$$\lim_{x\to\infty} \frac{(x+2)\cos^2(x)}{x}$$

does not exist, but

$$(x+2)\cos^2(x) = \mathcal{O}(x), \quad \text{since} \quad (x+2)\cos^2(x) \le x + 2 \le 2x \quad \text{for all } x \ge 2.$$

Example 2.16. Here are a few more examples of big-$\mathcal{O}$ notation. We leave the verification as an exercise.

(a) $x^2 + \sqrt{x} = \mathcal{O}\left(x^2\right)$.

(b) $5 + 6x^2 - 37x^5 = \mathcal{O}\left(x^5\right)$.

(c) $k^{300} = \mathcal{O}\left(2^k\right)$.

(d) $(\ln k)^{375} = \mathcal{O}\left(k^{0.001}\right)$.

(e) $k^2 2^k = \mathcal{O}\left(e^{2k}\right)$.

(f) $N^{10} 2^N = \mathcal{O}\left(e^N\right)$.

Order notation allows us to define several fundamental concepts that are used to get a rough handle on the computational complexity of mathematical problems.

Definition. Suppose that we are trying to solve a certain type of mathematical problem, where the input to the problem is a number whose size may vary. As an example, consider the *Integer Factorization Problem*, whose input is a number N and whose output is a prime factor of N. We are interested in knowing how long it takes to solve the problem in terms of the size of the input. Typically, one measures the size of the input by its number of bits, since that is how much storage it takes to record the input.

Suppose that there is a constant $A \ge 0$, independent of the size of the input, such that if the input is $\mathcal{O}(k)$ bits long, then it takes $\mathcal{O}(k^A)$ steps to solve the problem. Then the problem is said to be solvable in *polynomial time*.

If we can take $A = 1$, then the problem is solvable in *linear time*, and if we can take $A = 2$, then the problem is solvable in *quadratic time*. Polynomial-time algorithms are considered to be fast algorithms.

On the other hand, if there is a constant $c > 0$ such that for inputs of size $\mathcal{O}(k)$ bits, there is an algorithm to solve the problem in $\mathcal{O}\left(e^{ck}\right)$ steps, then the problem is solvable in *exponential time*. Exponential-time algorithms are considered to be slow algorithms.

Intermediate between polynomial-time algorithms and exponential-time algorithms are *subexponential-time* algorithms. These have the property that for every $\epsilon > 0$, they solve the problem in $\mathcal{O}_\epsilon\left(e^{\epsilon k}\right)$ steps. This notation means that the constants c and C appearing in the definition of order notation are allowed to depend on ϵ. For example, in Chap. 3 we will study a subexponential-time algorithm for the integer factorization problem whose running time is $\mathcal{O}\big(e^{c\sqrt{k\log k}}\big)$ steps.

As a general rule of thumb in cryptography, problems solvable in polynomial time are considered to be "easy" and problems that require exponential time are viewed as "hard," with subexponential time lying somewhere in between. However, bear in mind that these are asymptotic descriptions that are applicable only as the variables become very large. Depending on the big-$\mathcal{O}$ constants and on the size of the input, an exponential problem may be easier than a polynomial problem. We illustrate these general concepts by considering the discrete logarithm problem in various groups.

Example 2.17. We start with our original discrete logarithm problem $g^x = h$ in $G = \mathbb{F}_p^*$. If the prime p is chosen between 2^k and 2^{k+1}, then g, h, and p all require at most k bits, so the problem can be stated in $\mathcal{O}(k)$-bits. (Notice that $\mathcal{O}(k)$ is the same as $\mathcal{O}(\log_2 p)$.)

If we try to solve the DLP using the trial-and-error method mentioned earlier, then it takes $\mathcal{O}(p)$ steps to solve the problem. Since $\mathcal{O}(p) = \mathcal{O}(2^k)$, this algorithm takes exponential time. (If we consider instead multiplication or addition to be the basic operation, then the algorithm takes $\mathcal{O}(k \cdot 2^k)$ or $\mathcal{O}(k^2 \cdot 2^k)$ steps, but these distinctions are irrelevant; the running time is still exponential, since for example it is $\mathcal{O}(3^k)$.)

However, there are faster ways to solve the DLP in $\mathbb{F}_p^*$, some of which are very fast but work only for some primes, while others are less fast, but work for all primes. For example, the Pohlig–Hellman algorithm described in Sect. 2.9 shows that if $p - 1$ factors entirely into a product of small primes, then the DLP is quite easy. For arbitrary primes, the algorithm described in Sect. 2.7 solves the DLP in $\mathcal{O}(\sqrt{p}\log p)$ steps, which is much faster than $\mathcal{O}(p)$, but still exponential. Even better is the index calculus algorithm described in Sect. 3.8. The index calculus solves the DLP in $\mathcal{O}(e^{c\sqrt{(\log p)(\log\log p)}})$ steps, so it is a subexponential algorithm.

Example 2.18. We next consider the DLP in the group $G = \mathbb{F}_p$, where now the group operation is addition. The DLP in this context asks for a solution x to the congruence

$$x \cdot g \equiv h \pmod{p},$$

where g and h are given elements of $\mathbb{Z}/p\mathbb{Z}$. As described in Sect. 1.3, we can solve this congruence using the extended Euclidean algorithm (Theorem 1.11) to compute $g^{-1} \pmod{p}$ and setting $x \equiv g^{-1} \cdot h \pmod{p}$. This takes $\mathcal{O}(\log p)$ steps (see Remark 1.15), so there is a linear-time algorithm to solve the DLP in the additive group $\mathbb{F}_p$. This is a very fast algorithm, so the DLP in $\mathbb{F}_p$ with addition is not a good candidate for use as a one-way function in cryptography.

This is an important lesson to learn. The discrete logarithm problems in different groups may display different levels of difficulty for their solution. Thus the DLP in $\mathbb{F}_p$ with addition has a linear-time solution, while the best known general algorithm to solve the DLP in $\mathbb{F}_p^*$ with multiplication is subexponential. In Chap. 6 we discuss another sort of group called an elliptic curve. The discrete logarithm problem for elliptic curves is believed to be even more difficult than the DLP for $\mathbb{F}_p^*$. In particular, if the elliptic curve group is chosen carefully and has N elements, then the best known algorithm to solve the DLP requires $\mathcal{O}(\sqrt{N})$ steps. Thus it currently takes exponential time to solve the elliptic curve discrete logarithm problem (ECDLP).

2.7 A Collision Algorithm for the DLP

In this section we describe a discrete logarithm algorithm due to Shanks. It is an example of a collision, or meet-in-the-middle, algorithm. Algorithms of this type are discussed in more detail in Sects. 5.4 and 5.5. Shanks's algorithm works in any group, not just $\mathbb{F}_p^*$, and the proof that it works is no more difficult for arbitrary groups, so we state and prove it in full generality.

We begin by recalling the running time of the trivial brute-force algorithm to solve the DLP.

Proposition 2.19 (Trivial Bound for DLP). *Let G be a group and let $g \in G$ be an element of order N. (Recall that this means that $g^N = e$ and that no smaller positive power of g is equal to the identity element e.) Then the discrete logarithm problem*

$$g^x = h \tag{2.2}$$

can be solved in $\mathcal{O}(N)$ steps and $\mathcal{O}(1)$ storage, where each step consists of multiplication by g.

Proof. We simply compute $g, g^2, g^3, \ldots$, where each successive value is obtained by multiplying the previous value by g, so we only need to store two values at a time. If a solution to $g^x = h$ exists, then h will appear before we reach g^N. □

Remark 2.20. If we work in $\mathbb{F}_p^*$, then each computation of $g^x \pmod{p}$ requires $\mathcal{O}((\log p)^k)$ computer operations, where the constant k and the implied big-$\mathcal{O}$ constant depend on the computer and the algorithm used for modular multiplication. Then the total number of computer steps, or *running time*, is $\mathcal{O}(N(\log p)^k)$. In general, the factor contributed by the $\mathcal{O}((\log p)^k)$ is negligible, so we will suppress it and simply refer to the running time as $\mathcal{O}(N)$.

The idea behind a collision algorithm is to make two lists and look for an element that appears in both lists. For the discrete logarithm problem described in Proposition 2.19, the running time of a collision algorithm is a little more than $\mathcal{O}(\sqrt{N})$ steps, which is a huge savings over $\mathcal{O}(N)$ if N is large.

Proposition 2.21 (Shanks's Babystep–Giantstep Algorithm). *Let G be a group and let $g \in G$ be an element of order $N \geq 2$. The following algorithm solves the discrete logarithm problem $g^x = h$ in $\mathcal{O}(\sqrt{N} \cdot \log N)$ steps using $\mathcal{O}(\sqrt{N})$ storage.*

(1) *Let $n = 1 + \lfloor \sqrt{N} \rfloor$, so in particular, $n > \sqrt{N}$.*

(2) *Create two lists,*

$$\text{List 1:} \quad e,\ g,\ g^2,\ g^3, \ldots, g^n,$$
$$\text{List 2:} \quad h,\ h \cdot g^{-n},\ h \cdot g^{-2n},\ h \cdot g^{-3n}, \ldots, h \cdot g^{-n^2}.$$

(3) *Find a match between the two lists, say $g^i = hg^{-jn}$.*

(4) *Then $x = i + jn$ is a solution to $g^x = h$.*

Proof. We begin with a couple of observations. First, when creating List 2, we start by computing the quantity $u = g^{-n}$ and then compile List 2 by computing $h, h \cdot u, h \cdot u^2, \ldots, h \cdot u^n$. Thus creating the two lists takes approximately $2n$ multiplications.[8] Second, assuming that a match exists, we can find a match in a small multiple of $n \log(n)$ steps using standard sorting and searching algorithms, so Step (3) takes $\mathcal{O}(n \log n)$ steps. Hence the total running time for the algorithm is $\mathcal{O}(n \log n) = \mathcal{O}(\sqrt{N} \log N)$. For this last step we have used the fact that $n \approx \sqrt{N}$, so

$$n \log n \approx \sqrt{N} \log \sqrt{N} = \frac{1}{2} \sqrt{N} \log N.$$

Third, the lists in Step (2) have length n, so require $\mathcal{O}(\sqrt{N})$ storage.

In order to prove that the algorithm works, we must show that Lists 1 and 2 always have a match. To see this, let x be the unknown solution to $g^x = h$ and write x as

$$x = nq + r \quad \text{with } 0 \leq r < n.$$

[8]Multiplication by g is a "baby step" and multiplication by $u = g^{-n}$ is a "giant step," whence the name of the algorithm.

k	g^k	$h \cdot u^k$	k	g^k	$h \cdot u^k$	k	g^k	$h \cdot u^k$	k	g^k	$h \cdot u^k$
1	9704	347	9	15774	16564	17	10137	10230	25	4970	12260
2	6181	13357	10	12918	11741	18	17264	3957	26	9183	6578
3	5763	12423	11	16360	16367	19	4230	9195	27	10596	7705
4	1128	13153	12	13259	7315	20	9880	13628	28	2427	1425
5	8431	7928	13	4125	2549	21	9963	10126	29	6902	6594
6	16568	1139	14	16911	10221	22	15501	5416	30	11969	12831
7	**14567**	6259	15	4351	16289	23	6854	13640	31	6045	4754
8	2987	12013	16	1612	4062	24	15680	5276	32	7583	**14567**

Table 2.4: Babystep–giantstep to solve $9704^x \equiv 13896 \pmod{17389}$

We know that $1 \le x < N$, so

$$q = \frac{x - r}{n} < \frac{N}{n} < n \quad \text{since } n > \sqrt{N}.$$

Hence we can rewrite the equation $g^x = h$ as

$$g^r = h \cdot g^{-qn} \qquad \text{with } 0 \le r < n \text{ and } 0 \le q < n.$$

Thus g^r is in List 1 and $h \cdot g^{-qn}$ is in List 2, which shows that Lists 1 and 2 have a common element. □

Example 2.22. We illustrate Shanks's babystep–giantstep method by using it to solve the discrete logarithm problem

$$g^x = h \quad \text{in } \mathbb{F}_p^* \quad \text{with} \quad g = 9704, \quad h = 13896, \quad \text{and} \quad p = 17389.$$

The number 9704 has order 1242 in $\mathbb{F}_{17389}^*$.[9] Set $n = \lfloor\sqrt{1242}\rfloor + 1 = 36$ and $u = g^{-n} = 9704^{-36} = 2494$. Table 2.4 lists the values of g^k and $h \cdot u^k$ for $k = 1, 2, \ldots$. From the table we find the collision

$$9704^7 = 14567 = 13896 \cdot 2494^{32} \quad \text{in } \mathbb{F}_{17389}.$$

Using the fact that $2494 = 9704^{-36}$, we compute

$$13896 = 9704^7 \cdot 2494^{-32} = 9704^7 \cdot (9704^{36})^{32} = 9704^{1159} \quad \text{in } \mathbb{F}_{17389}.$$

Hence $x = 1159$ solves the problem $9704^x = 13896$ in $\mathbb{F}_{17389}$.

2.8 The Chinese Remainder Theorem

The Chinese remainder theorem describes the solutions to a system of simultaneous linear congruences. The simplest situation is a system of two congruences,

[9]Lagrange's theorem (Proposition 2.13) says that the order of g divides $17388 = 2^2 \cdot 3^3 \cdot 7 \cdot 23$. So we can determine the order of g by computing g^n for the 48 distinct divisors of 17388, although in practice there are more efficient methods.

$$x \equiv a \pmod{m} \qquad \text{and} \qquad x \equiv b \pmod{n}, \tag{2.3}$$

with $\gcd(m, n) = 1$, in which case the Chinese remainder theorem says that there is a unique solution modulo mn.

The first recorded instance of a problem of this type appears in a Chinese mathematical work from the late third or early fourth century. It actually deals with the harder problem of three simultaneous congruences.

> We have a number of things, but we do not know exactly how many. If we count them by threes, we have two left over. If we count them by fives, we have three left over. If we count them by sevens, we have two left over. How many things are there? [*Sun Tzu Suan Ching* (Master Sun's Mathematical Manual) circa 300 AD, volume 3, problem 26.]

The Chinese remainder theorem and its generalizations have many applications in number theory and other areas of mathematics. In Sect. 2.9 we will see how it can be used to solve certain instances of the discrete logarithm problem. We begin with an example in which we solve two simultaneous congruences. As you read this example, notice that it is not merely an abstract statement that a solution exists. The method that we describe is really an algorithm that allows us to find the solution.

Example 2.23. We look for an integer x that simultaneously solves both of the congruences

$$x \equiv 1 \pmod{5} \qquad \text{and} \qquad x \equiv 9 \pmod{11}. \tag{2.4}$$

The first congruence tells us that $x \equiv 1 \pmod{5}$, so the full set of solutions to the first congruence is the collection of integers

$$x = 1 + 5y, \qquad y \in \mathbb{Z}. \tag{2.5}$$

Substituting (2.5) into the second congruence in (2.4) gives

$$1 + 5y \equiv 9 \pmod{11}, \quad \text{and hence} \quad 5y \equiv 8 \pmod{11}. \tag{2.6}$$

We solve for y by multiplying both sides of (2.6) by the inverse of 5 modulo 11. This inverse exists because $\gcd(5, 11) = 1$ and can be computed using the procedure described in Proposition 1.13 (see also Remark 1.15). However, in this case the modulus is so small that we find it by trial and error; thus $5 \cdot 9 = 45 \equiv 1 \pmod{11}$.

In any case, multiplying both sides of (2.6) by 9 yields

$$y \equiv 9 \cdot 8 \equiv 72 \equiv 6 \pmod{11}.$$

Finally, substituting this value of y into (2.5) gives the solution

$$x = 1 + 5 \cdot 6 = 31$$

to the original problem.

The procedure outlined in Example 2.23 can be used to derive a general formula for the solution of two simultaneous congruences (see Exercise 2.20), but it is much better to learn the method, rather than memorizing a formula. This is especially true because the Chinese remainder theorem applies to systems of arbitrarily many simultaneous congruences.

Theorem 2.24 (Chinese Remainder Theorem). *Let $m_1, m_2, \ldots, m_k$ be a collection of pairwise relatively prime integers. This means that*

$$\gcd(m_i, m_j) = 1 \quad \textit{for all } i \neq j.$$

Let $a_1, a_2, \ldots, a_k$ be arbitrary integers. Then the system of simultaneous congruences

$$x \equiv a_1 \pmod{m_1}, \quad x \equiv a_2 \pmod{m_2}, \quad \ldots, \quad x \equiv a_k \pmod{m_k} \qquad (2.7)$$

has a solution $x = c$. Further, if $x = c$ and $x = c'$ are both solutions, then

$$c \equiv c' \pmod{m_1 m_2 \cdots m_k}. \qquad (2.8)$$

Proof. Suppose that for some value of i we have already managed to find a solution $x = c_i$ to the first i simultaneous congruences,

$$x \equiv a_1 \pmod{m_1}, \quad x \equiv a_2 \pmod{m_2}, \quad \ldots, \quad x \equiv a_i \pmod{m_i}. \qquad (2.9)$$

For example, if $i = 1$, then $c_1 = a_1$ works. We are going to explain how to find a solution to one more congruence,

$$x \equiv a_1 \pmod{m_1}, \quad x \equiv a_2 \pmod{m_2}, \quad \ldots, \quad x \equiv a_{i+1} \pmod{m_{i+1}}.$$

The idea is to look for a solution having the form

$$x = c_i + m_1 m_2 \cdots m_i y.$$

Notice that this value of x still satisfies all of the congruences (2.9), so we need merely choose y so that it also satisfies $x \equiv a_{i+1} \pmod{m_{i+1}}$. In other words, we need to find a value of y satisfying

$$c_i + m_1 m_2 \cdots m_i y \equiv a_{i+1} \pmod{m_{i+1}}.$$

Proposition 1.13(b) and the fact that $\gcd(m_{i+1}, m_1 m_2 \cdots m_i) = 1$ imply that we can always do this. This completes the proof of the existence of a solution. We leave to you the task of proving that different solutions satisfy (2.8); see Exercise 2.21. □

The proof of the Chinese remainder theorem (Theorem 2.24) is easily converted into an algorithm for finding the solution to a system of simultaneous congruences. An example suffices to illustrate the general method.

Example 2.25. We solve the three simultaneous congruences

$$x \equiv 2 \pmod{3}, \quad x \equiv 3 \pmod{7}, \quad x \equiv 4 \pmod{16}. \tag{2.10}$$

The Chinese remainder theorem says that there is a unique solution modulo 336, since $336 = 3 \cdot 7 \cdot 16$. We start with the solution $x = 2$ to the first congruence $x \equiv 2 \pmod{3}$. We use it to form the general solution $x = 2 + 3y$ and substitute it into the second congruence to get

$$2 + 3y \equiv 3 \pmod{7}.$$

This simplifies to $3y \equiv 1 \pmod{7}$, and we multiply both sides by 5 (since 5 is the inverse of 3 modulo 7) to get $y \equiv 5 \pmod{7}$. This gives the value

$$x = 2 + 3y = 2 + 3 \cdot 5 = 17$$

as a solution to the first two congruences in (2.10).

The general solution to the first two congruences is thus $x = 17 + 21z$. We substitute this into the third congruence to obtain

$$17 + 21z \equiv 4 \pmod{16}.$$

This simplifies to $5z \equiv 3 \pmod{16}$. We multiply by 13, which is the inverse of 5 modulo 16, to obtain

$$z \equiv 3 \cdot 13 \equiv 39 \equiv 7 \pmod{16}.$$

Finally, we substitute this into $x = 17 + 21z$ to get the solution

$$x = 17 + 21 \cdot 7 = 164.$$

All other solutions are obtained by adding and subtracting multiples of 336 to this particular solution.

2.8.1 Solving Congruences with Composite Moduli

It is usually easiest to solve a congruence with a composite modulus by first solving several congruences modulo primes (or prime powers) and then fitting together the solutions using the Chinese remainder theorem. We illustrate the principle in this section by discussing the problem of finding square roots modulo m. It turns out that it is relatively easy to compute square roots modulo a prime. Indeed, for primes congruent to 3 modulo 4, it is extremely easy to find square roots, as shown by the following proposition.

Proposition 2.26. *Let p be a prime satisfying $p \equiv 3 \pmod{4}$. Let a be an integer such that the congruence $x^2 \equiv a \pmod{p}$ has a solution, i.e., such that a has a square root modulo p. Then*

$$b \equiv a^{(p+1)/4} \pmod{p}$$

is a solution, i.e., it satisfies $b^2 \equiv a \pmod{p}$. *(N.B. This formula is valid only if a has a square root modulo p. In Sect. 3.9 we will describe an efficient method for checking which numbers have square roots modulo p.)*

Proof. Let g be a primitive root modulo p. Then a is equal to some power of g, and the fact that a has a square root modulo p means that a is an even power of g, say $a \equiv g^{2k} \pmod{p}$. (See Exercise 2.5.) Now we compute

$$\begin{aligned}
b^2 &\equiv a^{\frac{p+1}{2}} && \pmod{p} && \text{definition of } b,\\
&\equiv (g^{2k})^{\frac{p+1}{2}} && \pmod{p} && \text{since } a \equiv g^{2k} \pmod{p},\\
&\equiv g^{(p+1)k} && \pmod{p} &&\\
&\equiv g^{2k+(p-1)k} && \pmod{p} &&\\
&\equiv a \cdot (g^{p-1})^k && \pmod{p} && \text{since } a \equiv g^{2k} \pmod{p},\\
&\equiv a && \pmod{p} && \text{since } g^{p-1} \equiv 1 \pmod{p}.
\end{aligned}$$

Hence b is indeed a square root of a modulo p. □

Example 2.27. A square root of $a = 2201$ modulo the prime $p = 4127$ is

$$b \equiv a^{(p+1)/4} = 2201^{4128/4} \equiv 2201^{1032} \equiv 3718 \pmod{4127}.$$

To see that a does indeed have a square root modulo 4127, we simply square b and check that $3718^2 = 13823524 \equiv 2201 \pmod{4127}$.

Suppose now that we want to compute a square root modulo m, where m is not necessarily a prime. An efficient method is to factor m, compute the square root modulo each of the prime (or prime power) factors, and then combine the solutions using the Chinese remainder theorem. An example makes the idea clear.

Example 2.28. We look for a solution to the congruence

$$x^2 \equiv 197 \pmod{437}. \tag{2.11}$$

The modulus factors as $437 = 19 \cdot 23$, so we first solve the two congruences

$$y^2 \equiv 197 \equiv 7 \pmod{19} \quad \text{and} \quad z^2 \equiv 197 \equiv 13 \pmod{23}.$$

Since both 19 and 23 are congruent to 3 modulo 4, we can find these square roots using Proposition 2.26 (or by trial and error). In any case, we have

$$y \equiv \pm 8 \pmod{19} \quad \text{and} \quad z \equiv \pm 6 \pmod{23}.$$

We can pick either 8 or -8 for y and either 6 or -6 for z. Choosing the two positive solutions, we next use the Chinese remainder theorem to solve the simultaneous congruences

$$x \equiv 8 \pmod{19} \quad \text{and} \quad x \equiv 6 \pmod{23}. \tag{2.12}$$

We find that $x \equiv 236 \pmod{437}$, which gives the desired solution to (2.11).

Remark 2.29. The solution to Example 2.28 is not unique. In the first place, we can always take the negative,

$$-236 \equiv 201 \pmod{437},$$

to get a second square root of 197 modulo 437. If the modulus were prime, there would be only these two square roots (Exercise 1.36(a)). However, since $437 = 19 \cdot 23$ is composite, there are two others. In order to find them, we replace one of 8 and 6 with its negative in (2.12). This leads to the values $x = 144$ and $x = 293$, so 197 has four square roots modulo 437.

Remark 2.30. It is clear from Example 2.28 (see also Exercises 2.23 and 2.24) that it is relatively easy to compute square roots modulo m if one knows how to factor m into a product of prime powers. However, suppose that m is so large that we are not able to factor it. It is then a very difficult problem to find square roots modulo m. Indeed, in a certain reasonably precise sense, it is just as difficult to compute square roots modulo m as it is to factor m.

In fact, if m is a large composite number whose factorization is unknown, then it is a difficult problem to determine whether a given integer a has a square root modulo m, even without requiring that the square root be computed. The Goldwasser–Micali public key cryptosystem, which is described in Sect. 3.10, is based on the difficulty of identifying which numbers have square roots modulo a composite modulus m. The trapdoor information is knowledge of the factors of m.

2.9 The Pohlig–Hellman Algorithm

In addition to being a theorem and an algorithm, we would suggest to the reader that the Chinese remainder theorem is also a state of mind. If

$$m = m_1 \cdot m_2 \cdots m_t$$

is a product of pairwise relatively prime integers, then the Chinese remainder theorem says that solving an equation modulo m is more or less equivalent to solving the equation modulo m_i for each i, since it tells us how to knit the solutions together to get a solution modulo m.

In the discrete logarithm problem (DLP), we need to solve the equation

$$g^x \equiv h \pmod{p}.$$

In this case, the modulus p is prime, which suggests that the Chinese remainder theorem is irrelevant. However, recall that the solution x is determined only modulo $p-1$, so we can think of the solution as living in $\mathbb{Z}/(p-1)\mathbb{Z}$. This hints that the factorization of $p-1$ into primes may play a role in determining the difficulty of the DLP in $\mathbb{F}_p^*$. More generally, if G is any group and $g \in G$ is an element of order N, then solutions to $g^x = h$ in G are determined only modulo N, so the prime factorization of N would appear to be relevant. This idea is at the core of the Pohlig–Hellman algorithm.

As in Sect. 2.7 we state and prove results in this section for an arbitrary group G. But if you feel more comfortable working with integers modulo p, you may simply replace G by $\mathbb{F}_p^*$.

Theorem 2.31 (Pohlig–Hellman Algorithm). *Let G be a group, and suppose that we have an algorithm to solve the discrete logarithm problem in G for any element whose order is a power of a prime. To be concrete, if $g \in G$ has order q^e, suppose that we can solve $g^x = h$ in $\mathcal{O}(S_{q^e})$ steps. (For example, Proposition 2.21 says that we can take S_{q^e} to be $q^{e/2}$. See Remark 2.32 for a further discussion.)*

Now let $g \in G$ be an element of order N, and suppose that N factors into a product of prime powers as

$$N = q_1^{e_1} \cdot q_2^{e_2} \cdots q_t^{e_t}.$$

Then the discrete logarithm problem $g^x = h$ can be solved in

$$\mathcal{O}\left(\sum_{i=1}^{t} S_{q_i^{e_i}} + \log N\right) \text{ steps} \tag{2.13}$$

using the following procedure:

(1) *For each $1 \le i \le t$, let*

$$g_i = g^{N/q_i^{e_i}} \quad \text{and} \quad h_i = h^{N/q_i^{e_i}}.$$

Notice that g_i has prime power order $q_i^{e_i}$, so use the given algorithm to solve the discrete logarithm problem

$$g_i^y = h_i. \tag{2.14}$$

Let $y = y_i$ be a solution to (2.14).

(2) *Use the Chinese remainder theorem (Theorem 2.24) to solve*

$$x \equiv y_1 \pmod{q_1^{e_1}}, \quad x \equiv y_2 \pmod{q_2^{e_2}}, \quad \ldots, \quad x \equiv y_t \pmod{q_t^{e_t}}. \tag{2.15}$$

Proof. The running time is clear, since Step (1) takes $\mathcal{O}(\sum S_{q_i^{e_i}})$ steps, and Step (2), via the Chinese remainder theorem, takes $\mathcal{O}(\log N)$ steps. In practice, the Chinese remainder theorem computation is usually negligible compared to the discrete logarithm computations.

It remains to show that Steps (1) and (2) give a solution to $g^x = h$. Let x be a solution to the system of congruences (2.15). Then for each i we can write

$$x = y_i + q_i^{e_i} z_i \quad \text{for some } z_i. \tag{2.16}$$

This allows us to compute

$$\begin{aligned}
(g^x)^{N/q_i^{e_i}} &= \left(g^{y_i+q_i^{e_i}z_i}\right)^{N/q_i^{e_i}} && \text{from (2.16),}\\
&= \left(g^{N/q_i^{e_i}}\right)^{y_i}\cdot g^{Nz_i} \\
&= \left(g^{N/q_i^{e_i}}\right)^{y_i} && \text{since } g^N \text{ is the identity element,}\\
&= g_i^{y_i} && \text{by the definition of } g_i,\\
&= h_i && \text{from (2.14)}\\
&= h^{N/q_i^{e_i}} && \text{by the definition of } h_i.
\end{aligned}$$

In terms of discrete logarithms to the base g, we can rewrite this as

$$\frac{N}{q_i^{e_i}}\cdot x \equiv \frac{N}{q_i^{e_i}}\cdot \log_g(h) \pmod N, \tag{2.17}$$

where recall that the discrete logarithm to the base g is defined only modulo N, since g^N is the identity element.

Next we observe that the numbers

$$\frac{N}{q_1^{e_1}},\quad \frac{N}{q_2^{e_2}},\quad \dots\quad \frac{N}{q_t^{e_t}}$$

have no nontrivial common factor, i.e., their greatest common divisor is 1. Repeated application of the extended Euclidean theorem (Theorem 1.11) (see also Exercise 1.13) says that we can find integers $c_1, c_2, \dots, c_t$ such that

$$\frac{N}{q_1^{e_1}}\cdot c_1 + \frac{N}{q_2^{e_2}}\cdot c_2 + \cdots + \frac{N}{q_t^{e_t}}\cdot c_t = 1. \tag{2.18}$$

Now multiply both sides of (2.17) by c_i and sum over $i = 1, 2, \dots, t$. This gives

$$\sum_{i=1}^{t}\frac{N}{q_i^{e_i}}\cdot c_i\cdot x \equiv \sum_{i=1}^{t}\frac{N}{q_i^{e_i}}\cdot c_i\cdot \log_g(h) \pmod N,$$

and then (2.18) tells us that

$$x = \log_g(h) \pmod N.$$

This completes the proof that x satisfies $g^x \equiv h$. □

Remark 2.32. The Pohlig–Hellman algorithm more or less reduces the discrete logarithm problem for elements of arbitrary order to the discrete logarithm problem for elements of prime power order. A further refinement, which we discuss later in this section, essentially reduces the problem to elements of prime order. More precisely, in the notation of Theorem 2.31, the running time S_{q^e} for elements of order q^e can be reduced to $\mathcal{O}(eS_q)$. This is the content of Proposition 2.33.

The Pohlig–Hellman algorithm thus tells us that the discrete logarithm problem in a group G is not secure if the order of the group is a product of powers of small primes. More generally, $g^x = h$ is easy to solve if the order of the element g is a product of powers of small primes. This applies, in particular, to the discrete logarithm problem in $\mathbb{F}_p$ if $p-1$ factors into powers of small primes. Since $p-1$ is always even, the best that we can do is take $p = 2q+1$ with q prime and use an element g of order q. Then the running time of the collision algorithm described in Proposition 2.21 is $\mathcal{O}(\sqrt{q}) = \mathcal{O}(\sqrt{p})$. However, the index calculus method described in Sect. 3.8 has running time that is subexponential, so even if $p = 2q+1$, the prime q must be chosen to be quite large.

We now explain the algorithm that reduces the discrete logarithm problem for elements of prime power order to the discrete logarithm problem for elements of prime order. The idea is simple: if g has order q^e, then $g^{q^{e-1}}$ has order q. The trick is to repeat this process several times and then assemble the information into the final answer.

Proposition 2.33. *Let G be a group. Suppose that q is a prime, and suppose that we know an algorithm that takes S_q steps to solve the discrete logarithm problem $g^x = h$ in G whenever g has order q. Now let $g \in G$ be an element of order q^e with $e \geq 1$. Then we can solve the discrete logarithm problem*

$$g^x = h \quad \text{in } \mathcal{O}(eS_q) \text{ steps.} \tag{2.19}$$

Remark 2.34. Proposition 2.21 says that we can take $S_q = \mathcal{O}(\sqrt{q})$, so Proposition 2.33 says that we can solve the DLP (2.19) in $\mathcal{O}(e\sqrt{q})$ steps. Notice that if we apply Proposition 2.21 directly to the DLP (2.19), the running time is $\mathcal{O}(q^{e/2})$, which is much slower if $e \geq 2$.

Proof of Proposition 2.33. The key idea to proving the proposition is to write the unknown exponent x in the form

$$x = x_0 + x_1 q + x_2 q^2 + \cdots + x_{e-1} q^{e-1} \quad \text{with } 0 \leq x_i < q, \tag{2.20}$$

and then determine successively $x_0, x_1, x_2, \ldots$. We begin by observing that the element $g^{q^{e-1}}$ is of order q. This allows us to compute

$$\begin{aligned}
h^{q^{e-1}} &= (g^x)^{q^{e-1}} && \text{raising both sides of (2.19) to the } q^{e-1} \text{ power} \\
&= \left(g^{x_0 + x_1 q + x_2 q^2 + \cdots + x_{e-1} q^{e-1}}\right)^{q^{e-1}} && \text{from (2.20)} \\
&= g^{x_0 q^{e-1}} \cdot \left(g^{q^e}\right)^{x_1 + x_2 q + \cdots + x_{e-1} q^{e-2}} \\
&= \left(g^{q^{e-1}}\right)^{x_0} && \text{since } g^{q^e} = 1.
\end{aligned}$$

Since $g^{q^{e-1}}$ is an element of order q in G, the equation

$$\left(g^{q^{e-1}}\right)^{x_0} = h^{q^{e-1}}$$

is a discrete logarithm problem whose base is an element of order q. By assumption, we can solve this problem in S_q steps. Once this is done, we know an exponent x_0 with the property that

$$g^{x_0 q^{e-1}} = h^{q^{e-1}} \quad \text{in } G.$$

We next do a similar computation, this time raising both sides of (2.19) to the q^{e-2} power, which yields

$$\begin{aligned} h^{q^{e-2}} &= (g^x)^{q^{e-2}} \\ &= \left(g^{x_0+x_1 q+x_2 q^2+\cdots+x_{e-1}q^{e-1}}\right)^{q^{e-2}} \\ &= g^{x_0 q^{e-2}} \cdot g^{x_1 q^{e-1}} \cdot \left(g^{q^e}\right)^{x_2+x_3 q+\cdots+x_{e-1}q^{e-3}} \\ &= g^{x_0 q^{e-2}} \cdot g^{x_1 q^{e-1}}. \end{aligned}$$

Keep in mind that we have already determined the value of x_0 and that the element $g^{q^{e-1}}$ has order q in G. In order to find x_1, we must solve the discrete logarithm problem

$$\left(g^{q^{e-1}}\right)^{x_1} = \left(h \cdot g^{-x_0}\right)^{q^{e-2}}$$

for the unknown quantity x_1. Again applying the given algorithm, we can solve this in S_q steps. Hence in $\mathcal{O}(2S_q)$ steps, we have determined values for x_0 and x_1 satisfying

$$g^{(x_0+x_1 q)q^{e-2}} = h^{q^{e-2}} \quad \text{in } G.$$

Similarly, we find x_2 by solving the discrete logarithm problem

$$\left(g^{q^{e-1}}\right)^{x_2} = \left(h \cdot g^{-x_0-x_1 q}\right)^{q^{e-3}},$$

and in general, after we have determined $x_0, \ldots, x_{i-1}$, then the value of x_i is obtained by solving

$$\left(g^{q^{e-1}}\right)^{x_i} = \left(h \cdot g^{-x_0-x_1 q-\cdots-x_{i-1}q^{i-1}}\right)^{q^{e-i-1}} \quad \text{in } G.$$

Each of these is a discrete logarithm problem whose base is of order q, so each of them can be solved in S_q steps. Hence after $\mathcal{O}(eS_q)$ steps, we obtain an exponent $x = x_0 + x_1 q + \cdots + x_{e-1}q^{e-1}$ satisfying $g^x = h$, thus solving the original discrete logarithm problem. □

Example 2.35. We do an example to clarify the algorithm described in the proof of Proposition 2.33. We solve

$$5448^x = 6909 \quad \text{in } \mathbb{F}_{11251}^*. \tag{2.21}$$

The prime $p = 11251$ has the property that $p - 1$ is divisible by 5^4, and it is easy to check that 5448 has order exactly 5^4 in $\mathbb{F}_{11251}$. The first step is to solve

$$\left(5448^{5^3}\right)^{x_0} = 6909^{5^3},$$

which reduces to $11089^{x_0} = 11089$. This one is easy; the answer is $x_0 = 1$, so our initial value of x is $x = 1$.

The next step is to solve

$$\left(5448^{5^3}\right)^{x_1} = (6909 \cdot 5448^{-x_0})^{5^2} = (6909 \cdot 5448^{-1})^{5^2},$$

which reduces to $11089^{x_1} = 3742$. Note that we only need to check values of x_1 between 1 and 4, although if q were large, it would pay to use a faster algorithm such as Proposition 2.21 to solve this discrete logarithm problem. In any case, the solution is $x_1 = 2$, so the value of x is now $x = 11 = 1 + 2 \cdot 5$.

Continuing, we next solve

$$\left(5448^{5^3}\right)^{x_2} = (6909 \cdot 5448^{-x_0 - x_1 \cdot 5})^5 = (6909 \cdot 5448^{-11})^5,$$

which reduces to $11089^{x_2} = 1$. Thus $x_2 = 0$, which means that the value of x remains at $x = 11$.

The final step is to solve

$$\left(5448^{5^3}\right)^{x_3} = 6909 \cdot 5448^{-x_0 - x_1 \cdot 5 - x_2 \cdot 5^2} = 6909 \cdot 5448^{-11}.$$

This reduces to solving $11089^{x_3} = 6320$, which has the solution $x_3 = 4$. Hence our final answer is

$$x = 511 = 1 + 2 \cdot 5 + 4 \cdot 5^3.$$

As a check, we compute

$$5448^{511} = 6909 \quad \text{in } \mathbb{F}_{11251}. \checkmark$$

The Pohlig–Hellman algorithm (Theorem 2.31) for solving the discrete logarithm problem uses the Chinese remainder theorem (Theorem 2.24) to knot together the solutions for prime powers from Proposition 2.33. The following example illustrates the full Pohlig–Hellman algorithm.

Example 2.36. Consider the discrete logarithm problem

$$23^x = 9689 \quad \text{in } \mathbb{F}_{11251}. \tag{2.22}$$

The base 23 is a primitive root in $\mathbb{F}_{11251}$, i.e., it has order 11250. Since $11250 = 2 \cdot 3^2 \cdot 5^4$ is a product of small primes, the Pohlig–Hellman algorithm should work well. In the notation of Theorem 2.31, we set

$$p = 11251, \qquad g = 23, \qquad h = 9689, \qquad N = p - 1 = 2 \cdot 3^2 \cdot 5^4.$$

The first step is solve three subsidiary discrete logarithm problems, as indicated in the following table.

q	e	$g^{(p-1)/q^e}$	$h^{(p-1)/q^e}$	Solve $\left(g^{(p-1)/q^e}\right)^x = h^{(p-1)/q^e}$ for x
2	1	11250	11250	1
3	2	5029	10724	4
5	4	5448	6909	511

Notice that the first problem is trivial, while the third one is the problem that we solved in Example 2.35. In any case, the individual problems in this step of the algorithm may be solved as described in the proof of Proposition 2.33.

The second step is to use the Chinese remainder theorem to solve the simultaneous congruences

$$x \equiv 1 \pmod{2}, \qquad x \equiv 4 \pmod{3^2}, \qquad x \equiv 511 \pmod{5^4}.$$

The smallest solution is $x = 4261$. We check our answer by computing

$$23^{4261} = 9689 \quad \text{in } \mathbb{F}_{11251}. \quad \checkmark$$

2.10 Rings, Quotient Rings, Polynomial Rings, and Finite Fields

Note to the Reader: In this section we describe some topics that are typically covered in an introductory course in abstract algebra. This material is somewhat more mathematically sophisticated than the material that we have discussed up to this point. For cryptographic applications, the most important topics in this section are the theory of finite fields of prime power order, which in this book are used primarily in Sects. 6.7 and 6.8 in studying elliptic curve cryptography, and the theory of quotients of polynomial rings, which are used in Sect. 7.10 to describe the lattice-based NTRU public key cryptosystem. The reader interested in proceeding more rapidly to additional cryptographic topics may wish to omit this section at first reading and return to it when arriving at the relevant sections of Chaps. 6 and 7.

As we have seen, *groups* are fundamental objects that appear in many areas of mathematics. A group G is a set and an operation that allows us to "multiply" two elements to obtain a third element. We gave a brief overview of the theory of groups in Sect. 2.5. Another fundamental object in mathematics, called a *ring*, is a set having two operations. These two operations are analogous to ordinary addition and multiplication, and they are linked by the distributive law. In this section we begin with a brief discussion of the general theory of rings, then we discuss how to form one ring from another by taking quotients, and we conclude by examining in some detail the case of polynomial rings.

2.10.1 An Overview of the Theory of Rings

You are already familiar with many rings, for example the ring of integers with the operations of addition and multiplication. We abstract the fundamental properties of these operations and use them to formulate the following fundamental definition.

Definition. A *ring* is a set R that has two operations, which we denote by $+$ and $\star$,[10] having the following properties:

Properties of +

[*Identity Law*] There is an additive identity $0 \in R$ such that $0 + a = a + 0 = a$ for every $a \in R$.

[*Inverse Law*] For every element $a \in R$ there is an additive inverse $b \in R$ such that $a + b = b + a = 0$.

[*Associative Law*] $a + (b + c) = (a + b) + c$ for all $a, b, c \in R$.

[*Commutative Law*] $a + b = b + a$ for all $a, b \in R$,

Briefly, if we look at R with only the operation $+$, then it is a commutative group with (additive) identity element 0.

Properties of ⋆

[*Identity Law*] There is a multiplicative identity $1 \in R$ such that $1 \star a = a \star 1 = a$ for every $a \in R$.

[*Associative Law*] $a \star (b \star c) = (a \star b) \star c$ for all $a, b, c \in R$.

[*Commutative Law*] $a \star b = b \star a$ for all $a, b \in R$,

Thus if we look at R with only the operation $\star$, then it is *almost* a commutative group with (multiplicative) identity element 1, except that elements are not required to have multiplicative inverses.

Property Linking + and ⋆

[*Distributive Law*] $a \star (b + c) = a \star b + a \star c$ for all $a, b, c \in R$.

Remark 2.37. More generally, people sometimes work with rings that do not contain a multiplicative identity, and also with rings for which $\star$ is not commutative, i.e., $a \star b$ might not be equal to $b \star a$. So to be formal, our rings are really *commutative rings with (multiplicative) identity*. However, all of the rings that we use will be of this type, so we will just call them rings.

[10] Addition in a ring is virtually always denoted by $+$, but there are many different notations for multiplication. In this book use $a \star b$, $a \cdot b$, or simply ab, depending on the context.

Every element of a ring has an additive inverse, but there may be many nonzero elements that do not have multiplicative inverses. For example, in the ring of integers $\mathbb{Z}$, the only elements that have multiplicative inverses are 1 and -1.

Definition. A (commutative) ring in which every nonzero element has a multiplicative inverse is called a *field.*

Example 2.38. Here are a few examples of rings and fields with which you are probably already familiar.

(a) $R = \mathbb{Q}$, $\star =$ multiplication, and addition is as usual. The multiplicative identity element is 1. Every nonzero element has a multiplicative inverse, so $\mathbb{Q}$ is a field.

(b) $R = \mathbb{Z}$, $\star =$ multiplication, and addition is as usual. The multiplicative identity element is 1. The only elements that have multiplicative inverses are 1 and -1, so $\mathbb{Z}$ is a ring, but it is not a field.

(c) $R = \mathbb{Z}/n\mathbb{Z}$, n is any positive integer, $\star =$ multiplication, and addition is as usual. The multiplicative identity element is 1. Here R is always a ring, and it is a field if and only if n is prime.

(d) $R = \mathbb{F}_p$, p is any prime integer, $\star =$ multiplication, and addition is as usual. The multiplicative identity element is 1. By Proposition 1.21, every nonzero element has a multiplicative inverse, so $\mathbb{F}_p$ is a field.

(e) The collection of all polynomials with coefficients taken from $\mathbb{Z}$ forms a ring under the usual operations of polynomial addition and multiplication. This ring is denoted by $\mathbb{Z}[x]$. Thus we write

$$\mathbb{Z}[x] = \{a_0 + a_1x + a_2x^2 + \cdots + a_nx^n : n \ge 0 \text{ and } a_0, a_1, \ldots, a_n \in \mathbb{Z}\}.$$

For example, $1 + x^2$ and $3 - 7x^4 + 23x^9$ are polynomials in the ring $\mathbb{Z}[x]$, as are 17 and -203.

(f) More generally, if R is any ring, we can form a ring of polynomials whose coefficients are taken from the ring R. For example, the ring R might be $\mathbb{Z}/q\mathbb{Z}$ or a finite field $\mathbb{F}_p$. We discuss these general polynomial rings, denoted by $R[x]$, in Sect. 7.9.

2.10.2 Divisibility and Quotient Rings

The concept of divisibility, originally introduced for the integers $\mathbb{Z}$ in Sect. 1.2, can be generalized to any ring.

Definition. Let a and b be elements of a ring R with $b \ne 0$. We say that b *divides* a, or that a *is divisible by* b, if there is an element $c \in R$ such that

$$a = b \star c.$$

As before, we write $b \mid a$ to indicate that b divides a. If b does not divide a, then we write $b \nmid a$.

Remark 2.39. The basic properties of divisibility given in Proposition 1.4 apply to rings in general. The proof for $\mathbb{Z}$ works for any ring. Similarly, it is true in every ring that $b \mid 0$ for any $b \neq 0$. (See Exercise 2.30.) However, note that not every ring is as nice as $\mathbb{Z}$. For example, there are rings with nonzero elements a and b whose product $a \star b$ is 0. An example of such a ring is $\mathbb{Z}/6\mathbb{Z}$, in which 2 and 3 are nonzero, but $2 \cdot 3 = 6 = 0$.

Recall that an integer is called a prime if it has no nontrivial factors. What is a trivial factor? We can "factor" any integer by writing it as $a = 1 \cdot a$ and as $a = (-1)(-a)$, so these are trivial factorizations. What makes them trivial is the fact that 1 and -1 have multiplicative inverses. In general, if R is a ring and if $u \in R$ is an element that has a multiplicative inverse $u^{-1} \in R$, then we can factor any element $a \in R$ by writing it as $a = u^{-1} \cdot (ua)$. Elements that have multiplicative inverses and elements that have only trivial factorizations are special elements of a ring, so we give them special names.

Definition. Let R be a ring. An element $u \in R$ is called a *unit* if it has a multiplicative inverse, i.e., if there is an element $v \in R$ such that $u \star v = 1$.

An element a of a ring R is said to be *irreducible* if a is not itself a unit and if in every factorization of a as $a = b \star c$, either b is a unit or c is a unit.

Remark 2.40. The integers have the property that every integer factors uniquely into a product of irreducible integers, up to rearranging the order of the factors and throwing in some extra factors of 1 and -1. (Note that a positive irreducible integer is simply another name for a prime.) Not every ring has this important unique factorization property, but in the next section we prove that the ring of polynomials with coefficients in a field is a unique factorization ring.

We have seen that congruences are a very important and powerful mathematical tool for working with the integers. Using the definition of divisibility, we can extend the notion of congruence to arbitrary rings.

Definition. Let R be a ring and choose a nonzero element $m \in R$. We say that two elements a and b of R are *congruent modulo* m if their difference $a - b$ is divisible by m. We write

$$a \equiv b \pmod{m}$$

to indicate that a and b are congruent modulo m.

Congruences for arbitrary rings satisfy the same equation-like properties as they do in the original integer setting.

Proposition 2.41. *Let R be a ring and let $m \in R$ with $m \neq 0$. If*

$$a_1 \equiv a_2 \pmod{m} \qquad \textit{and} \qquad b_1 \equiv b_2 \pmod{m},$$

then

$$a_1 \pm b_1 \equiv a_2 \pm b_2 \pmod{m} \qquad \textit{and} \qquad a_1 \star b_1 \equiv a_2 \star b_2 \pmod{m}.$$

Proof. We leave the proof as an exercise; see Exercise 2.32. □

Remark 2.42. Our definition of congruence captures all of the properties that we need in this book. However, we must observe that there exists a more general notion of congruence modulo ideals. For our purposes, it is enough to work with congruences modulo principal ideals, which are ideals that are generated by a single element.

An important consequence of Proposition 2.41 is a method for creating new rings from old rings, just as we created $\mathbb{Z}/q\mathbb{Z}$ from $\mathbb{Z}$ by looking at congruences modulo q.

Definition. Let R be a ring and let $m \in R$ with $m \neq 0$. For any $a \in R$, we write $\overline{a}$ for the set of all $a' \in R$ such that $a' \equiv a \pmod{m}$. The set $\overline{a}$ is called the *congruence class of* a, and we denote the collection of all congruence classes by $R/(m)$ or R/mR. Thus

$$R/(m) = R/mR = \{\overline{a} : a \in R\}.$$

We add and multiply congruence classes using the obvious rules

$$\overline{a} + \overline{b} = \overline{a+b} \qquad \text{and} \qquad \overline{a} \star \overline{b} = \overline{a \star b}. \tag{2.23}$$

We call $R/(m)$ the *quotient ring of* R *by* m. This name is justified by the next proposition.

Proposition 2.43. *The formulas* (2.23) *give well-defined addition and multiplication rules on the set of congruence classes* $R/(m)$*, and they make* $R/(m)$ *into a ring.*

Proof. We leave the proof as an exercise; see Exercise 2.43. □

2.10.3 Polynomial Rings and the Euclidean Algorithm

In Example 2.38(f) we observed that if R is any ring, then we can create a polynomial ring with coefficients taken from R. This ring is denoted by

$$R[x] = \{a_0 + a_1x + a_2x^2 + \cdots + a_nx^n : n \geq 0 \text{ and } a_0, a_1, \ldots, a_n \in R\}.$$

The *degree* of a nonzero polynomial is the exponent of the highest power of x that appears. Thus if

$$\boldsymbol{a}(x) = a_0 + a_1x + a_2x^2 + \cdots + a_nx^n$$

with $a_n \neq 0$, then $\boldsymbol{a}(x)$ has degree n. We denote the degree of $\boldsymbol{a}$ by $\deg(\boldsymbol{a})$, and we call a_n the *leading coefficient* of $\boldsymbol{a}(x)$. A nonzero polynomial whose leading coefficient is equal to 1 is called a *monic polynomial.* For example, $3 + x^2$ is a monic polynomial, but $1 + 3x^2$ is not.

Especially important are those polynomial rings in which the ring R is a field; for example, R could be $\mathbb{Q}$ or $\mathbb{R}$ or $\mathbb{C}$ or a finite field $\mathbb{F}_p$. (For cryptography, by far the most important case is the last named one.) One reason why it is so useful to take R to be a field $\mathbb{F}$ is because virtually all of the properties of $\mathbb{Z}$ that we proved in Sect. 1.2 are also true for the polynomial ring $\mathbb{F}[x]$. This section is devoted to a discussion of the properties of $\mathbb{F}[x]$.

Back in high school you undoubtedly learned how to divide one polynomial by another. We recall the process by doing an example. Here is how one divides $x^5 + 2x^4 + 7$ by $x^3 - 5$:

$$\begin{array}{r l}
 & x^2 + 2x \qquad\qquad \text{R} \;\; 5x^2 + 10x + 7 \\
\cline{2-2}
x^3 - 5 \,) & x^5 + 2x^4 \qquad\qquad\qquad + 7 \\
 & x^5 \qquad\qquad - 5x^2 \\
\cline{2-2}
 & \quad 2x^4 \qquad + 5x^2 \qquad + 7 \\
 & \quad 2x^4 \qquad\quad - 10x \\
\cline{2-2}
 & \qquad\qquad 5x^2 + 10x + 7
\end{array}$$

In other words, $x^5 + 2x^4 + 7$ divided by $x^3 - 5$ gives a quotient of $x^2 + 2x$ with a remainder of $5x^2 + 10x + 7$. Another way to say this is to write[11]

$$x^5 + 2x^4 + 7 = (x^2 + 2x) \cdot (x^3 - 5) + (5x^2 + 10x + 7).$$

Notice that the degree of the remainder $5x^2 + 10x + 7$ is strictly smaller than the degree of the divisor $x^3 - 5$.

We can do the same thing for any polynomial ring $\mathbb{F}[x]$ as long as $\mathbb{F}$ is a field. Rings of this sort that have a "division with remainder" algorithm are called *Euclidean rings.*

Proposition 2.44 (The ring $\mathbb{F}[x]$ is Euclidean). *Let $\mathbb{F}$ be a field and let $\boldsymbol{a}$ and $\boldsymbol{b}$ be polynomials in $\mathbb{F}[x]$ with $\boldsymbol{b} \neq 0$. Then it is possible to write*

$$\boldsymbol{a} = \boldsymbol{b} \cdot \boldsymbol{k} + \boldsymbol{r} \qquad \textit{with } \boldsymbol{k} \textit{ and } \boldsymbol{r} \textit{ polynomials, and either } \boldsymbol{r} = 0 \textit{ or } \deg \boldsymbol{r} < \deg \boldsymbol{b}.$$

We say that $\boldsymbol{a}$ divided by $\boldsymbol{b}$ has quotient $\boldsymbol{k}$ and remainder $\boldsymbol{r}$.

Proof. We start with any values for $\boldsymbol{k}$ and $\boldsymbol{r}$ that satisfy

$$\boldsymbol{a} = \boldsymbol{b} \cdot \boldsymbol{k} + \boldsymbol{r}.$$

(For example, we could start with $\boldsymbol{k} = 0$ and $\boldsymbol{r} = \boldsymbol{a}$.) If $\deg \boldsymbol{r} < \deg \boldsymbol{b}$, then we're done. Otherwise we write

$$\boldsymbol{b} = b_0 + b_1 x + \cdots + b_d x^d \qquad \text{and} \qquad \boldsymbol{r} = r_0 + r_1 x + \cdots + r_e x^e$$

[11] For notational convenience, we drop the $\star$ for multiplication and just write $\boldsymbol{a} \cdot \boldsymbol{b}$, or even simply $\boldsymbol{ab}$.

with $b_d \neq 0$ and $r_e \neq 0$ and $e \geq d$. We rewrite the equation $\boldsymbol{a} = \boldsymbol{b} \cdot \boldsymbol{k} + \boldsymbol{r}$ as

$$\boldsymbol{a} = \boldsymbol{b} \cdot \left(\boldsymbol{k} + \frac{r_e}{b_d} x^{e-d} \right) + \left(\boldsymbol{r} - \frac{r_e}{b_d} x^{e-d} \cdot \boldsymbol{b} \right) = \boldsymbol{b} \cdot \boldsymbol{k}' + \boldsymbol{r}'.$$

Notice that we have canceled the top degree term of $\boldsymbol{r}$, so $\deg \boldsymbol{r}' < \deg \boldsymbol{r}$. If $\deg \boldsymbol{r}' < \deg \boldsymbol{b}$, then we're done. If not, we repeat the process. We can do this as long as the $\boldsymbol{r}$ term satisfies $\deg \boldsymbol{r} \geq \deg \boldsymbol{b}$, and every time we apply this process, the degree of our $\boldsymbol{r}$ term gets smaller. Hence eventually we arrive at an $\boldsymbol{r}$ term whose degree is strictly smaller than the degree of $\boldsymbol{b}$. □

We can now define common divisors and greatest common divisors in $\mathbb{F}[x]$.

Definition. A *common divisor* of two elements $\boldsymbol{a}, \boldsymbol{b} \in \mathbb{F}[x]$ is an element $\boldsymbol{d} \in \mathbb{F}[x]$ that divides both $\boldsymbol{a}$ and $\boldsymbol{b}$. We say that $\boldsymbol{d}$ is a *greatest common divisor of* $\boldsymbol{a}$ *and* $\boldsymbol{b}$ if every common divisor of $\boldsymbol{a}$ and $\boldsymbol{b}$ also divides $\boldsymbol{d}$.

We will see below that every pair of elements in $\mathbb{F}[x]$ has a greatest common divisor,[12] which is unique up to multiplying it by a nonzero element of $\mathbb{F}$. We write $\gcd(\boldsymbol{a}, \boldsymbol{b})$ for the unique monic polynomial that is a greatest common divisor of $\boldsymbol{a}$ and $\boldsymbol{b}$.

Example 2.45. The greatest common divisor of $x^2 - 1$ and $x^3 + 1$ is $x + 1$. Notice that

$$x^2 - 1 = (x+1)(x-1) \qquad \text{and} \qquad x^3 + 1 = (x+1)(x^2 - x + 1),$$

so $x+1$ is a common divisor. We leave it to you to check that it is the greatest common divisor.

It is not clear, a priori, that every pair of elements has a greatest common divisor. And indeed, there are many rings in which greatest common divisors do not exist, for example in the ring $\mathbb{Z}[x]$. But greatest common divisors do exist in the polynomial ring $\mathbb{F}[x]$ when $\mathbb{F}$ is a field.

Proposition 2.46 (The extended Euclidean algorithm for $\mathbb{F}[x]$). *Let* $\mathbb{F}$ *be a field and let* $\boldsymbol{a}$ *and* $\boldsymbol{b}$ *be polynomials in* $\mathbb{F}[x]$ *with* $\boldsymbol{b} \neq 0$. *Then the greatest common divisor* $\boldsymbol{d}$ *of* $\boldsymbol{a}$ *and* $\boldsymbol{b}$ *exists, and there are polynomials* $\boldsymbol{u}$ *and* $\boldsymbol{v}$ *in* $\mathbb{F}[x]$ *such that*

$$\boldsymbol{a} \cdot \boldsymbol{u} + \boldsymbol{b} \cdot \boldsymbol{v} = \boldsymbol{d}.$$

Proof. Just as in the proof of Theorem 1.7, the polynomial $\gcd(\boldsymbol{a}, \boldsymbol{b})$ can be computed by repeated application of Proposition 2.44, as described in Fig. 2.3. Similarly, the polynomials $\boldsymbol{u}$ and $\boldsymbol{v}$ can be computed by substituting one equation into another in Fig. 2.3, exactly as described in the proof of Theorem 1.11. □

[12] According to our definition, even if both $\boldsymbol{a}$ and $\boldsymbol{b}$ are 0, they have a greatest common divisor, namely 0. However, some authors prefer to leave $\gcd(0, 0)$ undefined.

$$\begin{array}{rl} \boldsymbol{a} = \boldsymbol{b} \cdot \boldsymbol{k}_1 + \boldsymbol{r}_2 & \text{with } 0 \le \deg \boldsymbol{r}_2 < \deg \boldsymbol{b}, \\ \boldsymbol{b} = \boldsymbol{r}_2 \cdot \boldsymbol{k}_2 + \boldsymbol{r}_3 & \text{with } 0 \le \deg \boldsymbol{r}_3 < \deg \boldsymbol{r}_2, \\ \boldsymbol{r}_2 = \boldsymbol{r}_3 \cdot \boldsymbol{k}_3 + \boldsymbol{r}_4 & \text{with } 0 \le \deg \boldsymbol{r}_4 < \deg \boldsymbol{r}_3, \\ \boldsymbol{r}_3 = \boldsymbol{r}_4 \cdot \boldsymbol{k}_4 + \boldsymbol{r}_5 & \text{with } 0 \le \deg \boldsymbol{r}_5 < \deg \boldsymbol{r}_4, \\ \vdots \qquad \vdots & \qquad \vdots \\ \boldsymbol{r}_{t-2} = \boldsymbol{r}_{t-1} \cdot \boldsymbol{k}_{t-2} + \boldsymbol{r}_t & \text{with } 0 \le \deg \boldsymbol{r}_t < \deg \boldsymbol{r}_{t-1}, \\ \boldsymbol{r}_{t-1} = \boldsymbol{r}_t \cdot \boldsymbol{k}_t & \\ \multicolumn{2}{c}{\text{Then } \boldsymbol{d} = \boldsymbol{r}_t = \gcd(\boldsymbol{a}, \boldsymbol{b}).} \end{array}$$

Figure 2.3: The Euclidean algorithm for polynomials

Example 2.47. We use the Euclidean algorithm in the ring $\mathbb{F}_{13}[x]$ to compute $\gcd(x^5 - 1, x^3 + 2x - 3)$:

$$\begin{aligned} x^5 - 1 &= (x^3 + 2x - 3) \cdot (x^2 + 11) + (3x^2 + 4x + 6) \\ x^3 + 2x - 3 &= (3x^2 + 4x + 6) \cdot (9x + 1) + (9x + 4) \quad \leftarrow \boxed{\gcd = 9x + 4} \\ 3x^2 + 4x + 6 &= (9x + 4) \cdot (9x + 8) + 0 \end{aligned}$$

Thus $9x + 4$ is a greatest common divisor of $x^5 - 1$ and $x^3 + 2x - 3$ in $\mathbb{F}_{13}[x]$. In order to get a monic polynomial, we multiply by $3 \equiv 9^{-1} \pmod{13}$. This gives

$$\gcd(x^5 - 1, x^3 + 2x - 3) = x - 1 \qquad \text{in } \mathbb{F}_{13}[x].$$

We recall from Sect. 2.10.2 that an element u of a ring is a unit if it has a multiplicative inverse u^{-1}, and that an element a of a ring is irreducible if it is not a unit and if the only way to factor a is as $a = bc$ with either b or c a unit. It is not hard to see that the units in a polynomial ring $\mathbb{F}[x]$ are precisely the nonzero constant polynomials, i.e., the nonzero elements of $\mathbb{F}$; see Exercise 2.34. The question of irreducibility is subtler, as shown by the following examples.

Example 2.48. The polynomial $x^5 - 4x^3 + 3x^2 - x + 2$ is irreducible as a polynomial in $\mathbb{Z}[x]$, but if we view it as an element of $\mathbb{F}_3[x]$, then it factors as

$$x^5 - 4x^3 + 3x^2 - x + 2 \equiv (x + 1)\left(x^4 + 2x^3 + 2\right) \pmod{3}.$$

It also factors if we view it as a polynomial in $\mathbb{F}_5[x]$, but this time as a product of a quadratic polynomial and a cubic polynomial,

$$x^5 - 4x^3 + 3x^2 - x + 2 \equiv \left(x^2 + 4x + 2\right)\left(x^3 + x^2 + 1\right) \pmod{5}.$$

On the other hand, if we work in $\mathbb{F}_{13}[x]$, then $x^5 - 4x^3 + 3x^2 - x + 2$ is irreducible.

Every integer has an essentially unique factorization as a product of primes. The same is true of polynomials with coefficients in a field. And just as for the integers, the key to proving unique factorization is the extended Euclidean algorithm.

Proposition 2.49. *Let $\mathbb{F}$ be a field. Then every nonzero polynomial in $\mathbb{F}[x]$ can be uniquely factored as a product of monic irreducible polynomials, in the following sense. If $\boldsymbol{a} \in \mathbb{F}[x]$ is factored as*

$$\boldsymbol{a} = \alpha \boldsymbol{p}_1 \cdot \boldsymbol{p}_2 \cdots \boldsymbol{p}_m \qquad \text{and} \qquad \boldsymbol{a} = \beta \boldsymbol{q}_1 \cdot \boldsymbol{q}_2 \cdots \boldsymbol{q}_n,$$

where $\alpha, \beta \in \mathbb{F}$ are constants and $\boldsymbol{p}_1, \ldots, \boldsymbol{p}_m, \boldsymbol{q}_1, \ldots, \boldsymbol{q}_n$ are monic irreducible polynomials, then after rearranging the order of $\boldsymbol{q}_1, \ldots, \boldsymbol{q}_n$, we have

$$\alpha = \beta, \qquad m = n, \qquad \text{and} \qquad \boldsymbol{p}_i = \boldsymbol{q}_i \quad \text{for all } 1 \le i \le m.$$

Proof. The existence of a factorization into irreducibles follows easily from the fact that if $\boldsymbol{a} = \boldsymbol{b} \cdot \boldsymbol{c}$, then $\deg \boldsymbol{a} = \deg \boldsymbol{b} + \deg \boldsymbol{c}$. (See Exercise 2.34.) The proof that the factorization is unique is exactly the same as the proof for integers, cf. Theorem 1.20. The key step in the proof is the statement that if $\boldsymbol{p} \in \mathbb{F}[x]$ is irreducible and divides the product $\boldsymbol{a} \cdot \boldsymbol{b}$, then either $\boldsymbol{p} \mid \boldsymbol{a}$ or $\boldsymbol{p} \mid \boldsymbol{b}$ (or both). This statement is the polynomial analogue of Proposition 1.19 and is proved in the same way, using the polynomial version of the extended Euclidean algorithm (Proposition 2.46). □

2.10.4 Quotients of Polynomial Rings and Finite Fields of Prime Power Order

In Sect. 2.10.3 we studied polynomial rings and in Sect. 2.10.2 we studied quotient rings. In this section we combine these two constructions and consider quotients of polynomial rings.

Recall that in working with the integers modulo m, it is often convenient to represent each congruence class modulo m by an integer between 0 and $m-1$. The division-with-remainder algorithm (Proposition 2.44) allows us to do something similar for the quotient of a polynomial ring.

Proposition 2.50. *Let $\mathbb{F}$ be field and let $\boldsymbol{m} \in \mathbb{F}[x]$ be a nonzero polynomial. Then every nonzero congruence class $\overline{\boldsymbol{a}} \in \mathbb{F}[x]/(\boldsymbol{m})$ has a unique representative $\boldsymbol{r}$ satisfying*

$$\deg \boldsymbol{r} < \deg \boldsymbol{m} \qquad \text{and} \qquad \boldsymbol{a} \equiv \boldsymbol{r} \pmod{\boldsymbol{m}}.$$

Proof. We use Proposition 2.44 to find polynomials $\boldsymbol{k}$ and $\boldsymbol{r}$ such that

$$\boldsymbol{a} = \boldsymbol{m} \cdot \boldsymbol{k} + \boldsymbol{r}$$

with either $\boldsymbol{r} = 0$ or $\deg \boldsymbol{r} < \deg \boldsymbol{m}$. If $\boldsymbol{r} = 0$, then $\boldsymbol{a} \equiv 0 \pmod{\boldsymbol{m}}$, so $\overline{\boldsymbol{a}} = 0$. Otherwise, reducing modulo $\boldsymbol{m}$ gives $\boldsymbol{a} \equiv \boldsymbol{r} \pmod{\boldsymbol{m}}$ with $\deg \boldsymbol{r} < \deg \boldsymbol{m}$. This shows that $\boldsymbol{r}$ exists. To show that it is unique, suppose that $\boldsymbol{r}'$ has the same properties. Then

$$\boldsymbol{r} - \boldsymbol{r}' \equiv \boldsymbol{a} - \boldsymbol{a} \equiv 0 \pmod{\boldsymbol{m}},$$

so $\boldsymbol{m}$ divides $\boldsymbol{r} - \boldsymbol{r}'$. But $\boldsymbol{r} - \boldsymbol{r}'$ has degree strictly smaller than the degree of $\boldsymbol{m}$, so we must have $\boldsymbol{r} - \boldsymbol{r}' = 0$. □

Example 2.51. Consider the ring $\mathbb{F}[x]/(x^2+1)$. Proposition 2.50 says that every element of this quotient ring is uniquely represented by a polynomial of the form

$$\overline{\alpha + \beta x} \quad \text{with } \alpha, \beta \in \mathbb{F}.$$

Addition is performed in the obvious way,

$$\overline{\alpha_1 + \beta_1 x} + \overline{\alpha_2 + \beta_2 x} = \overline{(\alpha_1 + \alpha_2) + (\beta_1 + \beta_2)x}.$$

Multiplication is similar, except that we have to divide the final result by x^2+1 and take the remainder. Thus

$$\begin{aligned} \overline{\alpha_1 + \beta_1 x} \cdot \overline{\alpha_2 + \beta_2 x} &= \overline{\alpha_1\alpha_2 + (\alpha_1\beta_2 + \alpha_2\beta_1)x + \beta_1\beta_2 x^2} \\ &= \overline{(\alpha_1\alpha_2 - \beta_1\beta_2) + (\alpha_1\beta_2 + \alpha_2\beta_1)x}. \end{aligned}$$

Notice that the effect of dividing by x^2+1 is the same as replacing x^2 with -1. The intuition is that in the quotient ring $\mathbb{F}[x]/(x^2+1)$, we have made the quantity x^2+1 equal to 0. Notice that if we take $\mathbb{F} = \mathbb{R}$ in this example, then $\mathbb{R}[x]/(x^2+1)$ is simply the field of complex numbers $\mathbb{C}$.

We can use Proposition 2.50 to count the number of elements in a polynomial quotient ring when $\mathbb{F}$ is a finite field.

Corollary 2.52. *Let $\mathbb{F}_p$ be a finite field and let $\boldsymbol{m} \in \mathbb{F}_p[x]$ be a nonzero polynomial of degree $d \geq 1$. Then the quotient ring $\mathbb{F}_p[x]/(\boldsymbol{m})$ contains exactly p^d elements.*

Proof. From Proposition 2.50 we know that every element of $\mathbb{F}_p[x]/(\boldsymbol{m})$ is represented by a unique polynomial of the form

$$a_0 + a_1x + a_2x^2 + \cdots + a_{d-1}x^{d-1} \quad \text{with } a_0, a_1, \ldots, a_{d-1} \in \mathbb{F}_p.$$

There are p choices for a_0, and p choices for a_1, and so on, leading to a total of p^d choices for $a_0, a_1, \ldots, a_d$. □

We next give an important characterization of the units in a polynomial quotient ring. This will allow us to construct new finite fields.

Proposition 2.53. *Let $\mathbb{F}$ be a field and let $\boldsymbol{a}, \boldsymbol{m} \in \mathbb{F}[x]$ be polynomials with $\boldsymbol{m} \neq 0$. Then $\overline{\boldsymbol{a}}$ is a unit in the quotient ring $\mathbb{F}[x]/(\boldsymbol{m})$ if and only if*

$$\gcd(\boldsymbol{a}, \boldsymbol{m}) = 1.$$

Proof. Suppose first that $\overline{\boldsymbol{a}}$ is a unit in $\mathbb{F}[x]/(\boldsymbol{m})$. By definition, this means that we can find some $\overline{\boldsymbol{b}} \in \mathbb{F}[x](\boldsymbol{m})$ satisfying $\overline{\boldsymbol{a}} \cdot \overline{\boldsymbol{b}} = \overline{1}$. In terms of congruences, this means that $\boldsymbol{a} \cdot \boldsymbol{b} \equiv 1 \pmod{\boldsymbol{m}}$, so there is some $\boldsymbol{c} \in \mathbb{F}[x]$ such that

$$\boldsymbol{a} \cdot \boldsymbol{b} - 1 = \boldsymbol{c} \cdot \boldsymbol{m}.$$

It follows that any common divisor of $\boldsymbol{a}$ and $\boldsymbol{m}$ must also divide 1. Therefore $\gcd(\boldsymbol{a}, \boldsymbol{m}) = 1$.

Next suppose that $\gcd(\boldsymbol{a}, \boldsymbol{m}) = 1$. Then Proposition 2.46 tells us that there are polynomials $\boldsymbol{u}, \boldsymbol{v} \in \mathbb{F}[x]$ such that

$$\boldsymbol{a} \cdot \boldsymbol{u} + \boldsymbol{m} \cdot \boldsymbol{v} = 1.$$

Reducing modulo $\boldsymbol{m}$ yields

$$\boldsymbol{a} \cdot \boldsymbol{u} \equiv 1 \pmod{\boldsymbol{m}},$$

so $\overline{\boldsymbol{u}}$ is an inverse for $\overline{\boldsymbol{a}}$ in $\mathbb{F}[x]/(\boldsymbol{m})$. □

An important instance of Proposition 2.53 is the case that the modulus is an irreducible polynomial.

Corollary 2.54. *Let $\mathbb{F}$ be a field and let $\boldsymbol{m} \in \mathbb{F}[x]$ be an irreducible polynomial. Then the quotient ring $\mathbb{F}[x]/(\boldsymbol{m})$ is a field, i.e., every nonzero element of $\mathbb{F}[x]/(\boldsymbol{m})$ has a multiplicative inverse.*

Proof. Replacing $\boldsymbol{m}$ by a constant multiple, we may assume that $\boldsymbol{m}$ is a monic polynomial. Let $\overline{\boldsymbol{a}} \in \mathbb{F}[x]/(\boldsymbol{m})$. There are two cases to consider. First, suppose that $\gcd(\boldsymbol{a}, \boldsymbol{m}) = 1$. Then Proposition 2.53 tells us that $\overline{\boldsymbol{a}}$ is a unit, so we are done. Second, suppose that $\boldsymbol{d} = \gcd(\boldsymbol{a}, \boldsymbol{m}) \neq 1$. Then in particular, we know that $\boldsymbol{d} \mid \boldsymbol{m}$. But $\boldsymbol{m}$ is monic and irreducible, and $\boldsymbol{d} \neq 1$, so we must have $\boldsymbol{d} = \boldsymbol{m}$. We also know that $\boldsymbol{d} \mid \boldsymbol{a}$, so $\boldsymbol{m} \mid \boldsymbol{a}$. Hence $\overline{\boldsymbol{a}} = 0$ in $\mathbb{F}[x]/(\boldsymbol{m})$. This completes the proof that every nonzero element of $\mathbb{F}[x]/(\boldsymbol{m})$ has a multiplicative inverse. □

Example 2.55. The polynomial $x^2 + 1$ is irreducible in $\mathbb{R}[x]$. The quotient ring $\mathbb{R}[x]/(x^2+1)$ is a field. Indeed, it is the field of complex numbers $\mathbb{C}$, where the "variable" $\overline{x}$ plays the role of $i = \sqrt{-1}$, since in the ring $\mathbb{R}[x]/(x^2+1)$ we have $\overline{x}^2 = -1$.

By way of contrast, the polynomial $x^2 - 1$ is clearly not irreducible in $\mathbb{R}[x]$. The quotient ring $\mathbb{R}[x]/(x^2 - 1)$ is not a field. In fact,

$$(\overline{x-1}) \cdot (\overline{x+1}) = 0 \qquad \text{in } \mathbb{R}[x]/(x^2 - 1).$$

Thus ring $\mathbb{R}[x]/(x^2-1)$ has nonzero elements whose product is 0, which means that they certainly cannot be units. (Nonzero elements of a ring whose product is 0 are called *zero divisors*.)

If we apply Corollary 2.54 to a polynomial ring with coefficients in a finite field $\mathbb{F}_p$, we can create new finite fields with a prime power number of elements.

Corollary 2.56. *Let $\mathbb{F}_p$ be a finite field and let $\boldsymbol{m} \in \mathbb{F}_p[x]$ be an irreducible polynomial of degree $d \geq 1$. Then $\mathbb{F}_p[x]/(\boldsymbol{m})$ is a field with p^d elements.*

Proof. We combine Corollary 2.54, which says that $\mathbb{F}_p[x]/(\boldsymbol{m})$ is a field, with Corollary 2.52, which says that $\mathbb{F}_p[x]/(\boldsymbol{m})$ has p^d elements. □

Example 2.57. It is not hard to check that the polynomial $x^3 + x + 1$ is irreducible in $\mathbb{F}_2[x]$ (see Exercise 2.37), so $\mathbb{F}_2[x]/(x^3 + x + 1)$ is a field with eight elements. Proposition 2.50 tells us that the following are representatives for the eight elements in this field:

$$0, 1, x, x^2, 1 + x, 1 + x^2, x + x^2, 1 + x + x^2.$$

Addition is easy as long as you remember to treat the coefficients modulo 2, so for example,

$$(1 + x) + (x + x^2) = 1 + x^2.$$

Multiplication is also easy, just multiply the polynomials, divide by x^3+x+1, and take the remainder. For example,

$$(1 + x) \cdot (x + x^2) = x + 2x^2 + x^3 = 1,$$

so $1 + x$ and $x + x^2$ are multiplicative inverses. The complete multiplication table for $\mathbb{F}_2[x]/(x^3 + x + 1)$ is described in Exercise 2.38.

Example 2.58. When is the polynomial $x^2 + 1$ irreducible in the ring $\mathbb{F}_p[x]$? If it is reducible, then it factors as

$$x^2 + 1 = (x + \alpha)(x + \beta) \qquad \text{for some } \alpha, \beta \in \mathbb{F}_p.$$

Comparing coefficients, we find that $\alpha + \beta = 0$ and $\alpha\beta = 1$; hence

$$\alpha^2 = \alpha \cdot (-\beta) = -\alpha\beta = -1.$$

In other words, the field $\mathbb{F}_p$ has an element whose square is -1. Conversely, if $\alpha \in \mathbb{F}_p$ satisfies $\alpha^2 = -1$, then $x^2 + 1 = (x - \alpha)(x + \alpha)$ factors in $\mathbb{F}_p[x]$. This proves that

$$x^2 + 1 \text{ is irreducible in } \mathbb{F}_p[x] \quad \text{if and only if} \quad -1 \text{ is not a square in } \mathbb{F}_p.$$

Quadratic reciprocity, which we study later in Sect. 3.9, then tells us that

$$x^2 + 1 \text{ is irreducible in } \mathbb{F}_p[x] \quad \text{if and only if} \quad p \equiv 3 \pmod{4}.$$

Let p be a prime satisfying $p \equiv 3 \pmod{4}$. Then the quotient field $\mathbb{F}_p[x]/(x^2 + 1)$ is a field containing p^2 elements. It contains an element $\overline{x}$ that is a square root of -1. So we can view $\mathbb{F}_p[x]/(x^2 + 1)$ as a sort of analogue of the complex numbers and can write its elements in the form

$$a + bi \quad \text{with } a, b \in \mathbb{F}_p,$$

where i is simply a symbol with the property that $i^2 = -1$. Addition, subtraction, multiplication, and division are performed just as in the complex numbers, with the understanding that instead of real numbers as coefficients, we are using integers modulo p. So for example, division is done by the usual "rationalizing the denominator" trick,

$$\frac{a+bi}{c+di} = \frac{a+bi}{c+di} \cdot \frac{c-di}{c-di} = \frac{(ac+bd)+(bc-ad)i}{c^2+d^2}.$$

Note that there is never a problem of 0 in the denominator, since the assumption that $p \equiv 3 \pmod 4$ ensures that $c^2 + d^2 \neq 0$ (as long as at least one of c and d is nonzero). These fields of order p^2 will be used in Sect. 6.9.3.

In order to construct a field with p^d elements, we need to find an irreducible polynomial of degree d in $\mathbb{F}_p[x]$. It is proven in more advanced texts that there is always such a polynomial, and indeed generally many such polynomials. Further, in a certain abstract sense it doesn't matter which irreducible polynomial we choose: we always get the same field. However, in a practical sense it does make a difference, because practical computations in $\mathbb{F}_p[x]/(\boldsymbol{m})$ are more efficient if $\boldsymbol{m}$ does not have very many nonzero coefficients.

We summarize some of the principal properties of finite fields in the following theorem.

Theorem 2.59. *Let $\mathbb{F}_p$ be a finite field.*

(a) *For every $d \geq 1$ there exists an irreducible polynomial $\boldsymbol{m} \in \mathbb{F}_p[x]$ of degree d.*
(b) *For every $d \geq 1$ there exists a finite field with p^d elements.*
(c) *If $\mathbb{F}$ and $\mathbb{F}'$ are finite fields with the same number of elements, then there is a way to match the elements of $\mathbb{F}$ with the elements of $\mathbb{F}'$ so that the addition and multiplication tables of $\mathbb{F}$ and $\mathbb{F}'$ are the same. (The mathematical terminology is that $\mathbb{F}$ and $\mathbb{F}'$ are* isomorphic.)

Proof. We know from Proposition 2.56 that (a) implies (b). For proofs of (a) and (c), see any basic algebra or number theory text, for example [40, §§13.5, 14.3], [53, Section 7.1], or [59, Chapter 7]. □

Definition. We write $\mathbb{F}_{p^d}$ for a field with p^d elements. Theorem 2.59 assures us that there is at least one such field and that any two fields with p^d elements are essentially the same, up to relabeling their elements. These fields are also sometimes called *Galois fields* and denoted by $\mathrm{GF}(p^d)$ in honor of the nineteenth-century French mathematician Évariste Galois, who studied them.

Remark 2.60. It is not difficult to prove that if $\mathbb{F}$ is a finite field, then $\mathbb{F}$ has p^d elements for some prime p and some $d \geq 1$. (The proof uses linear algebra; see Exercise 2.41.) So Theorem 2.59 describes all finite fields.

Remark 2.61. For cryptographic purposes, it is frequently advantageous to work in a field $\mathbb{F}_{2^d}$, rather than in a field $\mathbb{F}_p$ with p large. This is due to the fact that the binary nature of computers often enables them to work more efficiently with $\mathbb{F}_{2^d}$. A second reason is that sometimes it is useful to have a finite field that contains smaller fields. In the case of $\mathbb{F}_{p^d}$, one can show that every field $\mathbb{F}_{p^e}$ with $e \mid d$ is a subfield of $\mathbb{F}_{p^d}$. Of course, if one is going to use $\mathbb{F}_{2^d}$ for Diffie–Hellman key exchange or Elgamal encryption, it is necessary to choose 2^d to be of approximately the same size as one typically chooses p.

Let $\mathbb{F}$ be a finite field having q elements. Every nonzero element of $\mathbb{F}$ has an inverse, so the group of units $\mathbb{F}^*$ is a group of order $q-1$. Lagrange's theorem (Theorem 2.13) tells us that every element of $\mathbb{F}^*$ has order dividing $q-1$, so

$$a^{q-1} = 1 \quad \text{for all } a \in \mathbb{F}.$$

This is a generalization of Fermat's little theorem (Theorem 1.24) to arbitrary finite fields. The primitive root theorem (Theorem 1.30) is also true for all finite fields.

Theorem 2.62. *Let $\mathbb{F}$ be a finite field having q elements. Then $\mathbb{F}$ has a primitive root, i.e., there is an element $g \in \mathbb{F}$ such that*

$$\mathbb{F}^* = \{1, g, g^2, g^3, \ldots, g^{q-2}\}.$$

Proof. You can find a proof of this theorem in any basic number theory textbook; see for example [59, §4.1] or [137, Chapter 28]. □

Exercises

Section 2.1. Diffie–Hellman and RSA

2.1. Write a one page essay giving arguments, both pro and con, for the following assertion:

> If the government is able to convince a court that there is a valid reason for their request, then they should have access to an individual's private keys (even without the individual's knowledge), in the same way that the government is allowed to conduct court authorized secret wiretaps in cases of suspected criminal activity or threats to national security.

Based on your arguments, would you support or oppose the government being given this power? How about without court oversight? The idea that all private keys should be stored at a secure central location and be accessible to government agencies (with or without suitably stringent legal conditions) is called *key escrow*.

2.2. Research and write a one to two page essay on the classification of cryptographic algorithms as munitions under ITAR (International Traffic in Arms Regulations). How does that act define "export"? What are the potential fines and jail terms for those convicted of violating the Arms Export Control Act? Would teaching non-classified cryptographic algorithms to a college class that includes non-US citizens be considered a form of export? How has US government policy changed from the early 1990s to the present?

Section 2.2. The Discrete Logarithm Problem

2.3. Let g be a primitive root for $\mathbb{F}_p$.

(a) Suppose that $x = a$ and $x = b$ are both integer solutions to the congruence $g^x \equiv h \pmod{p}$. Prove that $a \equiv b \pmod{p-1}$. Explain why this implies that the map (2.1) on page 65 is well-defined.

(b) Prove that $\log_g(h_1 h_2) = \log_g(h_1) + \log_g(h_2)$ for all $h_1, h_2 \in \mathbb{F}_p^*$.
(c) Prove that $\log_g(h^n) = n \log_g(h)$ for all $h \in \mathbb{F}_p^*$ and $n \in \mathbb{Z}$.

2.4. Compute the following discrete logarithms.
(a) $\log_2(13)$ for the prime 23, i.e., $p = 23$, $g = 2$, and you must solve the congruence $2^x \equiv 13 \pmod{23}$.
(b) $\log_{10}(22)$ for the prime $p = 47$.
(c) $\log_{627}(608)$ for the prime $p = 941$. (*Hint.* Look in the second column of Table 2.1 on page 66.)

2.5. Let p be an odd prime and let g be a primitive root modulo p. Prove that a has a square root modulo p if and only if its discrete logarithm $\log_g(a)$ modulo $p-1$ is even.

Section 2.3. Diffie–Hellman Key Exchange

2.6. Alice and Bob agree to use the prime $p = 1373$ and the base $g = 2$ for a Diffie–Hellman key exchange. Alice sends Bob the value $A = 974$. Bob asks your assistance, so you tell him to use the secret exponent $b = 871$. What value B should Bob send to Alice, and what is their secret shared value? Can you figure out Alice's secret exponent?

2.7. Let p be a prime and let g be an integer. The *Decision Diffie–Hellman Problem* is as follows. Suppose that you are given three numbers A, B, and C, and suppose that A and B are equal to

$$A \equiv g^a \pmod{p} \qquad \text{and} \qquad B \equiv g^b \pmod{p},$$

but that you do not necessarily know the values of the exponents a and b. Determine whether C is equal to $g^{ab} \pmod{p}$. Notice that this is different from the Diffie–Hellman problem described on page 69. The Diffie–Hellman problem asks you to actually compute the value of g^{ab}.
(a) Prove that an algorithm that solves the Diffie–Hellman problem can be used to solve the decision Diffie–Hellman problem.
(b) Do you think that the decision Diffie–Hellman problem is hard or easy? Why? See Exercise 6.40 for a related example in which the decision problem is easy, but it is believed that the associated computational problem is hard.

Section 2.4. The Elgamal Public Key Cryptosystem

2.8. Alice and Bob agree to use the prime $p = 1373$ and the base $g = 2$ for communications using the Elgamal public key cryptosystem.
(a) Alice chooses $a = 947$ as her private key. What is the value of her public key A?
(b) Bob chooses $b = 716$ as his private key, so his public key is

$$B \equiv 2^{716} \equiv 469 \pmod{1373}.$$

Alice encrypts the message $m = 583$ using the random element $k = 877$. What is the ciphertext (c_1, c_2) that Alice sends to Bob?
(c) Alice decides to choose a new private key $a = 299$ with associated public key $A \equiv 2^{299} \equiv 34 \pmod{1373}$. Bob encrypts a message using Alice's public key and sends her the ciphertext $(c_1, c_2) = (661, 1325)$. Decrypt the message.

(d) Now Bob chooses a new private key and publishes the associated public key $B = 893$. Alice encrypts a message using this public key and sends the ciphertext $(c_1, c_2) = (693, 793)$ to Bob. Eve intercepts the transmission. Help Eve by solving the discrete logarithm problem $2^b \equiv 893 \pmod{1373}$ and using the value of b to decrypt the message.

2.9. Suppose that Eve is able to solve the Diffie–Hellman problem described on page 69. More precisely, assume that if Eve is given two powers g^u and g^v mod p, then she is able to compute g^{uv} mod p. Show that Eve can break the Elgamal PKC.

2.10. The exercise describes a public key cryptosystem that requires Bob and Alice to exchange several messages. We illustrate the system with an example.

Bob and Alice fix a publicly known prime $p = 32611$, and all of the other numbers used are private. Alice takes her message $m = 11111$, chooses a random exponent $a = 3589$, and sends the number $u = m^a \pmod{p} = 15950$ to Bob. Bob chooses a random exponent $b = 4037$ and sends $v = u^b \pmod{p} = 15422$ back to Alice. Alice then computes $w = v^{15619} \equiv 27257 \pmod{32611}$ and sends $w = 27257$ to Bob. Finally, Bob computes $w^{31883} \pmod{32611}$ and recovers the value 11111 of Alice's message.

(a) Explain why this algorithm works. In particular, Alice uses the numbers $a = 3589$ and 15619 as exponents. How are they related? Similarly, how are Bob's exponents $b = 4037$ and 31883 related?
(b) Formulate a general version of this cryptosystem, i.e., using variables, and show that it works in general.
(c) What is the disadvantage of this cryptosystem over Elgamal? (*Hint.* How many times must Alice and Bob exchange data?)
(d) Are there any advantages of this cryptosystem over Elgamal? In particular, can Eve break it if she can solve the discrete logarithm problem? Can Eve break it if she can solve the Diffie–Hellman problem?

Section 2.5. An Overview of the Theory of Groups

2.11. The group $\mathcal{S}_3$ consists of the following six distinct elements

$$e,\ \sigma,\ \sigma^2,\ \tau,\ \sigma\tau,\ \sigma^2\tau,$$

where e is the identity element and multiplication is performed using the rules

$$\sigma^3 = e, \qquad \tau^2 = e, \qquad \tau\sigma = \sigma^2\tau.$$

Compute the following values in the group $\mathcal{S}_3$:
(a) $\tau\sigma^2$ (b) $\tau(\sigma\tau)$ (c) $(\sigma\tau)(\sigma\tau)$ (d) $(\sigma\tau)(\sigma^2\tau)$.
Is $\mathcal{S}_3$ a commutative group?

2.12. Let G be a group, let $d \geq 1$ be an integer, and define a subset of G by

$$G[d] = \{g \in G : g^d = e\}.$$

(a) Prove that if g is in $G[d]$, then g^{-1} is in $G[d]$.
(b) Suppose that G is commutative. Prove that if g_1 and g_2 are in $G[d]$, then their product $g_1 \star g_2$ is in $G[d]$.
(c) Deduce that if G is commutative, then $G[d]$ is a group.

(d) Show by an example that if G is not a commutative group, then $G[d]$ need not be a group. (*Hint.* Use Exercise 2.11.)

2.13. Let G and H be groups. A function $\phi : G \to H$ is called a *(group) homomorphism* if it satisfies

$$\phi(g_1 \star g_2) = \phi(g_1) \star \phi(g_2) \qquad \text{for all } g_1, g_2 \in G.$$

(Note that the product $g_1 \star g_2$ uses the group law in the group G, while the product $\phi(g_1) \star \phi(g_2)$ uses the group law in the group H.)

(a) Let e_G be the identity element of G, let e_H be the identity element of H, and let $g \in G$. Prove that

$$\phi(e_G) = e_H \qquad \text{and} \qquad \phi(g^{-1}) = \phi(g)^{-1}.$$

(b) Let G be a commutative group. Prove that the map $\phi : G \to G$ defined by $\phi(g) = g^2$ is a homomorphism. Give an example of a noncommutative group for which this map is not a homomorphism.

(c) Same question as (b) for the map $\phi(g) = g^{-1}$.

2.14. Prove that each of the following maps is a group homomorphism.

(a) The map $\phi : \mathbb{Z} \to \mathbb{Z}/N\mathbb{Z}$ that sends $a \in \mathbb{Z}$ to a mod N in $\mathbb{Z}/N\mathbb{Z}$.

(b) The map $\phi : \mathbb{R}^* \to \mathrm{GL}_2(\mathbb{R})$ defined by $\phi(a) = \left(\begin{smallmatrix} a & 0 \\ 0 & a^{-1} \end{smallmatrix}\right)$.

(c) The discrete logarithm map $\log_g : \mathbb{F}_p^* \to \mathbb{Z}/(p-1)\mathbb{Z}$, where g is a primitive root modulo p.

2.15. (a) Prove that $\mathrm{GL}_2(\mathbb{F}_p)$ is a group.

(b) Show that $\mathrm{GL}_2(\mathbb{F}_p)$ is a noncommutative group for every prime p.

(c) Describe $\mathrm{GL}_2(\mathbb{F}_2)$ completely. That is, list its elements and describe the multiplication table.

(d) How many elements are there in the group $\mathrm{GL}_2(\mathbb{F}_p)$?

(e) How many elements are there in the group $\mathrm{GL}_n(\mathbb{F}_p)$?

Section 2.6. How Hard Is the Discrete Logarithm Problem?

2.16. Verify the following assertions from Example 2.16.

(a) $x^2 + \sqrt{x} = \mathcal{O}(x^2)$.

(b) $5 + 6x^2 - 37x^5 = \mathcal{O}(x^5)$.

(c) $k^{300} = \mathcal{O}(2^k)$.

(d) $(\ln k)^{375} = \mathcal{O}(k^{0.001})$.

(e) $k^2 2^k = \mathcal{O}(e^{2k})$.

(f) $N^{10} 2^N = \mathcal{O}(e^N)$.

Section 2.7. A Collision Algorithm for the DLP

2.17. Use Shanks's babystep–giantstep method to solve the following discrete logarithm problems. (For (b) and (c), you may want to write a computer program implementing Shanks's algorithm.)

(a) $11^x = 21$ in $\mathbb{F}_{71}$.

(b) $156^x = 116$ in $\mathbb{F}_{593}$.

(c) $650^x = 2213$ in $\mathbb{F}_{3571}$.

Section 2.8. The Chinese Remainder Theorem

2.18. Solve each of the following simultaneous systems of congruences (or explain why no solution exists).
(a) $x \equiv 3 \pmod{7}$ and $x \equiv 4 \pmod{9}$.
(b) $x \equiv 137 \pmod{423}$ and $x \equiv 87 \pmod{191}$.
(c) $x \equiv 133 \pmod{451}$ and $x \equiv 237 \pmod{697}$.
(d) $x \equiv 5 \pmod{9}$, $x \equiv 6 \pmod{10}$, and $x \equiv 7 \pmod{11}$.
(e) $x \equiv 37 \pmod{43}$, $x \equiv 22 \pmod{49}$, and $x \equiv 18 \pmod{71}$.

2.19. Solve the 1700-year-old Chinese remainder problem from the *Sun Tzu Suan Ching* stated on page 84.

2.20. Let a, b, m, n be integers with $\gcd(m, n) = 1$. Let

$$c \equiv (b - a) \cdot m^{-1} \pmod{n}.$$

Prove that $x = a + cm$ is a solution to

$$x \equiv a \pmod{m} \quad \text{and} \quad x \equiv b \pmod{n}, \tag{2.24}$$

and that every solution to (2.24) has the form $x = a + cm + ymn$ for some $y \in \mathbb{Z}$.

2.21. (a) Let a, b, c be positive integers and suppose that

$$a \mid c, \quad b \mid c, \quad \text{and} \quad \gcd(a, b) = 1.$$

Prove that $ab \mid c$.
(b) Let $x = c$ and $x = c'$ be two solutions to the system of simultaneous congruences (2.7) in the Chinese remainder theorem (Theorem 2.24). Prove that

$$c \equiv c' \pmod{m_1 m_2 \cdots m_k}.$$

2.22. For those who have studied ring theory, this exercise sketches a short proof of the Chinese remainder theorem. Let $m_1, \ldots, m_k$ be integers and let $m = m_1 m_2 \cdots m_k$ be their product.
(a) Prove that the map

$$\begin{array}{ccc} \dfrac{\mathbb{Z}}{m\mathbb{Z}} & \longrightarrow & \dfrac{\mathbb{Z}}{m_1\mathbb{Z}} \times \dfrac{\mathbb{Z}}{m_2\mathbb{Z}} \times \cdots \times \dfrac{\mathbb{Z}}{m_k\mathbb{Z}} \\ a \bmod m & \longmapsto & (a \bmod m_1, a \bmod m_2, \ldots, a \bmod m_k) \end{array} \tag{2.25}$$

is a well-defined homomorphism of rings. (*Hint.* First define a homomorphism from $\mathbb{Z}$ to the right-hand side of (2.25), and then show that $m\mathbb{Z}$ is in the kernel.)
(b) Assume that $m_1, \ldots, m_k$ are pairwise relatively prime. Prove that the map given by (2.25) is one-to-one. (*Hint.* What is the kernel?)
(c) Continuing with the assumption that the numbers $m_1, \ldots, m_k$ are pairwise relatively prime, prove that the map (2.25) is onto. (*Hint.* Use (b) and count the size of both sides.)
(d) Explain why the Chinese remainder theorem (Theorem 2.24) is equivalent to the assertion that (b) and (c) are true.

2.23. Use the method described in Sect. 2.8.1 to find square roots modulo the following composite moduli.

(a) Find a square root of 340 modulo 437. (Note that $437 = 19 \cdot 23$.)
(b) Find a square root of 253 modulo 3143.
(c) Find four square roots of 2833 modulo 4189. (The modulus factors as $4189 = 59 \cdot 71$. Note that your four square roots should be distinct modulo 4189.)
(d) Find eight square roots of 813 modulo 868.

2.24. Let p be an odd prime, let a be an integer that is not divisible by p, and let b be a square root of a modulo p. This exercise investigates the square root of a modulo powers of p.
(a) Prove that for some choice of k, the number $b + kp$ is a square root of a modulo p^2, i.e., $(b + kp)^2 \equiv a \pmod{p^2}$.
(b) The number $b = 537$ is a square root of $a = 476$ modulo the prime $p = 1291$. Use the idea in (a) to compute a square root of 476 modulo p^2.
(c) Suppose that b is a square root of a modulo p^n. Prove that for some choice of j, the number $b + jp^n$ is a square root of a modulo p^{n+1}.
(d) Explain why (c) implies the following statement: If p is an odd prime and if a has a square root modulo p, then a has a square root modulo p^n for every power of p. Is this true if $p = 2$?
(e) Use the method in (c) to compute the square root of 3 modulo 13^3, given that $9^2 \equiv 3 \pmod{13}$.

2.25. Suppose $n = pq$ with p and q distinct odd primes.
(a) Suppose that $\gcd(a, pq) = 1$. Prove that if the equation $x^2 \equiv a \pmod{n}$ has any solutions, then it has four solutions.
(b) Suppose that you had a machine that could find all four solutions for some given a. How could you use this machine to factor n?

Section 2.9. The Pohlig–Hellman Algorithm

2.26. Let $\mathbb{F}_p$ be a finite field and let $N \mid p - 1$. Prove that $\mathbb{F}_p^*$ has an element of order N. This is true in particular for any prime power that divides $p - 1$. (*Hint.* Use the fact that $\mathbb{F}_p^*$ has a primitive root.)

2.27. Write out your own proof that the Pohlig–Hellman algorithm works in the particular case that $p - 1 = q_1 \cdot q_2$ is a product of two distinct primes. This provides a good opportunity for you to understand how the proof works and to get a feel for how it was discovered.

2.28. Use the Pohlig–Hellman algorithm (Theorem 2.31) to solve the discrete logarithm problem

$$g^x = a \quad \text{in } \mathbb{F}_p$$

in each of the following cases.
(a) $p = 433$, $g = 7$, $a = 166$.
(b) $p = 746497$, $g = 10$, $a = 243278$.
(c) $p = 41022299$, $g = 2$, $a = 39183497$. (*Hint.* $p = 2 \cdot 29^5 + 1$.)
(d) $p = 1291799$, $g = 17$, $a = 192988$. (*Hint.* $p - 1$ has a factor of 709.)

Section 2.10. Rings, Quotient Rings, Polynomial Rings, and Finite Fields

2.29. Let R be a ring with the property that the only way that a product $a \cdot b$ can be 0 is if $a = 0$ or $b = 0$. (In the terminology of Example 2.55, the ring R has no zero divisors.) Suppose further that R has only finitely many elements. Prove that R is a field. (*Hint.* Let $a \in R$ with $a \neq 0$. What can you say about the map $R \to R$ defined by $b \mapsto a \cdot b$?)

2.30. Let R be a ring. Prove the following properties of R directly from the ring axioms described in Sect. 2.10.1.

(a) Prove that the additive identity element $0 \in R$ is unique, i.e., prove that there is only one element in R satisfying $0 + a = a + 0 = 0$ for every $a \in R$.
(b) Prove that the multiplicative identity element $1 \in R$ is unique.
(c) Prove that every element of R has a unique additive inverse.
(d) Prove that $0 \star a = a \star 0 = 0$ for all $a \in R$.
(e) We denote the additive inverse of a by $-a$. Prove that $-(-a) = a$.
(f) Let -1 be the additive inverse of the multiplicative identity element $1 \in R$. Prove that $(-1) \star (-1) = 1$.
(g) Prove that $b \mid 0$ for every nonzero $b \in R$.
(h) Prove that an element of R has at most one multiplicative inverse.

2.31. Let R and S be rings. A function $\phi : R \to S$ is called a (*ring*) *homomorphism* if it satisfies

$$\phi(a + b) = \phi(a) + \phi(b) \quad \text{and} \quad \phi(a \star a) = \phi(a) \star \phi(a) \qquad \text{for all } a, b, \in R.$$

(a) Let 0_R, 0_S, 1_R and 1_S denote the additive and multiplicative identities of R and S, respectively. Prove that

$$\phi(0_R) = 0_S, \quad \phi(1_R) = 1_S, \quad \phi(-a) = -\phi(a), \quad \phi(a^{-1}) = \phi(a)^{-1},$$

where the last equality holds for those $a \in R$ that have a multiplicative inverse.

(b) Let p be a prime, and let R be a ring with the property that $pa = 0$ for every $a \in R$. (Here pa means to add a to itself p times.) Prove that the map

$$\phi : R \longrightarrow R, \qquad \phi(a) = a^p$$

is a ring homomorphism. It is called the *Frobenius homomorphism*.

2.32. Prove Proposition 2.41.

2.33. Prove Proposition 2.43. (*Hint.* First use Exercise 2.32 to prove that the congruence classes $\overline{a + b}$ and $\overline{a \star b}$ depend only on the congruence classes of a and b.)

2.34. Let $\mathbb{F}$ be a field and let $\boldsymbol{a}$ and $\boldsymbol{b}$ be nonzero polynomials in $\mathbb{F}[x]$.

(a) Prove that $\deg(\boldsymbol{a} \cdot \boldsymbol{b}) = \deg(\boldsymbol{a}) + \deg(\boldsymbol{b})$.
(b) Prove that $\boldsymbol{a}$ has a multiplicative inverse in $\mathbb{F}[x]$ if and only if $\boldsymbol{a}$ is in $\mathbb{F}$, i.e., if and only if $\boldsymbol{a}$ is a constant polynomial.
(c) Prove that every nonzero element of $\mathbb{F}[x]$ can be factored into a product of irreducible polynomials. (*Hint.* Use (a), (b), and induction on the degree of the polynomial.)
(d) Let R be the ring $\mathbb{Z}/6\mathbb{Z}$. Give an example to show that (a) is false for some polynomials $\boldsymbol{a}$ and $\boldsymbol{b}$ in $R[x]$.

2.35. Let $\boldsymbol{a}$ and $\boldsymbol{b}$ be the polynomials

$$\boldsymbol{a} = x^5 + 3x^4 - 5x^3 - 3x^2 + 2x + 2,$$
$$\boldsymbol{b} = x^5 + x^4 - 2x^3 + 4x^2 + x + 5.$$

Use the Euclidean algorithm to compute $\gcd(\boldsymbol{a}, \boldsymbol{b})$ in each of the following rings.
(a) $\mathbb{F}_2[x]$ (b) $\mathbb{F}_3[x]$ (c) $\mathbb{F}_5[x]$ (d) $\mathbb{F}_7[x]$.

2.36. Continuing with the same polynomials $\boldsymbol{a}$ and $\boldsymbol{b}$ as in Exercise 2.35, for each of the polynomial rings (a)–(d) in Exercise 2.35, find polynomials $\boldsymbol{u}$ and $\boldsymbol{v}$ satisfying

$$\boldsymbol{a} \cdot \boldsymbol{u} + \boldsymbol{b} \cdot \boldsymbol{v} = \gcd(\boldsymbol{a}, \boldsymbol{b}).$$

2.37. Prove that the polynomial $x^3 + x + 1$ is irreducible in $\mathbb{F}_2[x]$. (*Hint.* Think about what a factorization would have to look like.)

2.38. The multiplication table for the field $\mathbb{F}_2[x]/(x^3 + x + 1)$ is given in Table 2.5, but we have omitted fourteen entries. Fill in the missing entries. (This is the field described in Example 2.57. You can download and print a copy of Table 2.5 at `www.math.brown.edu/~jhs/MathCrypto/Table2.5.pdf`.)

	0	1	x	x^2	$1+x$	$1+x^2$	$x+x^2$	$1+x+x^2$
0	0	0	0	0	0	0	0	0
1	0	1	x			$1+x^2$	$x+x^2$	$1+x+x^2$
x	0	x	x^2		$x+x^2$	1		$1+x^2$
x^2	0			$x+x^2$	$1+x+x^2$	x	$1+x^2$	1
$1+x$	0		$x+x^2$	$1+x+x^2$	$1+x^2$		1	x
$1+x^2$	0	$1+x^2$	1	x		$1+x+x^2$	$1+x$	
$x+x^2$	0	$x+x^2$		$1+x^2$	1	$1+x$	x	
$1+x+x^2$	0	$1+x+x^2$	$1+x^2$	1	x			$1+x$

Table 2.5: Multiplication table for the field $\mathbb{F}_2[x]/(x^3 + x + 1)$

2.39. The field $\mathbb{F}_7[x]/(x^2 + 1)$ is a field with 49 elements, which for the moment we denote by $\mathbb{F}_{49}$. (See Example 2.58 for a convenient way to work with $\mathbb{F}_{49}$.)
(a) Is $2 + 5x$ a primitive root in $\mathbb{F}_{49}$?
(b) Is $2 + x$ a primitive root in $\mathbb{F}_{49}$?
(c) Is $1 + x$ a primitive root in $\mathbb{F}_{49}$?
(*Hint.* Lagrange's theorem says that the order of $u \in \mathbb{F}_{49}$ must divide 48. So if $u^k \neq 1$ for all proper divisors k of 48, then u is a primitive root.)

2.40. Let p be a prime number and let $e \geq 2$. The quotient ring $\mathbb{Z}/p^e\mathbb{Z}$ and the finite field $\mathbb{F}_{p^e}$ are both rings and both have the same number of elements. Describe some ways in which they are intrinsically different.

2.41. Let $\mathbb{F}$ be a finite field.
(a) Prove that there is an integer $m \geq 1$ such that if we add 1 to itself m times,

$$\underbrace{1 + 1 + \cdots + 1}_{m \text{ ones}},$$

then we get 0. Note that here 1 and 0 are the multiplicative and additive identity elements of the field $\mathbb{F}$. If the notation is confusing, you can let u and z be the multiplicative and additive identity elements of $\mathbb{F}$, and then you need to prove that $u + u + \cdots + u = z$. (*Hint.* Since $\mathbb{F}$ is finite, the numbers 1, $1+1$, $1+1+1, \ldots$ cannot all be different.)

(b) Let m be the smallest positive integer with the property described in (a). Prove that m is prime. (*Hint.* If m factors, show that there are nonzero elements in $\mathbb{F}$ whose product is zero, so $\mathbb{F}$ cannot be a field.) This prime is called the *characteristic of the field* $\mathbb{F}$.

(c) Let p be the characteristic of $\mathbb{F}$. Prove that $\mathbb{F}$ is a finite-dimensional vector space over the field $\mathbb{F}_p$ of p elements.

(d) Use (c) to deduce that $\mathbb{F}$ has p^d elements for some $d \geq 1$.

Chapter 3

Integer Factorization and RSA

3.1 Euler's Formula and Roots Modulo pq

The Diffie–Hellman key exchange method and the Elgamal public key cryptosystem studied in Sects. 2.3 and 2.4 rely on the fact that it is easy to compute powers $a^n \bmod p$, but difficult to recover the exponent n if you know only the values of a and $a^n \bmod p$. An essential result that we used to analyze the security of Diffie–Hellman and Elgamal is Fermat's little theorem (Theorem 1.24),

$$a^{p-1} \equiv 1 \pmod{p} \qquad \text{for all } a \not\equiv 0 \pmod{p}.$$

Fermat's little theorem expresses a beautiful property of prime numbers. It is natural to ask what happens if we replace p with a number m that is not prime. Is it still true that $a^{m-1} \equiv 1 \pmod{m}$? A few computations such as Example 1.28 in Sect. 1.4 will convince you that the answer is no. In this section we investigate the correct generalization of Fermat's little theorem when $m = pq$ is a product of two distinct primes, since this is the case that is most important for cryptographic applications. We leave the general case for you to do in Exercises 3.4 and 3.5.

As usual, we begin with an example. What do powers modulo 15 look like? If we make a table of squares and cubes modulo 15, they do not look very interesting, but many fourth powers are equal to 1 modulo 15. More precisely, we find that

$$\begin{aligned} a^4 &\equiv 1 \pmod{15} \qquad \text{for } a = 1, 2, 4, 7, 8, 11, 13, \text{ and } 14; \\ a^4 &\not\equiv 1 \pmod{15} \qquad \text{for } a = 3, 5, 6, 9, 10, \text{ and } 12. \end{aligned}$$

J. Hoffstein et al., *An Introduction to Mathematical Cryptography*, Undergraduate Texts in Mathematics, DOI 10.1007/978-1-4939-1711-2_3

What distinguishes the list of numbers $1, 2, 4, 7, 8, 11, 13, 14$ whose fourth power is 1 modulo 15 from the list of numbers $3, 5, 6, 9, 10, 12, 15$ whose fourth power is not 1 modulo 15? A moment's reflection shows that each of the numbers $3, 5, 6, 9, 10, 12, 15$ has a nontrivial factor in common with the modulus 15, while the numbers $1, 2, 4, 7, 8, 11, 13, 14$ are relatively prime to 15. This suggests that some version of Fermat's little theorem should be true if the number a is relatively prime to the modulus m, but the correct exponent to use is not necessarily $m - 1$.

For $m = 15$ we found that the right exponent is 4. Why does 4 work? We could simply check each value of a, but a more enlightening argument would be better. In order to show that $a^4 \equiv 1 \pmod{15}$, it is enough to check the two congruences

$$a^4 \equiv 1 \pmod{3} \qquad \text{and} \qquad a^4 \equiv 1 \pmod{5}. \tag{3.1}$$

This is because the two congruences (3.1) say that

$$3 \text{ divides } a^4 - 1 \quad \text{and} \quad 5 \text{ divides } a^4 - 1,$$

which in turn imply that 15 divides $a^4 - 1$.

The two congruences in (3.1) are modulo primes, so we can use Fermat's little theorem to check that they are true. Thus

$$\begin{aligned} a^4 &= (a^2)^2 = (a^{(3-1)})^2 \equiv 1^2 \equiv 1 \pmod{3}, \\ a^4 &= a^{5-1} \equiv 1 \pmod{5}. \end{aligned}$$

If you think about these two congruences, you will see that the crucial property of the exponent 4 is that it is a multiple of $p - 1$ for both $p = 3$ and $p = 5$. Notice that this is not true of 14, which does not work as an exponent. With this observation, we are ready to state the fundamental formula that underlies the RSA public key cryptosystem.

Theorem 3.1 (Euler's Formula for pq). *Let p and q be distinct primes and let*

$$g = \gcd(p - 1, q - 1).$$

Then

$$a^{(p-1)(q-1)/g} \equiv 1 \pmod{pq} \qquad \textit{for all } a \textit{ satisfying } \gcd(a, pq) = 1.$$

In particular, if p and q are odd primes, then

$$a^{(p-1)(q-1)/2} \equiv 1 \pmod{pq} \qquad \textit{for all } a \textit{ satisfying } \gcd(a, pq) = 1.$$

Proof. By assumption we know that p does not divide a and that g divides $q - 1$, so we can compute

$$a^{(p-1)(q-1)/g} = \left(a^{(p-1)}\right)^{(q-1)/g} \qquad \text{since } (q-1)/g \text{ is an integer,}$$

$$\begin{aligned} &\equiv 1^{(q-1)/g} \pmod{p} && \text{since } a^{p-1} \equiv 1 \pmod{p} \\ & && \text{from Fermat's little theorem,} \\ &\equiv 1 \pmod{p} && \text{since 1 to any power is 1!} \end{aligned}$$

The exact same computation, reversing the roles of p and q, shows that

$$a^{(p-1)(q-1)/g} \equiv 1 \pmod{q}.$$

This proves that $a^{(p-1)(q-1)/g} - 1$ is divisible by both p and by q; hence it is divisible by pq, which completes the proof of Theorem 3.1. □

Diffie–Hellman key exchange and the Elgamal public key cryptosystem (Sects. 2.3 and 2.4) rely for their security on the difficulty of solving equations of the form

$$a^x \equiv b \pmod{p},$$

where a, b, and p are known quantities, p is a prime, and x is the unknown variable. The RSA public key cryptosystem, which we study in the next section, relies on the difficulty of solving equations of the form

$$x^e \equiv c \pmod{N},$$

where now the quantities e, c, and N are known and x is the unknown. In other words, the security of RSA relies on the assumption that it is difficult to take eth roots modulo N.

Is this a reasonable assumption? If the modulus N is prime, then it turns out that it is comparatively easy to compute eth roots modulo N, as described in the next proposition.

Proposition 3.2. *Let p be a prime and let $e \geq 1$ be an integer satisfying* $\gcd(e, p-1) = 1$. *Proposition 1.13 tells us that e has an inverse modulo $p-1$, say*

$$de \equiv 1 \pmod{p-1}.$$

Then the congruence

$$x^e \equiv c \pmod{p} \tag{3.2}$$

has the unique solution $x \equiv c^d \pmod{p}$.

Proof. If $c \equiv 0 \pmod{p}$, then $x \equiv 0 \pmod{p}$ is the unique solution and we are done. So we assume that $c \not\equiv 0 \pmod{p}$. The proof is then an easy application of Fermat's little theorem (Theorem 1.24). The congruence $de \equiv 1 \pmod{p-1}$ means that there is an integer k such that

$$de = 1 + k(p-1).$$

Now we check that c^d is a solution to $x^e \equiv c \pmod{p}$:

$$(c^d)^e \equiv c^{de} \pmod{p} \qquad \text{law of exponents,}$$

$$\begin{aligned}
&\equiv c^{1+k(p-1)} \pmod{p} && \text{since } de = 1 + k(p-1), \\
&\equiv c \cdot (c^{p-1})^k \pmod{p} && \text{law of exponents again,} \\
&\equiv c \cdot 1^k \pmod{p} && \text{from Fermat's little theorem,} \\
&\equiv c \pmod{p}.
\end{aligned}$$

This completes the proof that $x = c^d$ is a solution to $x^e \equiv c \pmod{p}$.

In order to see that the solution is unique, suppose that x_1 and x_2 are both solutions to the congruence (3.2). We've just proven that $z^{de} \equiv z \pmod{p}$ for any nonzero value z, so we find that

$$x_1 \equiv x_1^{de} \equiv (x_1^e)^d \equiv c^d \equiv (x_2^e)^d \equiv x_2^{de} \equiv x_2 \pmod{p}.$$

Thus x_1 and x_2 are the same modulo p, so (3.2) has at most one solution. □

Example 3.3. We solve the congruence

$$x^{1583} \equiv 4714 \pmod{7919},$$

where the modulus $p = 7919$ is prime. Proposition 3.2 says that first we need to solve the congruence

$$1583d \equiv 1 \pmod{7918}.$$

The solution, using the extended Euclidean algorithm (Theorem 1.11; see also Remark 1.15 and Exercise 1.12), is $d \equiv 5277 \pmod{7918}$. Then Proposition 3.2 tells us that

$$x \equiv 4714^{5277} \equiv 6059 \pmod{7919}$$

is a solution to $x^{1583} \equiv 4714 \pmod{7919}$.

Remark 3.4. Proposition 3.2 includes the assumption that $\gcd(e, p-1) = 1$. If this assumption is omitted, then the congruence $x^e \equiv c \pmod{p}$ will have a solution for some, but not all, values of c. Further, if it does have a solution, then it will have more than one. See Exercise 3.2 for further details.

Proposition 3.2 shows that it is easy to take eth roots if the modulus is a prime p. The situation for a composite modulus N looks similar, but there is a crucial difference. If we know how to factor N, then it is again easy to compute eth roots. The following proposition explains how to do this if $N = pq$ is a product of two primes. The general case is left for you to do in Exercise 3.6.

Proposition 3.5. *Let p and q be distinct primes and let $e \geq 1$ satisfy*

$$\gcd\bigl(e, (p-1)(q-1)\bigr) = 1.$$

Proposition 1.13 tells us that e has an inverse modulo $(p-1)(q-1)$, *say*

$$de \equiv 1 \pmod{(p-1)(q-1)}.$$

Then the congruence

$$x^e \equiv c \pmod{pq} \tag{3.3}$$

has the unique solution $x \equiv c^d \pmod{pq}$.

Proof. We assume that $\gcd(c, pq) = 1$; see Exercise 3.3 for the other cases. The proof of Proposition 3.5 is almost identical to the proof of Proposition 3.2, but instead of using Fermat's little theorem, we use Euler's formula (Theorem 3.1). The congruence $de \equiv 1 \pmod{(p-1)(q-1)}$ means that there is an integer k such that

$$de = 1 + k(p-1)(q-1).$$

Now we check that c^d is a solution to $x^e \equiv c \pmod{pq}$:

$$\begin{aligned}
(c^d)^e &\equiv c^{de} \pmod{pq} && \text{law of exponents,}\\
&\equiv c^{1+k(p-1)(q-1)} \pmod{pq} && \text{since } de = 1 + k(p-1)(q-1),\\
&\equiv c \cdot (c^{(p-1)(q-1)})^k \pmod{pq} && \text{law of exponents again,}\\
&\equiv c \cdot 1^k \pmod{pq} && \text{from Euler's formula (Theorem 3.1),}\\
&\equiv c \pmod{pq}.
\end{aligned}$$

This completes the proof that $x = c^d$ is a solution to the congruence (3.3). It remains to show that the solution is unique. Suppose that $x = u$ is a solution to (3.3). Then

$$\begin{aligned}
u &\equiv u^{de-k(p-1)(q-1)} \pmod{pq} && \text{since } de = 1 + k(p-1)(q-1),\\
&\equiv (u^e)^d \cdot (u^{(p-1)(q-1)})^{-k} \pmod{pq}\\
&\equiv (u^e)^d \cdot 1^{-k} \pmod{pq} && \text{using Euler's formula (Theorem 3.1),}\\
&\equiv c^d \pmod{pq} && \text{since } u \text{ is a solution to (3.3).}
\end{aligned}$$

Thus every solution to (3.3) is equal to $c^d \pmod{pq}$, so this is the unique solution. □

Remark 3.6. Proposition 3.5 gives an algorithm for solving $x^e \equiv c \pmod{pq}$ that involves first solving $de \equiv 1 \pmod{(p-1)(q-1)}$ and then computing $c^d \bmod pq$. We can often make the computation faster by using a smaller value of d. Let $g = \gcd(p-1, q-1)$ and suppose that we solve the following congruence for d:

$$de \equiv 1 \left(\bmod\ \frac{(p-1)(q-1)}{g}\right).$$

Euler's formula (Theorem 3.1) says that $a^{(p-1)(q-1)/g} \equiv 1 \pmod{pq}$. Hence just as in the proof of Proposition 3.5, if we write $de = 1 + k(p-1)(q-1)/g$, then

$$(c^d)^e = c^{de} = c^{1+k(p-1)(q-1)/g} = c \cdot (c^{(p-1)(q-1)/g})^k \equiv c \pmod{pq}.$$

Thus using this smaller value of d, we still find that $c^d \bmod pq$ is a solution to $x^e \equiv c \pmod{pq}$.

Example 3.7. We solve the congruence

$$x^{17389} \equiv 43927 \pmod{64349},$$

where the modulus $N = 64349 = 229 \cdot 281$ is a product of the two primes $p = 229$ and $q = 281$. The first step is to solve the congruence

$$17389d \equiv 1 \pmod{63840},$$

where $63840 = (p-1)(q-1) = 228 \cdot 280$. The solution, using the method described in Remark 1.15 or Exercise 1.12, is $d \equiv 53509 \pmod{63840}$. Then Proposition 3.5 tells us that

$$x \equiv 43927^{53509} \equiv 14458 \pmod{64349}$$

is the solution to $x^{17389} \equiv 43927 \pmod{64349}$.

We can save ourselves a little bit of work by using the idea described in Remark 3.6. We have

$$g = \gcd(p-1, q-1) = \gcd(228, 280) = 4,$$

so $(p-1)(q-1)/g = (228)(280)/4 = 15960$, which means that we can find a value of d by solving the congruence

$$17389d \equiv 1 \pmod{15960}.$$

The solution is $d \equiv 5629 \pmod{15960}$, and then

$$x \equiv 43927^{5629} \equiv 14458 \pmod{64349}$$

is the solution to $x^{17389} \equiv 43927 \pmod{64349}$. Notice that we obtained the same solution, as we should, but that we needed to raise 43927 to only the 5629th power, while using Proposition 3.5 directly required us to raise 43927 to the 53509th power. This saves some time, although not quite as much as it looks, since recall that computing $c^d \bmod N$ takes time $O(\ln d)$. Thus the faster method takes about 80 % as long as the slower method, since $\ln(5629)/\ln(53509) \approx 0.793$.

Example 3.8. Alice challenges Eve to solve the congruence

$$x^{9843} \equiv 134872 \pmod{30069476293}.$$

The modulus 30069476293 is not prime, since (cf. Example 1.28)

$$2^{30069476293-1} \equiv 18152503626 \not\equiv 1 \pmod{30069476293}.$$

It happens that 30069476293 is a product of two primes, but if Eve does not know the prime factors, she cannot use Proposition 3.5 to solve Alice's challenge. After accepting Eve's concession of defeat, Alice informs Eve that 30069476293 is equal to $104729 \cdot 287117$. With this new knowledge, Alice's challenge becomes easy. Eve computes $104728 \cdot 287116 = 30069084448$, solves the congruence $9843d \equiv 1 \pmod{30069084448}$ to find $d \equiv 18472798299 \pmod{30069084448}$, and computes the solution

$$x \equiv 134872^{18472798299} \equiv 25470280263 \pmod{30069476293}.$$

Bob	Alice
Key creation	
Choose secret primes p and q. Choose encryption exponent e with $\gcd(e,(p-1)(q-1)) = 1$. Publish $N = pq$ and e.	
Encryption	
	Choose plaintext m. Use Bob's public key (N, e) to compute $c \equiv m^e \pmod N$. Send ciphertext c to Bob.
Decryption	
Compute d satisfying $ed \equiv 1 \pmod{(p-1)(q-1)}$. Compute $m' \equiv c^d \pmod N$. Then m' equals the plaintext m.	

Table 3.1: RSA key creation, encryption, and decryption

3.2 The RSA Public Key Cryptosystem

Bob and Alice have the usual problem of exchanging sensitive information over an insecure communication line. We have seen in Chap. 2 various ways in which Bob and Alice can accomplish this task, based on the difficulty of solving the discrete logarithm problem. In this section we describe the RSA public key cryptosystem, the first invented and certainly best known such system. RSA is named after its (public) inventors, Ron Rivest, Adi Shamir, and Leonard Adleman.

The security of RSA depends on the following dichotomy:

- **Setup.** Let p and q be large primes, let $N = pq$, and let e and c be integers.

- **Problem.** Solve the congruence $x^e \equiv c \pmod{N}$ for the variable x.
- **Easy.** Bob, who knows the values of p and q, can easily solve for x as described in Proposition 3.5.
- **Hard.** Eve, who does not know the values of p and q, cannot easily find x.
- **Dichotomy.** Solving $x^e \equiv c \pmod{N}$ is easy for a person who possesses certain extra information, but it is apparently hard for all other people.

The *RSA public key cryptosystem* is summarized in Table 3.1. Bob's secret key is a pair of large primes p and q. His public key is the pair (N, e) consisting of the product $N = pq$ and an encryption exponent e that is relatively prime to $(p-1)(q-1)$. Alice takes her plaintext and converts it into an integer m between 1 and N. She encrypts m by computing the quantity

$$c \equiv m^e \pmod{N}.$$

The integer c is her ciphertext, which she sends to Bob. It is then a simple matter for Bob to solve the congruence $x^e \equiv c \pmod{N}$ to recover Alice's message m, because Bob knows the factorization $N = pq$. Eve, on the other hand, may intercept the ciphertext c, but unless she knows how to factor N, she presumably has a difficult time trying to solve $x^e \equiv c \pmod{N}$.

Example 3.9. We illustrate the RSA public key cryptosystem with a small numerical example. Of course, this example is not secure, since the numbers are so small that it would be easy for Eve to factor the modulus N. Secure implementations of RSA use moduli N with hundreds of digits.

RSA Key Creation

- Bob chooses two secret primes $p = 1223$ and $q = 1987$. Bob computes his public modulus

$$N = p \cdot q = 1223 \cdot 1987 = 2430101.$$

- Bob chooses a public encryption exponent $e = 948047$ with the property that

$$\gcd(e, (p-1)(q-1)) = \gcd(948047, 2426892) = 1.$$

RSA Encryption

- Alice converts her plaintext into an integer

$$m = 1070777 \qquad \text{satisfying} \quad 1 \le m < N.$$

- Alice uses Bob's public key $(N, e) = (2430101, 948047)$ to compute

$$c \equiv m^e \pmod{N}, \quad c \equiv 1070777^{948047} \equiv 1473513 \pmod{2430101}.$$

- Alice sends the ciphertext $c = 1473513$ to Bob.

RSA Decryption

- Bob knows $(p-1)(q-1) = 1222 \cdot 1986 = 2426892$, so he can solve

$$ed \equiv 1 \pmod{(p-1)(q-1)}, \qquad 948047 \cdot d \equiv 1 \pmod{2426892},$$

  for d and find that $d = 1051235$.
- Bob takes the ciphertext $c = 1473513$ and computes

$$c^d \pmod{N}, \qquad 1473513^{1051235} \equiv 1070777 \pmod{2430101}.$$

  The value that he computes is Alice's message $m = 1070777$.

Remark 3.10. The quantities N and e that form Bob's public key are called, respectively, the *modulus* and the *encryption exponent*. The number d that Bob uses to decrypt Alice's message, that is, the number d satisfying

$$ed \equiv 1 \pmod{(p-1)(q-1)}, \tag{3.4}$$

is called the *decryption exponent*. It is clear that encryption can be done more efficiently if the encryption exponent e is a small number, and similarly, decryption is more efficient if the decryption exponent d is small. Of course, Bob cannot choose both of them to be small, since once one of them is selected, the other is determined by the congruence (3.4). (This is not strictly true, since if Bob takes $e = 1$, then also $d = 1$, so both d and e are small. But then the plaintext and the ciphertext are identical, so taking $e = 1$ is a very bad idea!)

Notice that Bob cannot take $e = 2$, since he needs e to be relatively prime to $(p-1)(q-1)$. Thus the smallest possible value for e is $e = 3$. As far as is known, taking $e = 3$ is as secure as taking a larger value of e, although some doubts are raised in [22]. People who want fast encryption, but are worried that $e = 3$ is too small, often take $e = 2^{16} + 1 = 65537$, since it takes only sixteen squarings and one multiplication to compute m^{65537} via the square-and-multiply algorithm described in Sect. 1.3.2.

An alternative is for Bob to use a small value for d and use the congruence (3.4) to determine e, so e would be large. However, it turns out that this may lead to an insecure version of RSA. More precisely, if d is smaller than $N^{1/4}$, then the theory of continued fractions allows Eve to break RSA. See [17, 18, 19, 149] for details.

Remark 3.11. Bob's public key includes the number $N = pq$, which is a product of two secret primes p and q. Proposition 3.5 says that if Eve knows the value of $(p-1)(q-1)$, then she can solve $x^e \equiv c \pmod{N}$, and thus can decrypt messages sent to Bob.

Expanding $(p-1)(q-1)$ gives

$$(p-1)(q-1) = pq - p - q + 1 = N - (p+q) + 1. \tag{3.5}$$

Bob has published the value of N, so Eve already knows N. Thus if Eve can determine the value of the sum $p + q$, then (3.5) gives her the value of $(p-1)(q-1)$, which enables her to decrypt messages.

In fact, if Eve knows the values of $p+q$ and pq, then it is easy for her to compute the values of p and q. She simply uses the quadratic formula to find the roots of the polynomial

$$X^2 - (p+q)X + pq,$$

since this polynomial factors as $(X-p)(X-q)$, so its roots are p and q. Thus once Bob publishes the value of $N = pq$, it is no easier for Eve to find the value of $(p-1)(q-1)$ than it is for her to find p and q themselves.

We illustrate with an example. Suppose that Eve knows that

$$N = pq = 66240912547 \quad \text{and} \quad (p-1)(q-1) = 66240396760.$$

She first uses (3.5) to compute

$$p + q = N + 1 - (p-1)(q-1) = 515788.$$

Then she uses the quadratic formula to factor the polynomial

$$\begin{aligned} X^2 - (p+q)X + N &= X^2 - 515788X + 66240912547 \\ &= (X - 241511)(X - 274277). \end{aligned}$$

This gives her the factorization $N = 66240912547 = 241511 \cdot 274277$.

Remark 3.12. One final, but very important, observation. We have shown that it is no easier for Eve to determine $(p-1)(q-1)$ than it is for her to factor N. But this does not prove that that Eve must factor N in order to decrypt Bob's messages. The point is that what Eve really needs to do is to solve congruences of the form $x^e \equiv c \pmod{N}$, and conceivably there is an efficient algorithm to solve such congruences without knowing the value of $(p-1)(q-1)$. No one knows whether such a method exists, although see [22] for a suggestion that computing roots modulo N may be easier than factoring N.

3.3 Implementation and Security Issues

Our principal focus in this book is the mathematics of the hard problems underlying modern cryptography, but we would be remiss if we did not at least briefly mention some of the security issues related to implementation. The reader should be aware that we do not even scratch the surface of this vast and fascinating subject, but simply describe some examples to show that there is far more to creating a secure communications system than simply using a cryptosystem based on an intractable mathematical problem.

Example 3.13 (Woman-in-the-Middle Attack). Suppose that Eve is not simply an eavesdropper, but that she has full control over Alice and Bob's communication network. In this case, she can institute what is known as a *man-in-the-middle* attack. We describe this attack for Diffie–Hellman key exchange, but it exists for most public key constructions. (See Exercise 3.12.)

Recall that in Diffie–Hellman key exchange (Table 2.2), Alice sends Bob the value $A = g^a$ and Bob sends Alice the value $B = g^b$, where the computations take place in the finite field $\mathbb{F}_p$. What Eve does is to choose her own secret exponent e and compute the value $E = g^e$. She then intercepts Alice and Bob's communications, and instead of sending A to Bob and sending B to Alice, she sends both of them the number E. Notice that Eve has exchanged the value A^e with Alice and the value B^e with Bob, while Alice and Bob believe that they have exchanged values with each other. The man-in-the-middle attack is illustrated in Fig. 3.1.

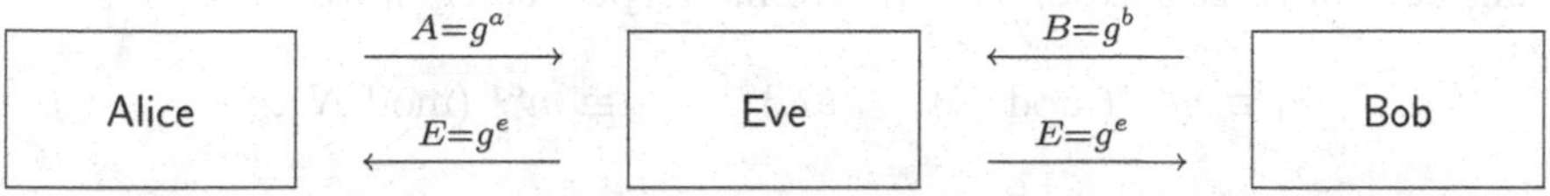

Figure 3.1: "Man-in-the-middle" attack on Diffie–Hellman key exchange

Suppose that Alice and Bob subsequently use their supposed secret shared value as the key for a symmetric cipher and send each other messages. For example, Alice encrypts a plaintext message m using E^a as the symmetric cipher key. Eve intercepts this message and is able to decrypt it using A^e as the symmetric cipher key, so she can read Alice's message. She then re-encrypts it using B^e as the symmetric cipher key and sends it to Bob. Since Bob is then able to decrypt it using E^b as the symmetric cipher key, he is unaware that there is a breach in security.

Notice the insidious nature of this attack. Eve does not solve the underlying hard problem (in this case, the discrete logarithm problem or the Diffie–Hellman problem), yet she is able to read Alice and Bob's communications, and they are not aware of her success.

Example 3.14. Suppose that Eve is able to convince Alice to decrypt "random" RSA messages using her (Alice's) private key. This is a plausible scenario, since one way for Alice to authenticate her identity as the owner of the public key (N, e) is to show that she knows how to decrypt messages. (One says that Eve has access to an *RSA oracle*.)

Eve can exploit Alice's generosity as follows. Suppose that Eve has intercepted a ciphertext c that Bob has sent to Alice. Eve chooses a random value k and sends Alice the "message"

$$c' \equiv k^e \cdot c \pmod{N}.$$

Alice decrypts c' and returns the resulting m' to Eve, where

$$m' \equiv (c')^d \equiv (k^e \cdot c)^d \equiv (k^e \cdot m^e)^d \equiv k \cdot m \pmod{N}.$$

Thus Eve knows the quantity $k \cdot m \pmod{N}$, and since she knows k, she immediately recovers Bob's plaintext m.

There are two important observations to make. First, Eve has decrypted Bob's message without knowing or gaining knowledge of how to factor N, so the difficulty of the underlying mathematical problem is irrelevant. Second, since Eve has used k to mask Bob's ciphertext, Alice has no way to tell that Eve's message is in any way related to Bob's message. Thus Alice sees only the values $k^e \cdot c \pmod{N}$ and $k \cdot m \pmod{N}$, which to her look random when compared to c and m.

Example 3.15. Suppose that Alice publishes two different exponents e_1 and e_2 for use with her public modulus N and that Bob encrypts a single plaintext m using both of Alice's exponents. If Eve intercepts the ciphertexts

$$c_1 \equiv m^{e_1} \pmod{N} \qquad \text{and} \qquad c_2 \equiv m^{e_2} \pmod{N},$$

she can take a solution to the equation

$$e_1 \cdot u + e_2 \cdot v = \gcd(e_1, e_2)$$

and use it to compute

$$c_1^u \cdot c_2^v \equiv (m^{e_1})^u \cdot (m^{e_2})^v \equiv m^{e_1 \cdot u + e_2 \cdot v} \equiv m^{\gcd(e_1,e_2)} \pmod{N}.$$

If it happens that $\gcd(e_1, e_2) = 1$, Eve has recovered the plaintext. (See Exercise 3.13 for a numerical example.) More generally, if Bob encrypts a single message using several exponents $e_1, e_2, \ldots, e_r$, then Eve can recover the plaintext if $\gcd(e_1, e_2, \ldots, e_r) = 1$. The moral is that Alice should use at most one encryption exponent for a given modulus.

3.4 Primality Testing

Bob has finished reading Sects. 3.2 and 3.3 and is now ready to communicate with Alice using his RSA public/private key pair. Or is he? In order to create an RSA key pair, Bob needs to choose two *very large* primes p and q. It's not enough for him to choose two very large, but possibly composite, numbers p and q. In the first place, if p and q are not prime, Bob will need to know how to factor them in order to decrypt Alice's message. But even worse, if p and q have small prime factors, then Eve may be able to factor pq and break Bob's system.

Bob is thus faced with the task of finding large prime numbers. More precisely, he needs a way of distinguishing between prime numbers and composite numbers, since if he knows how to do this, then he can choose large random numbers until he hits one that is prime. We discuss later (Sect. 3.4.1) the likelihood that a randomly chosen number is prime, but for now it is enough to know that he has a reasonably good chance of success. Hence what Bob really needs is an efficient way to tell whether a very large number is prime.

For example, suppose that Bob chooses the rather large number

$$n = 31987937737479355332620068643713101490952335301$$

and he wants to know whether n is prime. First Bob searches for small factors, but he finds that n is not divisible by any primes smaller than 1000000. So he begins to suspect that maybe n is prime. Next he computes the quantity $2^{n-1} \bmod n$ and he finds that

$$2^{n-1} \equiv 1281265953551359064133601216247151836053160074 \pmod{n}. \qquad (3.6)$$

The congruence (3.6) immediately tells Bob that n is a composite number, although it does not give him any indication of how to factor n. Why? Recall Fermat's little theorem, which says that if p is prime, then $a^{p-1} \equiv 1 \pmod{p}$ (unless p divides a). Thus if n were prime, then the right-hand side of (3.6) would equal 1; since it does not equal 1, Bob concludes that n is not prime.

Before continuing the saga of Bob's quest for large primes, we state a convenient version of Fermat's little theorem that puts no restrictions on a.

Theorem 3.16 (Fermat's Little Theorem, Version 2). *Let p be a prime number. Then*

$$a^p \equiv a \pmod{p} \qquad \textit{for every integer } a. \qquad (3.7)$$

Proof. If $p \nmid a$, then the first version of Fermat's little theorem (Theorem 1.24) implies that $a^{p-1} \equiv 1 \pmod{p}$. Multiplying both sides by a proves that (3.7) is true. On the other hand, if $p \mid a$, then both sides of (3.7) are 0 modulo p. □

Returning to Bob's quest, we find him undaunted as he randomly chooses another large number,

$$n = 2967952985951692762820418740138329004315165131. \qquad (3.8)$$

After checking for divisibility by small primes, Bob computes $2^n \bmod n$ and finds that

$$2^n \equiv 2 \pmod{n}. \qquad (3.9)$$

Does (3.9) combined with Fermat's little theorem 3.16 prove that n is prime? The answer is NO! Fermat's theorem works in only one direction:

$$\text{If } p \text{ is prime, then } a^p \equiv a \pmod{p}.$$

There is nothing to prevent an equality such as (3.9) being true for composite values of n, and indeed a brief search reveals examples such as

$$2^{341} \equiv 2 \pmod{341} \qquad \text{with} \quad 341 = 11 \cdot 31.$$

However, in some vague philosophical sense, the fact that $2^n \equiv 2 \pmod{n}$ makes it more likely that n is prime, since if the value of $2^n \bmod n$ had turned out differently, we would have known that n was composite. This leads us to make the following definition.

Definition. Fix an integer n. We say that an integer a is a *witness for* (*the compositeness of*) n if

$$a^n \not\equiv a \pmod{n}.$$

As we observed earlier, a single witness for n combined with Fermat's little theorem (Theorem 3.16) is enough to prove beyond a shadow of a doubt that n is composite.[1] Thus one way to assess the likelihood that n is prime is to try a lot of numbers $a_1, a_2, a_3, \ldots$. If any one of them is a witness for n, then Bob knows that n is composite; and if none of them is a witness for n, then Bob suspects, but does not know for certain, that n is prime.

Unfortunately, intruding on this idyllic scene are barbaric numbers such as 561. The number 561 is composite, $561 = 3 \cdot 11 \cdot 17$, yet 561 has no witnesses! In other words,

$$a^{561} \equiv a \pmod{561} \qquad \text{for every integer } a.$$

Composite numbers having no witnesses are called *Carmichael numbers*, after R.D. Carmichael, who in 1910 published a paper listing 15 such numbers. The fact that 561 is a Carmichael number can be verified by checking each value $a = 0, 1, 2, \ldots, 560$, but see Exercise 3.14 for an easier method and for more examples of Carmichael numbers. Although Carmichael numbers are rather rare, Alford, Granville, and Pomerance [5] proved in 1994 that there are infinitely many of them. So Bob needs something stronger than Fermat's little theorem in order to test whether a number is (probably) prime. What is needed is a better test for compositeness. The following property of prime numbers is used to formulate the *Miller–Rabin test*, which has the agreeable property that every composite number has a large number of witnesses.

Proposition 3.17. *Let p be an odd prime and write*

$$p - 1 = 2^k q \qquad \text{with } q \text{ odd.}$$

Let a be any number not divisible by p. Then one of the following two conditions is true:

(i) *a^q is congruent to 1 modulo p.*

(ii) *One of a^q, a^{2q}, $a^{4q}, \ldots, a^{2^{k-1}q}$ is congruent to -1 modulo p.*

Proof. Fermat's little theorem (Theorem 1.24) tells us that $a^{p-1} \equiv 1 \pmod{p}$. This means that when we look at the list of numbers

$$a^q,\ a^{2q},\ a^{4q}, \ldots,\ a^{2^{k-1}q},\ a^{2^k q},$$

we know that the last number in the list, which equals a^{p-1}, is congruent to 1 modulo p. Further, each number in the list is the square of the previous number. Therefore one of the following two possibilities must occur:

[1] In the great courthouse of mathematics, witnesses never lie!

(i) The first number in the list is congruent to 1 modulo p.

(ii) Some number in the list is not congruent to 1 modulo p, but when it is squared, it becomes congruent to 1 modulo p. But the only number satisfying both

$$b \not\equiv 1 \pmod{p} \quad \text{and} \quad b^2 \equiv 1 \pmod{p}$$

is -1, so one of the numbers in the list is congruent to -1 modulo p.

This completes the proof of Proposition 3.17. □

Input. Integer n to be tested, integer a as potential witness.

1. If n is even or $1 < \gcd(a, n) < n$, return **Composite**.
2. Write $n - 1 = 2^k q$ with q odd.
3. Set $a = a^q \pmod{n}$.
4. If $a \equiv 1 \pmod{n}$, return **Test Fails**.
5. Loop $i = 0, 1, 2, \ldots, k - 1$
   6. If $a \equiv -1 \pmod{n}$, return **Test Fails**.
   7. Set $a = a^2 \bmod n$.
8. End i loop.
9. Return **Composite**.

Table 3.2: Miller–Rabin test for composite numbers

Definition. Let n be an odd number and write $n - 1 = 2^k q$ with q odd. An integer a satisfying $\gcd(a, n) = 1$ is called a *Miller–Rabin witness for* (*the compositeness of*) n if both of the following conditions are true:
(a) $a^q \not\equiv 1 \pmod{n}$.
(b) $a^{2^i q} \not\equiv -1 \pmod{n}$ for all $i = 0, 1, 2, \ldots, k - 1$.

It follows from Proposition 3.17 that if there exists an a that is a Miller–Rabin witness for n, then n is definitely a composite number. This leads to the Miller–Rabin test for composite numbers described in Table 3.2.

Now suppose that Bob wants to check whether a large number n is probably a prime. To do this, he runs the Miller–Rabin test using a bunch of randomly selected values of a. Why is this better than using the Fermat's little theorem test? The answer is that there are no Carmichael-like numbers for the Miller–Rabin test, and in fact, every composite number has a lot of Miller–Rabin witnesses, as described in the following proposition.

Proposition 3.18. *Let n be an odd composite number. Then at least* 75 % *of the numbers a between* 1 *and $n - 1$ are Miller–Rabin witnesses for n.*

Proof. The proof is not hard, but we will not give it here. See for example [132, Theorem 10.6]. □

Consider now Bob's quest to identify large prime numbers. He takes his potentially prime number n and he runs the Miller–Rabin test on n for (say) 10 different values of a. If any a value is a Miller–Rabin witness for n, then Bob knows that n is composite. But suppose that none of his a values is a Miller–Rabin witness for n. Proposition 3.18 says that if n were composite, then each time Bob tries a value for a, he has at least a 75 % chance of getting a witness. Since Bob found no witnesses in 10 tries, it is reasonable[2] to conclude that the probability of n being composite is at most $(25\,\%)^{10}$, which is approximately 10^{-6}. And if this is not good enough, Bob can use 100 different values of a, and if none of them proves n to be composite, then the probability that n is actually composite is less than $(25\,\%)^{100} \approx 10^{-60}$.

Example 3.19. We illustrate the Miller–Rabin test with $a = 2$ and the number $n = 561$, which, you may recall, is a Carmichael number. We factor

$$n - 1 = 560 = 2^4 \cdot 35$$

and then compute

$$\begin{aligned}
2^{35} &\equiv 263 && \pmod{561},\\
2^{2\cdot 35} &\equiv 263^2 \equiv 166 && \pmod{561},\\
2^{4\cdot 35} &\equiv 166^2 \equiv 67 && \pmod{561},\\
2^{8\cdot 35} &\equiv 67^2 \equiv 1 && \pmod{561}.
\end{aligned}$$

The first number $2^{35} \bmod 561$ is neither 1 nor -1, and the other numbers in the list are not equal to -1, so 2 is a Miller–Rabin witness to the fact that 561 is composite.

Example 3.20. We do a second example, taking $n = 172947529$ and

$$n - 1 = 172947528 = 2^3 \cdot 21618441.$$

We apply the Miller–Rabin test with $a = 17$ and find that

$$17^{21618441} \equiv 1 \pmod{172947529}.$$

[2]Unfortunately, although this deduction seems reasonable, it is not quite accurate. In the language of probability theory, we need to compute the conditional probability that n is composite given that the Miller–Rabin test fails 10 times; and we know the conditional probability that the Miller–Rabin test succeeds at least 75 % of the time if n is composite. See Sect. 5.3.2 for a discussion of conditional probabilities and Exercise 5.30 for a derivation of the correct formula, which says that the probability $(25\,\%)^{10}$ must be approximately multiplied by $\ln(n)$.

Thus 17 is not a Miller–Rabin witness for n. Next we try $a = 3$, but unfortunately

$$3^{21618441} \equiv -1 \pmod{172947529},$$

so 3 also fails to be a Miller–Rabin witness. At this point we might suspect that n is prime, but if we try another value, say $a = 23$, we find that

$$\begin{aligned} 23^{21618441} &\equiv 40063806 \pmod{172947529}, \\ 23^{2\cdot 21618441} &\equiv 2257065 \pmod{172947529}, \\ 23^{4\cdot 21618441} &\equiv 1 \pmod{172947529}. \end{aligned}$$

Thus 23 is a Miller–Rabin witness and n is actually composite. In fact, n is a Carmichael number, but it's not so easy to factor (by hand).

3.4.1 The Distribution of the Set of Primes

If Bob picks a number at random, what is the likelihood that it is prime? The answer is provided by one of number theory's most famous theorems. In order to state the theorem, we need a definition.

Definition. For any number X, let

$$\pi(X) = (\# \text{ of primes } p \text{ satisfying } 2 \le p \le X).$$

For example, $\pi(10) = 4$, since the primes between 2 and 10 are 2, 3, 5, and 7.

Theorem 3.21 (The Prime Number Theorem).

$$\lim_{X\to\infty} \frac{\pi(X)}{X/\ln(X)} = 1.$$

Proof. The prime number theorem was proven independently by Hadamard and de la Vallée Poussin in 1896. The proof is unfortunately far beyond the scope of this book. The most direct proof uses complex analysis; see for example [7, Chapter 13]. □

Example 3.22. How many primes would we expect to find between 900000 and 1000000? The prime number theorem says that

$$\begin{aligned} &(\text{Number of primes between 900000 and 1000000}) \\ &\quad = \pi(1000000) - \pi(900000) \approx \frac{1000000}{\ln 1000000} - \frac{900000}{\ln 900000} = 6737.62\dots. \end{aligned}$$

In fact, it turns out that there are exactly 7224 primes between 900000 and 1000000.

For cryptographic purposes, we need even larger primes. For example, we might want to use primes having approximately 300 decimal digits, or almost equivalently, primes that are 1024 bits in length, since $2^{1024} \approx 10^{308.25}$.

How many primes p satisfy $2^{1023} < p < 2^{1024}$? The prime number theorem gives us an answer:

$$\# \text{ of 1024 bit primes} = \pi(2^{1024}) - \pi(2^{1023}) \approx \frac{2^{1024}}{\ln 2^{1024}} - \frac{2^{1023}}{\ln 2^{1023}} \approx 2^{1013.53}.$$

So there should be lots of primes in this interval.

Intuitively, the prime number theorem says that if we look at all of the numbers between 1 and X, then the proportion of them that are prime is approximately $1/\ln(X)$. Turning this statement around, the prime number theorem says:

$$\boxed{\text{A randomly chosen number } N \text{ has probability } 1/\ln(N) \text{ of being prime.}} \tag{3.10}$$

Of course, taken at face value, statement (3.10) is utter nonsense. A chosen number either is prime or is not prime; it cannot be partially prime and partially composite! A better interpretation of (3.10) is that it describes how many primes one expects to find in an interval around N. See Exercise 3.19 for a more precise statement of (3.10) that is both meaningful and mathematically correct.

Example 3.23. We illustrate statement (3.10) and the prime number theorem by searching for 1024-bit primes, i.e., primes that are approximately 2^{1024}. Statement (3.10) says that the probability that a random number $N \approx 2^{1024}$ is prime is approximately 0.14 %. Thus on average, Bob checks about 700 randomly chosen numbers of this size before finding a prime.

If he is clever, Bob can do better. He knows that he doesn't want a number that is even, nor does he want a number that is divisible by 3, nor divisible by 5, etc. Thus rather than choosing numbers completely at random, Bob might restrict attention (say) to numbers that are relatively prime to 2, 3, 5, 7 and 11. To do this, he first chooses a random number that is relatively prime to $2 \cdot 3 \cdot 5 \cdot 7 \cdot 11 = 2310$, say he chooses 1139. Then he considers only numbers N of the form

$$N = 2 \cdot 3 \cdot 5 \cdot 7 \cdot 11 \cdot K + 1139 = 2310K + 1139. \tag{3.11}$$

The probability that an N of this form is prime is approximately (see Exercise 3.20)

$$\frac{2}{1} \cdot \frac{3}{2} \cdot \frac{5}{4} \cdot \frac{7}{6} \cdot \frac{11}{10} \cdot \frac{1}{\ln(N)} \approx \frac{4.8}{\ln(N)}.$$

So if Bob chooses a random number N of the form (3.11) with $N \approx 2^{1024}$, then the probability that it is prime is approximately 0.67 %. Thus he only needs to check 150 numbers to have a good chance of finding a prime.

We used the Miller–Rabin test with 100 randomly chosen values of a to check the primality of

$$2310K + 1139 \qquad \text{for each} \qquad 2^{1013} \le K \le 2^{1013} + 1000.$$

We found that $2310(2^{1013}+J)+1139$ is probably prime for the following 12 values of J:

$$J \in \{41, 148, 193, 251, 471, 585, 606, 821, 851, 865, 910, 911\}.$$

This is a bit better than the 7 values predicted by the prime number theorem. The smallest probable prime that we found is $2310 \cdot (2^{1013} + 41) + 1139$, which is equal to the following 308 digit number:

20276714558261473373313940384387925462194955182405899331133959349334105522983
75121272248938548639688519470034484877532500936544755670421865031628734263599742737518719
78241831537235413710389881550750303525056818030281312537212445925881220354174468221605146
327969430834440565497127875070636801598203824198219369.

Remark 3.24. There are many deep open questions concerning the distribution of prime numbers, of which the most important and famous is certainly the *Riemann hypothesis*.[3] The usual way to state the Riemann hypothesis requires some complex analysis. The Riemann zeta function $\zeta(s)$ is defined by the series

$$\zeta(s) = \sum_{n=1}^{\infty} \frac{1}{n^s},$$

which converges when s is a complex number with real part greater than 1. It has an analytic continuation to the entire complex plane with a simple pole at $s = 1$ and no other poles. The Riemann hypothesis says that if $\zeta(\sigma + it) = 0$ with σ and t real and $0 \le \sigma \le 1$, then in fact $\sigma = \frac{1}{2}$.

At first glance, this somewhat bizarre statement appears to have little relation to prime numbers. However, it is not hard to show that $\zeta(s)$ is also equal to the product

$$\zeta(s) = \prod_{p \text{ prime}} \left(1 - \frac{1}{p^s}\right)^{-1},$$

so $\zeta(s)$ incorporates information about the set of prime numbers.

There are many statements about prime numbers that are equivalent to the Riemann hypothesis. For example, recall that the prime number theorem (Theorem 3.21) says that $\pi(X)$ is approximately equal to $X/\ln(X)$ for large values of X. The Riemann hypothesis is equivalent to the following more accurate statement:

$$\pi(X) = \int_2^X \frac{dt}{\ln t} + \mathcal{O}\left(\sqrt{X} \cdot \ln(X)\right). \tag{3.12}$$

This conjectural formula is stronger than the prime number theorem, since the integral is approximately equal to $X/\ln(X)$. (See Exercise 3.21.)

[3] The Riemann hypothesis is another of the $1,000,000 Millennium Prize problems.

3.4.2 Primality Proofs Versus Probabilistic Tests

The Miller–Rabin test is a powerful and practical method for finding large numbers that are "probably prime." Indeed, Proposition 3.18 says that every composite number has many Miller–Rabin witnesses, so 50 or 100 repetitions of the Miller–Rabin test provide solid evidence that n is prime. However, there is a difference between evidence for a statement and a rigorous proof that the statement is correct. Suppose that Bob is not satisfied with mere evidence. He wants to be completely certain that his chosen number n is prime.

In principle, nothing could be simpler. Bob checks to see whether n is divisible by any of the numbers $1, 2, 3, 4, \ldots$ up to $\sqrt{n}$. If none of these numbers divides n, then Bob knows, with complete certainty, that n is prime. Unfortunately, if n is large, say $n \approx 2^{1000}$, then the sun will have burnt out before Bob finishes his task. Notice that the running time of this naive algorithm is $O(\sqrt{n})$, which means that it is an exponential-time algorithm according to the definition in Sect. 2.6, since $\sqrt{n}$ is exponential in the *number of bits* required to write down the number n.

It would be nice if we could use the Miller–Rabin test to efficiently and conclusively prove that a number n is prime. More precisely, we would like a polynomial-time algorithm that proves primality. If a generalized version of the Riemann hypothesis is true, then the following proposition says that this can be done. (We discussed the Riemann hypothesis in Remark 3.24.)

Proposition 3.25. *If a generalized version of the Riemann hypothesis is true, then every composite number n has a Miller–Rabin witness a for its compositeness satisfying*

$$a \leq 2(\ln n)^2.$$

Proof. See [87] for a proof that every composite number n has a witness satisfying $a = O\big((\ln n)^2\big)$, and [9, 10] for the more precise estimate $a \leq 2(\ln n)^2$. □

Thus if the generalized Riemann hypothesis is true, then we can prove that n is prime by applying the Miller–Rabin test using every a smaller than $2(\ln n)^2$. If some a proves that n is composite, then n is composite, and otherwise, Proposition 3.25 tells us that n is prime. Unfortunately, the proof of Proposition 3.25 assumes that the generalized Riemann hypothesis is true, and no one has yet been able to prove even the original Riemann hypothesis, despite almost 150 years of work on the problem.

After the creation of public key cryptography, and especially after the publication of the RSA cryptosystem in 1978, it became of great interest to find a polynomial-time primality test that did not depend on any unproven hypotheses. Many years of research culminated in 2002, when M. Agrawal, N. Kayal, and N. Saxena [1] found such an algorithm. Subsequent improvements to their algorithm have given the following result.

Theorem 3.26 (AKS Primality Test). *For every $\epsilon > 0$, there is an algorithm that conclusively determines whether a given number n is prime in no more than $O\big((\ln n)^{6+\epsilon}\big)$ steps.*

Proof. The original algorithm was published in [1]. Further analysis and refinements may be found in [76]. The monograph [36] contains a nice description of primality testing, including the AKS test. □

Remark 3.27. The result described in Theorem 3.26 represents a triumph of modern algorithmic number theory. The significance for practical cryptography is less clear, since the AKS algorithm is much slower than the Miller–Rabin test. In practice, most people are willing to accept that a number is prime if it passes the Miller–Rabin test for (say) 50–100 randomly chosen values of a.

3.5 Pollard's $p-1$ Factorization Algorithm

We saw in Sect. 3.4 that it is relatively easy to check whether a large number is (probably) prime. This is good, since the RSA cryptosystem needs large primes in order to operate.

Conversely, the security of RSA relies on the apparent difficulty of factoring large numbers. The study of factorization dates back at least to ancient Greece, but it was only with the advent of computers that people started to develop algorithms capable of factoring very large numbers. The paradox of RSA is that in order to make RSA more efficient, we want to use a modulus $N = pq$ that is as small as possible. On the other hand, if an opponent can factor N, then our encrypted messages are not secure. It is thus vital to understand how hard it is to factor large numbers, and in particular, to understand the capabilities of the different algorithms that are currently used for factorization.

In the next few sections we discuss, with varying degrees of detail, some of the known methods for factoring large integers. A further method using elliptic curves is described in Sect. 6.6. Those readers interested in pursuing this subject might consult [28, 34, 109, 150] and the references cited in those works.

We begin with an algorithm called *Pollard's $p-1$ method.* Although not useful for all numbers, there are certain types of numbers for which it is quite efficient. Pollard's method demonstrates that there are insecure RSA moduli that at first glance appear to be secure. This alone warrants the study of Pollard's method. In addition, the $p-1$ method provides the inspiration for Lenstra's elliptic curve factorization method, which we study later, in Sect. 6.6.

We are presented with a number $N = pq$ and our task is to determine the prime factors p and q. Suppose that by luck or hard work or some other method, we manage to find an integer L with the property that

$$p-1 \text{ divides } L \qquad \text{and} \qquad q-1 \text{ does not divide } L.$$

This means that there are integers i, j, and k with $k \neq 0$ satisfying

$$L = i(p-1) \qquad \text{and} \qquad L = j(q-1)+k.$$

Consider what happens if we take a randomly chosen integer a and compute a^L. Fermat's little theorem (Theorem 1.24) tells us that[4]

$$\begin{aligned} a^L &= a^{i(p-1)} = (a^{p-1})^i \equiv 1^i \equiv 1 \pmod{p}, \\ a^L &= a^{j(q-1)+k} = a^k(a^{q-1})^j \equiv a^k \cdot 1^j \equiv a^k \pmod{q}. \end{aligned}$$

The exponent k is not equal to 0, so it is quite unlikely that a^k will be congruent to 1 modulo q. Thus with very high probability, i.e., for most choices of a, we find that

$$p \text{ divides } a^L - 1 \qquad \text{and} \qquad q \text{ does not divide } a^L - 1.$$

But this is wonderful, since it means that we can recover p via the simple gcd computation

$$p = \gcd(a^L - 1, N).$$

This is all well and good, but where, you may ask, can we find an exponent L that is divisible by $p-1$ and not by $q-1$? Pollard's observation is that if $p-1$ happens to be a product of many small primes, then it will divide $n!$ for some not-too-large value of n. So here is the idea. For each number $n = 2, 3, 4, \ldots$ we choose a value of a and compute

$$\gcd(a^{n!} - 1, N).$$

(In practice, we might simply take $a = 2$.) If the gcd is equal to 1, then we go on to the next value of n. If the gcd ever equals N, then we've been quite unlucky, but a different a value will probably work. And if we get a number strictly between 1 and N, then we have a nontrivial factor of N and we're done.

Remark 3.28. There are two important remarks to make before we put Pollard's idea into practice. The first concerns the quantity $a^{n!} - 1$. Even for $a = 2$ and quite moderate values of n, say $n = 100$, it is not feasible to compute $a^{n!} - 1$ exactly. Indeed, the number $2^{100!}$ has more than 10^{157} digits, which is larger than the number of elementary particles in the known universe! Luckily, there is no need to compute it exactly. We are interested only in the greatest common divisor of $a^{n!} - 1$ and N, so it suffices to compute

$$a^{n!} - 1 \pmod{N}$$

[4]We have assumed that $p \nmid a$ and $q \nmid a$, since if p and q are very large, this will almost certainly be the case. Further, if by some chance $p \mid a$ and $q \nmid a$, then we can recover p as $p = \gcd(a, N)$.

and then take the gcd with N. Thus we never need to work with numbers larger than N.

Second, we do not even need to compute the exponent $n!$. Instead, assuming that we have already computed $a^{n!}$ mod N in the previous step, we can compute the next value as

$$a^{(n+1)!} \equiv \left(a^{n!}\right)^{n+1} \pmod{N}.$$

This leads to the algorithm described in Table 3.3.

Remark 3.29. How long does it take to compute the value of $a^{n!}$ mod N? The fast exponentiation algorithm described in Sect. 1.3.2 gives a method for computing a^k mod N in at most $2\log_2 k$ steps, where each step is a multiplication modulo N. Stirling's formula[5] says that if n is large, then $n!$ is approximately equal to $(n/e)^n$. So we can compute $a^{n!}$ mod N in $2n\log_2(n)$ steps. Thus it is feasible to compute $a^{n!}$ mod N for reasonably large values of n.

Input. Integer N to be factored.

1. Set $a = 2$ (or some other convenient value).

2. Loop $j = 2, 3, 4, \ldots$ up to a specified bound.

  3. Set $a = a^j$ mod N.

  4. Compute $d = \gcd(a-1, N)^{\dagger}$.

  5. If $1 < d < N$ then **success**, return d.

6. Increment j and loop again at Step **2**.

† For added efficiency, choose an appropriate k and compute the gcd in Step **4** only every kth iteration.

Table 3.3: Pollard's $p-1$ factorization algorithm

Example 3.30. We use Pollard's $p-1$ method to factor $N = 13927189$. Starting with $\gcd(2^{9!}-1, N)$ and taking successively larger factorials in the exponent, we find that

$$\begin{array}{lll}
2^{9!}-1 \equiv 13867883 & \pmod{13927189}, & \gcd(2^{9!}-1, 13927189) = 1, \\
2^{10!}-1 \equiv 5129508 & \pmod{13927189}, & \gcd(2^{10!}-1, 13927189) = 1, \\
2^{11!}-1 \equiv 4405233 & \pmod{13927189}, & \gcd(2^{11!}-1, 13927189) = 1, \\
2^{12!}-1 \equiv 6680550 & \pmod{13927189}, & \gcd(2^{12!}-1, 13927189) = 1, \\
2^{13!}-1 \equiv 6161077 & \pmod{13927189}, & \gcd(2^{13!}-1, 13927189) = 1, \\
2^{14!}-1 \equiv 879290 & \pmod{13927189}, & \gcd(2^{14!}-1, 13927189) = 3823.
\end{array}$$

The final line gives us a nontrivial factor $p = 3823$ of N. This factor is prime, and the other factor $q = N/p = 13927189/3823 = 3643$ is also prime.

[5]Stirling's formula says more precisely that $\ln(n!) = n\ln(n) - n + \frac{1}{2}\ln(2\pi n) + O(1/n)$.

The reason that an exponent of 14! worked in this instance is that $p-1$ factors into a product of small primes,

$$p - 1 = 3822 = 2 \cdot 3 \cdot 7^2 \cdot 13.$$

The other factor satisfies $q - 1 = 3642 = 2 \cdot 3 \cdot 607$, which is not a product of small primes.

Example 3.31. We present one further example using larger numbers. Let $N = 168441398857$. Then

$$\begin{aligned}
2^{50!} - 1 &\equiv 114787431143 \pmod{N}, & \gcd(2^{50!} - 1, N) &= 1,\\
2^{51!} - 1 &\equiv 36475745067 \pmod{N}, & \gcd(2^{51!} - 1, N) &= 1,\\
2^{52!} - 1 &\equiv 67210629098 \pmod{N}, & \gcd(2^{52!} - 1, N) &= 1,\\
2^{53!} - 1 &\equiv 8182353513 \pmod{N}, & \gcd(2^{53!} - 1, N) &= 350437.
\end{aligned}$$

So using $2^{53!} - 1$ yields the prime factor $p = 350437$ of N, and the other (prime) factor is 480661. We were lucky, of course, that $p-1$ is a product of small factors,

$$p - 1 = 350436 = 2^2 \cdot 3 \cdot 19 \cdot 29 \cdot 53.$$

Remark 3.32. Notice that it is easy for Bob and Alice to avoid the dangers of Pollard's $p-1$ method when creating RSA keys. They simply check that their chosen secret primes p and q have the property that neither $p-1$ nor $q-1$ factors entirely into small primes. From a cryptographic perspective, the importance of Pollard's method lies in the following lesson. Most people would not expect, at first glance, that factorization properties of $p-1$ and $q-1$ should have anything to do with the difficulty of factoring pq. The moral is that even if we build a cryptosystem based on a seemingly hard problem such as integer factorization, we must be wary of special cases of the problem that, for subtle and nonobvious reasons, are easier to solve than the general case. We have already seen an example of this in the Pohlig–Hellman algorithm for the discrete logarithm problem (Sect. 2.9), and we will see it again later when we discuss elliptic curves and the elliptic curve discrete logarithm problem.

Remark 3.33. We have not yet discussed the likelihood that Pollard's $p-1$ algorithm succeeds. Suppose that p and q are randomly chosen primes of about the same size. Pollard's method works if at least one of $p-1$ or $q-1$ factors entirely into a product of small prime powers. Clearly $p-1$ is even, so we can pull off a factor of 2, but after that, the quantity $\frac{1}{2}(p-1)$ should behave more or less like a random number of size approximately $\frac{1}{2}p$. This leads to the following question:

> What is the probability that a randomly chosen integer of size approximately n divides $B!$ (B-factorial)?

Notice in particular that if n divides $B!$, then every prime ℓ dividing n must satisfy $\ell \le B$. A number whose prime factors are all less than or equal to B

is called a *B-smooth number*. It is thus natural to ask for the probability that a randomly chosen integer of size approximately n is a B-smooth number. Turning this question around, we can also ask:

> Given n, how large should we choose B so that a randomly chosen integer of size approximately n has a reasonably good probability of being a B-smooth number?

The efficiency (or lack thereof) of all modern methods of integer factorization is largely determined by the answer to this question. We study smooth numbers in Sect. 3.7.

3.6 Factorization via Difference of Squares

The most powerful factorization methods known today rely on one of the simplest identities in all of mathematics,

$$X^2 - Y^2 = (X+Y)(X-Y). \tag{3.13}$$

This beautiful formula says that a difference of squares is equal to a product. The potential applicability to factorization is immediate. In order to factor a number N, we look for an integer b such that the quantity $N + b^2$ is a perfect square, say equal to a^2. Then $N + b^2 = a^2$, so

$$N = a^2 - b^2 = (a+b)(a-b),$$

and we have effected a factorization of N.

Example 3.34. We factor $N = 25217$ by looking for an integer b making $N + b^2$ a perfect square:

$$\begin{aligned}
25217 + 1^2 &= 25218 && \text{not a square,}\\
25217 + 2^2 &= 25221 && \text{not a square,}\\
25217 + 3^2 &= 25226 && \text{not a square,}\\
25217 + 4^2 &= 25233 && \text{not a square,}\\
25217 + 5^2 &= 25242 && \text{not a square,}\\
25217 + 6^2 &= 25253 && \text{not a square,}\\
25217 + 7^2 &= 25266 && \text{not a square,}\\
25217 + 8^2 &= 25281 = 159^2 && \text{Eureka! ** square **.}
\end{aligned}$$

Then we compute

$$25217 = 159^2 - 8^2 = (159+8)(159-8) = 167 \cdot 151.$$

If N is large, then it is unlikely that a randomly chosen value of b will make $N + b^2$ into a perfect square. We need to find a clever way to select b. An important observation is that we don't necessarily need to write N itself as a difference of two squares. It often suffices to write some multiple kN of N as a difference of two squares, since if

$$kN = a^2 - b^2 = (a + b)(a - b),$$

then there is a reasonable chance that the factors of N are separated by the right-hand side of the equation, i.e., that N has a nontrivial factor in common with each of $a + b$ and $a - b$. It is then a simple matter to recover the factors by computing $\gcd(N, a + b)$ and $\gcd(N, a - b)$. We illustrate with an example.

Example 3.35. We factor $N = 203299$. If we make a list of $N + b^2$ for values of $b = 1, 2, 3, \ldots$, say up to $b = 100$, we do not find any square values. So next we try listing the values of $3N + b^2$ and we find

$$\begin{array}{ll}
3 \cdot 203299 + 1^2 = 609898 & \text{not a square,} \\
3 \cdot 203299 + 2^2 = 609901 & \text{not a square,} \\
3 \cdot 203299 + 3^2 = 609906 & \text{not a square,} \\
3 \cdot 203299 + 4^2 = 609913 & \text{not a square,} \\
3 \cdot 203299 + 5^2 = 609922 & \text{not a square,} \\
3 \cdot 203299 + 6^2 = 609933 & \text{not a square,} \\
3 \cdot 203299 + 7^2 = 609946 & \text{not a square,} \\
3 \cdot 203299 + 8^2 = 609961 = 781^2 & \text{Eureka! ** square **.}
\end{array}$$

Thus

$$3 \cdot 203299 = 781^2 - 8^2 = (781 + 8)(781 - 8) = 789 \cdot 773,$$

so when we compute

$$\gcd(203299, 789) = 263 \quad \text{and} \quad \gcd(203299, 773) = 773,$$

we find nontrivial factors of N. The numbers 263 and 773 are prime, so the full factorization of N is $203299 = 263 \cdot 773$.

Remark 3.36. In Example 3.35, we made a list of values of $3N + b^2$. Why didn't we try $2N + b^2$ first? The answer is that if N is odd, then $2N + b^2$ can never be a square, so it would have been a waste of time to try it. The reason that $2N + b^2$ can never be a square is as follows (cf. Exercise 1.23). We compute modulo 4,

$$2N + b^2 \equiv 2 + b^2 \equiv \begin{cases} 2 + 0 \equiv 2 \pmod{4} & \text{if } b \text{ is even,} \\ 2 + 1 \equiv 3 \pmod{4} & \text{if } b \text{ is odd.} \end{cases}$$

Thus $2N + b^2$ is congruent to either 2 or 3 modulo 4. But squares are congruent to either 0 or 1 modulo 4. Hence if N is odd, then $2N + b^2$ is never a square.

The multiples of N are the numbers that are congruent to 0 modulo N, so rather than searching for a difference of squares $a^2 - b^2$ that is a multiple of N, we may instead search for distinct numbers a and b satisfying

$$a^2 \equiv b^2 \pmod{N}. \tag{3.14}$$

This is exactly the same problem, of course, but the use of modular arithmetic helps to clarify our task.

In practice it is not feasible to search directly for integers a and b satisfying (3.14). Instead we use a three-step process as described in Table 3.4. This procedure, in one form or another, underlies most modern methods of factorization.

1. **Relation Building**: Find many integers $a_1, a_2, a_3, \ldots, a_r$ with the property that the quantity $c_i \equiv a_i^2 \pmod{N}$ factors as a product of small primes.
2. **Elimination**: Take a product $c_{i_1} c_{i_2} \cdots c_{i_s}$ of some of the c_i's so that every prime appearing in the product appears to an even power. Then $c_{i_1} c_{i_2} \cdots c_{i_s} = b^2$ is a perfect square.
3. **GCD Computation**: Let $a = a_{i_1} a_{i_2} \cdots a_{i_s}$ and compute the greatest common divisor $d = \gcd(N, a - b)$. Since

$$a^2 = (a_{i_1} a_{i_2} \cdots a_{i_s})^2 \equiv a_{i_1}^2 a_{i_2}^2 \cdots a_{i_s}^2 \equiv c_{i_1} c_{i_2} \cdots c_{i_s} \equiv b^2 \pmod{N},$$

there is a reasonable chance that d is a nontrivial factor of N.

Table 3.4: A three step factorization procedure

Example 3.37. We factor $N = 914387$ using the procedure described in Table 3.4. We first search for integers a with the property that $a^2 \bmod N$ is a product of small primes. For this example, we ask that each $a^2 \bmod N$ be a product of primes in the set $\{2, 3, 5, 7, 11\}$. Ignoring for now the question of how to find such a, we observe that

$$\begin{aligned} 1869^2 &\equiv 750000 \pmod{914387} &\quad\text{and}\quad&& 750000 &= 2^4 \cdot 3 \cdot 5^6,\\ 1909^2 &\equiv 901120 \pmod{914387} &\quad\text{and}\quad&& 901120 &= 2^{14} \cdot 5 \cdot 11,\\ 3387^2 &\equiv 499125 \pmod{914387} &\quad\text{and}\quad&& 499125 &= 3 \cdot 5^3 \cdot 11^3. \end{aligned}$$

None of the numbers on the right is a square, but if we multiply them together, then we do get a square. Thus

$$\begin{aligned} 1869^2 \cdot 1909^2 \cdot 3387^2 &\equiv 750000 \cdot 901120 \cdot 499125 \pmod{914387}\\ &\equiv (2^4 \cdot 3 \cdot 5^6)(2^{14} \cdot 5 \cdot 11)(3 \cdot 5^3 \cdot 11^3) \pmod{914387} \end{aligned}$$

$$\begin{aligned} &= (2^9 \cdot 3 \cdot 5^5 \cdot 11^2)^2 \\ &= 580800000^2 \\ &\equiv 164255^2 \pmod{914387}. \end{aligned}$$

We further note that $1869 \cdot 1909 \cdot 3387 \equiv 9835 \pmod{914387}$, so we compute

$$\gcd(914387, 9835 - 164255) = \gcd(914387, 154420) = 1103.$$

Hooray! We have factored $914387 = 1103 \cdot 829$.

Example 3.38. We do a second example to illustrate a potential pitfall in this method. We will factor $N = 636683$. After some searching, we find

$$\begin{aligned} 1387^2 &\equiv 13720 \pmod{636683} \quad \text{and} \quad 13720 = 2^3 \cdot 5 \cdot 7^3, \\ 2774^2 &\equiv 54880 \pmod{636683} \quad \text{and} \quad 54880 = 2^5 \cdot 5 \cdot 7^3. \end{aligned}$$

Multiplying these two values gives a square,

$$1387^2 \cdot 2774^2 \equiv 13720 \cdot 54880 = (2^4 \cdot 5 \cdot 7^3)^2 = 27440^2.$$

Unfortunately, when we compute the gcd, we find that

$$\gcd(636683, 1387 \cdot 2774 - 27440) = \gcd(636683, 3820098) = 636683.$$

Thus after all our work, we have made no progress! However, all is not lost. We can gather more values of a and try to find a different relation. Extending the above list, we discover that

$$3359^2 \equiv 459270 \pmod{636683} \qquad \text{and} \qquad 459270 = 2 \cdot 3^8 \cdot 5 \cdot 7.$$

Multiplying 1387^2 and 3359^2 gives

$$1387^2 \cdot 3359^2 \equiv 13720 \cdot 459270 = (2^2 \cdot 3^4 \cdot 5 \cdot 7^2)^2 = 79380^2,$$

and now when we compute the gcd, we obtain

$$\gcd(636683, 1387 \cdot 3359 - 79380) = \gcd(636683, 4579553) = 787.$$

This gives the factorization $N = 787 \cdot 809$.

Remark 3.39. How many solutions to $a^2 \equiv b^2 \pmod{N}$ are we likely to try before we find a factor of N? The most difficult case occurs when $N = pq$ is a product of two primes that are of roughly the same size. (This is because the smallest prime factor is $\mathcal{O}(\sqrt{N})$, while in any other case the smallest prime factor will be $\mathcal{O}(N^\alpha)$, with $\alpha < 1/2$. As α decreases, the difficulty of factoring N decreases.) Suppose that we can find more or less random values of a and b satisfying $a^2 \equiv b^2 \pmod{N}$. What are our chances of finding a nontrivial factor of N when we compute $\gcd(a - b, N)$? We know that

$$(a-b)(a+b) = a^2 - b^2 = kN = kpq \qquad \text{for some value of } k.$$

The prime p must divide at least one of $a-b$ and $a+b$, and it has approximately equal probability of dividing each. Similarly for q. We win if $a-b$ is divisible by exactly one of p and q, which happens approximately 50 % of the time. Hence if we can actually generate random a's and b's satisfying $a^2 \equiv b^2 \pmod{N}$, then it won't take us long to find a factor of N. Of course this leaves us with the question of just how hard it is to find these a's and b's.

Having given a taste of the process through several examples, we now do a more systematic analysis. The factorization procedure described in Table 3.4 consists of three steps:

1.	Relation Building
2.	Elimination
3.	GCD Computation

There is really nothing to say about Step 3, since the Euclidean algorithm (Theorem 1.7) tells us how to efficiently compute $\gcd(N, a-b)$ in $O(\ln N)$ steps. On the other hand, there is so much to say about relation building that we postpone our discussion until Sect. 3.7. Finally, what of Step 2, the elimination step?

We suppose that each of the numbers $a_1, \ldots, a_r$ found in Step 1 has the property that $c_i \equiv a_i^2 \pmod{m}$ factors into a product of small primes—say that each c_i is a product of primes chosen from the set of the first t primes $\{p_1, p_2, p_3, \ldots, p_t\}$. This means that there are exponents e_{ij} such that

$$\begin{aligned} c_1 &= p_1^{e_{11}} p_2^{e_{12}} p_3^{e_{13}} \cdots p_t^{e_{1t}}, \\ c_2 &= p_1^{e_{21}} p_2^{e_{22}} p_3^{e_{23}} \cdots p_t^{e_{2t}}, \\ &\vdots \qquad\qquad \vdots \\ c_r &= p_1^{e_{r1}} p_2^{e_{r2}} p_3^{e_{r3}} \cdots p_t^{e_{rt}}. \end{aligned}$$

Our goal is to take a product of some of the c_i's in order to make each prime on the right-hand side of the equation appear to an even power. In other words, our problem reduces to finding $u_1, u_2, \ldots, u_r \in \{0, 1\}$ such that

$$c_1^{u_1} \cdot c_2^{u_2} \cdots c_r^{u_r} \quad \text{is a perfect square.}$$

Here we take $u_i = 1$ if we want to include c_i in the product, and we take $u_i = 0$ if we do not want to include c_i in the product.

Writing out the product in terms of the prime factorizations of $c_1, \ldots, c_r$ gives the rather messy expression

$$\begin{aligned} &c_1^{u_1} \cdot c_2^{u_2} \cdots c_r^{u_r} \\ &= (p_1^{e_{11}} p_2^{e_{12}} p_3^{e_{13}} \cdots p_t^{e_{1t}})^{u_1} \cdot (p_1^{e_{21}} p_2^{e_{22}} p_3^{e_{23}} \cdots p_t^{e_{2t}})^{u_2} \cdots (p_1^{e_{r1}} p_2^{e_{r2}} p_3^{e_{r3}} \cdots p_t^{e_{rt}})^{u_r} \end{aligned}$$

$$= p_1^{e_{11}u_1+e_{21}u_2+\cdots+e_{r1}u_r} \cdot p_2^{e_{12}u_1+e_{22}u_2+\cdots+e_{r2}u_r} \cdots p_t^{e_{1t}u_1+e_{2t}u_2+\cdots+e_{rt}u_r}. \tag{3.15}$$

You may find this clearer if it is written using summation and product notation,

$$\prod_{i=1}^{r} c_i^{u_i} = \prod_{j=1}^{t} p_j^{\sum_{i=1}^{r} e_{ij}u_i}. \tag{3.16}$$

In any case, our goal is to choose $u_1, \ldots, u_r$ such that all of the exponents in (3.15), or equivalently in (3.16), are even.

To recapitulate, we are given integers

$$e_{11}, e_{12}, \ldots, e_{1t}, e_{21}, e_{22}, \ldots, e_{2t}, \ldots, e_{r1}, e_{r2}, \ldots, e_{rt}$$

and we are searching for integers $u_1, u_2, \ldots, u_r$ such that

$$\begin{aligned} e_{11}u_1 + e_{21}u_2 + \cdots + e_{r1}u_r &\equiv 0 \pmod{2}, \\ e_{12}u_1 + e_{22}u_2 + \cdots + e_{r2}u_r &\equiv 0 \pmod{2}, \\ \vdots \qquad\qquad & \quad \vdots \\ e_{1t}u_1 + e_{2t}u_2 + \cdots + e_{rt}u_r &\equiv 0 \pmod{2}. \end{aligned} \tag{3.17}$$

You have undoubtedly recognized that the system of congruences (3.17) is simply a system of linear equations over the finite field $\mathbb{F}_2$. Hence standard techniques from linear algebra, such as Gaussian elimination, can be used to solve these equations. In fact, doing linear algebra in the field $\mathbb{F}_2$ is much easier than doing linear algebra in the field $\mathbb{R}$, since there is no need to worry about round-off errors.

Example 3.40. We illustrate the linear algebra elimination step by factoring the number

$$N = 9788111.$$

We look for numbers a with the property that $a^2 \bmod N$ is 50-smooth, i.e., numbers a such that $a^2 \bmod N$ is equal to a product of primes in the set

$$\{2, 3, 5, 7, 11, 13, 17, 19, 23, 29, 31, 37, 41, 43, 47\}.$$

The top part of Table 3.5 lists the 20 numbers $a_1, a_2, \ldots, a_{20}$ between 3129 and 4700 having this property,[6] together with the factorization of each

$$c_i \equiv a_i^2 \pmod{N}.$$

The bottom part of Table 3.5 translates the requirement that a product $c_1^{u_1} c_2^{u_2} \cdots c_{20}^{u_{20}}$ be a square into a system of linear equation for $(u_1, u_2, \ldots, u_{20})$ as described by (3.17). For notational convenience, we have written the system of linear equations in Table 3.5 in matrix form.

[6] Why do we start with $a = 3129$? The answer is that unless a^2 is larger than N, then there is no reduction modulo N in $a^2 \bmod N$, so we cannot hope to gain any information. The value 3129 comes from the fact that $\sqrt{N} = \sqrt{9788111} \approx 3128.6$.

The next step is to solve the system of linear equations in Table 3.5. This can be done by standard Gaussian elimination, always keeping in mind that all computations are done modulo 2. The set of solutions turns out to be an $\mathbb{F}_2$-vector space of dimension 8. A basis for the set of solutions is given by the following 8 vectors, where we have written the vectors horizontally, rather than vertically, in order to save space:

$$\begin{aligned}
\boldsymbol{v}_1 &= (0,0,1,0,1,0,0,0,1,0,0,0,0,0,0,0,0,0,0,0),\\
\boldsymbol{v}_2 &= (0,1,1,1,1,1,0,0,0,0,0,0,0,0,0,0,0,0,0,0),\\
\boldsymbol{v}_3 &= (0,0,1,1,0,0,0,0,0,1,0,0,0,0,0,0,0,0,0,0),\\
\boldsymbol{v}_4 &= (1,0,1,0,0,0,0,0,0,0,1,0,0,0,0,0,0,0,0,0),\\
\boldsymbol{v}_5 &= (1,0,1,0,1,0,1,0,0,0,0,0,0,1,0,0,0,0,0,0),\\
\boldsymbol{v}_6 &= (1,0,0,0,0,0,0,1,0,0,0,0,1,0,1,0,0,0,0,0),\\
\boldsymbol{v}_7 &= (1,0,0,0,0,0,1,1,0,0,0,0,1,0,0,0,0,0,1,0),\\
\boldsymbol{v}_8 &= (1,0,0,0,0,0,0,1,0,0,0,1,1,0,0,0,0,0,0,1).
\end{aligned}$$

Each of the vectors $\boldsymbol{v}_1, \ldots, \boldsymbol{v}_8$ gives a congruence $a^2 \equiv b^2 \pmod{N}$ that has the potential to provide a factorization of N. For example, $\boldsymbol{v}_1$ says that if we multiply the 3rd, 5th, and 9th numbers in the list at the top of Table 3.5, we will get a square, and indeed we find that

$$\begin{aligned}
&3131^2 \cdot 3174^2 \cdot 3481^2\\
&\quad\equiv (2 \cdot 5^2 \cdot 7 \cdot 43)(5 \cdot 11^3 \cdot 43)(2 \cdot 5^3 \cdot 7 \cdot 11^3) \pmod{9788111}\\
&\quad= (2 \cdot 5^3 \cdot 7 \cdot 11^3 \cdot 43)^2\\
&\quad= 100157750^2.
\end{aligned}$$

Next we compute

$$\gcd(9788111, 3131 \cdot 3174 \cdot 3481 - 100157750) = 9788111,$$

which gives back the original number N. This is unfortunate, but all is not lost, since we have seven more independent solutions to our system of linear equations. Trying each of them in turn, we list the results in Table 3.6.

Seven of the eight solutions to the system of linear equations yield no useful information about N, the resulting gcd being either 1 or N. However, one solution, listed in the penultimate box of Table 3.6, leads to a nontrivial factorization of N. Thus 2741 is a factor of N, and dividing by it we obtain $N = 9788111 = 2741 \cdot 3571$. Since both 2741 and 3571 are prime, this gives the complete factorization of N.

Remark 3.41. In order to factor a large number N, it may be necessary to use a set $\{p_1, p_2, p_3, \ldots, p_t\}$ containing hundreds of thousands, or even millions, of primes. Then the system (3.17) contains millions of linear equations, and even working in the field $\mathbb{F}_2$, it can be very difficult to solve general systems

$$
\begin{array}{llcl}
3129^2 \equiv 2530 & \pmod{9788111} & \text{and} & 2530 = 2 \cdot 5 \cdot 11 \cdot 23 \\
3130^2 \equiv 8789 & \pmod{9788111} & \text{and} & 8789 = 11 \cdot 17 \cdot 47 \\
3131^2 \equiv 15050 & \pmod{9788111} & \text{and} & 15050 = 2 \cdot 5^2 \cdot 7 \cdot 43 \\
3166^2 \equiv 235445 & \pmod{9788111} & \text{and} & 235445 = 5 \cdot 7^2 \cdot 31^2 \\
3174^2 \equiv 286165 & \pmod{9788111} & \text{and} & 286165 = 5 \cdot 11^3 \cdot 43 \\
3215^2 \equiv 548114 & \pmod{9788111} & \text{and} & 548114 = 2 \cdot 7^3 \cdot 17 \cdot 47 \\
3313^2 \equiv 1187858 & \pmod{9788111} & \text{and} & 1187858 = 2 \cdot 7^2 \cdot 17 \cdot 23 \cdot 31 \\
3449^2 \equiv 2107490 & \pmod{9788111} & \text{and} & 2107490 = 2 \cdot 5 \cdot 7^2 \cdot 11 \cdot 17 \cdot 23 \\
3481^2 \equiv 2329250 & \pmod{9788111} & \text{and} & 2329250 = 2 \cdot 5^3 \cdot 7 \cdot 11^3 \\
3561^2 \equiv 2892610 & \pmod{9788111} & \text{and} & 2892610 = 2 \cdot 5 \cdot 7 \cdot 31^2 \cdot 43 \\
4394^2 \equiv 9519125 & \pmod{9788111} & \text{and} & 9519125 = 5^3 \cdot 7 \cdot 11 \cdot 23 \cdot 43 \\
4425^2 \equiv 4403 & \pmod{9788111} & \text{and} & 4403 = 7 \cdot 17 \cdot 37 \\
4426^2 \equiv 13254 & \pmod{9788111} & \text{and} & 13254 = 2 \cdot 3 \cdot 47^2 \\
4432^2 \equiv 66402 & \pmod{9788111} & \text{and} & 66402 = 2 \cdot 3^2 \cdot 7 \cdot 17 \cdot 31 \\
4442^2 \equiv 155142 & \pmod{9788111} & \text{and} & 155142 = 2 \cdot 3^3 \cdot 13^2 \cdot 17 \\
4468^2 \equiv 386802 & \pmod{9788111} & \text{and} & 386802 = 2 \cdot 3^3 \cdot 13 \cdot 19 \cdot 29 \\
4551^2 \equiv 1135379 & \pmod{9788111} & \text{and} & 1135379 = 7^2 \cdot 17 \cdot 29 \cdot 47 \\
4595^2 \equiv 1537803 & \pmod{9788111} & \text{and} & 1537803 = 3^2 \cdot 17 \cdot 19 \cdot 23^2 \\
4651^2 \equiv 2055579 & \pmod{9788111} & \text{and} & 2055579 = 3 \cdot 23 \cdot 31^3 \\
4684^2 \equiv 2363634 & \pmod{9788111} & \text{and} & 2363634 = 2 \cdot 3^3 \cdot 7 \cdot 13^2 \cdot 37
\end{array}
$$

Relation gathering step

$$
\begin{pmatrix}
1&0&1&0&0&1&1&1&1&1&0&0&1&1&1&1&0&0&0&1\\
0&0&0&0&0&0&0&0&0&0&0&0&1&0&1&1&0&0&1&1\\
1&0&0&1&1&0&0&1&1&1&1&0&0&0&0&0&0&0&0&0\\
0&0&1&0&0&1&0&0&1&1&1&1&0&1&0&0&0&0&0&1\\
1&1&0&0&1&0&0&1&1&0&1&0&0&0&0&0&0&0&0&0\\
0&0&0&0&0&0&0&0&0&0&0&0&0&0&0&1&0&0&0&0\\
0&1&0&0&0&1&1&1&0&0&0&1&0&1&1&0&1&1&0&0\\
0&0&0&0&0&0&0&0&0&0&0&0&0&0&0&1&0&1&0&0\\
1&0&0&0&0&0&1&1&0&0&1&0&0&0&0&0&0&0&1&0\\
0&0&0&0&0&0&0&0&0&0&0&0&0&0&0&1&1&0&0&0\\
0&0&0&0&0&0&1&0&0&0&0&0&0&1&0&0&0&0&1&0\\
0&0&0&0&0&0&0&0&0&0&0&1&0&0&0&0&0&0&0&1\\
0&0&0&0&0&0&0&0&0&0&0&0&0&0&0&0&0&0&0&0\\
0&0&1&0&1&0&0&0&0&1&1&0&0&0&0&0&0&0&0&0\\
0&1&0&0&0&1&0&0&0&0&0&0&0&0&0&0&1&0&0&0
\end{pmatrix}
\begin{pmatrix}
u_1\\u_2\\u_3\\u_4\\u_5\\u_6\\u_7\\u_8\\u_9\\u_{10}\\u_{11}\\u_{12}\\u_{13}\\u_{14}\\u_{15}\\u_{16}\\u_{17}\\u_{18}\\u_{19}\\u_{20}
\end{pmatrix}
\equiv
\begin{pmatrix}
0\\0\\0\\0\\0\\0\\0\\0\\0\\0\\0\\0\\0\\0\\0
\end{pmatrix}
\pmod{2}
$$

Linear algebra elimination step

Table 3.5: Factorization of $N = 9788111$

$\boldsymbol{v}_1 = (0,0,1,0,1,0,0,0,1,0,0,0,0,0,0,0,0,0,0,0)$ $3131^2 \cdot 3174^2 \cdot 3481^2 \equiv (2 \cdot 5^3 \cdot 7 \cdot 11^3 \cdot 43)^2$ $= 100157750^2$ $\gcd(9788111, 3131 \cdot 3174 \cdot 3481 - 100157750) = 9788111$
$\boldsymbol{v}_2 = (0,1,1,1,1,1,0,0,0,0,0,0,0,0,0,0,0,0,0,0)$ $3130^2 \cdot 3131^2 \cdot 3166^2 \cdot 3174^2 \cdot 3215^2 \equiv (2 \cdot 5^2 \cdot 7^3 \cdot 11^2 \cdot 17 \cdot 31 \cdot 43 \cdot 47)^2$ $= 2210173785050^2$ $\gcd(9788111, 3130 \cdot 3131 \cdot 3166 \cdot 3174 \cdot 3215 - 2210173785050) = 1$
$\boldsymbol{v}_3 = (0,0,1,1,0,0,0,0,1,0,0,0,0,0,0,0,0,0,0,0)$ $3131^2 \cdot 3166^2 \cdot 3561^2 \equiv (2 \cdot 5^2 \cdot 7^2 \cdot 31^2 \cdot 43)^2$ $= 101241350^2$ $\gcd(9788111, 3131 \cdot 3166 \cdot 3561 - 101241350) = 9788111$
$\boldsymbol{v}_4 = (1,0,1,0,0,0,0,0,0,0,1,0,0,0,0,0,0,0,0,0)$ $3129^2 \cdot 3131^2 \cdot 4394^2 \equiv (2 \cdot 5^3 \cdot 7 \cdot 11 \cdot 23 \cdot 43)^2$ $= 19038250^2$ $\gcd(9788111, 3129 \cdot 3131 \cdot 4394 - 19038250) = 9788111$
$\boldsymbol{v}_5 = (1,0,1,0,1,0,1,0,0,0,0,0,0,1,0,0,0,0,0,0)$ $3129^2 \cdot 3131^2 \cdot 3174^2 \cdot 3313^2 \cdot 4432^2 \equiv (2^2 \cdot 3 \cdot 5^2 \cdot 7^2 \cdot 11^2 \cdot 17 \cdot 23 \cdot 31 \cdot 43)^2$ $= 927063776100^2$ $\gcd(9788111, 3129 \cdot 3131 \cdot 3174 \cdot 3313 \cdot 4432 - 927063776100) = 1$
$\boldsymbol{v}_6 = (1,0,0,0,0,0,0,1,0,0,0,0,1,0,1,0,0,0,0,0)$ $3129^2 \cdot 3449^2 \cdot 4426^2 \cdot 4442^2 \equiv (2^2 \cdot 3^2 \cdot 5 \cdot 7 \cdot 11 \cdot 13 \cdot 17 \cdot 23 \cdot 47)^2$ $= 3311167860^2$ $\gcd(9788111, 3129 \cdot 3449 \cdot 4426 \cdot 4442 - 3311167860) = 1$
$\boldsymbol{v}_7 = (1,0,0,0,0,0,1,1,0,0,0,0,1,0,0,0,0,0,1,0)$ $3129^2 \cdot 3313^2 \cdot 3449^2 \cdot 4426^2 \cdot 4651^2 \equiv (2^2 \cdot 3 \cdot 5 \cdot 7^2 \cdot 11 \cdot 17 \cdot 23^2 \cdot 31^2 \cdot 47)^2$ $= 13136082114540^2$ $\gcd(9788111, 3129 \cdot 3313 \cdot 3449 \cdot 4426 \cdot 4651 - 13136082114540) = \mathbf{2741}$
$\boldsymbol{v}_8 = (1,0,0,0,0,0,0,1,0,0,0,1,1,0,0,0,0,0,0,1)$ $3129^2 \cdot 3449^2 \cdot 4425^2 \cdot 4426^2 \cdot 4684^2 \equiv (2^2 \cdot 3^2 \cdot 5 \cdot 7^2 \cdot 11 \cdot 13 \cdot 17 \cdot 23 \cdot 37 \cdot 47)^2$ $= 857592475740^2$ $\gcd(9788111, 3129 \cdot 3449 \cdot 4425 \cdot 4426 \cdot 4684 - 857592475740) = 1$

Table 3.6: Factorization of $N = 9788111$; computation of gcds

of this size. However, it turns out that the systems of linear equations used in factorization are quite *sparse*, which means that most of their coefficients are zero. (This is plausible because if a number A is a product of primes smaller than B, then one expects A to be a product of approximately $\ln(A)/\ln(B)$ distinct primes.) There are special techniques for solving sparse systems of linear equations that are much more efficient than ordinary Gaussian elimination; see for example [31, 72].

3.7 Smooth Numbers, Sieves, and Building Relations for Factorization

In this section we describe the two fastest known methods for doing "hard" factorization problems, i.e., factoring numbers of the form $N = pq$, where p and q are primes of approximately the same order of magnitude. We begin with a discussion of smooth numbers, which form the essential tool for building relations. Next we describe in some detail the quadratic sieve, which is a fast method for finding the necessary smooth numbers. Finally, we briefly describe the number field sieve, which is similar to the quadratic sieve in that it provides a fast method for finding smooth numbers of a certain form. However, when N is extremely large, the number field sieve is much faster than the quadratic sieve, because by working in a ring larger than $\mathbb{Z}$, it uses smaller auxiliary numbers in its search for smooth numbers.

3.7.1 Smooth Numbers

The relation building step in the three step factorization procedure described in Table 3.4 requires us to find many integers with the property that $a^2 \bmod N$ factors as a product of small primes. As noted at the end of Sect. 3.5, these highly factorizable numbers have a name.

Definition. An integer n is called *B-smooth* if all of its prime factors are less than or equal to B.

Example 3.42. Here are the first few 5-smooth numbers and the first few numbers that are not 5-smooth:

$$\begin{aligned} \text{5-smooth:} \quad & 2, 3, 4, 5, 6, 8, 9, 10, 12, 15, 16, 18, 20, 24, 25, 27, 30, 32, 36, \ldots \\ \text{Not 5-smooth:} \quad & 7, 11, 13, 14, 17, 19, 21, 22, 23, 26, 28, 29, 31, 33, 34, 35, 37, \ldots \end{aligned}$$

Definition. The function $\psi(X, B)$ counts B-smooth numbers,

$$\psi(X, B) = \text{Number of } B\text{-smooth integers } n \text{ such that } 1 < n \le X.$$

For example,

$$\psi(25, 5) = 15,$$

since the 5-smooth numbers between 1 and 25 are the 15 numbers

$$2, 3, 4, 5, 6, 8, 9, 10, 12, 15, 16, 18, 20, 24, 25.$$

In order to evaluate the efficiency of the three step factorization method, we need to understand how $\psi(X, B)$ behaves for large values of X and B. It turns out that in order to obtain useful results, the quantities B and X must increase together in *just* the right way. An important theorem in this direction was proven by Canfield, Erdős, and Pomerance [24].

Theorem 3.43 (Canfield, Erdős, Pomerance). *Fix a number* $0 < \epsilon < \frac{1}{2}$*, and let* X *and* B *increase together while satisfying*

$$(\ln X)^{\epsilon} < \ln B < (\ln X)^{1-\epsilon}.$$

For notational convenience, we let

$$u = \frac{\ln X}{\ln B}.$$

Then the number of B*-smooth numbers less than* X *satisfies*

$$\psi(X, B) = X \cdot u^{-u(1+o(1))}.$$

Remark 3.44. We've used *little-o* notation here for the first time. The expression $o(1)$ denotes a function that tends to 0 as X tends to infinity. More generally, we write

$$f(X) = o\big(g(X)\big)$$

if the ratio $f(X)/g(X)$ tends to 0 as X tends to infinity. Note that this is different from the big-$\mathcal{O}$ notation introduced in Sect. 2.6, where recall that $f(X) = \mathcal{O}\big(g(X)\big)$ means that $f(X)$ is smaller than a multiple of $g(X)$.

The question remains of how we should choose B in terms of X. It turns out that the following curious-looking function $L(X)$ is what we will need:

$$L(X) = e^{\sqrt{(\ln X)(\ln \ln X)}}. \tag{3.18}$$

Then, as an immediate consequence of Theorem 3.43, we obtain a fundamental estimate for ψ.

Corollary 3.45. *For any fixed value of* c *with* $0 < c < 1$,

$$\psi\big(X, L(X)^c\big) = X \cdot L(X)^{-(1/2c)(1+o(1))} \qquad \text{as } X \to \infty.$$

Proof. Note that if $B = L(X)^c$ and if we take any $\epsilon < \frac{1}{2}$, then

$$\ln B = c \ln L(X) = c\sqrt{(\ln X)(\ln \ln X)}$$

satisfies $(\ln X)^{\epsilon} < \ln B < (\ln X)^{1-\epsilon}$. So we can apply Theorem 3.43 with

$$u = \frac{\ln X}{\ln B} = \frac{1}{c} \cdot \sqrt{\frac{\ln X}{\ln \ln X}}$$

to deduce that $\psi(X, L(X)^c) = X \cdot u^{-u(1+o(1))}$. It is easily checked (see Exercise 3.32) that this value of u satisfies

$$u^{-u(1+o(1))} = L(X)^{-(1/2c)(1+o(1))},$$

which completes the proof of the corollary. □

The function $L(X) = e^{\sqrt{(\ln X)(\ln \ln X)}}$ and other similar functions appear prominently in the theory of factorization due to their close relationship to the distribution of smooth numbers. It is thus important to understand how fast $L(X)$ grows as a function of X.

Recall that in Sect. 2.6 we defined big-$\mathcal{O}$ notation and used it to discuss the notions of polynomial, exponential, and subexponential running times. What this meant was that the number of steps required to solve a problem was, respectively, polynomial, exponential, and subexponential in the number of bits required to describe the problem. As a supplement to big-$\mathcal{O}$ notation, it is convenient to introduce two other ways of comparing the rate at which functions grow.

Definition (Order Notation). Let $f(X)$ and $g(X)$ be functions of X whose values are positive. Recall that we write

$$f(X) = \mathcal{O}\big(g(X)\big)$$

if there are positive constants c and C such that

$$f(X) \le cg(X) \qquad \text{for all } X \ge C.$$

Similarly, we say that f *is big-*Ω *of* g and write

$$f(X) = \Omega\big(g(X)\big)$$

if there are positive constants c and C such that[7]

$$f(X) \ge cg(X) \qquad \text{for all } X \ge C.$$

Finally, if f is both big-$\mathcal{O}$ and big-Ω of g, we say that f *is big-*Θ *of* g and write $f(X) = \Theta\big(g(X)\big)$.

[7]Note: Big-Ω notation as used by computer scientists and cryptographers does not mean the same thing as the big-Ω notation of mathematicians. In mathematics, especially in the field of analytic number theory, the expression $f(n) = \Omega\big(g(n)\big)$ means that there is a constant c such that there are infinitely many integers n such that $f(n) \ge cg(n)$. In this book we use the computer science definition.

Remark 3.46. In analytic number theory there is an alternative version of order notation that is quite intuitive. For functions $f(X)$ and $g(X)$, we write

$$\begin{aligned} f(X) &\ll g(X) && \text{if } f(X) = \mathcal{O}(g(X)), \\ f(X) &\gg g(X) && \text{if } f(X) = \Omega(g(X)), \\ f(X) &\gg\ll g(X) && \text{if } f(X) = \Theta(g(X)). \end{aligned}$$

The advantage of this notation is that it is transitive, just as the usual "greater than" and "less than" relations are transitive. For example, if $f \gg g$ and $g \gg h$, then $f \gg h$.

Definition. With this notation in place, a function $f(X)$ is said to grow *exponentially* if there are positive constants α and β such that

$$\Omega(X^\alpha) = f(X) = \mathcal{O}(X^\beta),$$

and it is said to grow *polynomially* if there are positive constants α and β such that

$$\Omega\big((\ln X)^\alpha\big) = f(X) = \mathcal{O}\big((\ln X)^\beta\big).$$

In the alternative notation of Remark 3.46, exponential growth and polynomial growth are written, respectively, as

$$X^\alpha \ll f(X) \ll X^\beta \qquad \text{and} \qquad (\ln X)^\alpha \ll f(X) \ll (\ln X)^\beta.$$

A function that falls in between these two categories is called *subexponential*. Thus $f(X)$ is subexponential if for every positive constant α, no matter how large, and for every positive constant β, no matter how small,

$$\Omega\big((\ln X)^\alpha\big) = f(X) = \mathcal{O}(X^\beta). \tag{3.19}$$

(In the alternative notation, this becomes $(\ln X)^\alpha \ll f(X) \ll X^\beta$.)

Note that there is a possibility for confusion, since these definitions do not correspond to the usual meaning of exponential and polynomial growth that one finds in calculus. What is really happening is that "exponential" and "polynomial" refer to growth rates in the number of bits that it takes to write down X, i.e., exponential or polynomial functions of $\log_2(X)$.

Remark 3.47. The function $L(X)$ falls into the subexponential category. We leave this for you to prove in Exercise 3.30. See Table 3.7 for a rough idea of how fast $L(X)$ grows as X increases.

Suppose that we attempt to factor N by searching for values $a^2 \pmod N$ that are B-smooth. In order to perform the linear equation elimination step, we need (at least) as many B-smooth numbers as there are primes less than B. We need this many because in the elimination step, the smooth numbers correspond to the variables, while the primes less than B correspond to the equations, and we need more variables than equations. In order to ensure that

X	$\ln L(X)$	$L(X)$
2^{100}	17.141	$2^{24.73}$
2^{250}	29.888	$2^{43.12}$
2^{500}	45.020	$2^{64.95}$
2^{1000}	67.335	$2^{97.14}$
2^{2000}	100.145	$2^{144.48}$

Table 3.7: The growth of $L(X) = e^{\sqrt{(\ln X)(\ln\ln X)}}$

this is the case, we thus need there to be at least $\pi(B)$ B-smooth numbers, where $\pi(B)$ is the number of primes up to B. It will turn out that we can take $B = L(N)^c$ for a suitable value of c. In the next proposition we use the prime number theorem (Theorem 3.21) and the formula for $\psi(X, L(X)^c)$ given in Corollary 3.45 to choose the smallest value of c that gives us some chance of factoring N using this method.

Proposition 3.48. *Let* $L(X) = e^{\sqrt{(\ln X)(\ln\ln X)}}$ *be as in Corollary 3.45, let* N *be a large integer, and set* $B = L(N)^{1/\sqrt{2}}$.

(a) *We expect to check approximately* $L(N)^{\sqrt{2}}$ *random numbers modulo* N *in order to find* $\pi(B)$ *numbers that are* B*-smooth.*

(b) *We expect to check approximately* $L(N)^{\sqrt{2}}$ *random numbers of the form* $a^2 \pmod{N}$ *in order to find enough* B*-smooth numbers to factor* N.

Hence the factorization procedure described in Table 3.4 should have a subexponential running time.

Proof. We already explained why (a) and (b) are equivalent, assuming that the numbers $a^2 \pmod{N}$ are sufficiently random. We now prove (a).

The probability that a randomly chosen number modulo N is B-smooth is $\psi(N, B)/N$. In order to find $\pi(B)$ numbers that are B-smooth, we need to check approximately

$$\frac{\pi(B)}{\psi(N,B)/N} \quad \text{numbers.} \tag{3.20}$$

We want to choose B so as to minimize this function, since checking numbers for smoothness is a time-consuming process.

Corollary 3.45 says that

$$\psi(N, L(N)^c)/N \approx L(N)^{-1/2c},$$

so we set $B = L(N)^c$ and search for the value of c that minimizes (3.20). The prime number theorem (Theorem 3.21) tells us that $\pi(B) \approx B/\ln(B)$, so (3.20) is equal to

$$\frac{\pi(L(N)^c)}{\psi(N, L(N)^c)/N} \approx \frac{L(N)^c}{c \ln L(N)} \cdot \frac{1}{L(N)^{-1/2c}} = L(N)^{c+1/2c} \cdot \frac{1}{c \ln L(N)}.$$

The factor $L(N)^{c+1/2c}$ dominates this last expression, so we choose the value of c that minimizes the quantity $c + \frac{1}{2c}$. This is an elementary calculus problem. It is minimized when $c = \frac{1}{\sqrt{2}}$, and the minimum value is $\sqrt{2}$. Thus if we choose $B \approx L(N)^{1/\sqrt{2}}$, then we need to check approximately $L(N)^{\sqrt{2}}$ values in order to find $\pi(B)$ numbers that are B-smooth, and hence to find enough relations to factor N. □

Remark 3.49. Proposition 3.48 suggests that we need to check approximately $L(N)^{\sqrt{2}}$ randomly chosen numbers modulo N in order to find enough smooth numbers to factor N. There are various ways to decrease the search time. In particular, rather than using random values of a to compute numbers of the form $a^2 \pmod{N}$, we might instead select numbers a that are only a little bit larger than $\sqrt{N}$. Then $a^2 \pmod{N}$ is $\mathcal{O}(\sqrt{N}\,)$, so is more likely to be B-smooth than is a number that is $\mathcal{O}(N)$. Reworking the calculation in Proposition 3.48, one finds that it suffices to check approximately $L(N)$ random numbers of the form $a^2 \pmod{N}$ with a close to $\sqrt{N}$. This is a significant savings over $L(N)^{\sqrt{2}}$. See Exercise 3.33 for further details.

Remark 3.50. When estimating the effort needed to factor N, we have completely ignored the work required to check whether a given number is B-smooth. For example, if we check for B-smoothness using trial division, i.e., dividing by each prime less than B, then it takes approximately $\pi(B)$ trial divisions to check for B-smoothness. Taking this additional effort into account in the proof of Proposition 3.48, one finds that it takes approximately $L(N)^{\sqrt{2}}$ trial divisions to find enough smooth numbers to factor N, even using values of $a \approx \sqrt{N}$ as in Remark 3.49.

The quadratic sieve, which we describe in Sect. 3.7.2, uses a more efficient method for generating B-smooth numbers and thereby reduces the running time down to $L(N)$. (See Table 3.7 for a reminder of how $L(N)$ grows and why a running time of $L(N)$ is much better than a running time of $L(N)^{\sqrt{2}}$.) In Exercise 3.29 we ask you to estimate how long it takes to perform $L(N)$ operations on a moderately fast computer. For a number of years it was thought that no factorization algorithm could take fewer than a fixed power of $L(N)$ steps, but the invention of the number field sieve (Sect. 3.7.3) showed this to be incorrect. The number field sieve, whose running time of $e^{c\sqrt[3]{(\ln N)(\ln\ln N)^2}}$ is faster than $L(N)^\epsilon$ for every $\epsilon > 0$, achieves its speed by moving beyond the realm of the ordinary integers.

3.7.2 The Quadratic Sieve

In this section we address the final piece of the puzzle that must be solved in order to factor large numbers via the difference of squares method described in Sect. 3.6:

> How can we efficiently find many numbers $a > \sqrt{N}$ such that each $a^2 \pmod{N}$ is B-smooth?

From the discussion in Sect. 3.7.1 and the proof of Proposition 3.48, we know that we need to take $B \approx L(N)^{1/\sqrt{2}}$ in order to have a reasonable chance of factoring N.

An early approach to finding B-smooth squares modulo N was to look for fractions $\frac{a}{b}$ that are as close as possible to $\sqrt{kN}$ for $k = 1, 2, 3, \ldots$. Then

$$a^2 \approx b^2 kN,$$

so $a^2 \pmod{N}$ is reasonably small, and thus is more likely to be B-smooth. The theory of *continued fractions* gives an algorithm for finding such $\frac{a}{b}$. See [28, §10.1] for details.

An alternative approach that turns out to be much faster in practice is to allow slightly larger values of a and to use an efficient cancellation process called a *sieve* to simultaneously create a large number of values $a^2 \pmod{N}$ that are B-smooth. We next describe Pomerance's *quadratic sieve*, which is still the fastest known method for factoring large numbers $N = pq$ up to about 2^{350}. For numbers considerably larger than this, say larger than 2^{450}, the more complicated *number field sieve* holds the world record for quickest factorization. In the remainder of this section we describe the simplest version of the quadratic sieve as an illustration of modern factorization methods. For a description of the history of sieve methods and an overview of how they work, see Pomerance's delightful essay "A Tale of Two Sieves" [105].

We start with the simpler problem of rapidly finding many B-smooth numbers less than some bound X, without worrying whether the numbers have the form $a^2 \pmod{N}$. To do this, we adapt the *Sieve of Eratosthenes*, which is an ancient Greek method for making lists of prime numbers. Eratosthenes' idea for finding primes is as follows. Start by circling the first prime 2 and crossing off every larger multiple of 2. Then circle the next number, 3 (which must be prime) and cross off every larger multiple of 3. The smallest uncircled number is 5, so circle 5 and cross off all larger multiples of 5, and so on. At the end, the circled numbers are the primes.

This sieving process is illustrated in Fig. 3.2, where we have sieved all primes less than 10. (These are the boxed primes in the figure.) The remaining uncrossed numbers in the list are all remaining primes smaller than 100.

	[2]	[3]	~~4~~	[5]	~~6~~	[7]	~~8~~	~~9~~	~~10~~	11	~~12~~	13	~~14~~	~~15~~	~~16~~	17	~~18~~	19	~~20~~
~~21~~	~~22~~	23	~~24~~	~~25~~	~~26~~	~~27~~	~~28~~	29	~~30~~	31	~~32~~	~~33~~	~~34~~	~~35~~	~~36~~	37	~~38~~	~~39~~	~~40~~
41	~~42~~	43	~~44~~	~~45~~	~~46~~	47	~~48~~	~~49~~	~~50~~	~~51~~	~~52~~	53	~~54~~	~~55~~	~~56~~	~~57~~	~~58~~	59	~~60~~
61	~~62~~	~~63~~	~~64~~	~~65~~	~~66~~	67	~~68~~	~~69~~	~~70~~	71	~~72~~	73	~~74~~	~~75~~	~~76~~	~~77~~	~~78~~	79	~~80~~
~~81~~	~~82~~	83	~~84~~	~~85~~	~~86~~	~~87~~	~~88~~	89	~~90~~	~~91~~	~~92~~	~~93~~	~~94~~	~~95~~	~~96~~	97	~~98~~	~~99~~	

Figure 3.2: The sieve of Eratosthenes

Notice that some numbers are crossed off several times. For example, 6, 12 and 18 are crossed off twice, once because they are multiples of 2 and once because they are multiples of 3. Similarly, numbers such as 30 and 42 are crossed off three times. Suppose that rather than crossing numbers off, we instead divide. That is, we begin by dividing every even number by 2, then we divide every multiple of 3 by 3, then we divide every multiple of 5 by 5, and so on. If we do this for all primes less than B, which numbers end up being divided all the way down to 1? The answer is that these are the numbers that are a product of distinct primes less than B; in particular, they are B-smooth! So we end up with a list of many B-smooth numbers.

Unfortunately, we miss some B-smooth numbers, namely those divisible by powers of small primes, but it is easy to remedy this problem by sieving with prime powers. Thus after sieving by 3, rather than proceeding to 5, we first sieve by 4. To do this, we cancel an additional factor of 2 from every multiple of 4. (Notice that we've already canceled 2 from these numbers, since they are even, so we can cancel only one additional factor of 2.) If we do this, then at the end, the B-smooth numbers less than X are precisely the numbers that have been reduced to 1. One can show that the total number of divisions required is approximately $X \ln(\ln(B))$. The double logarithm function $\ln(\ln(B))$ grows extremely slowly, so the average number of divisions required to check each individual number for smoothness is approximately constant.

However, our goal is not to make a list of numbers from 1 to X that are B-smooth. What we need is a list of numbers of the form $a^2 \pmod{N}$ that are B-smooth. Our strategy for accomplishing this uses the polynomial

$$F(T) = T^2 - N.$$

We want to start with a value of a that is slightly larger than $\sqrt{N}$, so we set

$$a = \lfloor \sqrt{N} \rfloor + 1,$$

where $\lfloor x \rfloor$ denotes, as usual, the greatest integer less than or equal to x. We then look at the list of numbers

$$F(a), F(a+1), F(a+2), \ldots, F(b). \tag{3.21}$$

The idea is to find the B-smooth numbers in this list by sieving away the primes smaller than B and seeing which numbers in the list get sieved all the way down to 1. We choose B sufficiently large so that, by the end of the sieving process, we are likely to have found enough B-smooth numbers to factor N. The following definition is useful in describing this process.

Definition. The set of primes less than B (or sometimes the set of prime powers less than B) is called the *factor base*.

Suppose that p is a prime in our factor base. Which of the numbers in the list (3.21) are divisible by p? Equivalently, which numbers t between a and b satisfy

$$t^2 \equiv N \pmod{p}? \tag{3.22}$$

If the congruence (3.22) has no solutions, then we discard the prime p, since p divides none of the numbers in the list (3.21). Otherwise the congruence (3.22) has two solutions (see Exercise 1.36 on page 55), which we denote by

$$t = \alpha_p \quad \text{and} \quad t = \beta_p.$$

(If $p = 2$, there is only one solution α_p.) It follows that each of the numbers

$$F(\alpha_p), F(\alpha_p + p), F(\alpha_p + 2p), F(\alpha_p + 3p), \ldots$$

and each of the numbers

$$F(\beta_p), F(\beta_p + p), F(\beta_p + 2p), F(\beta_p + 3p), \ldots$$

is divisible by p. Thus we can sieve away a factor of p from every pth entry in the list (3.21), starting with the smallest a value satisfying $a \equiv \alpha_p \pmod{p}$, and similarly we can sieve away a factor of p from every pth entry in the list (3.21), starting with the smallest a value satisfying $a \equiv \beta_p \pmod{p}$.

Example 3.51. We illustrate the quadratic sieve applied to the composite number $N = 221$. The smallest number whose square is larger than N is $a = \lfloor\sqrt{221}\rfloor + 1 = 15$. We set

$$F(T) = T^2 - 221$$

and sieve the numbers from $F(15) = 4$ up to $F(30) = 679$ using successively the prime powers from 2 to 7. The initial list of numbers $T^2 - N$ is[8]

4 35 68 103 140 179 220 263 308 355 404 455 508 563 620 679.

We first sieve by $p = 2$, which means that we cancel 2 from every second entry in the list. This gives

4	35	68	103	140	179	220	263	308	355	404	455	508	563	620	679
$\downarrow^2$		$\downarrow^2$		$\downarrow^2$		$\downarrow^2$		$\downarrow^2$		$\downarrow^2$		$\downarrow^2$		$\downarrow^2$	
2	35	34	103	70	179	110	263	154	355	202	455	254	563	310	679

Next we sieve by $p = 3$. However, it turns out that the congruence

$$t^2 \equiv 221 \equiv 2 \pmod{3}$$

has no solutions, so none of the entries in our list are divisible by 3.

[8]In practice when N is large, the t values used in the quadratic sieve are close enough to $\sqrt{N}$ that the value of $t^2 - N$ is between 1 and N. For our small numerical example, this is not the case, so it would be more efficient to reduce our values of t^2 modulo N, rather than merely subtracting N from t^2. However, since our aim is illumination, not efficiency, we will pretend that there is no advantage to subtracting additional multiples of N from $t^2 - N$.

We move on to the prime power 2^2. Every odd number is a solution of the congruence

$$t^2 \equiv 221 \equiv 1 \pmod 4,$$

which means that we can sieve another factor of 2 from every second entry in our list. We put a small 4 next to the sieving arrows to indicate that in this step we are sieving by 4, although we cancel only a factor of 2 from each entry.

2	35	34	103	70	179	110	263	154	355	202	455	254	563	310	679
$\downarrow_4$		$\downarrow_4$		$\downarrow_4$		$\downarrow_4$		$\downarrow_4$		$\downarrow_4$		$\downarrow_4$		$\downarrow_4$	
1	35	17	103	35	179	55	263	77	355	101	455	127	563	155	679

Next we move on to $p = 5$. The congruence

$$t^2 \equiv 221 \equiv 1 \pmod 5$$

has two solutions, $\alpha_5 = 1$ and $\beta_5 = 4$ modulo 5. The first t value in our list that is congruent to 1 modulo 5 is $t = 16$, so starting with $F(16)$, we find that every fifth entry is divisible by 5. Sieving out these factors of 5 gives

1	35	17	103	35	179	55	263	77	355	101	455	127	563	155	679
	$\downarrow_5$					$\downarrow_5$					$\downarrow_5$				
1	7	17	103	35	179	11	263	77	355	101	91	127	563	155	679

Similarly, every fifth entry starting with $F(19)$ is divisible by 5, so we sieve out those factors

1	7	17	103	35	179	11	263	77	355	101	91	127	563	155	679
				$\downarrow_5$					$\downarrow_5$					$\downarrow_5$	
1	7	17	103	7	179	11	263	77	71	101	91	127	563	31	679

To conclude our example, we sieve the prime $p = 7$. The congruence

$$t^2 \equiv 221 = 4 \pmod 7$$

has the two solutions $\alpha_7 = 2$ and $\beta_7 = 5$. We can thus sieve 7 away from every seventh entry starting with $F(16)$, and also every seventh entry starting with $F(19)$. This yields

1	7	17	103	7	179	11	263	77	71	101	91	127	563	31	679
	$\downarrow_7$							$\downarrow_7$							$\downarrow_7$
1	1	17	103	7	179	11	263	11	71	101	91	127	563	31	97
				$\downarrow_7$							$\downarrow_7$				
1	1	17	103	1	179	11	263	11	71	101	13	127	563	31	97

Notice that the original entries

$$F(15) = 4, \quad F(16) = 35, \quad \text{and} \quad F(19) = 140$$

have been sieved all the way down to 1. This tells us that

$$F(15) = 15^2 - 221, \qquad F(16) = 16^2 - 221, \qquad \text{and} \qquad F(19) = 19^2 - 221$$

are each a product of small primes, so we have discovered several squares modulo 221 that are products of small primes:

$$\begin{aligned} 15^2 &\equiv 2^2 && \pmod{221}, \\ 16^2 &\equiv 5 \cdot 7 && \pmod{221}, \\ 19^2 &\equiv 2^2 \cdot 5 \cdot 7 && \pmod{221}. \end{aligned} \tag{3.23}$$

We can use the congruences (3.23) to obtain various relations between squares. For example,

$$(16 \cdot 19)^2 \equiv (2 \cdot 5 \cdot 7)^2 \pmod{221}.$$

Computing

$$\gcd(221, 16 \cdot 19 - 2 \cdot 5 \cdot 7) = \gcd(221, 234) = 13$$

gives a nontrivial factor of 221.[9]

We have successfully factored $N = 221$, but to illustrate the sieving process further, we continue sieving up to $B = 11$. The next prime power to sieve is 3^2. However, the fact that $t^2 \equiv 221 \pmod{3}$ has no solutions means that $t^2 \equiv 221 \pmod{9}$ also has no solutions, so we move on to the prime $p = 11$.

The congruence $t^2 \equiv 221 \equiv 1 \pmod{11}$ has the solutions $\alpha_{11} = 1$ and $\beta_{11} = 10$, which allows us to sieve a factor of 11 from $F(23)$ and from $F(21)$. We recapitulate the entire sieving process in Fig. 3.3, where the top row gives values of t and the subsequent rows sieve the values of $F(t) = t^2 - 221$ using prime powers up to 11.

Notice that two more entries, $F(21)$ and $F(23)$, have been sieved down to 1, which gives us two additional relations

$$F(21) \equiv 21^2 \equiv 2^2 \cdot 5 \cdot 11 \pmod{221} \quad \text{and} \quad F(23) \equiv 23^2 \equiv 2^2 \cdot 7 \cdot 11 \pmod{221}.$$

We can combine these relations with the earlier relations (3.23) to obtain new square equalities, for example

$$(19 \cdot 21 \cdot 23)^2 \equiv (2^3 \cdot 5 \cdot 7 \cdot 11)^2 \pmod{221}.$$

These give another way to factor 221:

$$\gcd(221, 19 \cdot 21 \cdot 23 - 2^3 \cdot 5 \cdot 7) = \gcd(221, 6097) = 13.$$

[9]Looking back at the congruences (3.23), you may have noticed that it is even easier to use the fact that 15^2 is itself congruent to a square modulo 221, yielding $\gcd(15 - 2, 221) = 13$. In practice, the true power of the quadratic sieve appears only when it is applied to numbers much too large to use in a textbook example.

Remark 3.52. If p is an odd prime, then the congruence $t^2 \equiv N \pmod{p}$ has either 0 or 2 solutions modulo p. More generally, congruences

$$t^2 \equiv N \pmod{p^e}$$

modulo powers of p have either 0 or 2 solutions. (See Exercises 1.36 and 1.37.) This makes sieving odd prime powers relatively straightforward. Sieving with

15	16	17	18	19	20	21	22	23	24	25	26	27	28	29	30
4	35	68	103	140	179	220	263	308	355	404	455	508	563	620	679
↓2		↓2		↓2		↓2		↓2		↓2		↓2		↓2	
2	35	34	103	70	179	110	263	154	355	202	455	254	563	310	679
↓4		↓4		↓4		↓4		↓4		↓4		↓4		↓4	
1	35	17	103	35	179	55	263	77	355	101	455	127	563	155	679
	↓5					↓5					↓5				
1	7	17	103	35	179	11	263	77	355	101	91	127	563	155	679
				↓5					↓5					↓5	
1	7	17	103	7	179	11	263	77	71	101	91	127	563	31	679
	↓7							↓7							↓7
1	1	17	103	7	179	11	263	11	71	101	91	127	563	31	97
				↓7							↓7				
1	1	17	103	1	179	11	263	11	71	101	13	127	563	31	97
								↓11							
1	1	17	103	1	179	11	263	1	71	101	13	127	563	31	97
						↓11									
1	1	17	103	1	179	1	263	1	71	101	13	127	563	31	97

Figure 3.3: Sieving $N = 221$ using prime powers up to $B = 11$

powers of 2 is a bit trickier, since the number of solutions may be different modulo 2, modulo 4, and modulo higher powers of 2. Further, there may be more than two solutions. For example, $t^2 \equiv N \pmod{8}$ has four different solutions modulo 8 if $N \equiv 1 \pmod{8}$. So although sieving powers of 2 is not intrinsically difficult, it must be dealt with as a special case.

Remark 3.53. There are many implementation ideas that can be used to greatly increase the practical speed of the quadratic sieve. Although the running time of the sieve remains a constant multiple of $L(N)$, the multiple can be significantly reduced.

A time-consuming part of the sieve is the necessity of dividing every pth entry by p, since if the numbers are large, division by p is moderately complicated. Of course, computers perform division quite rapidly, but the sieving process requires approximately $L(N)$ divisions, so anything that decreases this time will have an immediate effect. A key idea to speed up this step is to use approximate logarithms, which allows the slower division operations to be replaced by faster subtraction operations.

We explain the basic idea. Instead of using the list of values

$$F(a),\ F(a+1),\ F(a+2),\ \ldots,$$

we use a list of integer values that are approximately equal to

$$\log F(a),\ \log F(a+1),\ \log F(a+2),\ \log F(a+3),\ \ldots.$$

In order to sieve p from $F(t)$, we subtract an integer approximation of $\log p$ from the integer approximation to $\log F(t)$, since by the rule of logarithms,

$$\log F(t) - \log p = \log \frac{F(t)}{p}.$$

If we were to use exact values for the logarithms, then at the end of the sieving process, the entries that are reduced to 0 would be precisely the values of $F(t)$ that are B-smooth. However, since we use only approximate logarithm values, at the end we look for entries that have been reduced to a small number. Then we use division on only those few entries to find the ones that are actually B-smooth.

A second idea that can be used to speed the quadratic sieve is to use the polynomial $F(t) = t^2 - N$ only until t reaches a certain size, and then replace it with a new polynomial. For details of these two implementation ideas and many others, see for example [28, §10.4], [34], or [109] and the references that they list.

3.7.3 The Number Field Sieve

The *number field sieve* is a factorization method that works in a ring that is larger than the ordinary integers. The full details are very complicated, so in this section we are content to briefly explain some of the ideas that go into making the number field sieve the fastest known method for factoring large numbers of the form $N = pq$, where p and q are primes of approximately the same order of magnitude.

In order to factor N, we start by finding a nonzero integer m and an irreducible monic polynomial $f(x) \in \mathbb{Z}[x]$ of small degree satisfying

$$f(m) \equiv 0 \pmod{N}.$$

Example 3.54. Suppose that we want to factor the number $N = 2^{2^9} + 1$. Then we could take $m = 2^{103}$ and $f(x) = x^5 + 8$, since

$$f(m) = f(2^{103}) = 2^{515} + 8 = 8(2^{512} + 1) \equiv 0 \pmod{2^{2^9} + 1}.$$

Let d be the degree of $f(x)$ and let β be a root of $f(x)$. (Note that β might be a complex number.) We will work in the ring

$$\mathbb{Z}[\beta] = \{c_0 + c_1\beta + c_2\beta^2 + \cdots + c_{d-1}\beta^{d-1} \in \mathbb{C} : c_0, c_1, \ldots, c_{d-1} \in \mathbb{Z}\}.$$

Note that although we have written $\mathbb{Z}[\beta]$ as a subring of the complex numbers, it isn't actually necessary to deal with real or complex numbers. We can work with $\mathbb{Z}[\beta]$ purely algebraically, since it is equal to the quotient ring $\mathbb{Z}[x]/(f(x))$. (See Sect. 2.10.2 for information about quotient rings.)

Example 3.55. We give an example to illustrate how one performs addition and multiplication in the ring $\mathbb{Z}[\beta]$. Let $f(x) = 1 + 3x - 2x^3 + x^4$, let β be a root of $f(x)$, and consider the ring $\mathbb{Z}[\beta]$. In order to add the elements

$$u = 2 - 4\beta + 7\beta^2 + 3\beta^3 \quad \text{and} \quad v = 1 + 2\beta - 4\beta^2 - 2\beta^3,$$

we simply add their coefficients,

$$u + v = 3 - 2\beta + 3\beta^2 + \beta^3.$$

Multiplication is a bit more complicated. First we multiply u and v, treating β as if it were a variable,

$$uv = 2 - 9\beta^2 + 29\beta^3 - 14\beta^4 - 26\beta^5 - 6\beta^6.$$

Then we divide by $f(\beta) = 1 + 3\beta - 2\beta^3 + \beta^4$, still treating β as a variable, and keep the remainder,

$$uv = 92 + 308\beta + 111\beta^2 - 133\beta^3 \in \mathbb{Z}[\beta].$$

The next step in the number field sieve is to find a large number of pairs of integers $(a_1, b_1), \ldots, (a_k, b_k)$ that simultaneously satisfy

$$\prod_{i=1}^{k}(a_i - b_i m) \text{ is a square in } \mathbb{Z} \quad \text{and} \quad \prod_{i=1}^{k}(a_i - b_i\beta) \text{ is a square in } \mathbb{Z}[\beta].$$

Thus there is an integer $A \in \mathbb{Z}$ and an element $\alpha \in \mathbb{Z}[\beta]$ such that

$$\prod_{i=1}^{k}(a_i - b_i m) = A^2 \quad \text{and} \quad \prod_{i=1}^{k}(a_i - b_i\beta) = \alpha^2. \tag{3.24}$$

By definition of $\mathbb{Z}[\beta]$, we can find an expression for α of the form

$$\alpha = c_0 + c_1\beta + c_2\beta^2 + \cdots + c_{d-1}\beta^{d-1} \quad \text{with } c_0, c_1, \ldots, c_{d-1} \in \mathbb{Z}. \tag{3.25}$$

Recall our original assumption $f(m) \equiv 0 \pmod{N}$. This means that we have

$$m \equiv \beta \pmod{N} \quad \text{in the ring } \mathbb{Z}[\beta].$$

So on the one hand, (3.24) becomes

$$A^2 \equiv \alpha^2 \pmod{N} \quad \text{in the ring } \mathbb{Z}[\beta],$$

while on the other hand, (3.25) becomes

$$\alpha \equiv c_0 + c_1 m + c_2 m^2 + \cdots + c_{d-1}m^{d-1} \pmod{N} \quad \text{in the ring } \mathbb{Z}[\beta].$$

Hence

$$A^2 \equiv (c_0 + c_1 m + c_2 m^2 + \cdots + c_{d-1} m^{d-1})^2 \pmod{N}.$$

Thus we have created a congruence $A^2 \equiv B^2 \pmod{N}$ that is valid in the ring of integers $\mathbb{Z}$, and as usual, there is then a good chance that $\gcd(A - B, N)$ will yield a nontrivial factor of N.

How do we find the (a_i, b_i) pairs to make both of the products (3.24) into squares? For the first product, we can use a sieve-type algorithm, similar to the method used in the quadratic sieve, to find values of $a - bm$ that are smooth, and then use linear algebra to find a subset with the desired property.

Pollard's idea is to simultaneously do something similar for the second product while working in the ring $\mathbb{Z}[\beta]$. Thus we look for pairs of integers (a, b) such that the quantity $a - b\beta$ is "smooth" in $\mathbb{Z}[\beta]$. There are many serious issues that arise when we try to do this, including the following:

1. The ring $\mathbb{Z}[\beta]$ usually does not have unique factorization of elements into primes or irreducible elements. So instead, we factor the ideal $(a - b\beta)$ into a product of prime ideals. We say that $a - b\beta$ is smooth if the prime ideals appearing in the factorization are small.

2. Unfortunately, even ideals in the ring $\mathbb{Z}[\beta]$ may not have unique factorization as a product of prime ideals. However, there is a slightly larger ring, called the ring of integers of $\mathbb{Q}(\beta)$, in which unique factorization of ideals is true.

3. Suppose that we have managed to make the ideal $(\prod(a_i - b_i\beta))$ into the square of an ideal in $\mathbb{Z}[\beta]$. There are two further problems. First, it need not be the square of an ideal generated by a single element. Second, even if it is equal to an ideal of the form $(\gamma)^2$, we can conclude only that $\prod(a - i - b_i\beta) = u\gamma^2$ for some unit $u \in \mathbb{Z}[\beta]^*$, and generally the ring $\mathbb{Z}[\beta]$ has infinitely many units.

It would take us too far afield to explain how to deal with these potential difficulties. Suffice it to say that through a number of ingenious ideas due to Adleman, Buhler, H. Lenstra, Pomerance, and others, the obstacles were overcome, leading to a practical factorization method. (See [105] for a nice overview of the number field sieve and some of the ideas used to turn it from a theoretical construction into a working algorithm.)

However, we will comment further on the first step in the algorithm. In order to get started, we need an integer m and a monic irreducible polynomial $f(x)$ of small degree such that $f(m) \equiv 0 \pmod{N}$. The trick is first to choose the desired degree d of f, next to choose an integer m satisfying

$$(N/2)^{1/d} < m < N^{1/d},$$

and then to write N as a number to the base m,

$$N = c_0 + c_1 m + c_2 m^2 + \cdots + c_{d-1} m^{d-1} + c_d m^d \qquad \text{with } 0 \le c_i < m.$$

The condition on m ensures that $c_d = 1$, so we can take f to be the monic polynomial

$$f(x) = c_0 + c_1x + c_2x^2 + \cdots + c_{d-1}x^{d-1} + x^d.$$

We also need $f(x)$ to be irreducible, but if $f(x)$ factors in $\mathbb{Z}[x]$, say $f(x) = g(x)h(x)$, then $N = f(m) = g(m)h(m)$ gives a factorization of N and we are done. So now we have an $f(x)$ and an m, which allows us to get started using the number field sieve.

There is no denying the fact that the number field sieve is much more complicated than the quadratic sieve. So why is it useful? The reason has to do with the size of the numbers that must be considered. Recall that for the quadratic sieve, we sieved to find smooth numbers of the form

$$\left(\lfloor\sqrt{N}\rfloor + k\right)^2 - N \quad \text{for } k = 1, 2, 3, \ldots.$$

So we needed to pick out the smooth numbers from a set of numbers whose size is a little larger than $\sqrt{N}$. For the number field sieve one ends up looking for smooth numbers of the form

$$(a - mb) \cdot b^d f(a/b), \tag{3.26}$$

and it turns out that by a judicious choice of m and f, these numbers are much smaller than $\sqrt{N}$. In order to describe how much smaller, we use a generalization of the subexponential function $L(N)$ that was so useful in describing the running time of the quadratic sieve.

Definition. For any $0 < \epsilon < 1$, we define the function

$$L_\epsilon(X) = e^{(\ln X)^\epsilon (\ln\ln X)^{1-\epsilon}}.$$

Notice that with this notation, the function $L(X)$ defined in Sect. 3.7.1 is $L_{1/2}(X)$.

Then one can show that the numbers (3.26) used by the number field sieve have size a small power of $L_{2/3}(N)$. To put this into perspective, the quadratic sieve works with numbers having approximately half as many digits as N, while the number field sieve uses numbers K satisfying

$$(\text{Number of digits of } K) \approx (\text{Number of digits of } N)^{2/3}.$$

This leads to a vastly improved running time for sufficiently large values of N.

Theorem 3.56. *Under some reasonable assumptions, the expected running time of the number field sieve to factor the number N is $L_{1/3}(N)^c$ for a small value of c.*

For general numbers, the best known value of c in Theorem 3.56 is a bit less than 2, while for special numbers such as $2^{2^9} + 1$ it is closer to 1.5. Of course, the number field sieve is sufficiently complicated that it becomes faster than other methods only when N is sufficiently large. As a practical matter, the quadratic sieve is faster for numbers smaller than 10^{100}, while the number field sieve is faster for numbers larger than 10^{130}.

3.8 The Index Calculus Method for Computing Discrete Logarithms in $\mathbb{F}_p$

The *index calculus* is a method for solving the discrete logarithm problem in a finite field $\mathbb{F}_p$. The algorithm uses smooth numbers and bears some similarity to the sieve methods that we have studied in this chapter, which is why we cover it here, rather than in Chap. 2, where we originally discussed discrete logarithms.

The idea behind the index calculus is fairly simple. We want to solve the discrete logarithm problem

$$g^x \equiv h \pmod{p}, \tag{3.27}$$

where the prime p and the integers g and h are given. For simplicity, we will assume that g is a primitive root modulo p, so its powers give all of $\mathbb{F}_p^*$.

Rather than solving (3.27) directly, we instead choose a value B and solve the discrete logarithm problem

$$g^x \equiv \ell \pmod{p} \qquad \text{for all primes } \ell \leq B.$$

In other words, we compute the discrete logarithm $\log_g(\ell)$ for every prime satisfying $\ell \leq B$.

Having done this, we next look at the quantities

$$h \cdot g^{-k} \pmod{p} \qquad \text{for } k = 1, 2, \ldots$$

until we find a value of k such that $h \cdot g^{-k} \pmod{p}$ is B-smooth. For this value of k we have

$$h \cdot g^{-k} \equiv \prod_{\ell \leq B} \ell^{e_\ell} \pmod{p} \tag{3.28}$$

for certain exponents e_ℓ. We rewrite (3.28) in terms of discrete logarithms as

$$\log_g(h) \equiv k + \sum_{\ell \leq B} e_\ell \cdot \log_g(\ell) \pmod{p-1}, \tag{3.29}$$

where recall that discrete logarithms are defined only modulo $p-1$. But we are assuming that we already computed $\log_g(\ell)$ for all primes $\ell \leq B$. Hence (3.29) gives the value of $\log_g(h)$.

It remains to explain how to find $\log_g(\ell)$ for small primes ℓ. Again the idea is simple. For a random selection of exponents i we compute

$$g_i \equiv g^i \pmod{p} \qquad \text{with } 0 < g_i < p.$$

If g_i is not B-smooth, then we discard it, while if g_i is B-smooth, then we can factor it as

$$g_i = \prod_{\ell \leq B} \ell^{u_\ell(i)}.$$

In terms of discrete logarithms, this gives the relation

$$i \equiv \log_g(g_i) \equiv \sum_{\ell \le B} u_\ell(i) \cdot \log_g(\ell) \pmod{p-1}. \tag{3.30}$$

Notice that the only unknown quantities in the formula (3.30) are the discrete logarithm values $\log_g(\ell)$. So if we can find more than $\pi(B)$ equations like (3.30), then we can use linear algebra to solve for the $\log_g(\ell)$ "variables."

This method of solving the discrete logarithm problem in $\mathbb{F}_p$ is called the *index calculus*, where recall from Sect. 2.2 that *index* is an older name for *discrete logarithm*. The index calculus first appears in work of Western and Miller [148] in 1968, so it predates by a few years the invention of public key cryptography. The method was independently rediscovered by several cryptographers in the 1970s after the publication of the Diffie–Hellman paper [38].

Remark 3.57. A minor issue that we have ignored is the fact that the linear equations (3.30) are congruences modulo $p-1$. Standard linear algebra methods such as Gaussian elimination do not work well modulo composite numbers, because there are numbers that do not have multiplicative inverses. The Chinese remainder theorem (Theorem 2.24) solves this problem. First we solve the congruences (3.30) modulo q for each prime q dividing $p-1$. Then, if q appears in the factorization of $p-1$ to a power q^e, we lift the solution from $\mathbb{Z}/q\mathbb{Z}$ to $\mathbb{Z}/q^e\mathbb{Z}$. Finally, we use the Chinese remainder theorem to combine solutions modulo prime powers to obtain a solution modulo $p-1$. In cryptographic applications one should choose p such that $p-1$ is divisible by a large prime; otherwise, the Pohlig–Hellman algorithm (Sect. 2.9) solves the discrete logarithm problem. For example, if we select $p = 2q+1$ with q prime, then the index calculus requires us to solve simultaneous congruences (3.30) modulo q and modulo 2.

There are many implementation issues that arise and tricks that have been developed in practical applications of the index calculus. We do not pursue these matters here, but are content to present a small numerical example illustrating how the index calculus works.

Example 3.58. We let p be the prime $p = 18443$ and use the index calculus to solve the discrete logarithm problem

$$37^x \equiv 211 \pmod{18443}.$$

We note that $g = 37$ is a primitive root modulo $p = 18443$ We take $B = 5$, so our factor base is the set of primes $\{2, 3, 5\}$. We start by taking random powers of $g = 37$ modulo 18443 and pick out the ones that are B-smooth. A couple of hundred attempts gives four equations:

$$\begin{aligned} g^{12708} &\equiv 2^3 \cdot 3^4 \cdot 5 \pmod{18443}, & g^{11311} &\equiv 2^3 \cdot 5^2 \pmod{18443}, \\ g^{15400} &\equiv 2^3 \cdot 3^3 \cdot 5 \pmod{18443}, & g^{2731} &\equiv 2^3 \cdot 3 \cdot 5^4 \pmod{18443}. \end{aligned} \tag{3.31}$$

These in turn give linear relations for the discrete logarithms of 2, 3, and 5 to the base g. For example, the first one says that

$$12708 = 3 \cdot \log_g(2) + 4 \cdot \log_g(3) + \log_g(5).$$

To ease notation, we let

$$x_2 = \log_g(2), \quad x_3 = \log_g(3), \quad \text{and} \quad x_5 = \log_g(5).$$

Then the four congruences (3.31) become the following four linear relations:

$$\begin{aligned} 12708 &= 3x_2 + 4x_3 + x_5 && \pmod{18442}, \\ 11311 &= 3x_2 + 2x_5 && \pmod{18442}, \\ 15400 &= 3x_2 + 3x_3 + x_5 && \pmod{18442}, \\ 2731 &= 3x_2 + x_3 + 4x_5 && \pmod{18442}. \end{aligned} \tag{3.32}$$

Note that the formulas (3.32) are congruences modulo

$$p - 1 = 18442 = 2 \cdot 9221,$$

since discrete logarithms are defined only modulo $p-1$. The number 9221 is prime, so we need to solve the system of linear equations (3.32) modulo 2 and modulo 9221. This is easily accomplished by Gaussian elimination, i.e., by adding multiples of one equation to another to eliminate variables. The solutions are

$$\begin{aligned} (x_2, x_3, x_5) &\equiv (1, 0, 1) \pmod{2}, \\ (x_2, x_3, x_5) &\equiv (5733, 6529, 6277) \pmod{9221}. \end{aligned}$$

Combining these solutions yields

$$(x_2, x_3, x_5) \equiv (5733, 15750, 6277) \pmod{18442}.$$

We check the solutions by computing

$$37^{5733} \equiv 2 \pmod{18443}, \quad 37^{15750} \equiv 3 \pmod{18443}, \quad 37^{6277} \equiv 5 \pmod{18443}.$$

Recall that our ultimate goal is to solve the discrete logarithm problem

$$37^x \equiv 211 \pmod{18443}.$$

We compute the value of $211 \cdot 37^{-k} \pmod{18443}$ for random values of k until we find a value that is B-smooth. After a few attempts we find that

$$211 \cdot 37^{-9549} \equiv 2^5 \cdot 3^2 \cdot 5^2 \pmod{18443}.$$

Using the values of the discrete logs of 2, 3, and 5 from above, this yields

$$\begin{aligned} \log_g(211) &= 9549 + 5\log_g(2) + 2\log_g(3) + 2\log_g(5) \\ &= 9549 + 5 \cdot 5733 + 2 \cdot 15750 + 2 \cdot 6277 \equiv 8500 \pmod{18442}. \end{aligned}$$

Finally, we check our answer $\log_g(211) = 8500$ by computing

$$37^{8500} \equiv 211 \pmod{18443}. \quad \checkmark$$

Remark 3.59. We can roughly estimate the running time of the index calculus as follows. Using a factor base consisting of primes less than B, we need to find approximately $\pi(B)$ numbers of the form $g^i \pmod{p}$ that are B-smooth. Proposition 3.48 suggests that we should take $B = L(p)^{1/\sqrt{2}}$, and then we will have to check approximately $L(p)^{\sqrt{2}}$ values of i. There is also the issue of checking each value to see whether it is B-smooth, but sieve-type methods can be used to speed the process. Further, using ideas based on the number field sieve, the running time can be further reduced to a small power $L_{1/3}(p)$. In any case, the index calculus is a subexponential algorithm for solving the discrete logarithm problem in $\mathbb{F}_p^*$. This stands in marked contrast to the discrete logarithm problem in elliptic curve groups, which we study in Chap. 6. Currently, the best known algorithms to solve the general discrete logarithm problem in elliptic curve groups are fully exponential.

3.9 Quadratic Residues and Quadratic Reciprocity

Let p be a prime number. Here is a simple mathematical question:

How can Bob tell whether a given number a is equal to a square modulo p?

For example, suppose that Alice asks Bob whether 181 is a square modulo 1223. One way for Bob to answer Alice's question is by constructing a table of squares modulo 1223 as illustrated in Table 3.8, but this is a lot of work, so he gave up after computing $96^2 \bmod 1223$. Alice picked up the computation where Bob stopped and eventually found that $437^2 \equiv 181 \pmod{1223}$. Thus the answer to her question is that 181 is indeed a square modulo 1223. Similarly, if Alice is sufficiently motivated to continue the table all the way up to $1222^2 \bmod 1223$, she can verify that the number 385 is not a square modulo 1223, because it does not appear in her table. (In fact, Alice can save half her time by computing only up to $611^2 \bmod 1223$, since a^2 and $(p-a)^2$ have the same values modulo p.)

Our goal in this section is to describe a more much efficient way to check if a number is a square modulo a prime. We begin with a definition.

Definition. Let p be an odd prime number and let a be a number with $p \nmid a$. We say that a is a *quadratic residue modulo* p if a is a square modulo p, i.e., if there is a number c so that $c^2 \equiv a \pmod{p}$. If a is not a square modulo p, i.e., if there exists no such c, then a is called a *quadratic nonresidue modulo* p.

Example 3.60. The numbers 968 and 1203 are both quadratic residues modulo 1223, since

$$453^2 \equiv 968 \pmod{1223} \qquad \text{and} \qquad 375^2 \equiv 1203 \pmod{1223}.$$

$1^2 \equiv 1$	$2^2 \equiv 4$	$3^2 \equiv 9$	$4^2 \equiv 16$	$5^2 \equiv 25$	$6^2 \equiv 36$	$7^2 \equiv 49$	$8^2 \equiv 64$	$9^2 \equiv 81$
$10^2 \equiv 100$	$11^2 \equiv 121$	$12^2 \equiv 144$	$13^2 \equiv 169$	$14^2 \equiv 196$	$15^2 \equiv 225$	$16^2 \equiv 256$	$17^2 \equiv 289$	$18^2 \equiv 324$
$19^2 \equiv 361$	$20^2 \equiv 400$	$21^2 \equiv 441$	$22^2 \equiv 484$	$23^2 \equiv 529$	$24^2 \equiv 576$	$25^2 \equiv 625$	$26^2 \equiv 676$	$27^2 \equiv 729$
$28^2 \equiv 784$	$29^2 \equiv 841$	$30^2 \equiv 900$	$31^2 \equiv 961$	$32^2 \equiv 1024$	$33^2 \equiv 1089$	$34^2 \equiv 1156$	$35^2 \equiv 2$	$36^2 \equiv 73$
$37^2 \equiv 146$	$38^2 \equiv 221$	$39^2 \equiv 298$	$40^2 \equiv 377$	$41^2 \equiv 458$	$42^2 \equiv 541$	$43^2 \equiv 626$	$44^2 \equiv 713$	$45^2 \equiv 802$
$46^2 \equiv 893$	$47^2 \equiv 986$	$48^2 \equiv 1081$	$49^2 \equiv 1178$	$50^2 \equiv 54$	$51^2 \equiv 155$	$52^2 \equiv 258$	$53^2 \equiv 363$	$54^2 \equiv 470$
$55^2 \equiv 579$	$56^2 \equiv 690$	$57^2 \equiv 803$	$58^2 \equiv 918$	$59^2 \equiv 1035$	$60^2 \equiv 1154$	$61^2 \equiv 52$	$62^2 \equiv 175$	$63^2 \equiv 300$
$64^2 \equiv 427$	$65^2 \equiv 556$	$66^2 \equiv 687$	$67^2 \equiv 820$	$68^2 \equiv 955$	$69^2 \equiv 1092$	$70^2 \equiv 8$	$71^2 \equiv 149$	$72^2 \equiv 292$
$73^2 \equiv 437$	$74^2 \equiv 584$	$75^2 \equiv 733$	$76^2 \equiv 884$	$77^2 \equiv 1037$	$78^2 \equiv 1192$	$79^2 \equiv 126$	$80^2 \equiv 285$	$81^2 \equiv 446$
$82^2 \equiv 609$	$83^2 \equiv 774$	$84^2 \equiv 941$	$85^2 \equiv 1110$	$86^2 \equiv 58$	$87^2 \equiv 231$	$88^2 \equiv 406$	$89^2 \equiv 583$	$90^2 \equiv 762$
$91^2 \equiv 943$	$92^2 \equiv 1126$	$93^2 \equiv 88$	$94^2 \equiv 275$	$95^2 \equiv 464$	$96^2 \equiv 655$	....	....	....

Table 3.8: Bob's table of squares modulo 1223

On the other hand, the numbers 209 and 888 are quadratic nonresidues modulo 1223, since the congruences

$$c^2 \equiv 209 \pmod{1223} \qquad \text{and} \qquad c^2 \equiv 888 \pmod{1223}$$

have no solutions.

The next proposition describes what happens when quadratic residues and nonresidues are multiplied together.

Proposition 3.61. *Let p be an odd prime number.*

(a) *The product of two quadratic residues modulo p is a quadratic residue modulo p.*

(b) *The product of a quadratic residue and a quadratic nonresidue modulo p is a quadratic nonresidue modulo p.*

(c) *The product of two quadratic nonresidues modulo p is a quadratic residue modulo p.*

Proof. It is easy to prove (a) and (b) directly from the definition of quadratic residue, but we use a different approach that gives all three parts simultaneously. Let g be a primitive root modulo p as described in Theorem 1.30. This means that the powers $1, g, g^2, \ldots, g^{p-2}$ are all distinct modulo p.

Which powers of g are quadratic residues modulo p? Certainly if $m = 2k$ is even, then $g^m = g^{2k} = (g^k)^2$ is a square.

On the other hand, let m be odd, say $m = 2k + 1$, and suppose that g^m is a quadratic residue, say $g^m \equiv c^2 \pmod{p}$. Fermat's little theorem (Theorem 1.24) tells us that

$$c^{p-1} \equiv 1 \pmod{p}.$$

However, $c^{p-1} \pmod{p}$ is also equal to

$$c^{p-1} \equiv (c^2)^{\frac{p-1}{2}} \equiv (g^m)^{\frac{p-1}{2}} \equiv (g^{2k+1})^{\frac{p-1}{2}} \equiv g^{k(p-1)} \cdot g^{\frac{p-1}{2}} \pmod{p}.$$

Another application of Fermat's little theorem tells us that

$$g^{k(p-1)} \equiv (g^{p-1})^k \equiv 1^k \equiv 1 \pmod{p},$$

so we find that

$$g^{\frac{p-1}{2}} \equiv 1 \pmod{p}.$$

This contradicts the fact that g is a primitive root, which proves that every odd power of g is a quadratic nonresidue.

We have proven an important dichotomy. If g is a primitive root modulo p, then

$$g^m \text{ is a } \begin{cases} \text{quadratic residue} & \text{if } m \text{ is even,} \\ \text{quadratic nonresidue} & \text{if } m \text{ is odd.} \end{cases}$$

It is now a simple matter to prove Proposition 3.61. In each case we write a and b as powers of g, multiply a and b by adding their exponents, and read off the result.

(a) Suppose that a and b are quadratic residues. Then $a = g^{2i}$ and $b = g^{2j}$, so $ab = g^{2(i+j)}$ has even exponent, and hence ab is a quadratic residue.

(b) Let a be a quadratic residue and let b be a nonresidue. Then $a = g^{2i}$ and $b = g^{2j+1}$, so $ab = g^{2(i+j)+1}$ has odd exponent, and hence ab is a quadratic nonresidue.

(c) Finally, let a and b both be nonresidues. Then $a = g^{2i+1}$ and $b = g^{2j+1}$, so $ab = g^{2(i+j+1)}$ has even exponent, and hence ab is a quadratic residue. □

If we write QR to denote a quadratic residue and NR to denote a quadratic nonresidue, then Proposition 3.61 may be succinctly summarized by the three equations

$$\mathsf{QR} \cdot \mathsf{QR} = \mathsf{QR}, \qquad \mathsf{QR} \cdot \mathsf{NR} = \mathsf{NR}, \qquad \mathsf{NR} \cdot \mathsf{NR} = \mathsf{QR}.$$

Do these equations look familiar? They resemble the rules for multiplying 1 and -1. This observation leads to the following definition.

Definition. Let p be an odd prime. The *Legendre symbol* of a is the quantity $\left(\frac{a}{p}\right)$ defined by the rules

$$\left(\frac{a}{p}\right) = \begin{cases} 1 & \text{if } a \text{ is a quadratic residue modulo } p, \\ -1 & \text{if } a \text{ is a quadratic nonresidue modulo } p, \\ 0 & \text{if } p \mid a. \end{cases}$$

With this definition, Proposition 3.61 is summarized by the simple multiplication rule[10]

[10]Proposition 3.61 deals only with the case that $p \nmid a$ and $p \nmid b$. But if p divides a or b, then p also divides ab, so both sides of (3.33) are zero.

$$\left(\frac{a}{p}\right)\left(\frac{b}{p}\right) = \left(\frac{ab}{p}\right). \tag{3.33}$$

We also make the obvious, but useful, observation that

$$\text{If} \quad a \equiv b \pmod{p}, \quad \text{then} \quad \left(\frac{a}{p}\right) = \left(\frac{b}{p}\right). \tag{3.34}$$

Thus in computing $\left(\frac{a}{p}\right)$, we may reduce a modulo p into the interval from 0 to $p-1$. It is worth adding a cautionary note: The notation for the Legendre symbol resembles a fraction, but it is *not* a fraction!

Returning to our original question of determining whether a given number is a square modulo p, the following beautiful and powerful theorem provides a method for determining the answer.

Theorem 3.62 (Quadratic Reciprocity). *Let p and q be odd primes.*

(a) $$\left(\frac{-1}{p}\right) = \begin{cases} 1 & \text{if } p \equiv 1 \pmod{4}, \\ -1 & \text{if } p \equiv 3 \pmod{4}. \end{cases}$$

(b) $$\left(\frac{2}{p}\right) = \begin{cases} 1 & \text{if } p \equiv 1 \text{ or } 7 \pmod{8}, \\ -1 & \text{if } p \equiv 3 \text{ or } 5 \pmod{8}. \end{cases}$$

(c) $$\left(\frac{p}{q}\right) = \begin{cases} \left(\frac{q}{p}\right) & \text{if } p \equiv 1 \pmod{4} \text{ or } q \equiv 1 \pmod{4}, \\ -\left(\frac{q}{p}\right) & \text{if } p \equiv 3 \pmod{4} \text{ and } q \equiv 3 \pmod{4}. \end{cases}$$

Proof. We do not give a proof of quadratic reciprocity, but you will find a proof in any introductory number theory textbook, such as [35, 52, 59, 100, 111]. □

The name "quadratic reciprocity" comes from property (c), which tells us how $\left(\frac{p}{q}\right)$ is related to its "reciprocal" $\left(\frac{q}{p}\right)$. It is worthwhile spending some time contemplating Theorem 3.62, because despite the simplicity of its statement, quadratic reciprocity is saying something quite unexpected and profound. The value of $\left(\frac{p}{q}\right)$ tells us whether p is a square modulo q. Similarly, $\left(\frac{q}{p}\right)$ tells us whether q is a square modulo p. There is no a priori reason to suspect that these questions should have anything to do with one another. Quadratic reciprocity tells us that they are intimately related, and indeed, related by a very simple rule.

Similarly, parts (a) and (b) of quadratic reciprocity give us some surprising information. The first part says that the question whether -1 is a square modulo p is answered by the congruence class of p modulo 4, and the second part says that question whether 2 is a square modulo p is answered by the congruence class of p modulo 8.

We indicated earlier that quadratic reciprocity can be used to determine whether a is a square modulo p. The way to apply quadratic reciprocity is to use (c) to repeatedly flip the Legendre symbol, where each time that we flip, we're allowed to reduce the top number modulo the bottom number. This leads to a rapid reduction in the size of the numbers, as illustrated by the following example.

Example 3.63. We check whether -15750 is a quadratic residue modulo 37907 using quadratic reciprocity to compute the Legendre symbol $\left(\frac{-15750}{37907}\right)$.

$$\begin{aligned}
\left(\frac{-15750}{37907}\right) &= \left(\frac{-1}{37907}\right)\left(\frac{15750}{37907}\right) && \text{Multiplication rule (3.33)}\\
&= -\left(\frac{15750}{37907}\right) && \text{Quadratic Reciprocity 3.62(a)}\\
&= -\left(\frac{2\cdot 3^2\cdot 5^3\cdot 7}{37907}\right) && \text{Factor 15750}\\
&= -\left(\frac{2}{37907}\right)\left(\frac{3}{37907}\right)^2\left(\frac{5}{37907}\right)^3\left(\frac{7}{37907}\right) && \\
& && \text{Multiplication rule (3.33)}\\
&= -\left(\frac{2}{37907}\right)\left(\frac{5}{37907}\right)\left(\frac{7}{37907}\right) && \text{since } (-1)^2 = 1\\
&= \left(\frac{5}{37907}\right)\left(\frac{7}{37907}\right) && \text{Quadratic Reciprocity 3.62(b)}\\
&= \left(\frac{37907}{5}\right) \times -\left(\frac{37907}{7}\right) && \text{Quadratic Reciprocity 3.62(c)}\\
&= -\left(\frac{2}{5}\right)\left(\frac{2}{7}\right) && \text{since } 37907 \equiv 2 \pmod 5\\
& && \text{and } 37907 \equiv 2 \pmod 7\\
&= -(-1)\times 1 && \text{Quadratic Reciprocity 3.62(b)}\\
&= 1.
\end{aligned}$$

Thus $\left(\frac{-15750}{37907}\right) = 1$, so we conclude that -15750 is a square modulo 37907. Note that our computation using Legendre symbols does not tell us how to solve $c^2 \equiv -15750 \pmod{37907}$; it tells us only that there is a solution. For those who are curious, we mention that $c = 10982$ is a solution.

Example 3.63 shows how quadratic reciprocity can be used to evaluate the Legendre symbol. However, you may have noticed that in the middle of our calculation, we needed to factor the number 15750. We were lucky that 15750 is easy to factor, but suppose that we were faced with a more difficult factorization problem. For example, suppose that we want to determine whether $p = 228530738017$ is a square modulo $q = 9365449244297$. It turns out that both p and q are prime.[11] Hence we can use quadratic reciprocity to compute

$$\begin{aligned}
\left(\frac{228530738017}{9365449244297}\right) &= \left(\frac{9365449244297}{228530738017}\right) && \text{since } 228530738017 \equiv 1 \pmod 4,\\
&= \left(\frac{224219723617}{228530738017}\right) && \text{reducing } 9365449244297\\
& && \text{modulo } 228530738017.
\end{aligned}$$

[11] If you don't believe that p and q are prime, use Miller–Rabin (Table 3.2) to check.

Unfortunately, the number 224219723617 is not prime, so we cannot apply quadratic reciprocity directly, and even more unfortunately, it is not an easy number to factor (by hand). So it appears that quadratic reciprocity is useful only if the intermediate calculations lead to numbers that we are able to factor.

Luckily, there is a fancier version of quadratic reciprocity that completely eliminates this difficulty. In order to state it, we need to generalize the definition of the Legendre symbol.

Definition. Let a and b be integers and let b be *odd* and *positive*. Suppose that the factorization of b into primes is

$$b = p_1^{e_1} p_2^{e_2} p_3^{e_3} \cdots p_t^{e_t}.$$

The *Jacobi symbol* $\left(\frac{a}{b}\right)$ is defined by the formula

$$\left(\frac{a}{b}\right) = \left(\frac{a}{p_1}\right)^{e_1} \left(\frac{a}{p_2}\right)^{e_2} \left(\frac{a}{p_3}\right)^{e_3} \cdots \left(\frac{a}{p_t}\right)^{e_t}.$$

Notice that if b is itself prime, then $\left(\frac{a}{b}\right)$ is the original Legendre symbol, so the Jacobi symbol is a generalization of the Legendre symbol. Also note that we define the Jacobi symbol only for odd positive values of b.

Example 3.64. Here is a simple example of a Jacobi symbol, computed directly from the definition:

$$\left(\frac{123}{323}\right) = \left(\frac{123}{17 \cdot 19}\right) = \left(\frac{123}{17}\right)\left(\frac{123}{19}\right) = \left(\frac{4}{17}\right)\left(\frac{9}{19}\right) = 1.$$

Here is a more complicated example:

$$\begin{aligned}\left(\frac{171337608}{536134436237}\right) &= \left(\frac{171337608}{29^3 \cdot 59 \cdot 67^2 \cdot 83}\right) \\ &= \left(\frac{171337608}{29}\right)^3 \left(\frac{171337608}{59}\right) \left(\frac{171337608}{67}\right)^2 \left(\frac{171337608}{83}\right) \\ &= \left(\frac{171337608}{29}\right) \left(\frac{171337608}{59}\right) \left(\frac{171337608}{83}\right) \\ &= \left(\frac{11}{29}\right)\left(\frac{15}{59}\right)\left(\frac{44}{83}\right) = (-1) \cdot 1 \cdot 1 = -1.\end{aligned}$$

From the definition, it appears that we need to know how to factor b in order to compute the Jacobi symbol $\left(\frac{a}{b}\right)$, so we haven't gained anything. However, it turns out that the Jacobi symbol inherits most of the properties of the Legendre symbol, which will allow us to compute $\left(\frac{a}{b}\right)$ extremely rapidly without doing any factorization at all. We start with the basic multiplication and reduction properties.

Proposition 3.65. *Let a, a_1, a_2, b, b_1, b_2 be integers with b, b_1, and b_2 positive and odd.*

(a) $$\left(\frac{a_1 a_2}{b}\right) = \left(\frac{a_1}{b}\right)\left(\frac{a_2}{b}\right) \quad \textit{and} \quad \left(\frac{a}{b_1 b_2}\right) = \left(\frac{a}{b_1}\right)\left(\frac{a}{b_2}\right).$$

(b) $$\textit{If} \quad a_1 \equiv a_2 \pmod{b}, \quad \textit{then} \quad \left(\frac{a_1}{b}\right) = \left(\frac{a_2}{b}\right).$$

Proof. Both parts of Proposition 3.65 follow easily from the definition of the Jacobi symbol and the corresponding properties (3.33) and (3.34) of the Legendre symbol. □

Now we come to the amazing fact that the Jacobi symbol satisfies exactly the same reciprocity law as the Legendre symbol.

Theorem 3.66 (Quadratic Reciprocity: Version II). *Let a and b be integers that are* odd *and* positive.

(a) $$\left(\frac{-1}{b}\right) = \begin{cases} 1 & \text{if } b \equiv 1 \pmod{4}, \\ -1 & \text{if } b \equiv 3 \pmod{4}. \end{cases}$$

(b) $$\left(\frac{2}{b}\right) = \begin{cases} 1 & \text{if } b \equiv 1 \text{ or } 7 \pmod{8}, \\ -1 & \text{if } b \equiv 3 \text{ or } 5 \pmod{8}. \end{cases}$$

(c) $$\left(\frac{a}{b}\right) = \begin{cases} \left(\frac{b}{a}\right) & \text{if } a \equiv 1 \pmod{4} \text{ or } b \equiv 1 \pmod{4}, \\ -\left(\frac{b}{a}\right) & \text{if } a \equiv 3 \pmod{4} \text{ and } b \equiv 3 \pmod{4}. \end{cases}$$

Proof. It is not hard to use the original version of quadratic reciprocity for the Legendre symbol (Theorem 3.62) to prove the more general version for the Jacobi symbol. See for example [59, Proposition 5.2.2] or [137, Theorem 22.2]. □

Example 3.67. When we tried to use the original version of quadratic reciprocity (Theorem 3.62) to compute $\left(\frac{228530738017}{9365449244297}\right)$, we ran into the problem that we needed to factor the number 224219723617. Using the new and improved version of quadratic reciprocity (Theorem 3.66), we can perform the computation without doing any factoring:

$$\left(\frac{228530738017}{9365449244297}\right) = \left(\frac{9365449244297}{228530738017}\right) = \left(\frac{224219723617}{228530738017}\right) = \left(\frac{228530738017}{224219723617}\right)$$
$$= \left(\frac{4311014400}{224219723617}\right) = \left(\frac{2^{10} \cdot 4209975}{224219723617}\right) = \left(\frac{224219723617}{4209975}\right) = \left(\frac{665092}{4209975}\right)$$
$$= \left(\frac{2^2 \cdot 166273}{4209975}\right) = \left(\frac{4209975}{166273}\right) = \left(\frac{53150}{166273}\right) = \left(\frac{2 \cdot 26575}{166273}\right) = \left(\frac{26575}{166273}\right)$$
$$= \left(\frac{166273}{26575}\right) = \left(\frac{6823}{26575}\right) = -\left(\frac{26575}{6823}\right) = -\left(\frac{6106}{6823}\right) = -\left(\frac{2 \cdot 3053}{6823}\right)$$

$$= -\left(\frac{3053}{6823}\right) = -\left(\frac{6823}{3053}\right) = -\left(\frac{717}{3053}\right) = -\left(\frac{3053}{717}\right) = -\left(\frac{185}{717}\right) = -\left(\frac{717}{185}\right)$$
$$= -\left(\frac{162}{185}\right) = -\left(\frac{2\cdot 81}{185}\right) = -\left(\frac{81}{185}\right) = -\left(\frac{185}{81}\right) = -\left(\frac{23}{81}\right) = -\left(\frac{81}{23}\right)$$
$$= -\left(\frac{12}{23}\right) = -\left(\frac{2^2\cdot 3}{23}\right) = \left(\frac{23}{3}\right) = \left(\frac{2}{3}\right) = -1.$$

Hence 228530738017 is not a square modulo 9365449244297.

Remark 3.68. Suppose that $\left(\frac{a}{b}\right) = 1$, where b is some odd positive number. Does the fact that $\left(\frac{a}{b}\right) = 1$ tell us that a is a square modulo b? It does if b is prime, since that's how we defined the Legendre symbol, but what if b is composite? For example, suppose that $b = pq$ is a product of two primes. Then by definition,

$$\left(\frac{a}{b}\right) = \left(\frac{a}{pq}\right) = \left(\frac{a}{p}\right)\left(\frac{a}{q}\right).$$

We see that there are two ways in which $\left(\frac{a}{b}\right)$ can be equal to 1, namely $1 = 1\cdot 1$ and $1 = (-1)\cdot(-1)$. This leads to two different cases:

Case 1: $\left(\frac{a}{p}\right) = \left(\frac{a}{q}\right) = 1,$ so a is a square modulo pq.

Case 2: $\left(\frac{a}{p}\right) = \left(\frac{a}{q}\right) = -1,$ so a is not a square modulo pq.

We should justify our assertion that a is a square modulo pq in Case 1. Note that in Case 1, there are solutions to $c_1^2 \equiv a \pmod{p}$ and $c_2^2 \equiv a \pmod{q}$. We use the Chinese remainder theorem (Theorem 2.24) to find an integer c satisfying $c \equiv c_1 \pmod{p}$ and $c \equiv c_2 \pmod{q}$, and then $c^2 \equiv a \pmod{pq}$.

Our conclusion is that if $b = pq$ is a product of two primes, then although it is easy to compute the value of the Jacobi symbol $\left(\frac{a}{pq}\right)$, this value does not tell us whether a is a square modulo pq. This dichotomy can be exploited for cryptographic purposes as explained in the next section.

Example 3.69 (An application of quadratic reciprocity to the discrete logarithm problem). Let p be an odd prime, let $g \in \mathbb{F}_p^*$ be a primitive root, and let $h \in \mathbb{F}_p^*$. As we have discussed, it is in general a difficult problem to compute the discrete logarithm $\log_g(h)$, i.e., to solve $g^x = h$. But one might ask if it is possible to easily extract some information about $\log_g(h)$. The answer is yes, since we claim that

$$(-1)^{\log_g(h)} = \left(\frac{h}{p}\right). \tag{3.35}$$

Thus the Legendre symbol $\left(\frac{h}{p}\right)$ determines whether $\log_g(h)$ is odd or even, and quadratic reciprocity gives a fast algorithm to compute the value of $\left(\frac{h}{p}\right)$.

In order to prove (3.35), we note that while proving Proposition 3.61, we showed that g^r is a quadratic residue if r is even and that g^r is a quadratic nonresidue if r is odd. Taking $r = \log_g(h)$ gives (3.35). In fancier terminology, one says that the 0th bit of the discrete logarithm is insecure. See Exercise 3.40 for a generalization.

3.10 Probabilistic Encryption and the Goldwasser–Micali Cryptosystem

Suppose that Alice wants to use a public key cryptosystem to encrypt and send Bob 1 bit, i.e., Alice wants to send Bob one of the values 0 and 1. At first glance such an arrangement seems inherently insecure. All that Eve has to do is to encrypt the two possible plaintexts $m = 0$ and $m = 1$, and then she compares the encryptions with Alice's ciphertext. More generally, in any cryptosystem for which the set of possible plaintexts is small, Eve can encrypt every plaintext using Bob's public key until she finds the one that is Alice's.

Probabilistic encryption was invented by Goldwasser and Micali as a way around this problem. The idea is that Alice chooses both a plaintext m and a random string of data r, and then she uses Bob's public key to encrypt the pair (m, r). Ideally, as r varies over all of its possible values, the ciphertexts for (m, r) will vary "randomly" over the possible ciphertexts. More precisely, for any fixed m_1 and m_2 and for varying r, the distribution of values of the two quantities

$$e(m_1, r) = \text{the ciphertext for plaintext } m_1 \text{ and random string } r,$$
$$e(m_2, r) = \text{the ciphertext for plaintext } m_2 \text{ and random string } r,$$

should be essentially indistinguishable. Note that it is not necessary that Bob be able to recover the full pair (m, r) when he performs the decryption. He needs to recover only the plaintext m.

This abstract idea is clear, but how might one create a probabilistic encryption scheme in practice? Goldwasser and Micali describe one such scheme, which, although impractical, since it encrypts only 1 bit at a time, has the advantage of being quite simple to describe and analyze. The idea is based on the difficulty of the following problem.

> Let p and q be (secret) prime numbers and let $N = pq$ be given. For a given integer a, determine whether a is a square modulo N, i.e., determine whether there exists an integer u satisfying $u^2 \equiv a \pmod{N}$.

Note that Bob, who knows how to factor $N = pq$, is able to solve this problem very easily, since

$$a \text{ is a square modulo } pq \quad \text{if and only if} \quad \left(\frac{a}{p}\right) = 1 \quad \text{and} \quad \left(\frac{a}{q}\right) = 1.$$

Eve, on the other hand, has a harder time, since she knows only the value of N. Eve can compute $\left(\frac{a}{N}\right)$, but as we noted earlier (Remark 3.68), this does not tell her whether a is a square modulo N. Goldwasser and Micali exploit this fact[12] to create the probabilistic public key cryptosystem described in Table 3.9.

Bob	Alice
Key creation	
Choose secret primes p and q. Choose a with $\left(\frac{a}{p}\right) = \left(\frac{a}{q}\right) = -1$. Publish $N = pq$ and a.	
Encryption	
	Choose plaintext $m \in \{0,1\}$. Choose random r with $1 < r < N$. Use Bob's public key (N, a) to compute $c = \begin{cases} r^2 \bmod N & \text{if } m = 0, \\ ar^2 \bmod N & \text{if } m = 1. \end{cases}$ Send ciphertext c to Bob.
Decryption	
Compute $\left(\frac{c}{p}\right)$. Decrypt to $m = \begin{cases} 0 & \text{if } \left(\frac{c}{p}\right) = 1, \\ 1 & \text{if } \left(\frac{c}{p}\right) = -1. \end{cases}$	

Table 3.9: Goldwasser–Micali probabilistic public key cryptosystem

It is easy to check that the Goldwasser–Micali cryptosystem works as advertised, since

$$\left(\frac{c}{p}\right) = \begin{cases} \left(\frac{r^2}{p}\right) = \left(\frac{r}{p}\right)^2 = 1 & \text{if } m = 0, \\ \left(\frac{ar^2}{p}\right) = \left(\frac{a}{p}\right)\left(\frac{r}{p}\right)^2 = \left(\frac{a}{p}\right) = -1 & \text{if } m = 1. \end{cases}$$

Further, since Alice chooses r randomly, the set of values that Eve sees when Alice encrypts $m = 0$ consists of all possible squares modulo N, and the set of values that Eve sees when Alice encrypts $m = 1$ consists of all possible numbers c satisfying $\left(\frac{c}{N}\right) = 1$ that are not squares modulo N.

[12]Goldwasser and Micali were not the first to use the problem of squares modulo pq for cryptography. Indeed, an early public key cryptosystem due to Rabin that is provably secure against chosen plaintext attacks (assuming the hardness of factorization) relies on this problem.

What information does Eve obtain if she computes the Jacobi symbol $\left(\frac{c}{N}\right)$, which she can do since N is a public quantity? If $m = 0$, then $c \equiv r^2 \pmod{N}$, so

$$\left(\frac{c}{N}\right) = \left(\frac{r^2}{N}\right) = \left(\frac{r}{N}\right)^2 = 1.$$

On the other hand, if $m = 1$, then $c \equiv ar^2 \pmod{N}$, so

$$\left(\frac{c}{N}\right) = \left(\frac{ar^2}{N}\right) = \left(\frac{a}{N}\right) = \left(\frac{a}{pq}\right) = \left(\frac{a}{p}\right)\left(\frac{a}{q}\right) = (-1) \cdot (-1) = 1$$

is also equal to 1. (Note that Bob chose a to satisfy $\left(\frac{a}{p}\right) = \left(\frac{a}{q}\right) = -1$.) Thus $\left(\frac{c}{N}\right)$ is equal to 1, regardless of the value of N, so the Jacobi symbol gives Eve no useful information.

Example 3.70. Bob creates a Goldwasser–Micali public key by choosing

$$p = 2309, \qquad q = 5651, \qquad N = pq = 13048159, \qquad a = 6283665.$$

Note that a has the property that $\left(\frac{a}{p}\right) = \left(\frac{a}{q}\right) = -1$. He publishes the pair (N, a) and keeps the values of the primes p and q secret.

Alice begins by sending Bob the plaintext bit $m = 0$. To do this, she chooses $r = 1642087$ at random from the interval 1 to 13048158. She then computes

$$c \equiv r^2 \equiv 1642087^2 \equiv 8513742 \pmod{13048159},$$

and sends the ciphertext $c = 8513742$ to Bob. Bob decrypts the ciphertext $c = 8513742$ by computing $\left(\frac{8513742}{2309}\right) = 1$, which gives the plaintext bit $m = 0$.

Next Alice decides to send Bob the plaintext bit $m = 1$. She chooses a random value $r = 11200984$ and computes

$$c \equiv ar^2 \equiv 6283665 \cdot 11200984^2 \equiv 2401627 \pmod{13048159}.$$

Bob decrypts $c = 2401627$ by computing $\left(\frac{2401627}{2309}\right) = -1$, which tells him that the plaintext bit $m = 1$.

Finally, Alice wants to send Bob another plaintext bit $m = 1$. She chooses the random value $r = 11442423$ and computes

$$c \equiv ar^2 \equiv 6283665 \cdot 11442423^2 \equiv 4099266 \pmod{13048159}.$$

Notice that the ciphertext for this encryption of $m = 1$ is completely unrelated to the previous encryption of $m = 1$. Bob decrypts $c = 4099266$ by computing $\left(\frac{4099266}{2309}\right) = -1$ to conclude that the plaintext bit is $m = 1$.

Remark 3.71. The Goldwasser–Micali public key cryptosystem is not practical, because each bit of the plaintext is encrypted with a number modulo N. For it to be secure, it is necessary that Eve be unable to factor the number $N = pq$, so in practice N will be (at least) a 1000-bit number. Thus if Alice wants to send k bits of plaintext to Bob, her ciphertext will be $1000k$ bits long.

Thus the Goldwasswer–Micali public key cryptosystem has a *message expansion ratio* of 1000, since the ciphertext is 1000 times as long as the plaintext. In general, the Goldwasswer–Micali public key cryptosystem expands a message by a factor of $\log_2(N)$.

There are other probabilistic public key cryptosystems whose message expansion is much smaller. Indeed, we have already seen one: the random element k used by the Elgamal public key cryptosystem (Sect. 2.4) makes Elgamal a probabilistic cryptosystem. Elgamal has a message expansion ratio of 2, as explained in Remark 2.9. Later, in Sect. 7.10, we will see another probabilistic cryptosystem called NTRU. More generally, it is possible, and indeed usually desirable, to take a deterministic cryptosystem such as RSA and turn it into a probabilistic system, even at the cost of increasing its message expansion ratio. (See Exercise 3.43 and Sect. 8.6.)

Exercises

Section 3.1. Euler's Theorem and Roots Modulo pq

3.1. Solve the following congruences.
(a) $x^{19} \equiv 36 \pmod{97}$.
(b) $x^{137} \equiv 428 \pmod{541}$.
(c) $x^{73} \equiv 614 \pmod{1159}$.
(d) $x^{751} \equiv 677 \pmod{8023}$.
(e) $x^{38993} \equiv 328047 \pmod{401227}$. (*Hint.* $401227 = 607 \cdot 661$.)

3.2. This exercise investigates what happens if we drop the assumption that $\gcd(e, p-1) = 1$ in Proposition 3.2. So let p be a prime, let $c \not\equiv 0 \pmod{p}$, let $e \geq 1$, and consider the congruence

$$x^e \equiv c \pmod{p}. \tag{3.36}$$

(a) Prove that if (3.36) has one solution, then it has exactly $\gcd(e, p-1)$ distinct solutions. (*Hint.* Use primitive root theorem (Theorem 1.30), combined with the extended Euclidean algorithm (Theorem 1.11) or Exercise 1.27.)
(b) For how many non-zero values of $c \pmod{p}$ does the congruence (3.36) have a solution?

3.3. Let p and q be distinct primes and let e and d be positive integers satisfying

$$de \equiv 1 \pmod{(p-1)(q-1)}.$$

Suppose further that c is an integer with $\gcd(c, pq) > 1$. Prove that

$$x \equiv c^d \pmod{pq} \quad \text{is a solution to the congruence} \quad x^e \equiv c \pmod{pq},$$

thereby completing the proof of Proposition 3.5.

3.4. Recall from Sect. 1.3 that *Euler's phi function* $\phi(N)$ is the function defined by

$$\phi(N) = \#\{0 \le k < N : \gcd(k, N) = 1\}.$$

In other words, $\phi(N)$ is the number of integers between 0 and $N-1$ that are relatively prime to N, or equivalently, the number of elements in $\mathbb{Z}/N\mathbb{Z}$ that have inverses modulo N.

(a) Compute the values of $\phi(6)$, $\phi(9)$, $\phi(15)$, and $\phi(17)$.

(b) If p is prime, what is the value of $\phi(p)$?

(c) Prove *Euler's formula*

$$a^{\phi(N)} \equiv 1 \pmod{N} \quad \text{for all integers } a \text{ satisfying } \gcd(a, N) = 1.$$

(*Hint.* Mimic the proof of Fermat's little theorem (Theorem 1.24), but instead of looking at all of the multiples of a as was done in (1.8), just take the multiples ka of a for values of k satisfying $\gcd(k, N) = 1$.)

3.5. Euler's phi function has many beautiful properties.

(a) If p and q are distinct primes, how is $\phi(pq)$ related to $\phi(p)$ and $\phi(q)$?

(b) If p is prime, what is the value of $\phi(p^2)$? How about $\phi(p^j)$? Prove that your formula for $\phi(p^j)$ is correct. (*Hint.* Among the numbers between 0 and $p^j - 1$, remove the ones that have a factor of p. The ones that are left are relatively prime to p.)

(c) Let M and N be integers satisfying $\gcd(M, N) = 1$. Prove the multiplication formula

$$\phi(MN) = \phi(M)\phi(N).$$

(d) Let $p_1, p_2, \ldots, p_r$ be the distinct primes that divide N. Use your results from (b) and (c) to prove the following formula:

$$\phi(N) = N \prod_{i=1}^{r} \left(1 - \frac{1}{p_i}\right).$$

(e) Use the formula in (d) to compute the following values of $\phi(N)$.

(i) $\phi(1728)$. (ii) $\phi(1575)$. (iii) $\phi(889056)$ (*Hint.* $889056 = 2^5 \cdot 3^4 \cdot 7^3$).

3.6. Let N, c, and e be positive integers satisfying the conditions $\gcd(N, c) = 1$ and $\gcd(e, \phi(N)) = 1$.

(a) Explain how to solve the congruence

$$x^e \equiv c \pmod{N},$$

assuming that you know the value of $\phi(N)$. (*Hint.* Use the formula in Exercise 3.4(c).)

(b) Solve the following congruences. (The formula in Exercise 3.5(d) may be helpful for computing the value of $\phi(N)$.)

(i) $x^{577} \equiv 60 \pmod{1463}$.

(ii) $x^{959} \equiv 1583 \pmod{1625}$.

(iii) $x^{133957} \equiv 224689 \pmod{2134440}$.

Section 3.2. The RSA Public Key Cryptosystem

3.7. Alice publishes her RSA public key: modulus $N = 2038667$ and exponent $e = 103$.

(a) Bob wants to send Alice the message $m = 892383$. What ciphertext does Bob send to Alice?

(b) Alice knows that her modulus factors into a product of two primes, one of which is $p = 1301$. Find a decryption exponent d for Alice.

(c) Alice receives the ciphertext $c = 317730$ from Bob. Decrypt the message.

3.8. Bob's RSA public key has modulus $N = 12191$ and exponent $e = 37$. Alice sends Bob the ciphertext $c = 587$. Unfortunately, Bob has chosen too small a modulus. Help Eve by factoring N and decrypting Alice's message. (*Hint.* N has a factor smaller than 100.)

3.9. For each of the given values of $N = pq$ and $(p-1)(q-1)$, use the method described in Remark 3.11 to determine p and q.

(a) $N = pq = 352717$ and $(p-1)(q-1) = 351520$.

(b) $N = pq = 77083921$ and $(p-1)(q-1) = 77066212$.

(c) $N = pq = 109404161$ and $(p-1)(q-1) = 109380612$.

(d) $N = pq = 172205490419$ and $(p-1)(q-1) = 172204660344$.

3.10. A *decryption exponent* for an RSA public key (N, e) is an integer d with the property that $a^{de} \equiv a \pmod{N}$ for all integers a that are relatively prime to N.

(a) Suppose that Eve has a magic box that creates decryption exponents for (N, e) for a fixed modulus N and for a large number of different encryption exponents e. Explain how Eve can use her magic box to try to factor N.

(b) Let $N = 38749709$. Eve's magic box tells her that the encryption exponent $e = 10988423$ has decryption exponent $d = 16784693$ and that the encryption exponent $e = 25910155$ has decryption exponent $d = 11514115$. Use this information to factor N.

(c) Let $N = 225022969$. Eve's magic box tells her the following three encryption/decryption pairs for N:

$$(70583995, 4911157), \quad (173111957, 7346999), \quad (180311381, 29597249).$$

Use this information to factor N.

(d) Let $N = 1291233941$. Eve's magic box tells her the following three encryption/decryption pairs for N:

$$(1103927639, 76923209), \quad (1022313977, 106791263), \quad (387632407, 7764043).$$

Use this information to factor N.

3.11. Here is an example of a public key system that was proposed at a cryptography conference. It was designed to be more efficient than RSA.

Alice chooses two large primes p and q and she publishes $N = pq$. It is assumed that N is hard to factor. Alice also chooses three random numbers g, r_1, and r_2 modulo N and computes

$$g_1 \equiv g^{r_1(p-1)} \pmod{N} \quad \text{and} \quad g_2 \equiv g^{r_2(q-1)} \pmod{N}.$$

Her public key is the triple (N, g_1, g_2) and her private key is the pair of primes (p, q).

Now Bob wants to send the message m to Alice, where m is a number modulo N. He chooses two random integers s_1 and s_2 modulo N and computes

$$c_1 \equiv m g_1^{s_1} \pmod{N} \qquad \text{and} \qquad c_2 \equiv m g_2^{s_2} \pmod{N}.$$

Bob sends the ciphertext (c_1, c_2) to Alice.

Decryption is extremely fast and easy. Alice uses the Chinese remainder theorem to solve the pair of congruences

$$x \equiv c_1 \pmod{p} \qquad \text{and} \qquad x \equiv c_2 \pmod{q}.$$

(a) Prove that Alice's solution x is equal to Bob's plaintext m.
(b) Explain why this cryptosystem is not secure.

Section 3.3. Implementation and Security Issues

3.12. Formulate a man-in-the-middle attack, similar to the attack described in Example 3.13 on page 126, for the following public key cryptosystems.
(a) The Elgamal public key cryptosystem (Table 2.3 on page 72).
(b) The RSA public key cryptosystem (Table 3.1 on page 123).

3.13. Alice decides to use RSA with the public key $N = 1889570071$. In order to guard against transmission errors, Alice has Bob encrypt his message twice, once using the encryption exponent $e_1 = 1021763679$ and once using the encryption exponent $e_2 = 519424709$. Eve intercepts the two encrypted messages

$$c_1 = 1244183534 \quad \text{and} \quad c_2 = 732959706.$$

Assuming that Eve also knows N and the two encryption exponents e_1 and e_2, use the method described in Example 3.15 to help Eve recover Bob's plaintext without finding a factorization of N.

Section 3.4. Primality Testing

3.14. We stated that the number 561 is a Carmichael number, but we never checked that $a^{561} \equiv a \pmod{561}$ for every value of a.
(a) The number 561 factors as $3 \cdot 11 \cdot 17$. First use Fermat's little theorem to prove that

$$a^{561} \equiv a \pmod{3}, \quad a^{561} \equiv a \pmod{11}, \quad \text{and} \quad a^{561} \equiv a \pmod{17}$$

for every value of a. Then explain why these three congruences imply that $a^{561} \equiv a \pmod{561}$ for every value of a.
(b) Mimic the idea used in (a) to prove that each of the following numbers is a Carmichael number. (To assist you, we have factored each number into primes.)
  (i) $1729 = 7 \cdot 13 \cdot 19$
  (ii) $10585 = 5 \cdot 29 \cdot 73$
  (iii) $75361 = 11 \cdot 13 \cdot 17 \cdot 31$
  (iv) $1024651 = 19 \cdot 199 \cdot 271$
(c) Prove that a Carmichael number must be odd.

(d) Prove that a Carmichael number must be a product of *distinct* primes.
(e) Look up Korselt's criterion in a book or online, write a brief description of how it works, and use it to show that $29341 = 13 \cdot 37 \cdot 61$ and $172947529 = 307 \cdot 613 \cdot 919$ are Carmichael numbers.

3.15. Use the Miller–Rabin test on each of the following numbers. In each case, either provide a Miller–Rabin witness for the compositeness of n, or conclude that n is probably prime by providing 10 numbers that are not Miller–Rabin witnesses for n.

(a) $n = 1105$. (Yes, 5 divides n, but this is just a warm-up exercise!)
(b) $n = 294409$ (c) $n = 294439$
(d) $n = 118901509$ (e) $n = 118901521$
(f) $n = 118901527$ (g) $n = 118915387$

3.16. Looking back at Exercise 3.10, let's suppose that for a given N, the magic box can produce only one decryption exponent. Equivalently, suppose that an RSA key pair has been compromised and that the private decryption exponent corresponding to the public encryption exponent has been discovered. Show how the basic idea in the Miller–Rabin primality test can be applied to use this information to factor N.

3.17. The function $\pi(X)$ counts the number of primes between 2 and X.
(a) Compute the values of $\pi(20)$, $\pi(30)$, and $\pi(100)$.
(b) Write a program to compute $\pi(X)$ and use it to compute $\pi(X)$ and the ratio $\pi(X)/(X/\ln(X))$ for $X = 100$, $X = 1000$, $X = 10000$, and $X = 100000$. Does your list of ratios make the prime number theorem plausible?

3.18. Let

$$\pi_1(X) = (\#\text{ of primes } p \text{ between 2 and } X \text{ satisfying } p \equiv 1 \pmod 4),$$
$$\pi_3(X) = (\#\text{ of primes } p \text{ between 2 and } X \text{ satisfying } p \equiv 3 \pmod 4).$$

Thus every prime other than 2 gets counted by either $\pi_1(X)$ or by $\pi_3(X)$.
(a) Compute the values of $\pi_1(X)$ and $\pi_3(X)$ for each of the following values of X.
(i) $X = 10$. (ii) $X = 25$. (iii) $X = 100$.
(b) Write a program to compute $\pi_1(X)$ and $\pi_3(X)$ and use it to compute their values and the ratio $\pi_3(X)/\pi_1(X)$ for $X = 100$, $X = 1000$, $X = 10000$, and $X = 100000$.
(c) Based on your data from (b), make a conjecture about the relative sizes of $\pi_1(X)$ and $\pi_3(X)$. Which one do you think is larger? What do you think is the limit of the ratio $\pi_3(X)/\pi_1(X)$ as $X \to \infty$?

3.19. We noted in Sect. 3.4 that it really makes no sense to say that the number n has probability $1/\ln(n)$ of being prime. Any particular number that you choose either will be prime or will not be prime; there are no numbers that are 35 % prime and 65 % composite! In this exercise you will prove a result that gives a more sensible meaning to the statement that a number has a certain probability of being prime. You may use the prime number theorem (Theorem 3.21) for this problem.
(a) Fix a (large) number N and suppose that Bob chooses a random number n in the interval $\frac{1}{2}N \le n \le \frac{3}{2}N$. If he repeats this process many times, prove that approximately $1/\ln(N)$ of his numbers will be prime. More precisely, define

$$P(N) = \frac{\text{number of primes between } \frac{1}{2}N \text{ and } \frac{3}{2}N}{\text{number of integers between } \frac{1}{2}N \text{ and } \frac{3}{2}N}$$
$$= \begin{bmatrix}\text{Probability that an integer } n \text{ in the} \\ \text{interval } \frac{1}{2}N \le n \le \frac{3}{2}N \text{ is a prime number}\end{bmatrix},$$

and prove that

$$\lim_{N\to\infty} \frac{P(N)}{1/\ln(N)} = 1.$$

This shows that if N is large, then $P(N)$ is approximately $1/\ln(N)$.

(b) More generally, fix two numbers c_1 and c_2 satisfying $c_2 > c_1 > 0$. Bob chooses random numbers n in the interval $c_1N \le n \le c_2N$. Keeping c_1 and c_2 fixed, let

$$P(c_1, c_2; N) = \begin{bmatrix}\text{Probability that an integer } n \text{ in the inter-} \\ \text{val } c_1N \le n \le c_2N \text{ is a prime number}\end{bmatrix}.$$

In the following formula, fill in the box with a simple function of N so that the statement is true:

$$\lim_{N\to\infty} \frac{P(c_1, c_2; N)}{\boxed{}} = 1.$$

3.20. Continuing with the previous exercise, explain how to make mathematical sense of the following statements.

(a) A randomly chosen *odd* number N has probability $2/\ln(N)$ of being prime. (What is the probability that a randomly chosen even number is prime?)

(b) A randomly chosen number N satisfying $N \equiv 1 \pmod 3$ has probability $3/(2\ln(N))$ of being prime.

(c) A randomly chosen number N satisfying $N \equiv 1 \pmod 6$ has probability $3/\ln(N)$ of being prime.

(d) Let $m = p_1p_2\cdots p_r$ be a product of distinct primes and let k be a number satisfying $\gcd(k, m) = 1$. What number should go into the box to make statement (3.37) correct? Why?

A randomly chosen number N satisfying
$N \equiv k \pmod m$ has probability $\boxed{}/\ln(N)$ (3.37)
of being prime.

(e) Same question, but for arbitrary m, not just for m that are products of distinct primes.

3.21. The *logarithmic integral function* $\mathrm{Li}(X)$ is defined to be

$$\mathrm{Li}(X) = \int_2^X \frac{dt}{\ln t}.$$

(a) Prove that

$$\mathrm{Li}(X) = \frac{X}{\ln X} + \int_2^X \frac{dt}{(\ln t)^2} + O(1).$$

(*Hint.* Integration by parts.)

(b) Compute the limit

$$\lim_{X\to\infty} \frac{\mathrm{Li}(X)}{X/\ln X}.$$

(*Hint.* Break the integral in (a) into two pieces, $2 \le t \le \sqrt{X}$ and $\sqrt{X} \le t \le X$, and estimate each piece separately.)

(c) Use (b) to show that formula (3.12) on page 135 implies the prime number theorem (Theorem 3.21).

Section 3.5. Pollard's $p-1$ Factorization Algorithm

3.22. Use Pollard's $p-1$ method to factor each of the following numbers.

(a) $n = 1739$ (b) $n = 220459$ (c) $n = 48356747$

Be sure to show your work and to indicate which prime factor p of n has the property that $p-1$ is a product of small primes.

3.23. A prime of the form $2^n - 1$ is called a *Mersenne prime*.

(a) Factor each of the numbers $2^n - 1$ for $n = 2, 3, \ldots, 10$. Which ones are Mersenne primes?
(b) Find the first seven Mersenne primes. (You may need a computer.)
(c) If n is even and $n > 2$, prove that $2^n - 1$ is not prime.
(d) If $3 \mid n$ and $n > 3$, prove that $2^n - 1$ is not prime.
(e) More generally, prove that if n is a composite number, then $2^n - 1$ is not prime. Thus all Mersenne primes have the form $2^p - 1$ with p a prime number.
(f) What is the largest known Mersenne prime? Are there any larger primes known? (You can find out at the "Great Internet Mersenne Prime Search" web site `www.mersenne.org/prime.htm`.)
(g) Write a one page essay on Mersenne primes, starting with the discoveries of Father Mersenne and ending with GIMPS.

Section 3.6. Factorization via Difference of Squares

3.24. For each of the following numbers N, compute the values of

$$N + 1^2, \quad N + 2^2, \quad N + 3^2, \quad N + 4^2, \quad \ldots$$

as we did in Example 3.34 until you find a value $N + b^2$ that is a perfect square a^2. Then use the values of a and b to factor N.

(a) $N = 53357$ (b) $N = 34571$ (c) $N = 25777$ (d) $N = 64213$

3.25. For each of the listed values of N, k, and b_{init}, factor N by making a list of values of $k \cdot N + b^2$, starting at $b = b_{\text{init}}$ and incrementing b until $k \cdot N + b^2$ is a perfect square. Then take greatest common divisors as we did in Example 3.35.

(a)	$N = 143041$	$k = 247$	$b_{\text{init}} = 1$
(b)	$N = 1226987$	$k = 3$	$b_{\text{init}} = 36$
(c)	$N = 2510839$	$k = 21$	$b_{\text{init}} = 90$

3.26. For each part, use the data provided to find values of a and b satisfying $a^2 \equiv b^2 \pmod{N}$, and then compute $\gcd(N, a-b)$ in order to find a nontrivial factor of N, as we did in Examples 3.37 and 3.38.

(a) $N = 61063$

$$1882^2 \equiv 270 \pmod{61063} \quad \text{and} \quad 270 = 2 \cdot 3^3 \cdot 5$$
$$1898^2 \equiv 60750 \pmod{61063} \quad \text{and} \quad 60750 = 2 \cdot 3^5 \cdot 5^3$$

(b) $N = 52907$

$$399^2 \equiv 480 \pmod{52907} \quad \text{and} \quad 480 = 2^5 \cdot 3 \cdot 5$$
$$763^2 \equiv 192 \pmod{52907} \quad \text{and} \quad 192 = 2^6 \cdot 3$$
$$773^2 \equiv 15552 \pmod{52907} \quad \text{and} \quad 15552 = 2^6 \cdot 3^5$$
$$976^2 \equiv 250 \pmod{52907} \quad \text{and} \quad 250 = 2 \cdot 5^3$$

(c) $N = 198103$

$$1189^2 \equiv 27000 \pmod{198103} \quad \text{and} \quad 27000 = 2^3 \cdot 3^3 \cdot 5^3$$
$$1605^2 \equiv 686 \pmod{198103} \quad \text{and} \quad 686 = 2 \cdot 7^3$$
$$2378^2 \equiv 108000 \pmod{198103} \quad \text{and} \quad 108000 = 2^5 \cdot 3^3 \cdot 5^3$$
$$2815^2 \equiv 105 \pmod{198103} \quad \text{and} \quad 105 = 3 \cdot 5 \cdot 7$$

(d) $N = 2525891$

$$1591^2 \equiv 5390 \pmod{2525891} \quad \text{and} \quad 5390 = 2 \cdot 5 \cdot 7^2 \cdot 11$$
$$3182^2 \equiv 21560 \pmod{2525891} \quad \text{and} \quad 21560 = 2^3 \cdot 5 \cdot 7^2 \cdot 11$$
$$4773^2 \equiv 48510 \pmod{2525891} \quad \text{and} \quad 48510 = 2 \cdot 3^2 \cdot 5 \cdot 7^2 \cdot 11$$
$$5275^2 \equiv 40824 \pmod{2525891} \quad \text{and} \quad 40824 = 2^3 \cdot 3^6 \cdot 7$$
$$5401^2 \equiv 1386000 \pmod{2525891} \quad \text{and} \quad 1386000 = 2^4 \cdot 3^2 \cdot 5^3 \cdot 7 \cdot 11$$

Section 3.7. Smooth Numbers, Sieves, and Building Relations for Factorization

3.27. Compute the following values of $\psi(X, B)$, the number of B-smooth numbers between 2 and X (see page 150).

(a) $\psi(25, 3)$ (b) $\psi(35, 5)$ (c) $\psi(50, 7)$ (d) $\psi(100, 5)$ (e) $\psi(100, 7)$

3.28. An integer M is called *B-power-smooth* if every prime power p^e dividing M satisfies $p^e \le B$. For example, $180 = 2^2 \cdot 3^2 \cdot 5$ is 10-power-smooth, since the largest prime power dividing 180 is 9, which is smaller than 10.

(a) Suppose that M is B-power-smooth. Prove that M is also B-smooth.

(b) Suppose that M is B-smooth. Is it always true that M is also B-power-smooth? Either prove that it is true or give an example for which it is not true.

(c) The following is a list of 20 randomly chosen numbers between 1 and 1000, sorted from smallest to largest. Which of these numbers are 10-power-smooth? Which of them are 10-smooth?

$$\{84, 141, 171, 208, 224, 318, 325, 366, 378, 390, 420, 440,$$
$$504, 530, 707, 726, 758, 765, 792, 817\}$$

(d) Prove that M is B-power-smooth if and only if M divides the least common multiple of $[1, 2, \ldots, B]$. (The *least common multiple* of a list of numbers $k_1, \ldots, k_r$ is the smallest number K that is divisible by every number in the list.)

3.29. Let $L(N) = e^{\sqrt{(\ln N)(\ln\ln N)}}$ as usual. Suppose that a computer does one billion operations per second.
(a) How many seconds does it take to perform $L(2^{100})$ operations?
(b) How many hours does it take to perform $L(2^{250})$ operations?
(c) How many days does it take to perform $L(2^{350})$ operations?
(d) How many years does it take to perform $L(2^{500})$ operations?
(e) How many years does it take to perform $L(2^{750})$ operations?
(f) How many years does it take to perform $L(2^{1000})$ operations?
(g) How many years does it take to perform $L(2^{2000})$ operations?
(For simplicity, you may assume that there are 365.25 days in a year.)

3.30. Prove that the function $L(X) = e^{\sqrt{(\ln X)(\ln\ln X)}}$ is subexponential. That is, prove the following two statements.
(a) For every positive constant α, no matter how large, $L(X) = \Omega\bigl((\ln X)^\alpha\bigr)$.
(b) For every positive constant β, no matter how small, $L(X) = \mathcal{O}(X^\beta)$.

3.31. For any fixed positive constants a and b, define the function

$$F_{a,b}(X) = e^{(\ln X)^{1/a}(\ln\ln X)^{1/b}}.$$

Prove the following properties of $F_{a,b}(X)$.
(a) If $a > 1$, prove that $F_{a,b}(X)$ is subexponential.
(b) If $a = 1$, prove that $F_{a,b}(X) = \Omega(X^\alpha)$ for every $\alpha > 0$. Thus $F_{a,b}(X)$ grows faster than every exponential function, so one says that $F_{a,b}(X)$ has *superexponential growth*.
(c) What happens if $a < 1$?

3.32. This exercise asks you to verify an assertion in the proof of Corollary 3.45. Let $L(X)$ be the usual function $L(X) = e^{\sqrt{(\ln X)(\ln\ln X)}}$.
(a) Prove that there is a value of $\epsilon > 0$ such that

$$(\ln X)^\epsilon < \ln L(X) < (\ln X)^{1-\epsilon} \qquad \text{for all } X > 10.$$

(b) Let $c > 0$, let $Y = L(X)^c$, and let $u = (\ln X)/(\ln Y)$. Prove that

$$u^{-u} = L(X)^{-\frac{1}{2c}(1+o(1))}.$$

3.33. Proposition 3.48 assumes that we choose random numbers a modulo N, compute $a^2 \pmod{N}$, and check whether the result is B-smooth. We can achieve better results if we take values for a of the form

$$a = \lfloor\sqrt{N}\rfloor + k \qquad \text{for } 1 \le k \le K.$$

(For simplicity, you may treat K as a fixed integer, independent of N. More rigorously, it is necessary to take K equal to a power of $L(N)$, which has a small effect on the final answer.)

(a) Prove that $a^2 - N \le 2K\sqrt{N} + K^2$, so in particular, $a^2 \pmod N$ is smaller than a multiple of $\sqrt{N}$.

(b) Prove that $L(\sqrt{N}) \approx L(N)^{1/\sqrt{2}}$ by showing that

$$\lim_{N\to\infty} \frac{\log L(\sqrt{N})}{\log L(N)^{1/\sqrt{2}}} = 1.$$

More generally, prove that in the same sense, $L(N^{1/r}) \approx L(N)^{1/\sqrt{r}}$ for any fixed $r > 0$.

(c) Re-prove Proposition 3.48 using this better choice of values for a. Set $B = L(N)^c$ and find the optimal value of c. Approximately how many relations are needed to factor N?

3.34. Illustrate the quadratic sieve, as was done in Fig. 3.3 (page 161), by sieving prime powers up to B on the values of $F(T) = T^2 - N$ in the indicated range.

(a) Sieve $N = 493$ using prime powers up to $B = 11$ on values from $F(23)$ to $F(38)$. Use the relation(s) that you find to factor N.

(b) Extend the computations in (a) by using prime powers up to $B = 16$ and sieving values from $F(23)$ to $F(50)$. What additional value(s) are sieved down to 1 and what additional relation(s) do they yield?

3.35. Let $\mathbb{Z}[\beta]$ be the ring described in Example 3.55, i.e., β is a root of $f(x) = 1 + 3x - 2x^3 + x^4$. For each of the following pairs of elements $u, v \in \mathbb{Z}[\beta]$, compute the sum $u + v$ and the product uv. Your answers should involve only powers of β up to β^3.

(a) $u = -5 - 2\beta + 9\beta^2 - 9\beta^3$ and $v = 2 + 9\beta - 7\beta^2 + 7\beta^3$.

(b) $u = 9 + 9\beta + 6\beta^2 - 5\beta^3$ and $v = -4 - 6\beta - 2\beta^2 - 5\beta^3$.

(c) $u = 6 - 5\beta + 3\beta^2 + 3\beta^3$ and $v = -2 + 7\beta + 6\beta^2$.

Section 3.8. The Index Calculus and Discrete Logarithms

3.36. This exercise asks you to use the index calculus to solve a discrete logarithm problem. Let $p = 19079$ and $g = 17$.

(a) Verify that $g^i \pmod p$ is 5-smooth for each of the values $i = 3030$, $i = 6892$, and $i = 18312$.

(b) Use your computations in (a) and linear algebra to compute the discrete logarithms $\log_g(2)$, $\log_g(3)$, and $\log_g(5)$. (Note that $19078 = 2 \cdot 9539$ and that 9539 is prime.)

(c) Verify that $19 \cdot 17^{-12400} \pmod p$ is 5-smooth.

(d) Use the values from (b) and the computation in (c) to solve the discrete logarithm problem

$$17^x \equiv 19 \pmod{19079}.$$

Section 3.9. Quadratic Residues and Quadratic Reciprocity

3.37. Let p be an odd prime and let a be an integer with $p \nmid a$.

(a) Prove that $a^{(p-1)/2}$ is congruent to either 1 or -1 modulo p.

(b) Prove that $a^{(p-1)/2}$ is congruent to 1 modulo p if and only if a is a quadratic residue modulo p. (*Hint.* Let g be a primitive root for p and use the fact, proven during the course of proving Proposition 3.61, that g^m is a quadratic residue if and only if m is even.)

(c) Prove that $a^{(p-1)/2} \equiv \left(\frac{a}{p}\right) \pmod{p}$. (This holds even if $p \mid a$.)
(d) Use (c) to prove Theorem 3.62(a), that is, prove that

$$\left(\frac{-1}{p}\right) = \begin{cases} 1 & \text{if } p \equiv 1 \pmod 4, \\ -1 & \text{if } p \equiv 3 \pmod 4. \end{cases}$$

3.38. Prove that the three parts of the quadratic reciprocity theorem (Theorem 3.62) are equivalent to the following three concise formulas, where p and q are odd primes:

(a) $\left(\frac{-1}{p}\right) = (-1)^{\frac{p-1}{2}}$ (b) $\left(\frac{2}{p}\right) = (-1)^{\frac{p^2-1}{8}}$ (c) $\left(\frac{p}{q}\right)\left(\frac{q}{p}\right) = (-1)^{\frac{p-1}{2}\cdot\frac{q-1}{2}}$

3.39. Let p be a prime satisfying $p \equiv 3 \pmod 4$.
(a) Let a be a quadratic residue modulo p. Prove that the number

$$b \equiv a^{\frac{p+1}{4}} \pmod{p}$$

has the property that $b^2 \equiv a \pmod{p}$. (*Hint.* Write $\frac{p+1}{2}$ as $1 + \frac{p-1}{2}$ and use Exercise 3.37.) This gives an easy way to take square roots modulo p for primes that are congruent to 3 modulo 4.
(b) Use (a) to compute the following square roots modulo p. Be sure to check your answers.
  (i) Solve $b^2 \equiv 116 \pmod{587}$.
  (ii) Solve $b^2 \equiv 3217 \pmod{8627}$.
  (iii) Solve $b^2 \equiv 9109 \pmod{10663}$.

3.40. Let p be an odd prime, let $g \in \mathbb{F}_p^*$ be a primitive root, and let $h \in \mathbb{F}_p^*$. Write $p - 1 = 2^s m$ with m odd and $s \geq 1$, and write the binary expansion of $\log_g(h)$ as

$$\log_g(h) = \epsilon_0 + 2\epsilon_2 + 4\epsilon_2 + 8\epsilon_3 + \cdots \quad \text{with} \quad \epsilon_0, \epsilon_1, \ldots \in \{0, 1\}.$$

Give an algorithm that generalizes Example 3.69 and allows you to rapidly compute $\epsilon_0, \epsilon_1, \ldots, \epsilon_{s-1}$, thereby proving that the first s bits of the discrete logarithm are insecure. You may assume that you have a fast algorithm to compute square roots in $\mathbb{F}_p^*$, as provided for example by Exercise 3.39(a) if $p \equiv 3 \pmod 4$. (*Hint.* Use Example 3.69 to compute the 0th bit, take the square root of either h or $g^{-1}h$, and repeat.)

3.41. Let p be a prime satisfying $p \equiv 1 \pmod 3$. We say that a is a *cubic residue modulo p* if $p \nmid a$ and there is an integer c satisfying $a \equiv c^3 \pmod{p}$.
(a) Let a and b be cubic residues modulo p. Prove that ab is a cubic residue modulo p.
(b) Give an example to show that (unlike the case with quadratic residues) it is possible for none of a, b, and ab to be a cubic residue modulo p.
(c) Let g be a primitive root modulo p. Prove that a is a cubic residue modulo p if and only if $3 \mid \log_g(a)$, where $\log_g(a)$ is the discrete logarithm of a to the base g.

(d) Suppose instead that $p \equiv 2 \pmod 3$. Prove that for every integer a there is an integer c satisfying $a \equiv c^3 \pmod p$. In other words, if $p \equiv 2 \pmod 3$, show that every number is a cube modulo p.

Section 3.10. Probabilistic Encryption and the Goldwasser–Micali Cryptosystem

3.42. Perform the following encryptions and decryptions using the Goldwasser–Micali public key cryptosystem (Table 3.9).

(a) Bob's public key is the pair $N = 1842338473$ and $a = 1532411781$. Alice encrypts 3 bits and sends Bob the ciphertext blocks

$$1794677960, \quad 525734818, \quad \text{and} \quad 420526487.$$

Decrypt Alice's message using the factorization

$$N = pq = 32411 \cdot 56843.$$

(b) Bob's public key is $N = 3149$ and $a = 2013$. Alice encrypts 3 bits and sends Bob the ciphertext blocks 2322, 719, and 202. Unfortunately, Bob used primes that are much too small. Factor N and decrypt Alice's message.

(c) Bob's public key is $N = 781044643$ and $a = 568980706$. Encrypt the 3 bits 1, 1, 0 using, respectively, the three random values

$$r = 705130839, \quad r = 631364468, \quad r = 67651321.$$

3.43. Suppose that the plaintext space $\mathcal{M}$ of a certain cryptosystem is the set of bit strings of length $2b$. Let e_k and d_k be the encryption and decryption functions associated with a key $k \in \mathcal{K}$. This exercise describes one method of turning the original cryptosystem into a probabilistic cryptosystem. Most practical cryptosystems that are currently in use rely on more complicated variants of this idea in order to thwart certain types of attacks. (See Sect. 8.6 for further details.)

Alice sends Bob an encrypted message by performing the following steps:

1. Alice chooses a b-bit message m' to be encrypted.
2. Alice chooses a string r consisting of b random bits.
3. Alice sets $m = r \,\|\, (r \oplus m')$, where $\|$ denotes concatenation[13] and $\oplus$ denotes exclusive or (see Sect. 1.7.4). Notice that m has length $2b$ bits.
4. Alice computes $c = e_k(m)$ and sends the ciphertext c to Bob.

(a) Explain how Bob decrypts Alice's message and recovers the plaintext m'. We assume, of course, that Bob knows the decryption function d_k.

(b) If the plaintexts and the ciphertexts of the original cryptosystem have the same length, what is the message expansion ratio of the new probabilistic cryptosystem?

(c) More generally, if the original cryptosystem has a message expansion ratio of μ, what is the message expansion ratio of the new probabilistic cryptosystem?

---

[13]The *concatenation* of 2 bit strings is formed by placing the first string before the second string. For example, $1101 \,\|\, 1001$ is the bit string 11011001.

Chapter 4

Digital Signatures

4.1 What Is a Digital Signature?

Encryption schemes, whether symmetric or asymmetric, solve the problem of secure communications over an insecure network. *Digital signatures* solve a different problem, analogous to the purpose of a pen-and-ink signature on a physical document. It is thus interesting that the tools used to construct digital signatures are very similar to the tools used to construct asymmetric ciphers.

Here is the exact problem that a digital signature is supposed to solve. Samantha[1] has a (digital) document D, for example a computer file, and she wants to create some additional piece of information D^{Sam} that can be used to prove conclusively that Samantha herself approves of the document. So you might view Samantha's digital signature D^{Sam} as analogous to her actual signature on an ordinary paper document.

To contrast the purpose and functionality of public key (asymmetric) cryptosystems versus digital signatures, we consider an analogy using bank deposit vaults and signet rings. A bank deposit vault has a narrow slot (the "public encryption key") into which anyone can deposit an envelope, but only the owner of the combination (the "private decryption key") to the vault's lock is able to open the vault and read the message. Thus a public key cryptosystem is a digital version of a bank deposit vault. A signet ring (the "private signing key") is a ring that has a recessed image. The owner drips some wax from a candle onto his document and presses the ring into the wax to make an impression (the "public signature"). Anyone who looks at the document can verify that the wax impression was made by the owner of the signet ring, but

[1] In this chapter we give Alice and Bob a well deserved rest and let Samantha, the signer, and Victor, the verifier, take over cryptographic duties.

J. Hoffstein et al., *An Introduction to Mathematical Cryptography*, Undergraduate Texts in Mathematics, DOI 10.1007/978-1-4939-1711-2_4

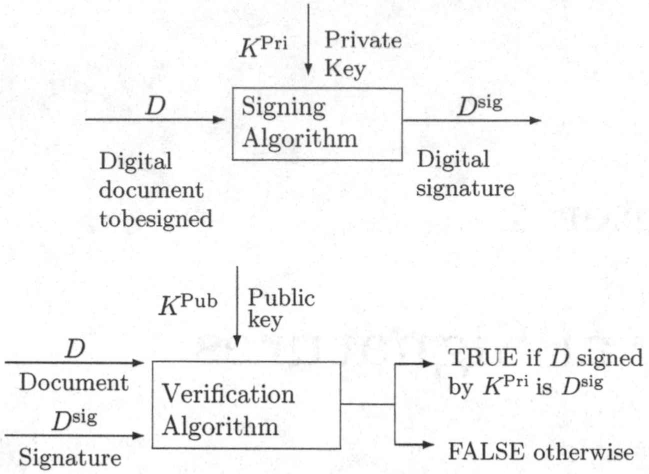

Figure 4.1: The two components of a digital signature scheme

only the owner of the ring is able to create valid impressions.[2] Thus one may view a digital signature system as a modern version of a signet ring.

Despite their different purposes, digital signature schemes are similar to asymmetric cryptosystems in that they involve public and private keys and invoke algorithms that use these keys. Here is an abstract description of the pieces that make up a digital signature scheme:

K^{Pri} A private signing key.

K^{Pub} A public verification key.

Sign A *signing algorithm* that takes as input a digital document D and a private key K^{Pri} and returns a signature D^{sig} for D.

Verify A *verification algorithm* that takes as input a digital document D, a signature D^{sig}, and a public key K^{Pub}. The algorithm returns True if D^{sig} is a signature for D associated to the private key K^{Pri}, and otherwise it returns False.

The operation of a digital signature scheme is depicted in Fig. 4.1. An important point to observe in Fig. 4.1 is that the verification algorithm does not know the private key K^{Pri} when it determines whether D signed by K^{Pri} is equal to D^{sig}. The verification algorithm has access only to the public key K^{Pub}.

It is not difficult to produce (useless) algorithms that satisfy the digital signature properties. For example, let $K^{Pub} = K^{Pri}$. What is difficult is to

[2]Back in the days when interior illumination was by candlelight, sealing documents with signet rings was a common way to create unforgeable signatures. In today's world, with its plentiful machine tools, signet rings and wax images obviously would not provide much security.

create a digital signature scheme in which the owner of the private key K^{Pri} is able to create valid signatures, but knowledge of the public key K^{Pub} does not reveal the private key K^{Pri}. Necessary general conditions for a secure digital signature scheme include the following:

- Given K^{Pub}, an attacker cannot feasibly determine K^{Pri}, nor can she determine any other private key that produces the same signatures as K^{Pri}.
- Given K^{Pub} and a list of signed documents $D_1, \ldots, D_n$ with their signatures $D_1^{\text{sig}}, \ldots, D_n^{\text{sig}}$, an attacker cannot feasibly determine a valid signature on any document D that is not in the list $D_1, \ldots, D_n$.

The second condition is rather different from the situation for encryption schemes. In public key encryption, an attacker can create as many ciphertext/plaintext pairs as she wants, since she can create them using the known public key. However, each time a digital signature scheme is used to sign a new document, it is revealing a new document/signature pair, which provides new information to an attacker. The second condition says that the attacker gains nothing beyond knowledge of that new pair. An attack on a digital signature scheme that makes use of a large number of known signatures is called a *transcript attack*. (See Sect. 7.12 for further discussion.)

Remark 4.1. Digital signatures are at least as important as public key cryptosystems for the conduct of business in a digital age, and indeed, one might argue that they are of greater importance. To take a significant instance, your computer undoubtedly receives program and system upgrades over the Internet. How can your computer tell that an upgrade comes from a legitimate source, in this case the company that wrote the program in the first place? The answer is a digital signature. The original program comes equipped with the company's public verification key. The company uses its private signing key to sign the upgrade and sends your computer both the new program and the signature. Your computer can use the public key to verify the signature, thereby verifying that the program comes from a trusted source, before installing it on your system.

We must stress, however, that although this conveys the idea of how a digital signature might be used, it is a vastly oversimplified explanation. Real-world applications of digital signature schemes require considerable care to avoid a variety of subtle, but fatal, security problems. In particular, as digital signatures proliferate, it can become problematic to be sure that a purported public verification key actually belongs to the supposed owner. And clearly an adversary who tricks you into using her verification key, instead of the real one, will then be able to convince you to accept all of her forged documents.

Remark 4.2. The natural capability of most digital signature schemes is to sign only a small amount of data, say b bits, where b is between 80 and 1000. It is thus quite inefficient to sign a large digital document D, both because it takes a lot of time to sign each b bits of D and because the resulting digital signature is likely to be as large as the original document.

The standard solution to this problem is to use a *hash function*, which is an easily computable function

$$\mathsf{Hash} : (\text{arbitrary size documents}) \longrightarrow \{0,1\}^k$$

that is very hard to invert. (More generally, one wants it to be very difficult to find two distinct inputs D and D' whose outputs $\mathsf{Hash}(D)$ and $\mathsf{Hash}(D')$ are the same.) Then, rather than signing her document D, Samantha instead computes and signs the hash $\mathsf{Hash}(D)$. For verification, Victor computes and verifies the signature on $\mathsf{Hash}(D)$.

There are also security advantages to signing a hash of D, including intrinsically linking the signature to the entire document, and preventing an adversary from choosing random signatures and determining which documents they sign. For a brief introduction to hash functions and references for further reading, see Sect. 8.1. We will not concern ourselves further with such issues in this chapter.

Remark 4.3. There are many variants of the basic digital signature paradigm. For example, a *blinded signature* is one in which the signer does not know the contents of the document being signed. This could be useful, for example, if voters want an election official to sign their votes without revealing what those votes are. Further material on blinded signatures, with an RSA-style example and applications to digital cash systems, are given in Sect. 8.8.

In this chapter we discuss digital signature schemes whose underlying hard problems are integer factorization and the discrete logarithm problem in $\mathbb{F}_p^*$. Subsequent chapters include descriptions of digital signature schemes based on the discrete logarithm problem in elliptic curve groups (Sect. 6.4.3) and on hard lattice problems (Sect. 7.12).

4.2 RSA Digital Signatures

The original RSA paper described both the RSA encryption scheme and an RSA digital signature scheme. The idea is very simple. The setup is the same as for RSA encryption, Samantha chooses two large secret primes p and q and she publishes their product $N = pq$ and a public verification exponent e. Samantha uses her knowledge of the factorization of N to solve the congruence

$$de \equiv 1 \pmod{(p-1)(q-1)}. \tag{4.1}$$

Note that if Samantha were doing RSA encryption, then e would be her encryption exponent and d would be her decryption exponent. However, in the present setup d is her *signing exponent* and e is her *verification exponent*.

In order to sign a digital document D, which we assume to be an integer in the range $1 < D < N$, Samantha computes

$$S \equiv D^d \pmod{N}.$$

Samantha	Victor
Key creation	
Choose secret primes p and q. Choose verification exponent e with $\gcd(e,(p-1)(q-1)) = 1$. Publish $N = pq$ and e.	
Signing	
Compute d satisfying $de \equiv 1 \pmod{(p-1)(q-1)}$. Sign document D by computing $S \equiv D^d \pmod{N}$.	
Verification	
	Compute $S^e \bmod N$ and verify that it is equal to D.

Table 4.1: RSA digital signatures

Victor verifies the validity of the signature S on D by computing

$$S^e \bmod N$$

and checking that it is equal to D. This process works because Euler's formula (Theorem 3.1) tells us that

$$S^e \equiv D^{de} \equiv D \pmod{N}.$$

The RSA digital signature scheme is summarized in Table 4.1.

If Eve can factor N, then she can solve (4.1) for Samantha's secret signing key d. However, just as with RSA encryption, the hard problem underlying RSA digital signatures is not directly the problem of factorization. In order to forge a signature on a document D, Eve needs to find a eth root of D modulo N. This is identical to the hard problem underlying RSA decryption, in which the plaintext is the eth root of the ciphertext.

Remark 4.4. As with RSA encryption, one can gain a bit of efficiency by choosing d and e to satisfy

$$de \equiv 1 \left(\bmod \frac{(p-1)(q-1)}{\gcd(p-1,q-1)} \right).$$

Theorem 3.1 ensures that the verification step still works.

Example 4.5. We illustrate the RSA digital signature scheme with a small numerical example.

RSA Signature Key Creation

- Samantha chooses two secret primes $p = 1223$ and $q = 1987$ and computes her public modulus
$$N = p \cdot q = 1223 \cdot 1987 = 2430101.$$
- Samantha chooses a public verification exponent $e = 948047$ with the property that
$$\gcd\big(e, (p-1)(q-1)\big) = \gcd(948047, 2426892) = 1.$$

RSA Signing

- Samantha computes her private signing key d using the secret values of p and q to compute $(p-1)(q-1) = 1222 \cdot 1986 = 2426892$ and then solving the congruence
$$ed \equiv 1 \ \big(\mathrm{mod}\,(p-1)(q-1)\big), \qquad 948047 \cdot d \equiv 1 \pmod{2426892}.$$
She finds that $d = 1051235$.
- Samantha selects a digital document to sign,
$$D = 1070777 \quad \text{with} \quad 1 \le D < N.$$
She computes the digital signature
$$S \equiv D^d \pmod{N}, \quad S \equiv 1070777^{1051235} \equiv 153337 \pmod{2430101}.$$
- Samantha publishes the document and signature
$$D = 1070777 \qquad \text{and} \qquad S = 153337.$$

RSA Verification

- Victor uses Samantha's public modulus N and verification exponent e to compute
$$S^e \bmod N, \qquad 153337^{948047} \equiv 1070777 \pmod{2430101}.$$
He verifies that the value of S^e modulo N is the same as the value of the digital document $D = 1070777$.

4.3 Elgamal Digital Signatures and DSA

The transition from RSA encryption to RSA digital signatures, as described in Sect. 4.2, is quite straightforward. This is not true for discrete logarithm based encryption schemes such as Elgamal (Sect. 2.4).

An Elgamal-style digital signature scheme was put forward in 1985, and a modified version called the *Digital Signature Algorithm* (DSA), which allows shorter signatures, was proposed in 1991 and officially published as a national *Digital Signature Standard* (DSS) in 1994; see [98]. We start with the Elgamal scheme, which is easier to understand, and then explain how DSA works.

Samantha, or some trusted third party, chooses a large prime p and a primitive root g modulo p. Samantha next chooses a secret signing exponent a and computes

$$A \equiv g^a \pmod{p}.$$

The quantity a, together with the public parameters p and g, form Samantha's public verification key.

Suppose now that Samantha wants to sign a digital document D, where D is an integer satisfying $1 < D < p$. She chooses a random element $1 < k < p$ satisfying $\gcd(k, p-1) = 1$ and computes the two quantities

$$S_1 \equiv g^k \pmod{p} \qquad \text{and} \qquad S_2 \equiv (D - aS_1)k^{-1} \pmod{p-1}. \tag{4.2}$$

Notice that S_2 is computed modulo $p-1$, not modulo p. Samantha's digital signature on the document D is the pair (S_1, S_2).

Victor verifies the signature by checking that

$$A^{S_1} S_1^{S_2} \bmod p \quad \text{is equal to} \quad g^D \bmod p.$$

The Elgamal digital signature algorithm is illustrated in Table 4.2.

Why does Elgamal work? When Victor computes $A^{S_1} S_1^{S_2}$, he is actually computing

$$\begin{aligned} A^{S_1} \cdot S_1^{S_2} &\equiv g^{aS_1} \cdot g^{kS_2} \\ &\equiv g^{aS_1 + kS_2} \equiv g^{aS_1 + k(D - aS_1)k^{-1}} \equiv g^{aS_1 + (D - aS_1)} \equiv g^D \pmod{p}, \end{aligned}$$

so verification returns TRUE for a valid signature.

Notice the significance of choosing S_2 modulo $p - 1$. The quantity S_2 appears as an exponent of g, and we know that $g^{p-1} \equiv 1 \pmod{p}$, so in the expression $g^{S_2} \bmod p$, we may replace S_2 by any quantity that is congruent to S_2 modulo $p - 1$.

If Eve knows how to solve the discrete logarithm problem, then she can solve $g^a \equiv A \pmod{p}$ for Samantha's private signing key a, and thence can forge Samantha's signature. However, it is not at all clear that this is the only way to forge an Elgamal signature. Eve's task is as follows. Given the values of A and g^D, Eve must find integers x and y satisfying

$$A^x x^y \equiv g^D \pmod{p}. \tag{4.3}$$

The congruence (4.3) is a rather curious one, because the variable x appears as both a base and an exponent. Using discrete logarithms to the base g, we can rewrite (4.3) as

$$\log_g(A)x + y\log_g(x) \equiv D \pmod{p-1}. \tag{4.4}$$

<table>
<tr><th colspan="2">Public parameter creation</th></tr>
<tr><td colspan="2">A trusted party chooses and publishes a large prime p
and primitive root g modulo p.</td></tr>
<tr><th>Samantha</th><th>Victor</th></tr>
<tr><th colspan="2">Key creation</th></tr>
<tr><td>Choose secret signing key
$1 \le a \le p-1$.
Compute $A = g^a \pmod{p}$.
Publish the verification key A.</td><td></td></tr>
<tr><th colspan="2">Signing</th></tr>
<tr><td>Choose document $D \bmod p$.
Choose random element $1 < k < p$
satisfying $\gcd(k, p-1) = 1$.
Compute signature
$S_1 \equiv g^k \pmod{p}$ and
$S_2 \equiv (D - aS_1)k^{-1} \pmod{p-1}$.</td><td></td></tr>
<tr><th colspan="2">Verification</th></tr>
<tr><td></td><td>Compute $A^{S_1} S_1^{S_2} \bmod p$.
Verify that it is equal to $g^D \bmod p$.</td></tr>
</table>

Table 4.2: The Elgamal digital signature algorithm

If Eve can solve the discrete logarithm problem, she can take an arbitrary value for x, compute $\log_g(A)$ and $\log_g(x)$, and then solve (4.4) for y. At present, this is the only known method for finding a solution to (4.4).

Remark 4.6. There are many subtleties associated to using an ostensibly secure digital signature scheme such as Elgamal. See Exercises 4.7 and 4.8 for some examples of what can go wrong.

Example 4.7. Samantha chooses the prime $p = 21739$ and primitive root $g = 7$. She selects the secret signing key $a = 15140$ and computes her public verification key

$$A \equiv g^a \equiv 7^{15140} \equiv 17702 \pmod{21739}.$$

She signs the digital document $D = 5331$ using the random element $k = 10727$ by computing

$$S_1 \equiv g^k \equiv 7^{10727} \equiv 15775 \pmod{21739},$$
$$S_2 \equiv (D - aS_1)k^{-1} \equiv (5331 - 15140 \cdot 15775) \cdot 6353 \equiv 791 \pmod{21738}.$$

Samantha publishes the signature $(S_1, S_2) = (15775, 791)$ and the digital document $D = 5331$. Victor verifies the signature by computing

$$A^{S_1} S_1^{S_2} \equiv 17702^{15775} \cdot 15775^{791} \equiv 13897 \pmod{21739}$$

and verifying that it agrees with

$$g^D \equiv 7^{5331} \equiv 13897 \pmod{21739}.$$

An Elgamal signature (S_1, S_2) consists of one number modulo p and one number modulo $p-1$, so has length approximately $2\log_2(p)$ bits. In order to be secure against index calculus attacks on the discrete logarithm problem, the prime p is generally taken to be between 1000 and 2000 bits, so signatures are between 2000 and 4000 bits.

The *Digital Signature Algorithm* (DSA) significantly shortens the signature by working in a subgroup of $\mathbb{F}_p^*$ of prime order q. The underlying assumption is that using the index calculus to solve the discrete logarithm problem in the subgroup is no easier than solving it in $\mathbb{F}_p^*$. So it suffices to take a subgroup in which it is infeasible to solve the discrete logarithm problem using a collision algorithm. We now describe the details of DSA.

Samantha, or some trusted third party, chooses two primes p and q with

$$p \equiv 1 \pmod{q}.$$

(In practice, typical choices satisfy $2^{1000} < p < 2^{2000}$ and $2^{160} < q < 2^{320}$.) She also chooses an element $g \in \mathbb{F}_p^*$ of exact order q. This is easy to do. For example, she can take

$$g = g_1^{(p-1)/q} \quad \text{for a primitive root } g_1 \text{ in } \mathbb{F}_p.$$

Samantha chooses a secret exponent a and computes

$$A \equiv g^a \pmod{p}.$$

The quantity A, together with the public parameters (p, q, g), form Samantha's public verification key.

Suppose now that Samantha wants to sign a digital document D, where D is an integer satisfying $1 \le D < q$. She chooses a random element k in the range $1 < k < q$ and computes the two quantities

$$S_1 = (g^k \bmod p) \bmod q \qquad \text{and} \qquad S_2 \equiv (D + aS_1)k^{-1} \pmod{q}. \tag{4.5}$$

Notice the similarity between (4.5) and the Elgamal signature (4.2). However, there is an important difference, since when computing S_1 in (4.5), Samantha first computes $g^k \bmod p$ as an integer in the range from 1 to $p-1$, and then she reduces modulo q to obtain an integer in the range from 1 to $q-1$. Samantha's digital signature on the document D is the pair (S_1, S_2), so the signature consists of two numbers modulo q.

Victor verifies the signature by first computing

$$V_1 \equiv DS_2^{-1} \pmod{q} \qquad \text{and} \qquad V_2 \equiv S_1 S_2^{-1} \pmod{q}.$$

He then checks that

Public parameter creation	
A trusted party chooses and publishes large primes p and q satisfying $p \equiv 1 \pmod{q}$ and an element g of order q modulo p.	
Samantha	**Victor**
Key creation	
Choose secret signing key $1 \le a \le q-1$. Compute $A = g^a \pmod{p}$. Publish the verification key A.	
Signing	
Choose document $D \bmod q$. Choose random element $1 < k < q$. Compute signature $S_1 \equiv (g^k \bmod p) \bmod q$ and $S_2 \equiv (D + aS_1)k^{-1} \pmod{q}$.	
Verification	
	Compute $V_1 \equiv DS_2^{-1} \pmod{q}$ and $V_2 \equiv S_1S_2^{-1} \pmod{q}$. Verify that $(g^{V_1}A^{V_2} \bmod p) \bmod q = S_1$.

Table 4.3: The digital signature algorithm (DSA)

$$(g^{V_1}A^{V_2} \bmod p) \bmod q \quad \text{is equal to} \quad S_1.$$

The digital signature algorithm (DSA) is illustrated in Table 4.3.

DSA seems somewhat complicated, but it is easy to check that it works. Thus Victor computes

$$\begin{aligned} g^{V_1}A^{V_2} \pmod{p} &\equiv g^{DS_2^{-1}} g^{aS_1S_2^{-1}} && \text{since } V_1 \equiv DS_2^{-1} \text{ and } V_2 \equiv S_1S_2^{-1} \\ & && \text{and } A \equiv g^a, \\ &\equiv g^{(D+aS_1)S_2^{-1}} \pmod{p} \\ &\equiv g^k \pmod{p} && \text{since } S_2 \equiv (D+aS_1)k^{-1}. \end{aligned}$$

Hence

$$(g^{V_1}A^{V_2} \bmod p) \bmod q = (g^k \bmod p) \bmod q = S_1.$$

Example 4.8. We illustrate DSA with a small numerical example. Samantha uses the public parameters

$$p = 48731, \quad q = 443, \quad \text{and} \quad g = 5260.$$

(The element g was computed as $g \equiv 7^{48730/443} \pmod{48731}$, where 7 is a primitive root modulo 48731.) Samantha chooses the secret signing key $a = 242$ and publishes her public verification key

$$A \equiv 5260^{242} \equiv 3438 \pmod{48731}.$$

She signs the document $D = 343$ using the random element $k = 427$ by computing the two quantities

$$S_1 = (5260^{427} \bmod 48731) \bmod 443 = 2727 \bmod 443 = 59,$$
$$S_2 \equiv (343 + 343 \cdot 59)427^{-1} \equiv 166 \pmod{443}.$$

Samantha publishes the signature $(S_1, S_2) = (59, 166)$ for the document $D = 343$.

Victor verifies the signature by first computing

$$V_1 \equiv 343 \cdot 166^{-1} \equiv 357 \pmod{443} \quad \text{and} \quad V_2 \equiv 59 \cdot 166^{-1} \equiv 414 \pmod{443}.$$

He then computes

$$g^{V_1} A^{V_2} \equiv 5260^{357} \cdot 3438^{414} \equiv 2717 \pmod{48731}$$

and checks that

$$(g^{V_1} A^{V_2} \bmod 48731) \bmod 443 = 2717 \bmod 443 = 59$$

is equal to $S_1 = 59$.

Both the Elgamal digital signature scheme and DSA can be adapted to other groups in which the discrete logarithm problem is ostensibly more difficult to solve. In particular, the use of elliptic curve groups leads to the *Elliptic Curve Digital Signature Algorithm* (ECDSA), which is described in Sect. 6.4.3.

Exercises

Section 4.2. RSA Digital Signatures

4.1. Samantha uses the RSA signature scheme with primes $p = 541$ and $q = 1223$ and public verification exponent $e = 159853$.
(a) What is Samantha's public modulus? What is her private signing key?
(b) Samantha signs the digital document $D = 630579$. What is the signature?

4.2. Samantha uses the RSA signature scheme with public modulus $N = 1562501$ and public verification exponent $e = 87953$. Adam claims that Samantha has signed each of the documents

$$D = 119812, \quad D' = 161153, \quad D'' = 586036,$$

and that the associated signatures are

$$S = 876453, \quad S' = 870099, \quad S'' = 602754.$$

Which of these are valid signatures?

4.3. Samantha uses the RSA signature scheme with public modulus and public verification exponent

$$N = 27212325191 \quad \text{and} \quad e = 22824469379.$$

Use whatever method you want to factor N, and then forge Samantha's signature on the document $D = 12910258780$.

4.4. Suppose that Alice and Bob communicate using the RSA PKC. This means that Alice has a public modulus $N_A = p_A q_A$, a public encryption exponent e_A, and a private decryption exponent d_A, where p_A and q_A are primes and e_A and d_A satisfy

$$e_A d_A \equiv 1 \pmod{(p_A - 1)(q_A - 1)}.$$

Similarly, Bob has a public modulus $N_B = p_B q_B$, a public encryption exponent e_B, and a private decryption exponent d_B.

In this situation, Alice can simultaneously encrypt and sign a message in the following way. Alice chooses her plaintext m and computes the usual RSA ciphertext

$$c \equiv m^{e_B} \pmod{N_B}.$$

She next applies a hash function to her plaintext and uses her private decryption key to compute

$$s \equiv \mathsf{Hash}(m)^{d_A} \pmod{N_A}.$$

She sends the pair (c, s) to Bob.

Bob first decrypts the ciphertext using his private decryption exponent d_B,

$$m \equiv c^{d_B} \pmod{N_B}.$$

He then uses Alice's public encryption exponent e_A to verify that

$$\mathsf{Hash}(m) \equiv s^{e_A} \pmod{N_A}.$$

Explain why verification works, and why it would be difficult for anyone other than Alice to send Bob a validly signed message.

Section 4.3. Discrete Logarithm Digital Signatures

4.5. Samantha uses the Elgamal signature scheme with prime $p = 6961$ and primitive root $g = 437$.

(a) Samantha's private signing key is $a = 6104$. What is her public verification key?

(b) Samantha signs the digital document $D = 5584$ using the random element $k = 4451$. What is the signature?

4.6. Samantha uses the Elgamal signature scheme with prime $p = 6961$ and primitive root $g = 437$. Her public verification key is $A = 4250$. Adam claims that Samantha has signed each of the documents

$$D = 1521, \quad D' = 1837, \quad D'' = 1614,$$

and that the associated signatures are

$$(S_1, S_2) = (4129, 5575), \quad (S_1', S_2') = (3145, 1871), \quad (S_1'', S_2'') = (2709, 2994).$$

Which of these are valid signatures?

4.7. Let p be a prime, let i and j be integers with $\gcd(j, p-1) = 1$, and let A be arbitrary. Set

$$S_1 \equiv g^i A^j \pmod{p}, \quad S_2 \equiv -S_1 j^{-1} \pmod{p-1}, \quad D \equiv -S_1 i j^{-1} \pmod{p-1}.$$

Prove that (S_1, S_2) is a valid Elgamal signature on the document D for the verification key A. Thus Eve can produce signatures on random documents.

4.8. Suppose that Samantha is using the Elgamal signature scheme and that she is careless and uses the same random element k to sign two documents D and D'.

(a) Explain how Eve can tell at a glance whether Samantha has made this mistake.

(b) If the signature on D is (S_1, S_2) and the signature on D' is (S'_1, S'_2), explain how Eve can recover a, Samantha's private signing key.

(c) Apply your method from (b) to the following example and recover Samantha's signing key a, where Samantha is using the prime $p = 348149$, base $g = 113459$, and verification key $A = 185149$.

$$\begin{aligned} D &= 153405, & S_1 &= 208913, & S_2 &= 209176, \\ D' &= 127561, & S'_1 &= 208913, & S'_2 &= 217800. \end{aligned}$$

4.9. Samantha uses DSA with public parameters $(p, q, g) = (22531, 751, 4488)$. She chooses the secret signing key $a = 674$.

(a) What is Samantha's public verification key?

(b) Samantha signs the document $D = 244$ using the random element $k = 574$. What is the signature?

4.10. Samantha uses DSA with public parameters $(p, q, g) = (22531, 751, 4488)$. Her public verification key is $A = 22476$.

(a) Is $(S_1, S_2) = (183, 260)$ a valid signature on the document $D = 329$?

(b) Is $(S_1, S_2) = (211, 97)$ a valid signature on the document $D = 432$?

4.11. Samantha's DSA public parameters are $(p, q, g) = (103687, 1571, 21947)$, and her public verification key is $A = 31377$. Use whatever method you prefer (brute-force, collision, index calculus,...) to solve the DLP and find Samantha's private signing key. Use her key to sign the document $D = 510$ using the random element $k = 1105$.

4.[illegible] [illegible] integers [illegible] k [illegible] and [illegible]

[illegible]

[illegible] and the final digit [illegible] is [illegible]

4.[illegible] [illegible] and [illegible]

[illegible] (1984) [illegible]

[illegible]

[illegible]

4.16 [illegible]

[illegible]

4.17 [illegible] RSA [illegible]

[illegible]

Chapter 5

Combinatorics, Probability, and Information Theory

In considering the usefulness and practicality of a cryptographic system, it is necessary to measure its resistance to various forms of attack. Such attacks include simple brute-force searches through the key or message space, somewhat faster searches via *collision* or *meet-in-the-middle* algorithms, and more sophisticated methods that are used to compute discrete logarithms, factor integers, and find short vectors in lattices. We have already studied some of these algorithms in Chaps. 2 and 3, and we will see the others in this and later chapters. In studying these algorithms, it is important to be able to analyze how long they take to solve the targeted problem. Such an analysis generally requires tools from combinatorics, probability theory, and information theory. In this chapter we present, in a largely self-contained form, an introduction to these topics.

We start with basic principles of counting, and continue with the development of the foundations of probability theory, primarily in the discrete setting. Subsequent sections introduce (discrete) random variables, probability density functions, conditional probability and Bayes's formula. The applications of probability theory to cryptography are legion. We cover in some detail Monte Carlo algorithms and collision algorithms and their uses in cryptography. We also include a section on the statistical cryptanalysis of a historically interesting polyalphabetic substitution cipher called the Vigenère cipher, but we note that the material on the Vigenère cipher is not used elsewhere in the book, so it may be omitted by the reader who wishes to proceed more rapidly to the more modern cryptographic material.

J. Hoffstein et al., *An Introduction to Mathematical Cryptography*,
Undergraduate Texts in Mathematics, DOI 10.1007/978-1-4939-1711-2_5

The chapter concludes with a very short introduction to the concept of complexity and the notions of polynomial-time and nondeterministic polynomial-time algorithms. This section, if properly developed, would be a book in itself, and we can only give a hint of the powerful ideas and techniques used in this subject.

5.1 Basic Principles of Counting

As I was going to St. Ives,
I met a man with seven wives,
Each wife had seven sacks,
Each sack had seven cats.
Each cat had seven kits.
Kits, cats, sacks, and wives,
How many were going to St. Ives?

The trick answer to this ancient riddle is that there is only one person going to St. Ives, namely the narrator, since all of the other people and animals and objects that he meets in the rhyme are not traveling *to* St. Ives, they are traveling *away* from St. Ives! However, if we are in a pedantic, rather than a clever, frame of mind, we might instead ask the natural question: How many people, animals, and objects does the narrator meet?

The answer is

$$2801 = \underbrace{1}_{\text{man}} + \underbrace{7}_{\text{wives}} + \underbrace{7^2}_{\text{sacks}} + \underbrace{7^3}_{\text{cats}} + \underbrace{7^4}_{\text{kits}}.$$

The computation of this number employs basic counting principles that are fundamental to the probability calculations used in cryptography and in many other areas of mathematics. We have already seen an example in Sect. 1.1.1, where we computed the number of different simple substitution ciphers.

A cipher is said to be *combinatorially secure* if it is not feasible to break the system by exhaustively checking every possible key.[1] This depends to some extent on how long it takes to check each key, but more importantly, it depends on the number of keys. In this section we develop some basic counting techniques that are used in a variety of ways to analyze the security of cryptographic constructions.

Example 5.1 (A Basic Counting Principle). Bob is at a restaurant that features two appetizers, egg rolls and fried wontons, and 20 main dishes. Assuming that he plans to order one appetizer and one main dish, how many possible meals could Bob order?

We need to count the number of pairs (x, y), where x is either "egg roll" or "fried wonton" and y is a main dish. The total number is obtained by letting x

[1] Sometimes the length of the search can be significantly shortened by matching pieces of keys taken from two or more lists. Such an attack is called a collision or meet-in-the-middle attack; see Sect. 5.4.

vary over the 2 possibilities and letting y vary over the 20 possibilities, and then counting up the total number of pairs

$$(ER, 1), (ER, 2), \ldots, (ER, 20), (FW, 1), (FW, 2), \ldots, (FW, 20).$$

The answer is that there are 40 possibilities, which we compute as

$$40 = \underbrace{2}_{\text{appetizers}} \cdot \underbrace{20}_{\text{main dishes}} .$$

In this example, we first counted the number of ways of assigning an appetizer (egg roll or fried wonton) to the variable x. It is convenient to view this assignment as the outcome of an *experiment*. That is, we perform an experiment whose outcome is either "egg roll" or "fried wonton," and we assign the outcome's value to x. Similarly, we perform a second independent experiment whose possible outcomes are any one of the 20 main courses, and we assign that value to y. The total number of outcomes of the two experiments is the product of the number of outcomes for each one individually. This leads to the following basic counting principle:

Basic Counting Principle

If two experiments are performed, one of which has n possible outcomes and the other of which has m possible outcomes, then there are nm possible outcomes of performing both experiments.

More generally, if k independent experiments are performed and if the number of possible outcomes of the ith experiment is n_i, then the total number of outcomes for all of the experiments is the product $n_1 n_2 \cdots n_k$. It is easy to derive this result by writing x_i for the outcome of the ith experiment. Then the outcome of all k experiments is the value of the k-tuple $(x_1, x_2, \ldots, x_k)$, and the total number of possible k-tuples is the product $n_1 n_2 \cdots n_k$.

Example 5.2. Suppose that Bob also wants to order dessert, and that there are five desserts on the menu. We are now counting triples (x, y, z), where x is one of the two appetizers, y is one of the 20 main dishes, and z is one of the five desserts. Hence the total number of meals is

$$200 = \underbrace{2}_{\text{appetizers}} \cdot \underbrace{20}_{\text{main courses}} \cdot \underbrace{5}_{\text{desserts}} .$$

The basic counting principle is used in the solution of the pedantic version of the St. Ives problem. For example, the number of cats traveling from St. Ives is

$$\#\text{ of cats} = 343 = 7^3 = \underbrace{1}_{\text{man}} \cdot \underbrace{7}_{\text{wives}} \cdot \underbrace{7}_{\text{sacks}} \cdot \underbrace{7}_{\text{cats}} .$$

The earliest published version of the St. Ives riddle dates to around 1730, but similar problems date back to antiquity; see Exercise 5.1.

5.1.1 Permutations

The numbers $1, 2, \ldots, 10$ are typically listed in increasing order, but suppose instead we allow the order to be mixed. Then how many different ways are there to list these ten integers? Each possible configuration is called a *permutation* of $1, 2, \ldots, 10$. The problem of counting the number of possible permutations of a given list of objects occurs in many forms and contexts throughout mathematics.

Each permutation of $1, 2, \ldots, 10$ is a sequence of all ten distinct integers in some order. For example, here is a random choice: $8, 6, 10, 3, 9, 2, 4, 7, 5, 1$. How can we create all of the possibilities? It's easiest to create them by listing the numbers one at a time, say from left to right. We thus start by assigning a number to the first position. There are ten choices. Next we assign a number to the second position, but for the second position there are only nine choices, because we already used up one of the integers in the first position. (Remember that we are not allowed to use an integer twice.) Then there are eight integers left as possibilities for the third position, because we already used two integers in the first two positions. And so on. Hence the total number of permutations of $1, 2, \ldots, 10$ is

$$10! = 10 \cdot 9 \cdot 8 \cdots 2 \cdot 1.$$

The value of $10!$ is 3628800, so between three and four million.

Notice how we are using the basic counting principle. The only subtlety is that the outcome of the first experiment reduces the number of possible outcomes of the second experiment, the results of the first two experiments further reduce the number of possible outcomes of the third experiment, and so on.

Definition. Let S be a set containing n distinct objects. A *permutation of S* is an ordered list of the objects in S. A permutation of the set $\{1, 2, \ldots, n\}$ is simply called a *permutation of n*.

Proposition 5.3. *Let S be a set containing n distinct objects. Then there are exactly $n!$ different permutations of S.*

Proof. Our discussion of the permutations of $\{1, \ldots, 10\}$ works in general. Thus suppose that S contains n objects and that we want to create a permutation of S. There are n choices for the first entry, then $n-1$ choices for the second entry, then $n-2$ choices for the third entry, etc. This leads to a total of $n \cdot (n-1) \cdot (n-2) \cdots 2 \cdot 1$ possible permutations. □

Remark 5.4 (Permutations and Simple Substitution Ciphers). By definition, a permutation of the set $\{a_1, a_2, \ldots, a_n\}$ is a list consisting of the a_i's in some order. We can also describe a permutation by using a bijective (i.e., one-to-one and onto) function

$$\pi : \{1, 2, \ldots, n\} \longrightarrow \{1, 2, \ldots, n\}.$$

The function π determines the permutation

$$(a_{\pi(1)}, a_{\pi(2)}, \ldots, a_{\pi(n)}),$$

and given a permutation, it is easy to write down the corresponding function.

Now suppose that we take the set of letters $\{\mathtt{A}, \mathtt{B}, \mathtt{C}, \ldots, \mathtt{Z}\}$. A permutation π of this set is just another name for a simple substitution cipher, where π acts as the encryption function. Thus π tells us that A gets sent to the $\pi(1)$st letter, and B gets sent to the $\pi(2)$nd letter, and so on. In order to decrypt, we use the inverse function π^{-1}.

Example 5.5. Sometimes one needs to count the number of possible permutations of n objects when some of the objects are indistinguishable. For example, there are six permutations of three distinct objects A, B, C,

$$ABC, \quad CAB, \quad BCA, \quad ACB, \quad BAC, \quad \text{and} \quad CBA,$$

but if two of them are indistinguishable, say A, A, B, then there are only three different arrangements,

$$AAB, \quad ABA, \quad \text{and} \quad BAA.$$

To illustrate the idea in a more complicated case, we count the number of different letter arrangements of the five letters A, A, A, B, B. If the five letters were distinguishable, say they were labeled A_1, A_2, A_3, B_1, B_2, then there would be 5! permutations. However, permutations such as

$$A_1A_2B_1B_2A_3 \qquad \text{and} \qquad A_2A_3B_2B_1A_1$$

become the same when the subscripts are dropped, so we have overcounted in arriving at the number 5!. How many different arrangements have been counted more than once?

For example, in any particular permutation, the two B's have been placed into specific positions, but we can always switch them and get the same unsubscripted list. This means that we need to divide 5! by 2 to compensate for overcounting the placement of the B's. Similarly, once the three A's have been placed into specific positions, we can permute them among themselves in 3! ways, so we need to divide 5! by 3! to compensate for overcounting the placement of the A's. Hence there are $\frac{5!}{3!\cdot 2!} = 10$ different letter arrangements of the five letters A, A, A, B, B.

5.1.2 Combinations

A permutation is a way of arranging a set of objects into a list. A combination is similar, except that now the order of the list no longer matters. We start with an example that is typical of problems involving combinations.

Example 5.6. Five people (Alice, Bob, Carl, Dave, and Eve[2]) are ordering a meal at a Chinese restaurant. The menu contains 20 different items. Each person gets to choose one dish, no dish may be ordered twice, and they plan to share the food. How many different meals are possible?

Alice orders first and she has 20 choices for her dish. Then Bob orders from the remaining 19 dishes, and then Carl chooses from the remaining 18 dishes, and so on. It thus appears that there are $20 \cdot 19 \cdot 18 \cdot 17 \cdot 16 = 1860480$ possible meals. However, the order in which the dishes are ordered is immaterial. If Alice orders fried rice and Bob orders egg rolls, or if Alice orders egg rolls and Bob orders fried rice, the meal is the same. Unfortunately, we did not take this into account when we arrived at the number 1860480.

Let's number the dishes $D_1, D_2, \ldots, D_{20}$. Then, for example, we want to count the two possible dinners

$$D_1, D_5, D_7, D_{18}, D_{20} \qquad \text{and} \qquad D_5, D_{18}, D_{20}, D_7, D_1$$

as being the same, although the order of the dishes is different. To correct the overcount, note that in the computation $20 \cdot 19 \cdot 18 \cdot 17 \cdot 16 = 1860480$, every permutation of any set of five dishes was counted separately, but we really want to count these permutations as giving the same meal. Thus we should divide 1860480 by the number of ways to permute the five distinct dishes in each possible order, i.e., we should divide by 5!. Hence the total number of different meals is

$$\frac{20 \cdot 19 \cdot 18 \cdot 17 \cdot 16}{5!} = 15504.$$

It is often convenient to rewrite this quantity entirely in terms of factorials by multiplying the numerator and the denominator by 15! to get

$$\frac{20 \cdot 19 \cdot 18 \cdot 17 \cdot 16}{5!} = \frac{(20 \cdot 19 \cdot 18 \cdot 17 \cdot 16) \cdot (15 \cdot 14 \cdots 3 \cdot 2 \cdot 1)}{5! \cdot 15!} = \frac{20!}{5! \cdot 15!}.$$

Definition. Let S be a set containing n distinct objects. A *combination* of r objects of S is a subset consisting of exactly r distinct elements of S, where the order of the objects in the subset does not matter.

Proposition 5.7. *The number of possible combinations of r objects chosen from a set of n objects is equal to*

$$\binom{n}{r} = \frac{n!}{r!(n-r)!}.$$

Remark 5.8. The symbol $\binom{n}{r}$ is called a *combinatorial symbol* or a *binomial coefficient.* It is read as "n choose r." Note that by convention, zero factorial is set equal to 1, so $\binom{n}{0} = \frac{n!}{n! \cdot 0!} = 1$. This makes sense, since there is only one way to choose zero objects from a set.

[2]You may wonder why Alice and Bob, those intrepid exchangers of encrypted secret messages, are sitting down for a meal with their cryptographic adversary Eve. In the real world, this happens all the time, especially at cryptography conferences!

Proof of Proposition 5.7. If you understand the discussion in Example 5.6, then the proof of the general case is clear. The number of ways to make an *ordered* list of r distinct elements from the set S is

$$n(n-1)(n-2)\cdots(n-r+1),$$

since there are n choices for the first element, then $n-1$ choices for the second element, and so on until we have selected r elements. Then we need to divide by $r!$ in order to compensate for the ways to permute the r elements in our subset. Dividing by $r!$ accounts for the fact that we do not care in which order the r elements were chosen. Hence the total number of combinations is

$$\frac{n(n-1)(n-2)\cdots(n-r+1)}{r!} = \frac{n!}{r!(n-r)!}. \qquad \square$$

Example 5.9. Returning to the five people ordering a meal at the Chinese restaurant, suppose that they want the order to consist of two vegetarian dishes and three meat dishes, and suppose that the menu contains 5 vegetarian choices and 15 meat choices. Now how many possible meals can they order? There are $\binom{5}{2}$ possibilities for the two vegetarian dishes and there are $\binom{15}{3}$ choices for the three meat dishes. Hence by our basic counting principle, there are

$$\binom{5}{2}\cdot\binom{15}{3} = 10\cdot 455 = 4550$$

possible meals.

5.1.3 The Binomial Theorem

You may have seen the combinatorial numbers $\binom{n}{r}$ appearing in the binomial theorem,[3] which gives a formula for the nth power of the sum of two numbers.

Theorem 5.10 (The Binomial Theorem).

$$(x+y)^n = \sum_{j=0}^{n}\binom{n}{j}x^j y^{n-j}. \tag{5.1}$$

Proof. Let's start with a particular case, say $n=3$. If we multiply out the product

[3]The binomial theorem's fame extends beyond mathematics. Moriarty, Sherlock Holmes's arch enemy, "wrote a treatise upon the Binomial Theorem," on the strength of which he won a mathematical professorship. And Major General Stanley, that very Model of a Modern Major General, proudly informs the Pirate King and his cutthroat band:

About Binomial Theorem I'm teeming with a lot o' news—
With many cheerful facts about the square of the hypotenuse.
(*The Pirates of Penzance*, W.S. Gilbert and A. Sullivan 1879)

$$(x+y)^3 = (x+y)\cdot(x+y)\cdot(x+y), \tag{5.2}$$

the result is a sum of terms x^3, x^2y, xy^2, and y^3. There is only one x^3 term, since to get x^3 we must take x from each of the three factors in (5.2). How many copies of x^2y are there? We can get x^2y in several ways. For example, we could take x from the first two factors and y from the last factor. Or we could take x from the first and third factors and take y from the second factor. Thus we get x^2y by choosing two of the three factors in (5.2) to give x (note that the order doesn't matter), and then the remaining factor gives y. There are thus $\binom{3}{2} = 3$ ways to get x^2y. Similarly, there are $\binom{3}{1} = 3$ ways to get xy^2 and only one way to get y^3. Hence

$$(x+y)^3 = \binom{3}{3}x^3 + \binom{3}{2}x^2y + \binom{3}{1}xy^2 + \binom{3}{0}y^3 = x^3 + 3x^2y + 3xy^2 + y^3.$$

The general case is exactly the same. When multiplied out, the product

$$(x+y)^n = (x+y)\cdot(x+y)\cdot(x+y)\cdots(x+y) \tag{5.3}$$

is a sum of terms $x^n, x^{n-1}y, \ldots, xy^{n-1}, y^n$. We get copies of x^jy^{n-j} by choosing x from any j of the factors in (5.3) and then taking y from the other $n-j$ factors. Thus we get $\binom{n}{j}$ copies of x^jy^{n-j}. Summing over the possible values of j gives (5.1), which completes the proof of the binomial theorem. □

Example 5.11. We use the binomial theorem to compute

$$\begin{aligned}(2t+3)^4 &= \binom{4}{4}(2t)^4 + \binom{4}{3}(2t)^3\cdot 3 + \binom{4}{2}(2t)^2\cdot 3^2 + \binom{4}{1}2t\cdot 3^3 + \binom{4}{0}3^4\\ &= 16t^4 + 4\cdot 8t^3\cdot 3 + 6\cdot 4t^2\cdot 9 + 4\cdot 2t\cdot 27 + 81\\ &= 16t^4 + 96t^3 + 216t^2 + 216t + 81.\end{aligned}$$

5.2 The Vigenère Cipher

The simple substitution ciphers that we studied in Sect. 1.1 are examples of *monoalphabetic ciphers*, since every plaintext letter is encrypted using only one cipher alphabet. As cryptanalytic methods became more sophisticated in Renaissance Italy, correspondingly more sophisticated ciphers were invented (although it seems that they were seldom used in practice). Consider how much more difficult a task is faced by the cryptanalyst if every plaintext letter is encrypted using a different ciphertext alphabet. This ideal resurfaces in modern cryptography in the form of the one-time pad, which we discuss in Sect. 5.6, but in this section we discuss a less complicated *polyalphabetic cipher* called the Vigenère cipher[4] dating back to the sixteenth century.

[4]This cipher is named after Blaise de Vigenère (1523–1596), whose 1586 book *Traicté des Chiffres* describes the known ciphers of his time. These include polyalphabetic ciphers such as the "Vigenère cipher," which according to [63] Vigenère did not invent, and an ingenious autokey system (see Exercise 5.19), which he did.

The Vigenère cipher works by using different shift ciphers to encrypt different letters. In order to decide how far to shift each letter, Bob and Alice first agree on a keyword or phrase. Bob then uses the letters of the keyword, one by one, to determine how far to shift each successive plaintext letter. If the keyword letter is `a`, there is no shift, if the keyword letter is `b`, he shifts by 1, if the keyword letter is `c`, he shifts by 2, and so on. An example illustrates the process:

Example 5.12. Suppose that the keyword is `dog` and the plaintext is `yellow`. The first letter of the keyword is `d`, which gives a shift of 3, so Bob shifts the first plaintext letter `y` forward by 3, which gives the ciphertext letter `b`. (Remember that `a` follows `z`.) The second letter of the keyword is `o`, which gives a shift of 14, so Bob shifts the second plaintext letter `e` forward by 14, which gives the ciphertext letter `s`. The third letter of the keyword is `g`, which gives a shift of 6, so Bob shifts the third plaintext letter `l` forward by 6, which gives the ciphertext letter `r`.

Bob has run out of keyword letters, so what does he do now? He simply starts again with the first letter of the keyword. The first letter of the keyword is `d`, which again gives a shift of 3, so Bob shifts the fourth plaintext letter `l` forward by 3, which gives the ciphertext letter `o`. Then the second keyword letter `o` tells him to shift the fifth plaintext letter `o` forward by 14, giving the ciphertext letter `c`, and finally the third keyword letter `g` tells him to shift the sixth plaintext letter `w` forward by 6, giving the ciphertext letter `c`.

In conclusion, Bob has encrypted the plaintext `yellow` using the keyword `dog` and obtained the ciphertext `bsrocc`.

Even this simple example illustrates two important characteristics of the Vigenère cipher. First, the repeated letters `ll` in the plaintext lead to non-identical letters `ro` in the ciphertext, and second, the repeated letters `cc` in the ciphertext correspond to different letters `ow` of the plaintext. Thus a straightforward frequency analysis as we used to cryptanalyze simple substitution ciphers (Sect. 1.1.1) is not going to work for the Vigenère cipher.

A useful tool for doing Vigenère encryption and decryption, at least if no computer is available (as was typically the case in the sixteenth century!), is the so-called *Vigenère tableau* illustrated in Table 5.1. The Vigenère tableau consists of 26 alphabets arranged in a square, with each alphabet shifted one further than the alphabet to its left. In order to use a given keyword letter to encrypt a given plaintext letter, Bob finds the plaintext letter in the top row and the keyword letter in the first column. He then looks for the letter in the tableau lying below the plaintext letter and to the right of the keyword letter. That is, he locates the encrypted letter at the intersection of the row beginning with the keyword letter and the column with the plaintext letter on top.

For example, if the keyword letter is `d` and the plaintext letter is `y`, Bob looks in the fourth row (which is the one that starts with `d`) and in the next

a	b	c	d	e	f	g	h	i	j	k	l	m	n	o	p	q	r	s	t	u	v	w	x	y	z
b	c	d	e	f	g	h	i	j	k	l	m	n	o	p	q	r	s	t	u	v	w	x	y	z	a
c	d	e	f	g	h	i	j	k	l	m	n	o	p	q	r	s	t	u	v	w	x	y	z	a	b
d	e	f	g	h	i	j	k	l	m	n	o	p	q	r	s	t	u	v	w	x	y	z	a	b	c
e	f	g	h	i	j	k	l	m	n	o	p	q	r	s	t	u	v	w	x	y	z	a	b	c	d
f	g	h	i	j	k	l	m	n	o	p	q	r	s	t	u	v	w	x	y	z	a	b	c	d	e
g	h	i	j	k	l	m	n	o	p	q	r	s	t	u	v	w	x	y	z	a	b	c	d	e	f
h	i	j	k	l	m	n	o	p	q	r	s	t	u	v	w	x	y	z	a	b	c	d	e	f	g
i	j	k	l	m	n	o	p	q	r	s	t	u	v	w	x	y	z	a	b	c	d	e	f	g	h
j	k	l	m	n	o	p	q	r	s	t	u	v	w	x	y	z	a	b	c	d	e	f	g	h	i
k	l	m	n	o	p	q	r	s	t	u	v	w	x	y	z	a	b	c	d	e	f	g	h	i	j
l	m	n	o	p	q	r	s	t	u	v	w	x	y	z	a	b	c	d	e	f	g	h	i	j	k
m	n	o	p	q	r	s	t	u	v	w	x	y	z	a	b	c	d	e	f	g	h	i	j	k	l
n	o	p	q	r	s	t	u	v	w	x	y	z	a	b	c	d	e	f	g	h	i	j	k	l	m
o	p	q	r	s	t	u	v	w	x	y	z	a	b	c	d	e	f	g	h	i	j	k	l	m	n
p	q	r	s	t	u	v	w	x	y	z	a	b	c	d	e	f	g	h	i	j	k	l	m	n	o
q	r	s	t	u	v	w	x	y	z	a	b	c	d	e	f	g	h	i	j	k	l	m	n	o	p
r	s	t	u	v	w	x	y	z	a	b	c	d	e	f	g	h	i	j	k	l	m	n	o	p	q
s	t	u	v	w	x	y	z	a	b	c	d	e	f	g	h	i	j	k	l	m	n	o	p	q	r
t	u	v	w	x	y	z	a	b	c	d	e	f	g	h	i	j	k	l	m	n	o	p	q	r	s
u	v	w	x	y	z	a	b	c	d	e	f	g	h	i	j	k	l	m	n	o	p	q	r	s	t
v	w	x	y	z	a	b	c	d	e	f	g	h	i	j	k	l	m	n	o	p	q	r	s	t	u
w	x	y	z	a	b	c	d	e	f	g	h	i	j	k	l	m	n	o	p	q	r	s	t	u	v
x	y	z	a	b	c	d	e	f	g	h	i	j	k	l	m	n	o	p	q	r	s	t	u	v	w
y	z	a	b	c	d	e	f	g	h	i	j	k	l	m	n	o	p	q	r	s	t	u	v	w	x
z	a	b	c	d	e	f	g	h	i	j	k	l	m	n	o	p	q	r	s	t	u	v	w	x	y

- Find the plaintext letter in the top row.
- Find the keyword letter in the first column.
- The ciphertext letter lies below the plaintext letter and to the right of the keyword letter.

Table 5.1: The Vigenère Tableau

to last column (which is the one headed by y). This row and column intersect at the letter b, so the corresponding ciphertext letter is b.

Decryption is just as easy. Alice uses the row containing the keyword letter and looks in that row for the ciphertext letter. Then the top of that column is the plaintext letter. For example, if the keyword letter is g and the ciphertext letter is r, Alice looks in the row starting with g until she finds r and then she moves to the top of that column to find the plaintext letter l.

Example 5.13. We illustrate the use of the Vigenère tableau by encrypting the plaintext message

`The rain in Spain stays mainly in the plain,`

using the keyword `flamingo`. Since the key word has eight letters, the first step is to split the plaintext into eight-letter blocks,

`theraini | nspainst | aysmainl | yinthepl | ain.`

Next we write the keyword beneath each block of plaintext, where for convenience we label lines $\mathcal{P}$, $\mathcal{K}$, and $\mathcal{C}$ to indicate, respectively, the plaintext, the keyword, and the ciphertext.

$\mathcal{P}$	t h e r a i n i	n s p a i n s t	a y s m a i n l	y i n t h e p l	a i n
$\mathcal{K}$	f l a m i n g o	f l a m i n g o	f l a m i n g o	f l a m i n g o	f l a

Finally, we encrypt each letter using the Vigenère tableau. The initial plaintext letter t and initial keyword letter f combine in the Vigenère tableau to yield the ciphertext letter y, the second plaintext letter h and second keyword letter l combine in the Vigenère tableau to yield the ciphertext letter s, and so on. Continuing in this fashion, we complete the encryption process.

$\mathcal{P}$	t h e r a i n i	n s p a i n s t	a y s m a i n l	y i n t h e p l	a i n
$\mathcal{K}$	f l a m i n g o	f l a m i n g o	f l a m i n g o	f l a m i n g o	f l a
$\mathcal{C}$	y s e d i v t w	s d p m q a y h	f j s y i v t z	d t n f p r v z	f t n

Splitting the ciphertext into convenient blocks of five letters each, we are ready to transmit our encrypted message

`ysedi vtwsd pmqay hfjsy ivtzd tnfpr vzftn.`

Remark 5.14. As we already pointed out, the same plaintext letter in a Vigenère cipher is represented in the ciphertext by many different letters. However, if the keyword is short, there will be a tendency for repetitive parts of the plaintext to end up aligned at the same point in the keyword, in which case they will be identically enciphered. This occurs in Example 5.13, where the ain in rain and in mainly are encrypted using the same three keyword letters ing, so they yield the same ciphertext letters ivt. This repetition in the ciphertext, which appears separated by 16 letters, suggests that the keyword has length dividing 16. Of course, not every occurrence of ain in the plaintext yields the same ciphertext. It is only when two occurrences line up with the same part of the keyword that repetition occurs.

In the next section we develop the idea of using ciphertext repetitions to guess the length of the keyword, but here we simply want to make the point that short keywords are less secure than long keywords.[5] On the other hand, Bob and Alice find it easier to remember a short keyword than a long one. We thus see the beginnings of the eternal struggle in practical (as opposed to purely theoretical) cryptography, namely the battle between

$$\textbf{Efficiency (and ease of use)} \longleftarrow \text{versus} \longrightarrow \textbf{Security}.$$

As a further illustration of this dichotomy, we consider ways in which Bob and Alice might make their Vigenère-type cipher more secure. They can certainly make Eve's job harder by mixing up the letters in the first row of their Vigenère tableau and then rotating this "mixed alphabet" in the subsequent rows. Unfortunately, a mixed alphabet makes encryption and decryption more cumbersome, plus it means that Bob and Alice must remember (or write down for safekeeping!) not only their keyword, but also the mixed alphabet. And if they want to be even more secure, they can use different randomly mixed alphabets in every row of their Vigenère tableau. But if they do that, then they will certainly need to keep a written copy of the tableau, which is a serious security risk.

5.2.1 Cryptanalysis of the Vigenère Cipher: Theory

At various times in history it has been claimed that Vigenère-type ciphers, especially with mixed alphabets, are "unbreakable." In fact, nothing could be further from the truth. If Eve knows Bob and Alice, she may be able to guess part of the keyword and proceed from there. (How many people do you know who use some variation of their name and birthday as an Internet password?) But even without lucky guesses, elementary statistical methods developed in the nineteenth century allow for a straightforward cryptanalysis of Vigenère-type ciphers. In the interest of simplicity, we stick with the original Vigenère, i.e., we do not allow mixed alphabets in the tableau.

You may wonder why we take the time to cryptanalyze the Vigenère cipher, since no one these days uses the Vigenère for secure communications. The answer is that our exposition is designed principally to introduce you to the use of statistical tools in cryptanalysis. This builds on and extends the elementary application of frequency tables as we used them in Sect. 1.1.1 to cryptanalyze simple substitution ciphers. In this section we describe the theoretical tools used to cryptanalyze the Vigenère, and in the next section we apply those tools to decrypt a sample ciphertext. If at any point you find that the theory in this section becomes confusing, it may help to turn to Sect. 5.2.2 and see how the theory is applied in practice.

[5]More typically one uses a key phrase consisting of several words, but for simplicity we use the term "keyword" to cover both single keywords and longer key phrases.

The first goal in cryptanalyzing a Vigenère cipher is to find the length of the keyword, which is sometimes called the *blocksize* or the *period*. We already saw in Remark 5.14 how this might be accomplished by looking for repeated fragments in the ciphertext. The point is that certain plaintext fragments such as `the` occur quite frequently, while other plaintext fragments such as `ugw` occur infrequently or not at all. Among the many occurrences of the letters `the` in the plaintext, a certain percentage of them will line up with exactly the same part of the keyword.

This leads to the *Kasiski method*, first described by a German military officer named Friedrich Kasiski in his book *Die Geheimschriften und die Dechiffrir-kunst*[6] published in 1863. One looks for repeated fragments within the ciphertext and compiles a list of the distances that separate the repetitions. The key length is likely to divide many of these distances. Of course, a certain number of repetitions will occur by pure chance, but these are random, while the ones coming from repeated plaintext fragments are always divisible by the key length. It is generally not hard to pick out the key length from this data.

There is another method of guessing the key length that works with individual letters, rather than with fragments consisting of several letters. The underlying idea can be traced all the way back to the frequency table of English letters (Table 1.3), which shows that some letters are more likely to occur than others. Suppose now that you are presented with a ciphertext encrypted using a Vigenère cipher and that you guess that it was encrypted using a keyword of length 5. This means that every fifth letter was encrypted using the same rotation, so if you pull out every fifth letter and form them into a string, this entire string was encrypted using a single substitution cipher. Hence the string's letter frequencies should look more or less as they do in English, with some letters much more frequent and some much less frequent. And the same will be true of the string consisting of the 2nd, 7th, 12th,... letters of the ciphertext, and so on. On the other hand, if you guessed wrong and the key length is not five, then the string consisting of every fifth letter should be more or less random, so its letter frequencies should look different from the frequencies in English.

How can we quantify the following two statements so as to be able to distinguish between them?

String 1 has letter frequencies similar to those in Table 1.3. (5.4)

String 2 has letter frequencies that look more or less random. (5.5)

One method is to use the following device.

Definition. Let $\boldsymbol{s} = c_1c_2c_3\cdots c_n$ be a string of n alphabetic characters. The *index of coincidence* of $\boldsymbol{s}$, denoted by $\operatorname{IndCo}(\boldsymbol{s})$, is the probability that two randomly chosen characters in the string $\boldsymbol{s}$ are identical.

[6] *Cryptography and the Art of Decryption.*

We are going to derive a formula for the index of coincidence. It is convenient to identify the letters a,...,z with the numbers $0, 1, \ldots, 25$ respectively. For each value $i = 0, 1, 2, \ldots, 25$, let F_i be the frequency with which letter i appears in the string $\boldsymbol{s}$. For example, if the letter h appears 23 times in the string $\boldsymbol{s}$, then $F_7 = 23$, since h $= 7$ in our labeling of the alphabet.

For each i, there are $\binom{F_i}{2} = \frac{F_i(F_i-1)}{2}$ ways to select two instances of the ith letter of the alphabet from $\boldsymbol{s}$, so the total number of ways to get a repeated letter is the sum of $\frac{F_i(F_i-1)}{2}$ for $i = 0, 1, \ldots, 25$. On the other hand, there are $\binom{n}{2} = \frac{n(n-1)}{2}$ ways to select two arbitrary characters from $\boldsymbol{s}$. The probability of selecting two identical letters is the total number of ways to choose two identical letters divided by the total number of ways to choose any two letters. That is,

$$\mathrm{IndCo}(\boldsymbol{s}) = \frac{1}{n(n-1)} \sum_{i=0}^{25} F_i(F_i - 1). \tag{5.6}$$

Example 5.15. Let $\boldsymbol{s}$ be the string

$\boldsymbol{s} =$ "A bird in hand is worth two in the bush."

Ignoring the spaces between words, $\boldsymbol{s}$ consists of 30 characters. The following table counts the frequencies of each letter that appears at least once:

	A	B	D	E	H	I	N	O	R	S	T	U	W
i	0	1	3	4	7	8	13	14	17	18	19	20	22
F_i	2	2	2	1	4	4	3	2	2	2	3	1	2

Then the index of coincidence of $\boldsymbol{s}$, as given by (5.6), is

$$\mathrm{IndCo}(\boldsymbol{s}) = \frac{1}{30 \cdot 29}(2{\cdot}1+2{\cdot}1+2{\cdot}1+4{\cdot}3+4{\cdot}3+3{\cdot}2+\cdots+3{\cdot}2+2{\cdot}1) \approx 0.0575.$$

We return to our two statements (5.4) and (5.5). Suppose first that the string $\boldsymbol{s}$ consists of random characters. Then the probability that $c_i = c_j$ is exactly $\frac{1}{26}$, so we would expect $\mathrm{IndCo}(\boldsymbol{s}) \approx \frac{1}{26} \approx 0.0385$. On the other hand, if $\boldsymbol{s}$ consists of English text, then we would expect the relative frequencies to be as in Table 1.3. So for example, if $\boldsymbol{s}$ consists of 10,000 characters, we would expect approximately 815 A's, approximately 144 B's, approximately 276 C's, and so on. Thus the index of coincidence for a string of English text should be approximately

$$\frac{815 \cdot 814 + 144 \cdot 143 + 276 \cdot 275 + \cdots + 8 \cdot 7}{10000 \cdot 9999} \approx 0.0685.$$

The disparity between 0.0385 and 0.0685, as small as it may seem, provides the means to distinguish between Statement 5.4 and Statement 5.5. More precisely:

$$\text{If } \operatorname{IndCo}(\boldsymbol{s}) \approx 0.068, \text{ then } \boldsymbol{s} \text{ looks like simple substitution English.} \tag{5.7}$$

$$\text{If } \operatorname{IndCo}(\boldsymbol{s}) \approx 0.038, \text{ then } \boldsymbol{s} \text{ looks like random letters.} \tag{5.8}$$

Of course, the value of $\operatorname{IndCo}(\boldsymbol{s})$ will tend to fluctuate, especially if $\boldsymbol{s}$ is fairly short. But the moral of (5.7) and (5.8) is that larger values of $\operatorname{IndCo}(\boldsymbol{s})$ make it more likely that $\boldsymbol{s}$ is English encrypted with some sort of simple substitution, while smaller values of $\operatorname{IndCo}(\boldsymbol{s})$ make it more likely that $\boldsymbol{s}$ is random.

Now suppose that Eve intercepts a message $\boldsymbol{s}$ that she believes was encrypted using a Vigenère cipher and wants to check whether the keyword has length k. Her first step is to break the string $\boldsymbol{s}$ into k pieces $\boldsymbol{s}_1, \boldsymbol{s}_2, \ldots, \boldsymbol{s}_k$, where $\boldsymbol{s}_1$ consists of every kth letter starting from the first letter, $\boldsymbol{s}_2$ consists of every kth letter starting from the second letter, and so on. In mathematical terms, if we write $\boldsymbol{s} = c_1c_2c_3\ldots c_n$, then

$$\boldsymbol{s}_i = c_i c_{i+k} c_{i+2k} c_{i+3k} \cdots .$$

Notice that if Eve's guess is correct and the keyword has length k, then each $\boldsymbol{s}_i$ consists of characters that were encrypted using the same shift amount, so although they do not decrypt to form actual words (remember that $\boldsymbol{s}_i$ is every kth letter of the text), the pattern of their letter frequencies will look like English. On the other hand, if Eve's guess is incorrect, then the $\boldsymbol{s}_i$ strings will be more or less random.

Thus for each k, Eve computes $\operatorname{IndCo}(\boldsymbol{s}_i)$ for $i = 1, 2, \ldots, k$ and checks whether these numbers are closer to 0.068 or closer to 0.038. She does this for $k = 3, 4, 5, \ldots$ until she finds a value of k for which the average value of $\operatorname{IndCo}(\boldsymbol{s}_1), \operatorname{IndCo}(\boldsymbol{s}_2), \ldots, \operatorname{IndCo}(\boldsymbol{s}_k)$ is large, say greater than 0.06. Then this k is probably the correct blocksize.

We assume now that Eve has used the Kasiski test or the index of coincidence test to determine that the keyword has length k. That's a good start, but she's still quite far from her goal of finding the plaintext. The next step is to compare the strings $\boldsymbol{s}_1, \boldsymbol{s}_2, \ldots, \boldsymbol{s}_k$ to one another. The tool she uses to compare different strings is called the mutual index of coincidence. The general idea is that each of the k strings has been encrypted using a different shift cipher. If the string $\boldsymbol{s}_i$ is shifted by β_i and the string $\boldsymbol{s}_j$ is shifted by β_j, then one would expect the frequencies of $\boldsymbol{s}_i$ to best match those of $\boldsymbol{s}_j$ when the symbols in $\boldsymbol{s}_i$ are shifted by an additional amount

$$\sigma \equiv \beta_j - \beta_i \pmod{26}.$$

This leads to the following useful definition.

Definition. Let

$$\boldsymbol{s} = c_1c_2c_3\ldots c_n \qquad \text{and} \qquad \boldsymbol{t} = d_1d_2d_3\ldots d_m$$

be strings of alphabetic characters. The *mutual index of coincidence* of $\boldsymbol{s}$ and $\boldsymbol{t}$, denoted by $\operatorname{MutIndCo}(\boldsymbol{s}, \boldsymbol{t})$, is the probability that a randomly chosen character from $\boldsymbol{s}$ and a randomly chosen character from $\boldsymbol{t}$ will be the same.

If we let $F_i(\boldsymbol{s})$ denote the number of times the ith letter of the alphabet appears in the string $\boldsymbol{s}$, and similarly for $F_i(\boldsymbol{t})$, then the probability of choosing the ith letter from both is the product of the probabilities $\frac{F_i(\boldsymbol{s})}{n}$ and $\frac{F_i(\boldsymbol{t})}{m}$. In order to obtain a formula for the mutual index of coincidence of $\boldsymbol{s}$ and $\boldsymbol{t}$, we add these probabilities over all possible letters,

$$\text{MutIndCo}(\boldsymbol{s},\boldsymbol{t}) = \frac{1}{nm}\sum_{i=0}^{25} F_i(\boldsymbol{s})F_i(\boldsymbol{t}). \tag{5.9}$$

Example 5.16. Let $\boldsymbol{s}$ and $\boldsymbol{t}$ be the strings

$$\begin{aligned} \boldsymbol{s} &= \text{“A bird in hand is worth two in the bush,”} \\ \boldsymbol{t} &= \text{“A stitch in time saves nine.”} \end{aligned}$$

Using formula (5.9) to compute the mutual index of coincidence of $\boldsymbol{s}$ and $\boldsymbol{t}$ yields $\text{MutIndCo}(\boldsymbol{s},\boldsymbol{t}) = 0.0773$.

The mutual index of coincidence has very similar properties to the index of coincidence. For example, there are analogues of the two statements (5.7) and (5.8). The value of $\text{MutIndCo}(\boldsymbol{s},\boldsymbol{t})$ can be used to confirm that a guessed shift amount is correct. Thus if two strings $\boldsymbol{s}$ and $\boldsymbol{t}$ are encrypted using the *same* simple substitution cipher, then $\text{MutIndCo}(\boldsymbol{s},\boldsymbol{t})$ tends to be large, because of the uneven frequency with which letters appear. On the other hand, if $\boldsymbol{s}$ and $\boldsymbol{t}$ are encrypted using *different* substitution ciphers, then they have no relation to one another, and the mutual index of coincidence $\text{MutIndCo}(\boldsymbol{s},\boldsymbol{t})$ will be much smaller.

We return now to Eve's attack on a Vigenère cipher. She knows the key length k and has split the ciphertext into k blocks, $\boldsymbol{s}_1, \boldsymbol{s}_2, \ldots, \boldsymbol{s}_k$, as usual. The characters in each block have been encrypted using the same shift amount, say

$$\beta_i = \text{Amount that block } \boldsymbol{s}_i \text{ has been shifted.}$$

Eve's next step is to compare $\boldsymbol{s}_i$ with the string obtained by shifting the characters in $\boldsymbol{s}_j$ by different amounts. As a notational convenience, we write

$$\boldsymbol{s}_j + \sigma = \begin{pmatrix} \text{The string } \boldsymbol{s}_j \text{ with every character} \\ \text{shifted } \sigma \text{ spots down the alphabet.} \end{pmatrix}$$

Suppose that σ happens to equal $\beta_i - \beta_j$. Then $\boldsymbol{s}_j + \sigma$ has been shifted a total of $\beta_j + \sigma = \beta_i$ from the plaintext, so $\boldsymbol{s}_j + \sigma$ and $\boldsymbol{s}_i$ have been encrypted using the same shift amount. Hence, as noted above, their mutual index of coincidence will be fairly large. On the other hand, if σ is not equal to $\beta_i - \beta_j$, then $\boldsymbol{s}_j + \sigma$ and $\boldsymbol{s}_i$ have been encrypted using different shift amounts, so $\text{MutIndCo}(\boldsymbol{s},\boldsymbol{t})$ will tend to be small.

To put this concept into action, Eve computes all of the mutual indices of coincidence

$$\text{MutIndCo}(\boldsymbol{s}_i, \boldsymbol{s}_j + \sigma) \qquad \text{for } 1 \le i < j \le k \text{ and } 0 \le \sigma \le 25.$$

Scanning the list of values, she picks out the ones that are large, say larger than 0.065. Each large value of $\text{MutIndCo}(\boldsymbol{s}_i, \boldsymbol{s}_j + \sigma)$ makes it likely that

$$\beta_i - \beta_j \equiv \sigma \pmod{26}. \tag{5.10}$$

(Note that (5.10) is only a congruence modulo 26, since a shift of 26 is the same as a shift of 0.) The leads to a system of equations of the form (5.10) for the variables $\beta_1, \ldots, \beta_k$. In practice, some of these equations will be spurious, but after a certain amount of trial and error, Eve will end up with values $\gamma_2, \ldots, \gamma_k$ satisfying

$$\beta_2 = \beta_1 + \gamma_2, \quad \beta_3 = \beta_1 + \gamma_3, \quad \beta_4 = \beta_1 + \gamma_4, \quad \ldots, \quad \beta_k = \beta_1 + \gamma_k.$$

Thus if the keyword happens to start with A, then the second letter of the keyword would be A shifted by γ_2, the third letter of the keyword would be A shifted by γ_3, and so on. Similarly, if the keyword happens to start with B, then its second letter would be B shifted by γ_2, its third letter would be B shifted by γ_3, etc. So all that Eve needs to do is try each of the 26 possible starting letters and decrypt the message using each of the 26 corresponding keywords. Looking at the first few characters of the 26 putative plaintexts, it is easy for her to pick out the correct one.

Remark 5.17. We make one final remark before doing an example. We noted earlier that among the many occurrences of the letters the in the plaintext, a certain percentage of them will line up with exactly the same part of the keyword. It turns out that these repeated encryptions occur much more frequently than one might guess. This is an example of the "birthday paradox," which says that the probability of getting a match (e.g. of trigrams or birthdays or colors) is quite high. We discuss the birthday paradox and some of its many applications to cryptography in Sect. 5.4.

5.2.2 Cryptanalysis of the Vigenère Cipher: Practice

In this section we illustrate how to cryptanalyze a Vigenère ciphertext by decrypting the message given in Table 5.2.

```
zpgdl rjlaj kpylx zpyyg lrjgd lrzhz qyjzq repvm swrzy rigzh
zvreg kwivs saolt nliuw oldie aqewf iiykh bjowr hdogc qhkwa
jyagg emisr zqoqh oavlk bjofr ylvps rtgiu avmsw lzgms evwpc
dmjsv jqbrn klpcf iowhv kxjbj pmfkr qthtk ozrgq ihbmq sbivd
ardym qmpbu nivxm tzwqv gefjh ucbor vwpcd xuwft qmoow jipds
fluqm oeavl jgqea lrkti wvext vkrrg xani
```

Table 5.2: A Vigenère ciphertext to cryptanalyze

Trigram	Appears at places	Difference
avl	117 and 258	$141 = 3 \cdot 47$
bjo	86 and 121	$35 = 5 \cdot 7$
dlr	4 and 25	$21 = 3 \cdot 7$
gdl	3 and 24	$16 = 2^4$
lrj	5 and 21	$98 = 2 \cdot 7^2$
msw	40 and 138	$84 = 2^2 \cdot 3 \cdot 7$
pcd	149 and 233	$13 = 13$
qmo	241 and 254	$98 = 2 \cdot 7^2$
vms	39 and 137	$84 = 2^2 \cdot 3 \cdot 7$
vwp	147 and 231	$84 = 2^2 \cdot 3 \cdot 7$
wpc	148 and 232	$21 = 3 \cdot 7$
zhz	28 and 49	$21 = 3 \cdot 7$

Table 5.3: Repeated trigrams in the ciphertext given in Table 5.2

Key length	Average index	Individual indices of coincidence
4	0.038	0.034, 0.042, 0.039, 0.035
5	0.037	0.038, 0.039, 0.043, 0.027, 0.036
6	0.036	0.038, 0.038, 0.039, 0.038, 0.032, 0.033
7	0.062	0.062, 0.057, 0.065, 0.059, 0.060, 0.064, 0.064
8	0.038	0.037, 0.029, 0.038, 0.030, 0.034, 0.057, 0.040, 0.039
9	0.037	0.032, 0.036, 0.028, 0.030, 0.026, 0.032, 0.045, 0.047, 0.056

Table 5.4: Indices of coincidence of Table 5.2 for various key lengths

We begin by applying the Kasiski test. A list of repeated trigrams is given in Table 5.3, together with their location within the ciphertext and the number of letters that separates them. Most of the differences in the last column are divisible by 7, and 7 is the largest number with this property, so we guess that the keyword length is 7.

Although the Kasiski test shows that the period is probably 7, we also apply the index of coincidence test in order to illustrate how it works. Table 5.4 lists the indices of coincidence for various choices of key length and the average index of coincidence for each key length. We see from Table 5.4 that key length 7 has far higher average index of coincidence than the other potential key lengths, which confirms the conclusion from the Kasiski test.

Now that Eve knows that the key length is 7, she compares the blocks with one another as described in Sect. 5.2.1. She first breaks the ciphertext into seven blocks by taking every seventh letter. (Notice how the first seven letters of the ciphertext run down the first column, the second seven down the second column, and so on.)

Blocks		Shift amount												
i	j	0	1	2	3	4	5	6	7	8	9	10	11	12
1	2	0.025	0.034	0.045	0.049	0.025	0.032	0.037	0.042	0.049	0.031	0.032	0.037	0.043
1	3	0.023	**0.067**	0.055	0.022	0.034	0.049	0.036	0.040	0.040	0.046	0.025	0.031	0.046
1	4	0.032	0.041	0.027	0.040	0.045	0.037	0.045	0.028	0.049	0.042	0.042	0.030	0.039
1	5	0.043	0.021	0.031	0.052	0.027	0.049	0.037	0.050	0.033	0.033	0.035	0.044	0.030
1	6	0.037	0.036	0.030	0.037	0.037	0.055	0.046	0.038	0.035	0.031	0.032	0.037	0.032
1	7	0.054	0.063	0.034	0.030	0.034	0.040	0.035	0.032	0.042	0.025	0.019	0.061	0.054
2	3	0.041	0.029	0.036	0.041	0.045	0.038	0.060	0.031	0.020	0.045	0.056	0.029	0.030
2	4	0.028	0.043	0.042	0.032	0.032	0.047	0.035	0.048	0.037	0.040	0.028	0.051	0.037
2	5	0.047	0.037	0.032	0.044	0.059	0.029	0.017	0.044	0.060	0.034	0.037	0.046	0.039
2	6	0.033	0.035	0.052	0.040	0.032	0.031	0.031	0.029	0.055	0.052	0.043	0.028	0.023
2	7	0.038	0.037	0.035	0.046	0.046	0.054	0.037	0.018	0.029	0.052	0.041	0.026	0.037
3	4	0.029	0.039	0.033	0.048	0.044	0.043	0.030	0.051	0.033	0.034	0.034	0.040	0.038
3	5	0.021	0.041	0.041	0.037	0.051	0.035	0.036	0.038	0.025	0.043	0.034	0.039	0.036
3	6	0.037	0.034	0.042	0.034	0.051	0.029	0.027	0.041	0.034	0.040	0.037	0.046	0.036
3	7	0.046	0.023	0.028	0.040	0.031	0.040	0.045	0.039	0.020	0.030	**0.069**	0.042	0.037
4	5	0.041	0.033	0.041	0.038	0.036	0.031	0.056	0.032	0.026	0.034	0.049	0.029	0.054
4	6	0.035	0.037	0.032	0.039	0.041	0.033	0.032	0.039	0.042	0.031	0.049	0.039	0.058
4	7	0.031	0.032	0.046	0.038	0.039	0.042	0.033	0.056	0.046	0.027	0.027	0.036	0.036
5	6	0.048	0.036	0.026	0.031	0.033	0.039	0.037	0.027	0.037	0.045	0.032	0.040	0.041
5	7	0.030	0.051	0.043	0.031	0.034	0.041	0.048	0.032	0.053	0.037	0.024	0.029	0.045
6	7	0.032	0.033	0.030	0.038	0.032	0.035	0.047	0.050	0.049	0.033	0.057	0.050	0.021

Blocks		Shift amount												
i	j	13	14	15	16	17	18	19	20	21	22	23	24	25
1	2	0.034	0.052	0.037	0.030	0.037	0.054	0.021	0.018	0.052	0.052	0.043	0.042	0.046
1	3	0.031	0.037	0.038	0.050	0.039	0.040	0.026	0.037	0.044	0.043	0.023	0.045	0.032
1	4	0.039	0.040	0.032	0.041	0.028	0.019	**0.071**	0.038	0.040	0.034	0.045	0.026	0.052
1	5	0.042	0.032	0.038	0.037	0.032	0.045	0.045	0.033	0.041	0.043	0.035	0.028	0.063
1	6	0.040	0.030	0.028	**0.071**	0.051	0.033	0.036	0.047	0.029	0.037	0.046	0.041	0.027
1	7	0.040	0.032	0.049	0.037	0.035	0.035	0.039	0.023	0.043	0.035	0.041	0.042	0.027
2	3	0.054	0.040	0.028	0.031	0.039	0.033	0.052	0.046	0.037	0.026	0.028	0.036	0.048
2	4	0.047	0.034	0.027	0.038	0.047	0.042	0.026	0.038	0.029	0.046	0.040	0.061	0.025
2	5	0.034	0.026	0.035	0.038	0.048	0.035	0.033	0.032	0.040	0.041	0.045	0.033	0.036
2	6	0.033	0.034	0.036	0.036	0.048	0.040	0.041	0.049	0.058	0.028	0.021	0.043	0.049
2	7	0.042	0.037	0.041	0.059	0.031	0.027	0.043	0.046	0.028	0.021	0.044	0.048	0.040
3	4	0.037	0.045	0.033	0.028	0.029	**0.073**	0.026	0.040	0.040	0.026	0.043	0.042	0.043
3	5	0.035	0.029	0.036	0.044	0.055	0.034	0.033	0.046	0.041	0.024	0.041	**0.067**	0.037
3	6	0.023	0.043	**0.074**	0.047	0.033	0.043	0.030	0.026	0.042	0.045	0.032	0.035	0.040
3	7	0.035	0.035	0.035	0.028	0.048	0.033	0.035	0.041	0.038	0.052	0.038	0.029	0.062
4	5	0.032	0.041	0.036	0.032	0.046	0.035	0.039	0.042	0.038	0.034	0.043	0.036	0.048
4	6	0.034	0.034	0.036	0.029	0.043	0.037	0.039	0.036	0.039	0.033	**0.066**	0.037	0.028
4	7	0.043	0.032	0.039	0.034	0.029	**0.071**	0.037	0.039	0.030	0.044	0.037	0.030	0.041
5	6	0.052	0.035	0.019	0.036	0.063	0.045	0.030	0.039	0.049	0.029	0.036	0.052	0.041
5	7	0.040	0.031	0.034	0.052	0.026	0.034	0.051	0.044	0.041	0.039	0.034	0.046	0.029
6	7	0.029	0.035	0.039	0.032	0.028	0.039	0.026	0.036	**0.069**	0.052	0.035	0.034	0.038

Table 5.5: Mutual indices of coincidence of Table 5.2 for shifted blocks

$$\begin{aligned}
s_1 &= \texttt{zlxrhrrhwloehdweoklilwvlhphqbynwhwfjulrxx}\\
s_2 &= \texttt{pazjzezzitlwboamqbvuzpjpvmtiimiquptiqjkta}\\
s_3 &= \texttt{gjpgqpyvvndfjgjihjpagcqckfkhvqvvccqpmgtvn}\\
s_4 &= \texttt{dkydyvrrsliiocysoosvmdbfxkobdmxgbdmdoqiki}\\
s_5 &= \texttt{lpyljmiesieiwqarafrmsmrijrzmapmeoxoseewr}\\
s_6 &= \texttt{rygrzsggauayrhgzvrtsejnobqrqrbtfruofaavr}\\
s_7 &= \texttt{jllzqwzkowqkhkgqlygwvskwjtgsduzjvwwlvleg}
\end{aligned}$$

She then compares the ith block $\boldsymbol{s}_i$ to the jth block shifted by σ, which we denote by $\boldsymbol{s}_j + \sigma$, taking successively $\sigma = 0, 1, 2, \ldots, 25$. Table 5.5 gives a complete list of the 546 mutual indices of coincidence

$$\operatorname{MutIndCo}(\boldsymbol{s}_i, \boldsymbol{s}_j + \sigma) \qquad \text{for } 1 \le i < j \le 7 \text{ and } 0 \le \sigma \le 25.$$

In Table 5.5, the entry in the row corresponding to (i, j) and the column corresponding to the shift σ is equal to

$$\operatorname{MutIndCo}(\boldsymbol{s}_i, \boldsymbol{s}_j + \sigma) = \operatorname{MutIndCo}(\text{Block } \boldsymbol{s}_i, \text{Block } \boldsymbol{s}_j \text{ shifted by } \sigma). \quad (5.11)$$

If this quantity is large, it suggests that $\boldsymbol{s}_j$ has been shifted σ further than $\boldsymbol{s}_i$. As in Sect. 5.2.1 we let

i	j	Shift	MutIndCo	Shift relation
1	3	1	0.067	$\beta_1 - \beta_3 = 1$
3	7	10	0.069	$\beta_3 - \beta_7 = 10$
1	4	19	0.071	$\beta_1 - \beta_4 = 19$
1	6	16	0.071	$\beta_1 - \beta_6 = 16$
3	4	18	0.073	$\beta_3 - \beta_4 = 18$
3	5	24	0.067	$\beta_3 - \beta_5 = 24$
3	6	15	0.074	$\beta_3 - \beta_6 = 15$
4	6	23	0.066	$\beta_4 - \beta_6 = 23$
4	7	18	0.071	$\beta_4 - \beta_7 = 18$
6	7	21	0.069	$\beta_6 - \beta_7 = 21$

Table 5.6: Large indices of coincidence and shift relations

$$\beta_i = \text{Amount that the block } \boldsymbol{s}_i \text{ has been shifted.}$$

Then a large value for (5.11) makes it likely that

$$\beta_i - \beta_j = \sigma. \tag{5.12}$$

We have underlined the large values (those greater than 0.065) in Table 5.5 and compiled them, with the associated shift relation (5.12), in Table 5.6.

Eve's next step is to solve the system of linear equations appearing in the final column of Table 5.6, keeping in mind that all values are modulo 26, since a shift of 26 is the same as no shift at all. Notice that there are 10 equations for the six variables $\beta_1, \beta_3, \beta_4, \beta_5, \beta_6, \beta_7$. (Unfortunately, β_2 does not appear, so we'll deal with it later). In general, a system of 10 equations in 6 variables has no solutions,[7] but in this case a little bit of algebra shows that not only is there a solution, there is actually one solution for each value of β_1. In other words, the full set of solutions is obtained by expressing each of the variables $\beta_3, \ldots, \beta_7$ in terms of β_1:

$$\beta_3 = \beta_1 + 25, \quad \beta_4 = \beta_1 + 7, \quad \beta_5 = \beta_1 + 1, \quad \beta_6 = \beta_1 + 10, \quad \beta_7 = \beta_1 + 15. \tag{5.13}$$

What should Eve do about β_2? She could just ignore it for now, but instead she picks out the largest values in Table 5.5 that relate to block 2 and uses those. The largest such values are $(i, j) = (2, 3)$ with shift 6 and index 0.060 and $(i, j) = (2, 4)$ with shift 24 and index 0.061, which give the relations

$$\beta_2 - \beta_3 = 6 \qquad \text{and} \qquad \beta_2 - \beta_4 = 24.$$

[7]We were a little lucky in that every relation in Table 5.6 is correct. Sometimes there are erroneous relations, but it is not hard to eliminate them with some trial and error.

Substituting in from (5.13), these both yield $\beta_2 = \beta_1 + 5$, and the fact that they give the same value gives Eve confidence that they are correct.

Shift	Keyword	Decrypted text
0	AFZHBKP	zkhwkhulvkdoowxuqrxwwrehwkhkhurripbrzqolih
1	BGAICLQ	yjgvjgtkujcnnvwtpqwvvqdgvjgjgtqqhoaqypnkhg
2	CHBJDMR	xifuifsjtibmmuvsopvuupcfuififsppgnzpxomjgf
3	DICKENS	whetherishallturnouttobetheheroofmyownlife
4	EJDLFOT	vgdsgdqhrgzkkstqmntssnadsgdgdqnnelxnvmkhed
5	FKEMGPU	ufcrfcpgqfyjjrsplmsrrmzcrfcfcpmmdkwmuljgdc
6	GLFNHQV	tebqebofpexiiqroklrqqlybqebebollcjvltkifcb
7	HMGOIRW	sdapdaneodwhhpqnjkqppkxapdadankkbiuksjheba
8	INHPJSX	rczoczmdncvggopmijpoojwzoczczmjjahtjrigdaz
⋮	⋮	⋮

Table 5.7: Decryption of Table 5.2 using shifts of the keyword AFZHBKP

To summarize, Eve now knows that however much the first block s_1 is rotated, blocks $s_2, s_3, \ldots, s_7$ are rotated, respectively, 5, 25, 7, 1, 10, and 15 steps further than s_1. So for example, if s_1 is not rotated at all (i.e., if $\beta_1 = 0$ and the first letter of the keyword is A), then the full keyword is AFZHBKP. Eve uses the keyword AFZHBKP to decrypt the first few blocks of the ciphertext, finding the "plaintext"

zkhwkhulvkdoowxuqrxwwrehwkhkhurripbrzqolihruzkhwkh.

That doesn't look good! So next she tries $\beta_1 = 1$ and a keyword starting with the letter B. Continuing in this fashion, she need only check the 26 possibilities for β_1. The results are listed in Table 5.7.

Taking $\beta_1 = 3$ yields the keyword DICKENS and an acceptable plaintext. Completing the decryption using this keyword and supplying the appropriate word breaks, punctuation, and capitalization, Eve recovers the full plaintext:

> Whether I shall turn out to be the hero of my own life, or whether that station will be held by anybody else, these pages must show. To begin my life with the beginning of my life, I record that I was born (as I have been informed and believe) on a Friday, at twelve o'clock at night. It was remarked that the clock began to strike, and I began to cry, simultaneously.[8]

[8] *David Copperfield*, 1850, Charles Dickens.

5.3 Probability Theory

5.3.1 Basic Concepts of Probability Theory

In this section we introduce the basic ideas of probability theory in the discrete setting. A *probability space* consists of two pieces. The first is a finite set Ω consisting of all possible outcomes of an experiment and the second is a method for assigning a probability to each possible outcome. In mathematical terms, a probability space is a finite set of outcomes Ω, called the *sample space*, and a function

$$\mathrm{Pr} : \Omega \longrightarrow \mathbb{R}.$$

We want the function Pr to satisfy our intuition that

$$\mathrm{Pr}(\omega) = \text{“probability that event } \omega \text{ occurred.”}$$

In particular, the value of $\mathrm{Pr}(\omega)$ should be between 0 and 1.

Example 5.18. Consider the toss of a single coin. There are two outcomes, heads and tails, so we let Ω be the set $\{H, T\}$. Assuming that it is a fair coin, each outcome is equally likely, so $\mathrm{Pr}(H) = \mathrm{Pr}(T) = \frac{1}{2}$.

Example 5.19. Consider the roll of two dice. The sample space Ω is the following set of 36 pairs of numbers:

$$\Omega = \{(n, m) : n, m \in \mathbb{Z} \text{ with } 1 \le n, m \le 6\}.$$

As in Example 5.18, each possible outcome is equally likely. For example, the probability of rolling $(6, 6)$ is the same as the probability of rolling $(3, 4)$. Hence

$$\mathrm{Pr}\big((n, m)\big) = \frac{1}{36}$$

for any choice of (n, m). Note that order matters in this scenario. We might imagine that one die is red and the other is blue, so “red 3 and blue 5” is a different outcome from “red 5 and blue 3.”

Example 5.20. Suppose that an urn contains 100 balls, of which 21 are red and the rest are blue. If we pick 10 balls at random (without replacement), what is the probability that exactly 3 of them are red?

The total number of ways of selecting 10 balls from among 100 is $\binom{100}{10}$. Similarly, there are $\binom{21}{3}$ ways to select 3 red balls from among the 21 that are red, and there are $\binom{79}{7}$ ways to pick the other 7 balls from among the 79 that are blue. There are thus $\binom{21}{3}\binom{79}{7}$ ways to select exactly 3 red balls and exactly 7 blue balls. Hence the probability of picking exactly 3 red balls in 10 tries is

$$\mathrm{Pr}\begin{pmatrix}\text{exactly 3 red balls in}\\ \text{10 attempts}\end{pmatrix} = \frac{\binom{21}{3}\binom{79}{7}}{\binom{100}{10}} = \frac{20271005}{91015876} \approx 0.223.$$

We are typically more interested in computing the probability of *compound events*. These are subsets of the sample space that may include more than one outcome. For example, in the roll of two dice in Example 5.19, we might be interested in the probability that at least one of the dice shows a 6. This compound event is the subset of Ω consisting of all outcomes that include the number six, which is the set

$$\{(1,6),(2,6),(3,6),(4,6),(5,6),(6,6),(6,1),(6,2),(6,3),(6,4),(6,5)\}.$$

Suppose that we know the probability of each particular outcome. How then do we compute the probability of compound events or of events consisting of repeated independent trials of an experiment? Analyzing this problem leads to the idea of *independence of events*, a concept that gives probability theory much of its complexity and richness.

The formal theory of probability is an axiomatic theory. You have probably seen such theories when you studied Euclidean geometry and when you studied abstract vector spaces. In an axiomatic theory, one starts with a small list of basic axioms and derives from them additional interesting facts and formulas. The axiomatic theory of probability allows us to derive formulas to compute the probabilities of compound events. In this book we are content with an informal presentation of the theory, but for those who are interested in a more rigorous axiomatic treatment of probability theory, see for example [112, §2.3].

We begin with some definitions.

Definition. A *sample space* (or *set of outcomes*) is a finite[9] set Ω. Each outcome $\omega \in \Omega$ is assigned a probability $\Pr(\omega)$, where we require that the probability function

$$\Pr : \Omega \longrightarrow \mathbb{R}$$

satisfy the following two properties:

$$\text{(a)}\quad 0 \le \Pr(\omega) \le 1 \quad \text{for all } \omega \in \Omega \qquad \text{and} \qquad \text{(b)}\quad \sum_{\omega \in \Omega} \Pr(\omega) = 1. \tag{5.14}$$

Notice that (5.14)(a) corresponds to our intuition that every outcome has a probability between 0 (if it never occurs) and 1 (if it always occurs), while (5.14)(b) says that some outcome must occur, so Ω contains all possible outcomes for the experiment.

Definition. An *event* is any subset of Ω. We assign a probability to an event $E \subset \Omega$ by setting

$$\Pr(E) = \sum_{\omega \in E} \Pr(\omega). \tag{5.15}$$

In particular, $\Pr(\emptyset) = 0$ by convention, and $\Pr(\Omega) = 1$ from (5.14)(b).

[9]General (continuous) probability theory also deals with infinite sample spaces Ω, in which case only certain subsets of Ω are allowed to be events and are assigned probabilities. There are also further restrictions on the probability function $\Pr : \Omega \to \mathbb{R}$. For our study of cryptography in this book, it suffices to use discrete (finite) sample spaces.

Definition. We say that two events E and F are *disjoint* if $E \cap F = \emptyset$.

It is clear that

$$\Pr(E \cup F) = \Pr(E) + \Pr(F) \qquad \text{if } E \text{ and } F \text{ are disjoint},$$

since then $E \cup F$ is the collection of all outcomes in either E or F. When E and F are not disjoint, the probability of the event $E \cup F$ is not the sum of $\Pr(E)$ and $\Pr(F)$, since the outcomes common to both E and F should not be counted twice. Thus we need to subtract the outcomes common to E and F, which gives the useful formula

$$\Pr(E \cup F) = \Pr(E) + \Pr(F) - \Pr(E \cap F). \tag{5.16}$$

(See Exercise 5.20.)

Definition. The *complement* of an event E is the event E^c consisting of all outcomes that are not in E, i.e.,

$$E^c = \{\omega \in \Omega : \omega \notin E\}.$$

The probability of the complementary event is given by

$$\Pr(E^c) = 1 - \Pr(E). \tag{5.17}$$

It is sometimes easier to compute the probability of the complement of an event E and then use (5.17) to find $\Pr(E)$.

Example 5.21. We continue with Example 5.19 in which Ω consists of the possible outcomes of rolling two dice. Let E be the event

$$E = \{\text{at least one six is rolled}\}.$$

We can write down E explicitly; it is the set

$$E = \bigl\{(1,6),(6,1),(2,6),(6,2),(3,6),(6,3),(4,6),(6,4),(5,6),(6,5),(6,6)\bigr\}.$$

Each of these 11 outcomes has probability $\frac{1}{36}$, so

$$\Pr(E) = \sum_{\omega \in E} \Pr(\omega) = \frac{11}{36}.$$

We can then compute the probability of not rolling a six as

$$\Pr(\text{no sixes are rolled}) = \Pr(E^c) = 1 - \Pr(E) = \frac{25}{36}.$$

Next consider the event F defined by

$$F = \{\text{no number higher than two is rolled}\}.$$

Notice that

$$F = \{(1,1),(1,2),(2,1),(2,2)\}$$

is disjoint from E, so the probability of either rolling a six or else rolling no number higher than two is

$$\Pr(E \cup F) = \Pr(E) + \Pr(F) = \frac{11}{36} + \frac{4}{36} = \frac{15}{36}.$$

For nondisjoint events, the computation is more complicated, since we need to avoid double counting outcomes. Consider the event G defined by

$$G = \{\text{doubles}\},$$

i.e., $G = \{(1,1),(2,2),(3,3),(4,4),(5,5),(6,6)\}$. Then E and G both contain the outcome $(6,6)$, so their union $E \cup G$ only contains 16 outcomes, not 17. Thus the probability of rolling either a six or doubles is $\frac{16}{36}$. We can also compute this probability using formula (5.16),

$$\Pr(E \cup G) = \Pr(E) + \Pr(G) - \Pr(E \cap G) = \frac{11}{36} + \frac{6}{36} - \frac{1}{36} = \frac{16}{36} = \frac{4}{9}.$$

To conclude this example, let H be the event

$$H = \{\text{the sum of the two dice is at least } 4\}.$$

We could compute $\Pr(H)$ directly, but it is easier to compute the probability of H^c. Indeed, there are only three outcomes that give a sum smaller than 4, namely

$$H^c = \{(1,1),(1,2),(2,1)\}.$$

Thus $\Pr(H^c) = \frac{3}{36} = \frac{1}{12}$, and then $\Pr(H) = 1 - \Pr(H^c) = \frac{11}{12}$.

Suppose now that E and F are events. The event consisting of both E and F is the intersection $E \cap F$, so the probability that both E and F occur is

$$\Pr(E \text{ and } F) = \Pr(E \cap F).$$

As the next example makes clear, the probability of the intersection of two events is not a simple function of the probabilities of the individual events.

Example 5.22. Consider the experiment consisting of drawing two cards from a deck of cards, where the second card is drawn without replacing the first card. Let E and F be the following events:

$$E = \{\text{the first card drawn is a king}\},$$
$$F = \{\text{the second card drawn is a king}\}.$$

Clearly $\Pr(E) = \frac{1}{13}$. It is also true that $\Pr(F) = \frac{1}{13}$, since with no information about the value of the first card, there's no difference between events E and F. (If this seems unclear, suppose instead that the deck of cards were dealt to 52 people. Then the probability that any particular person gets a king is $\frac{1}{13}$, regardless of whether they received the first card or the second card or....) However, it is also clear that if we know whether event E has occurred, then that knowledge does affect the probability of F occurring. More precisely, if E occurs, then there are only 3 kings left in the remaining 51 cards, so F is less likely, while if E does not occur, then there are 4 kings left and F is more likely. Mathematically we find that

$$\Pr(F \text{ if } E \text{ has occurred}) = \frac{3}{51} \quad \text{and} \quad \Pr(F \text{ if } E \text{ has not occurred}) = \frac{4}{51}.$$

Thus the probability of both E and F occurring, i.e., the probability of drawing two consecutive kings, is smaller than the product of $\Pr(E)$ and $\Pr(F)$, because the occurrence of the event E makes the event F less likely. The correct computation is

$$\begin{aligned}\Pr(\text{drawing two kings}) &= \Pr(E \cap F) \\ &= \Pr(E) \cdot \Pr(F \text{ given that } E \text{ has occurred}) \\ &= \frac{1}{13} \cdot \frac{3}{51} = \frac{1}{221} \approx 0.0045.\end{aligned}$$

Let

$$G = \{\text{the second card drawn is an ace}\}.$$

Then the occurrence of E makes G more likely, since if the first card is known to be a king, then there are still four aces left. Thus if we know that E occurs, then the probability of G increases from $\frac{4}{52}$ to $\frac{4}{51}$.

Notice, however, that if we change the experiment and require that the first card be replaced in the deck before the second card is drawn, then whether E occurs has no effect at all on F. Thus using this card replacement scenario, the probability that E and F both occur is simply the product

$$\Pr(E)\Pr(F) = \left(\frac{1}{13}\right)^2 \approx 0.006.$$

We learn two things from the discussion in Example 5.22. First, we see that the probability of one event can depend on whether another event has occurred. Second, we develop some probabilistic intuitions that lead to the mathematical definition of independence.

Definition. Two events E and F are said to be *independent* if

$$\Pr(E \cap F) = \Pr(E) \cdot \Pr(F),$$

where recall that the probability of the intersection $\Pr(E \cap F)$ is the probability that both E and F occur. In other words, E and F are independent if the probability of their both occurring is the product of their individual probabilities of occurring.

Example 5.23. A coin is tossed 10 times and the results recorded. What are the probabilities of the following events?

$$E_1 = \{\text{the first five tosses are all heads}\}.$$
$$E_2 = \{\text{the first five tosses are heads and the rest are tails}\}.$$
$$E_3 = \{\text{exactly five of the ten tosses are heads}\}.$$

The result of any one toss is *independent* of the result of any other toss, so we can compute the probability of getting H on the first five tosses by multiplying together the probability of getting H on any one of these tosses. Assuming that it is a fair coin, the answer to our first question is thus

$$\Pr(E_1) = \left(\frac{1}{2}\right)^5 = \frac{1}{32} \approx 0.031.$$

In order to compute the probability of E_2, note that we are now asking for the probability that our sequence of tosses is exactly $\mathsf{HHHHHTTTTT}$. Again using the independence of the individual tosses, we see that

$$\Pr(E_2) = \left(\frac{1}{2}\right)^{10} = \frac{1}{1024} \approx 0.00098.$$

The computation of $\Pr(E_3)$ is a little trickier, because it asks for exactly five H's to occur, but places no restriction on when they occur. If we were to specify exactly when the five H's and the five T's occur, then the probability would be $\frac{1}{2^{10}}$, just as it was for E_2. So all that we need to do is to count how many ways we can distribute five H's and five T's into ten spots, or equivalently, how many different sequences we can form consisting of five H's and five T's. This is simply the number of ways of choosing five locations from ten possible locations, which is given by the combinatorial symbol $\binom{10}{5}$. Hence dividing the number of outcomes satisfying E_3 by the total number of outcomes, we find that

$$\Pr(E_3) = \binom{10}{5} \cdot \frac{1}{2^{10}} = \frac{252}{1024} = \frac{63}{256} \approx 0.246.$$

Thus there is just under a 25 % chance of getting exactly five heads in ten tosses of a coin.

5.3.2 Bayes's Formula

As we saw in Example 5.22, there is a connection between the probability that two events E and F occur simultaneously and the probability that one of them occurs if we know that the other one has occurred. The former quantity is simply $\Pr(E \cap F)$. The latter quantity is called the *conditional probability* of F on E.

Definition. The *conditional probability of F on E* is denoted by

$$\Pr(F \mid E) = \Pr(F \text{ given that } E \text{ has occurred}).$$

The probability that both E and F occur is related to the conditional probability of F on E by the formula

$$\Pr(F \mid E) = \frac{\Pr(F \cap E)}{\Pr(E)}. \tag{5.18}$$

The intuition behind (5.18), which is usually taken as the definition of the conditional probability $\Pr(F \mid E)$, is simple. On the left-hand side, we are assuming that E occurs, so our sample space or universe is now E instead of Ω. We are asking for the probability that the event F occurs in this smaller universe of outcomes, so we should compute the proportion of the event F that is included in the event E, divided by the total size of the event E itself. This gives the right-hand side of (5.18).

Formula (5.18) immediately implies that

$$\Pr(F \mid E)\Pr(E) = \Pr(F \cap E) = \Pr(E \cap F) = \Pr(E \mid F)\Pr(F).$$

Dividing both sides by $\Pr(F)$ gives a preliminary version of *Bayes's formula:*

$$\Pr(E \mid F) = \frac{\Pr(F \mid E)\Pr(E)}{\Pr(F)} \qquad \text{(Bayes's formula).} \tag{5.19}$$

This formula is useful if we know the conditional probability of F on E and want to know the reverse conditional probability of E on F.

Sometimes it is easier to compute the probability of an event by dividing it into a union of disjoint events, as in the next proposition, which includes another version of Bayes's formula.

Proposition 5.24. *Let E and F be events.*

(a) $$\Pr(E) = \Pr(E \mid F)\Pr(F) + \Pr(E \mid F^c)\Pr(F^c). \tag{5.20}$$

(b) $$\Pr(E \mid F) = \frac{\Pr(F \mid E)\Pr(E)}{\Pr(F \mid E)\Pr(E) + \Pr(F \mid E^c)\Pr(E^c)} \qquad \text{(Bayes's formula).} \tag{5.21}$$

Proof. The proof of (a) illustrates how one manipulates basic probability formulas.

$$\begin{aligned}
&\Pr(E \mid F)\Pr(F) + \Pr(E \mid F^c)\Pr(F^c) \\
&\qquad = \Pr(E \cap F) + \Pr(E \cap F^c) && \text{from (5.18),} \\
&\qquad = \Pr\big((E \cap F) \cup (E \cap F^c)\big) && \text{since } E \cap F \text{ and } E \cap F^c \text{ are disjoint,} \\
&\qquad = \Pr(E) && \text{since } F \cup F^c = \Omega.
\end{aligned}$$

This completes the proof of (a).

In order to prove (b), we reverse the roles of E and F in (a) to get

$$\Pr(F) = \Pr(F \mid E)\Pr(E) + \Pr(F \mid E^c)\Pr(E^c), \tag{5.22}$$

and then substitute (5.22) into the denominator of (5.19) to obtain (5.21). □

Here are some examples that illustrate the use of conditional probabilities. Bayes's formula will be applied in the next section.

Example 5.25. We are given two urns[10] containing gold and silver coins. Urn #1 contains 10 gold coins and 5 silver coins, and Urn #2 contains 2 gold coins and 8 silver coins. An urn is chosen at random, and then a coin is picked at random. What is the probability of choosing a gold coin?

Let

$$E = \{\text{a gold coin is chosen}\}.$$

The probability of E depends first on which urn was chosen, and then on which coin is chosen in that urn. It is thus natural to break E up according to the outcome of the event

$$F = \{\text{Urn \#1 is chosen}\}.$$

Notice that F^c is the event that Urn #2 is chosen. The decomposition formula (5.20) says that

$$\Pr(E) = \Pr(E \mid F)\Pr(F) + \Pr(E \mid F^c)\Pr(F^c).$$

The key point here is that it is easy to compute the conditional probabilities on the right-hand side, and similarly easy to compute $\Pr(F)$ and $\Pr(F^c)$. Thus

$$\Pr(E \mid F) = \frac{10}{15} = \frac{2}{3}, \qquad \Pr(E \mid F^c) = \frac{2}{10} = \frac{1}{5}, \qquad \Pr(F) = \Pr(F^c) = \frac{1}{2}.$$

Using these values, we can compute

$$\Pr(E) = \Pr(E \mid F)\Pr(F) + \Pr(E \mid F^c)\Pr(F^c) = \frac{2}{3}\cdot\frac{1}{2} + \frac{1}{5}\cdot\frac{1}{2} = \frac{13}{30} \approx 0.433.$$

Example 5.26 (The Three Prisoners Problem). The three prisoners problem is a classical problem about conditional probability. Three prisoners, Alice, Bob, and Carl, are informed by their jailer that the next day, one of them will be released from prison, but that the other two will have to serve life sentences. The jailer says that he will not tell any prisoner what will happen to him or her. But Alice, who reasons that her chances of going free are now $\frac{1}{3}$, asks the jailer to give her the name of one prisoner, other than herself, who will

[10]The authors of [51, chapter 1] explain the ubiquity of urns in the field of probability theory as being connected with the French phrase *aller aux urnes* (to vote).

not go free. The jailer tells Alice that Bob will remain in jail. Now what are Alice's chances of going free? Has the probability changed? Alice could argue that she now has a $\frac{1}{2}$ chance of going free, since Bob will definitely remain behind. On the other hand, it also seems reasonable to argue that since one of Bob or Carl had to stay in jail, this new information could not possibly change the odds for Alice.

In fact, either answer may be correct. It depends on the strategy that the jailer follows in deciding which name to give to Alice (assuming that Alice knows which strategy is being used). If the jailer picks a name at random whenever both Bob and Carl are possible choices, then Alice's chances of freedom have not changed. However, if the jailer names Bob whenever possible, and otherwise names Carl, then the new information does indeed change Alice's probability of release to $\frac{1}{2}$. See Exercise 5.26.

There are many other versions of the three prisoners problem, including the "Monty Hall problem" that is a staple of popular culture. Exercise 5.27 describes the Monty Hall problem and other fun applications of these ideas.

5.3.3 Monte Carlo Algorithms

There are many algorithms whose output is not guaranteed to be correct. For example, Table 3.2 in Sect. 3.4 describes the Miller–Rabin algorithm, which is used to check whether a given large number is prime. In practice, one runs the algorithm many times to obtain an output that is "probably" correct. In applying these so-called *Monte Carlo* or *probabilistic algorithms*, it is important to be able to compute a confidence level, which is the probability that the output is indeed correct. In this section we describe how to use Bayes's formula to do such a computation.

The basic scenario consists of a large (possibly infinite) set of integers $\mathcal{S}$ and an interesting property A. For example, $\mathcal{S}$ could be the set of all integers, or more realistically $\mathcal{S}$ might be the set of all integers between, say, 2^{1024} and 2^{1025}. An example of an interesting property A is the property of being composite.

Now suppose that we are looking for numbers that do not have property A. Using the Miller–Rabin test, we might be looking for integers between 2^{1024} and 2^{1025} that are not composite, i.e., that are prime. In general, suppose that we are given an integer m in $\mathcal{S}$ and that we want to know whether m has property A. Usually we know approximately how many of the integers in $\mathcal{S}$ have property A. For example, we might know that 99 % of elements have property A and that the other 1 % do not. However, it may be difficult to determine with certainty that any particular $m \in \mathcal{S}$ does not have property A. So instead we settle for a faster algorithm that is not absolutely certain to be correct.

A *Monte Carlo algorithm* for property A takes as its input both a number $m \in \mathcal{S}$ to be tested and a randomly chosen number r and returns as output either Yes or No according to the following rules:

(1) If the algorithm returns Yes, then m definitely has property A. In conditional probability notation, this says that

$$\Pr(m \text{ has property } A \mid \text{algorithm returns Yes}) = 1.$$

(2) If m has property A, then the algorithm returns Yes for at least 50 % of the choices for r.[11] Using conditional probability notation,

$$\Pr(\text{algorithm returns Yes} \mid m \text{ has property } A) \geq \frac{1}{2}.$$

Now suppose that we run the algorithm N times on an integer $m \in \mathcal{S}$, using N different randomly chosen values for r. If even a single trial returns Yes, then we know that m has property A. But suppose instead that all N trials return the answer No. How confident can we be that our integer does not have property A? In probability terminology, we want to estimate

$$\Pr(m \text{ does not have property } A \mid \text{algorithm returns No } N \text{ times}).$$

More precisely, we want to show that if N is large, then this probability is close to 1.

We define two events:

$$\begin{aligned} E &= \{\text{an integer in } \mathcal{S} \text{ does not have property } A\}, \\ F &= \{\text{the algorithm returns No } N \text{ times in a row}\}. \end{aligned}$$

We are interested in the conditional probability $\Pr(E \mid F)$, that is, the probability that m does not have property A, given the fact that the algorithm returned No N times. We can compute this probability using Bayes's formula (5.21),

$$\Pr(E \mid F) = \frac{\Pr(F \mid E)\Pr(E)}{\Pr(F \mid E)\Pr(E) + \Pr(F \mid E^c)\Pr(E^c)}.$$

We are given that 99 % of the elements in $\mathcal{S}$ have property A, so

$$\Pr(E) = 0.01 \qquad \text{and} \qquad \Pr(E^c) = 0.99.$$

Next consider $\Pr(F \mid E)$. If m does not have property A, which is our assumption on this conditional probability, then the algorithm always returns No, since Property (1) of the Monte Carlo method tells us that a Yes output forces m to have property A. In symbols, Property (1) says that

$$\Pr(\text{No} \mid \text{not } A) = \Pr(A \mid \text{Yes}) = 1.$$

[11] More generally, the success rate in a Monte Carlo algorithm need not be 50 %, but may instead be any positive probability that is not too small. For the Miller–Rabin test described in Sect. 3.4, the corresponding probability is 75 %. See Exercise 5.28 for details.

It follows that $\Pr(F \mid E) = \Pr(\textsf{No} \mid \text{not } A)^N = 1$.

Finally, we must compute the value of $\Pr(F \mid E^c)$. Since the algorithm is run N independent times, we have

$$\begin{aligned}\Pr(F \mid E^c) &= \Pr(\text{Output is } \textsf{No} \mid m \text{ has property } A)^N \\ &= \bigl(1 - \Pr(\text{Output is } \textsf{Yes} \mid m \text{ has property } A)\bigr)^N \\ &\le \left(1 - \frac{1}{2}\right)^N \quad \text{from Property (2) of the Monte Carlo method,} \\ &= \frac{1}{2^N}.\end{aligned}$$

Substituting these values into Bayes's formula, we find that if the algorithm returns No N times in a row, then the probability that the integer m does not have property A is

$$\Pr(E \mid F) \ge \frac{1 \cdot (0.01)}{1 \cdot (0.01) + 2^{-N} \cdot (0.99)} = \frac{1}{1 + 99 \cdot 2^{-N}} = 1 - \frac{99}{2^N + 99}.$$

Notice that if N is large, the lower bound is very close to 1.

For example, if we run the algorithm 100 times and get 100 No answers, then the probability that m does not have property A is at least

$$\frac{99}{99 + 2^{-100}} \approx 1 - 10^{-32.1}.$$

So for most practical purposes, it is safe to conclude that m does not have property A.

5.3.4 Random Variables

We are generally more interested in the consequences of an experiment, for example the net loss or gain from a game of chance, than in the experiment itself. Mathematically, this means that we are interested in functions that are defined on events and that take values in some set.

Definition. A *random variable* is a function

$$X : \Omega \longrightarrow \mathbb{R}$$

whose domain is the sample space Ω and that takes values in the real numbers. More generally, a random variable is a function $X : \Omega \to \mathcal{S}$ whose range may be any set; for example, $\mathcal{S}$ could be a set of keys or a set of plaintexts.

We note that since our sample spaces are finite, a random variable takes on only finitely many values. Random variables are useful for defining events. For example, if $X : \Omega \to \mathbb{R}$ is a random variable, then any real number x defines three interesting events,

$$\{\omega \in \Omega : X(\omega) \le x\}, \qquad \{\omega \in \Omega : X(\omega) = x\}, \qquad \{\omega \in \Omega : X(\omega) > x\}.$$

Definition. Let $X : \Omega \to \mathbb{R}$ be a random variable. The *probability density function of* X, denoted by $f_X(x)$, is defined to be

$$f_X(x) = \Pr(X = x).$$

In other words, $f_X(x)$ is the probability that X takes on the value x. Sometimes we write $f(x)$ if the random variable is clear.

Remark 5.27. In probability theory, people often use the *distribution function of* X, which is the function

$$F_X(x) = \Pr(X \le x),$$

instead of the density function. Indeed, when studying probability theory for infinite sample spaces, it is essential to use F_X. However, since our sample spaces are finite, and thus our random variables are finite and discrete, the two notions are essentially interchangeable. For simplicity, we will stick to density functions.

There are a number of standard density functions that occur frequently in discrete probability calculations. We briefly describe a few of the more common ones.

Example 5.28 (Uniform Distribution). Let S be a set containing N elements; for example, S could be the set $S = \{0, 1, \ldots, N-1\}$. Let X be a random variable satisfying

$$f_X(j) = \Pr(X = j) = \begin{cases} \dfrac{1}{N} & \text{if } j \in S, \\ 0 & \text{if } j \notin S. \end{cases}$$

This random variable X is said to be *uniformly distributed* or to have *uniform density*, since each of the outcomes in S is equally likely.

Example 5.29 (Binomial Distribution). Suppose that an experiment has two outcomes, success or failure. Let p denote the probability of success. The experiment is performed n times and the random variable X records the number of successes. The sample space Ω consists of all binary strings $\omega = b_1 b_2 \ldots b_n$ of length n, where $b_i = 0$ if the i'th experiment is a failure and $b_i = 1$ if the i'th experiment is a success. Then the value of the random variable X at ω is simply $X(\omega) = b_1 + b_2 + \cdots + b_n$, which is the number of successes. Using the random variable X, we can express the probability of a single event ω as

$$\Pr(\{\omega\}) = p^{X(\omega)}(1-p)^{n-X(\omega)}.$$

(Do you see why this is the correct formula?) This allows us to compute the probability of exactly k successes as

$$\begin{aligned} f_X(k) &= \Pr(X = k) \\ &= \sum_{\omega \in \Omega,\, X(\omega)=k} \Pr(\{\omega\}) \end{aligned}$$

$$
\begin{aligned}
&= \sum_{\omega\in\Omega,\, X(\omega)=k} p^{X(\omega)}(1-p)^{n-X(\omega)} \\
&= \sum_{\omega\in\Omega,\, X(\omega)=k} p^{k}(1-p)^{n-k} \\
&= \#\{\omega\in\Omega : X(\omega)=k\}p^{k}(1-p)^{n-k} \\
&= \binom{n}{k} p^{k}(1-p)^{n-k}.
\end{aligned}
$$

Here the last line follows from the fact that there are $\binom{n}{k}$ ways to select exactly k of the n experiments to be successes. The function

$$f_X(k) = \binom{n}{k} p^k(1-p)^{n-k} \tag{5.23}$$

is called the *binomial density function.*

Example 5.30 (Hypergeometric Distribution). An urn contains N balls of which m are red and $N-m$ are blue. From this collection, n balls are chosen at random without replacement. Let X denote the number of red balls chosen. Then X is a random variable taking on the integer values

$$0 \le X(\omega) \le \min\{m,n\}.$$

In the case that $n \le m$, an argument similar to the one that we gave in Example 5.20 shows that the density function of X is given by the formula

$$f_X(i) = \Pr(X=i) = \frac{\binom{m}{i}\binom{N-m}{n-i}}{\binom{N}{n}}. \tag{5.24}$$

This is called the *hypergeometric density function.*

Example 5.31 (Geometric Distribution). We give one example of an infinite probability space. Suppose that we repeatedly toss an unfair coin, where the probability of getting heads is some number $0 < p < 1$. Let X be the random variable giving the total number of coin tosses required before heads appears for the first time. Note that it is possible for X to take on any positive integer value, since it is possible (although unlikely) that we could have a tremendously long string of tails.[12]

The sample space Ω consists of all binary strings $\omega = b_1b_2b_3\ldots$, where $b_i = 0$ if the i'th toss is tails and $b_i = 1$ if the i'th toss is heads. Note that Ω is

[12]For an amusing commentary on long strings of heads, see Act I of Tom Stoppard's *Rosencrantz and Guildenstern Are Dead.*

an infinite set. We assign probabilities to certain events, i.e. to certain subsets of Ω, by specifying the values of some initial tosses. So for any given finite binary string $\gamma_1\gamma_2\ldots\gamma_n$, we assign a probability

$$\Pr\bigl(\{\omega \in \Omega : \omega \text{ starts } \gamma_1\gamma_2\ldots\gamma_n\}\bigr) = p^{(\#\text{ of }\gamma_i\text{ equal to }1)}(1-p)^{(\#\text{ of }\gamma_i\text{ equal to }0)}.$$

The random variable X is defined by

$$X(\omega) = X(b_1b_2b_3\ldots) = (\text{smallest } i \text{ such that } b_i = 1).$$

Then

$$\{X = n\} = \{\omega \in \Omega : X(\omega) = n\} = \{\underbrace{000\ldots00}_{n-1\text{ zeros}}1b_{n+1}b_{n+2}\ldots\},$$

which gives the formula

$$f_X(n) = \Pr(X = n) = (1-p)^{n-1}p \qquad \text{for } n = 1, 2, 3, \ldots. \tag{5.25}$$

A random variable with the density function (5.25) is said to have a *geometric density*, because the sequence of probabilities $f_X(1), f_X(2), f_X(3), \ldots$ form a geometric progression.[13] Later, in Example 5.37, we compute the expected value of this X by summing an infinite geometric series.

Earlier we studied aspects of probability theory involving two or more events interacting in various ways. We now discuss material that allows us study the interaction of two or more random variables.

Definition. Let X and Y be two random variables. The *joint density function of X and Y*, denoted by $f_{X,Y}(x, y)$, is the probability that X takes the value x and Y takes the value y. Thus[14]

$$f_{X,Y}(x, y) = \Pr(X = x \text{ and } Y = y).$$

Similarly, the *conditional density function*, denoted by $f_{X|Y}(x \mid y)$, is the probability that X takes the value x, given that Y takes the value y:

$$f_{X|Y}(x \mid y) = \Pr(X = x \mid Y = y).$$

We say that X and Y are *independent* if

[13] A sequence $a_1, a_2, a_3, \ldots$ is called a *geometric progression* if all of the ratios a_{n+1}/a_n are the same. Similarly, the sequence is an *arithmetic progression* if all of the differences $a_{n+1} - a_n$ are the same.

[14] Note that the expression $\Pr(X = x \text{ and } Y = y)$ is really shorthand for the probability of the event

$$\{\omega \in \Omega : X(\omega) = x \text{ and } Y(\omega) = y\}.$$

If you find yourself becoming confused about probabilities expressed in terms of values of random variables, it often helps to write them out explicitly in terms of an event, i.e., as the probability of a certain subset of Ω.

$$f_{X,Y}(x,y) = f_X(x) f_Y(y) \quad \text{for all } x \text{ and } y.$$

This is equivalent to the events $\{X = x\}$ and $\{Y = y\}$ being independent in the earlier sense of independence that is defined on page 232. If there is no chance for confusion, we sometimes write $f(x,y)$ and $f(x \mid y)$ for $f_{X,Y}(x,y)$ and $f_{X|Y}(x \mid y)$, respectively.

Example 5.32. An urn contains four gold coins and three silver coins. A coin is drawn at random, examined, and returned to the urn, and then a second coin is randomly drawn and examined. Let X be the number of gold coins drawn and let Y be the number of silver ones. To find the joint density function $f_{X,Y}(x,y)$, we need to compute the probability of the event $\{X = x \text{ and } Y = y\}$. To help explain the calculation, we define two additional random variables. Let

$$F = \begin{cases} 1 & \text{if first pick is gold,} \\ 0 & \text{if first pick is silver,} \end{cases} \quad \text{and} \quad S = \begin{cases} 1 & \text{if second pick is gold,} \\ 0 & \text{if second pick is silver.} \end{cases}$$

Notice that $X = F + S$ and $Y = 2 - X = 2 - F - S$. Further, the random variables F and S are independent, and $\Pr(F = 1) = \Pr(S = 1) = \frac{4}{7}$. We can compute $f_{X,Y}(1,1)$ as follows:

$$\begin{aligned} f_{X,Y}(1,1) &= \Pr(X = 1 \text{ and } Y = 1) \\ &= \Pr(F = 1 \text{ and } S = 0) + \Pr(F = 0 \text{ and } S = 1) \\ &= \Pr(F = 1) \cdot \Pr(S = 0) + \Pr(F = 0) \cdot \Pr(S = 1) \\ &= \frac{4}{7} \cdot \frac{3}{7} + \frac{3}{7} \cdot \frac{4}{7} = \frac{24}{49} \approx 0.4898. \end{aligned}$$

In other words, the probability of drawing one gold coin and one silver coin is about 0.4898. The computation of the other values of $f_{X,Y}$ is similar.

These computations were easy because F and S are independent. How do our computations change if the first coin is not replaced before the second coin is selected? Then the probability of getting a silver coin on the second pick depends on whether the first pick was gold or silver. For example, the earlier computation of $f_{X,Y}(1,1)$ changes to

$$\begin{aligned} f_{X,Y}(1,1) &= \Pr(X = 1 \text{ and } Y = 1) \\ &= \Pr(F = 1 \text{ and } S = 0) + \Pr(F = 0 \text{ and } S = 1) \\ &= \Pr(S = 0 \mid F = 1) \Pr(F = 1) + \Pr(S = 1 \mid F = 0) \Pr(F = 0) \\ &= \frac{3}{6} \cdot \frac{4}{7} + \frac{4}{6} \cdot \frac{3}{7} = \frac{4}{7} \approx 0.5714. \end{aligned}$$

Thus the chance of getting exactly one gold coin and exactly one silver coin is somewhat larger if the coins are not replaced after each pick.

We remark that this last computation is a special case of the hypergeometric distribution; see Example 5.30. Thus the value $f_{X,Y}(1,1) = \frac{4}{7}$ may be computed using (5.24) with $N = 7$, $m = 4$, $n = 2$, and $i = 1$, which yields $\binom{4}{1}\binom{3}{1} / \binom{7}{2} = \frac{4}{7}$.

The following restatement of Bayes's formula is often convenient for calculations involving conditional probabilities.

Theorem 5.33 (Bayes's formula). *Let X and Y be random variables and assume that $f_Y(y) > 0$. Then*

$$f_{X|Y}(x \mid y) = \frac{f_X(x) f_{Y|X}(y \mid x)}{f_Y(y)}.$$

In particular,

$$X \text{ and } Y \text{ are independent} \iff f_{X|Y}(x \mid y) = f_X(x) \quad \text{for all } x \text{ and } y.$$

Example 5.34. In this example we use Bayes's formula to explore the independence of pairs of random variables taken from a triple (X, Y, Z). Let X and Y be independent random variables taking on values $+1$ and -1 with probability $\frac{1}{2}$ each, and let $Z = XY$. Then Z also takes on the values $+1$ and -1, and we have

$$f_Z(1) = \sum_{x \in \{-1,+1\}} \sum_{y \in \{-1,+1\}} \Pr(Z = 1 \mid X = x \text{ and } Y = y) \cdot f_{X,Y}(x, y). \tag{5.26}$$

If $(X, Y) = (+1, -1)$ or $(X, Y) = (-1, +1)$, then $Z \neq 1$, so only the two terms with $(x, y) = (1, 1)$ and $(x, y) = (-1, -1)$ appear in the sum (5.26). For these two terms, we have $\Pr(Z = 1 \mid X = x \text{ and } Y = y) = 1$, so

$$\begin{aligned} f_Z(1) &= \Pr(X = 1 \text{ and } Y = 1) + \Pr(X = -1 \text{ and } Y = -1) \\ &= \frac{1}{2} \cdot \frac{1}{2} + \frac{1}{2} \cdot \frac{1}{2} = \frac{1}{2}. \end{aligned}$$

It follows that $f_Z(-1) = 1 - f_Z(1)$ is also equal to $\frac{1}{2}$.

Next we compute the joint probability density of Z and X. For example,

$$\begin{aligned} f_{Z,X}(1, 1) &= \Pr(Z = 1 \text{ and } X = 1) \\ &= \Pr(X = 1 \text{ and } Y = 1) \\ &= \frac{1}{4} \qquad \text{since } X \text{ and } Y \text{ are independent,} \\ &= f_Z(1) f_X(1). \end{aligned}$$

Similar computations show that

$$f_{Z,X}(z, x) = f_Z(z) f_X(x) \qquad \text{for all } z, x \in \{-1, +1\},$$

so by Theorem 5.33, Z and X are independent. The argument works equally well for Z and Y, so Z and Y are also independent. Thus among the three random variables X, Y, and Z, any pair of them are independent. Yet we would not want to call the three of them together an independent family, since the value of Z is determined by the values of X and Y. This prompts the following definition.

Definition. A family of two or more random variables $\{X_1, X_2, \dots, X_n\}$ is *independent* if the events

$$\{X_1 = x_1 \text{ and } X_2 = x_2 \text{ and } \cdots \text{ and } X_n = x_n\}$$

are independent for every choice of $x_1, x_2, \dots, x_n$.

Notice that the random variables X, Y and $Z = XY$ in Example 5.34 are not an independent family, since

$$\Pr(Z = 1 \text{ and } X = 1 \text{ and } Y = -1) = 0,$$

while

$$\Pr(Z = 1) \cdot \Pr(X = 1) \cdot \Pr(Y = -1) = \frac{1}{8}.$$

5.3.5 Expected Value

The *expected value* of a random variable X is the average of its values weighted by their probability of occurrence. The expected value thus provides a rough initial indication of the behavior of X.

Definition. Let X be a random variable that takes on the values $x_1, \dots, x_n$. The *expected value* (or *mean*) of X is the quantity

$$E(X) = \sum_{i=1}^{n} x_i \cdot f_X(x_i) = \sum_{i=1}^{n} x_i \cdot \Pr(X = x_i). \tag{5.27}$$

Example 5.35. Let X be the random variable whose value is the sum of the numbers appearing on two tossed dice. The possible values of X are the integers between 2 and 12, so

$$E(X) = \sum_{i=2}^{12} i \cdot \Pr(X = i).$$

There are 36 ways for the two dice to fall, as indicated in Table 5.8a. We read off from that table the number of ways that the sum can equal i for each value of i between 2 and 12 and compile the results in Table 5.8b. The probability that $X = i$ is $\frac{1}{36}$ times the total number of ways that two dice can sum to i, so we can use Table 5.8b to compute

$$\begin{aligned} E(X) &= 2 \cdot \frac{1}{36} + 3 \cdot \frac{2}{36} + 4 \cdot \frac{3}{36} + 5 \cdot \frac{4}{36} + 6 \cdot \frac{5}{36} + 7 \cdot \frac{6}{36} \\ &\quad + 8 \cdot \frac{5}{36} + 9 \cdot \frac{4}{36} + 10 \cdot \frac{3}{36} + 11 \cdot \frac{2}{36} + 12 \cdot \frac{1}{36} \\ &= 7. \end{aligned}$$

This answers makes sense, since the middle value is 7, and for any integer j, the value of X is just as likely to be $7 + j$ as it is to be $7 - j$.

	1	2	3	4	5	6
1	2	3	4	5	6	7
2	3	4	5	6	7	8
3	4	5	6	7	8	9
4	5	6	7	8	9	10
5	6	7	8	9	10	11
6	7	8	9	10	11	12

(a) Sum of two dice

Sum	# of ways
2 or 12	1
3 or 11	2
4 or 10	3
5 or 9	4
6 or 8	5
7	6

(b) Number of ways to make a sum

Table 5.8: Outcome of rolling two dice

The name "expected" value is somewhat misleading, since the fact that the expectation $E(X)$ is a weighted average means that it may take on a value that is not actually attained, as the next example shows.

Example 5.36. Suppose that we choose an integer at random from among the integers $\{1,2,3,4,5,6\}$ and let X be the value of our choice. Then $\Pr(X = i) = \frac{1}{6}$ for each $1 \le i \le 6$, i.e., X is uniformly distributed. The expected value of X is

$$E(X) = \frac{1}{6}(1+2+3+4+5+6) = \frac{7}{2}.$$

Thus the expectation of X is a value that X does not actually attain. More generally, the expected value of a random variable uniformly distributed on $\{1,2,\ldots,N\}$ is $(N+1)/2$.

Example 5.37. We return to our coin tossing experiment (Example 5.31), where the probability of getting H on any one coin toss is equal to p. Let X be the random variable that is equal to n if H appears for the first time at the nth coin toss. Then X has a geometric density, and its density function $f_X(n)$ is given by the formula (5.25). We compute $E(X)$, which is the expected number of tosses before the first H appears:

$$\begin{aligned} E(X) &= \sum_{n=1}^{\infty} np(1-p)^{n-1} = -p\sum_{n=1}^{\infty}\frac{d}{dp}((1-p)^n) \\ &= -p\frac{d}{dp}\left(\sum_{n=1}^{\infty}(1-p)^n\right) = -p\frac{d}{dp}\left(\frac{1}{p}-1\right) = \frac{p}{p^2} = \frac{1}{p}. \end{aligned}$$

This answer seems plausible, since the smaller the value of p, the more tosses we expect to need before obtaining our first H. The computation of $E(X)$ uses a very useful trick with derivatives followed by the summation of a geometric series. See Exercise 5.33 for further applications of this method.

5.4 Collision Algorithms and Meet-in-the-Middle Attacks

A simple, yet surprisingly powerful, search method is based on the observation that it is usually much easier to find matching objects than it is to find a particular object. Methods of this sort go by many names, including meet-in-the-middle attacks and collision algorithms.

5.4.1 The Birthday Paradox

The fundamental idea behind collision algorithms is strikingly illustrated by the famous birthday paradox. In a random group of 40 people, consider the following two questions:

(1) What is the probability that someone has the same birthday as you?

(2) What is the probability that at least two people share the same birthday?

It turns out that the answers to (1) and (2) are very different. As a warm-up, we start by answering the easier first question.

A rough answer is that since any one person has a 1-in-365 chance of sharing your birthday, then in a crowd of 40 people, the probability of someone having your birthday is approximately $\frac{40}{365} \approx 11\,\%$. However, this is an overestimate, since it double counts the occurrences of more than one person in the crowd sharing your birthday.[15] The exact answer is obtained by computing the probability that none of the people share your birthday and then subtracting that value from 1.

$$\begin{aligned}\Pr\begin{pmatrix}\text{someone has}\\ \text{your birthday}\end{pmatrix} &= 1 - \Pr\begin{pmatrix}\text{none of the 40 people}\\ \text{has your birthday}\end{pmatrix}\\ &= 1 - \prod_{i=1}^{40} \Pr\begin{pmatrix}i\text{th person does not}\\ \text{have your birthday}\end{pmatrix}\\ &= 1 - \left(\frac{364}{365}\right)^{40}\\ &\approx 10.4\,\%.\end{aligned}$$

Thus among 40 strangers, there is only slightly better than a 10 % chance that one of them shares your birthday.

Now consider the second question, in which you win if any two of the people in the group have the same birthday. Again it is easier to compute the probability that all 40 people have different birthdays. However, the computation

[15]If you think that $\frac{40}{365}$ is the right answer, think about the same situation with 366 people. The probability that someone shares your birthday cannot be $\frac{366}{365}$, since that's larger than 1.

changes because we now require that the ith person have a birthday that is different from all of the previous $i-1$ people's birthdays. Hence the calculation is

$$\begin{aligned}
\Pr\binom{\text{two people have}}{\text{the same birthday}} &= 1 - \Pr\binom{\text{all 40 people have}}{\text{different birthdays}} \\
&= 1 - \prod_{i=1}^{40} \Pr\begin{pmatrix} i\text{th person does not have} \\ \text{the same birthday as any} \\ \text{of the previous } i-1 \text{ people} \end{pmatrix} \\
&= 1 - \prod_{i=1}^{40} \frac{365-(i-1)}{365} \\
&= 1 - \frac{365}{365} \cdot \frac{364}{365} \cdot \frac{363}{365} \cdots \frac{326}{365} \\
&\approx 89.1\,\%.
\end{aligned}$$

Thus among 40 strangers, there is almost a 90 % chance that two of them share a birthday.

The only part of this calculation that merits some comment is the formula for the probability that the ith person has a birthday different from any of the previous $i-1$ people. Among the 365 possible birthdays, note that the previous $i-1$ people have taken up $i-1$ of them. Hence the probability that the ith person has his or her birthday among the remaining $365-(i-1)$ days is

$$\frac{365-(i-1)}{365}.$$

Most people tend to assume that questions (1) and (2) have essentially the same answer. The fact that they do not is called the *birthday paradox.* In fact, it requires only 23 people to have a better than 50 % chance of a matched birthday, while it takes 253 people to have better than a 50 % chance of finding someone who has your birthday.

5.4.2 A Collision Theorem

Cryptographic applications of collision algorithms are generally based on the following setup. Bob has a box that contains N numbers. He chooses n distinct numbers from the box and puts them in a list. He then makes a second list by choosing m (not necessarily distinct) numbers from the box. The remarkable fact is that if n and m are each slightly larger than $\sqrt{N}$, then it is very likely that the two lists contain a common element.

We start with an elementary result that illustrates the sort of calculation that is used to quantify the probability of success of a collision algorithm.

Theorem 5.38 (Collision Theorem). *An urn contains N balls, of which n are red and $N-n$ are blue. Bob randomly selects a ball from the urn, replaces* *it in the urn, randomly selects a second ball, replaces it, and so on. He does this until he has looked at a total of* m *balls.*

(a) *The probability that Bob selects at least one red ball is*

$$\Pr(\textit{at least one red}) = 1 - \left(1 - \frac{n}{N}\right)^m. \tag{5.28}$$

(b) *A lower bound for the probability* (5.28) *is*

$$\Pr(\textit{at least one red}) \geq 1 - e^{-mn/N}. \tag{5.29}$$

If N *is large and if* m *and* n *are not too much larger than* $\sqrt{N}$ *(e.g.,* $m, n < 10\sqrt{N}$*), then* (5.29) *is almost an equality.*

Proof. Each time Bob selects a ball, his probability of choosing a red one is $\frac{n}{N}$, so you might think that since he chooses m balls, his probability of getting a red one is $\frac{mn}{N}$. However, a small amount of thought shows that this must be incorrect. For example, if m is large, this would lead to a probability that is larger than 1. The difficulty, just as in the birthday example in Sect. 5.4.1, is that we are overcounting the times that Bob happens to select more than one red ball. The correct way to calculate is to compute the probability that Bob chooses only blue balls and then subtract this complementary probability from 1. Thus

$$\begin{aligned}
\Pr\begin{pmatrix}\text{at least one red} \\ \text{ball in } m \text{ attempts}\end{pmatrix} &= 1 - \Pr(\text{all } m \text{ choices are blue}) \\
&= 1 - \prod_{i=1}^{m} \Pr(i\text{th choice is blue}) \\
&= 1 - \prod_{i=1}^{m} \left(\frac{N-n}{N}\right) \\
&= 1 - \left(1 - \frac{n}{N}\right)^m.
\end{aligned}$$

This completes the proof of (a).

For (b), we use the inequality

$$e^{-x} \geq 1 - x \qquad \text{for all } x \in \mathbb{R}.$$

(See Exercise 5.38(a) for a proof.) Setting $x = n/N$ and raising both sides of the inequality to the mth power shows that

$$1 - \left(1 - \frac{n}{N}\right)^m \geq 1 - (e^{-n/N})^m = 1 - e^{-mn/N},$$

which proves the important inequality in (b). We leave it to the reader (Exercise 5.38(b)) to prove that the inequality is close to being an equality if m and n is not too large compared to $\sqrt{N}$. $\square$

In order to connect Theorem 5.38 with the problem of finding a match in two lists of numbers, we view the list of numbers as an urn containing N numbered blue balls. After making our first list of n different numbered balls, we repaint those n balls with red paint and return them to the box. The second list is constructed by drawing m balls out of the urn one at a time, noting their number and color, and then replacing them. The probability of selecting at least one red ball is the same as the probability of a matched number on the two lists.

Example 5.39. A deck of cards is shuffled and eight cards are dealt face up. Bob then takes a second deck of cards and chooses eight cards at random, replacing each chosen card before making the next choice. What is Bob's probability of matching one of the cards from the first deck?

We view the eight dealt cards from the first deck as "marking" those same cards in the second deck. So our "urn" is the second deck, the "red balls" are the eight marked cards in the second deck, and the "blue balls" are the other 44 cards in the second deck. Theorem 5.38(a) tells us that

$$\Pr(\text{a match}) = 1 - \left(1 - \frac{8}{52}\right)^8 \approx 73.7\,\%.$$

The approximation in Theorem 5.38(b) gives a lower bound of 70.8 %.

Suppose instead that Bob deals ten cards from the first deck and chooses only five cards from the second deck. Then

$$\Pr(\text{a match}) = 1 - \left(1 - \frac{10}{52}\right)^5 \approx 65.6\,\%.$$

Example 5.40. A box contains 10 billion labeled objects. Bob randomly selects 100,000 distinct objects from the box, makes a list of which objects he's chosen, and returns them to the box. If he next randomly selects another 100,000 objects (with replacement) and makes a second list, what is the probability that the two lists contain a match? Formula (5.28) in Theorem 5.38(a) says that

$$\Pr(\text{a match}) = 1 - \left(1 - \frac{100{,}000}{10^{10}}\right)^{100{,}000} \approx 0.632122.$$

The approximate lower bound given by the formula (5.29) in Theorem 5.38(b) is 0.632121. As you can see, the approximation is quite accurate.

It is interesting to observe that if Bob doubles the number of objects in his lists to 200,000, then his probability of getting a match increases quite substantially to 98.2 %. And if he triples the number of elements in each list to 300,000, then the probability of a match is 99.988 %. This rapid increase reflects that fact that the exponential function in (5.29) decreases very rapidly as soon as mn becomes larger than N.

Example 5.41. A set contains N objects. Bob randomly chooses n of them, makes a list of his choices, replaces them, and then chooses another n of them. How large should he choose n to give himself a 50 % chance of getting a match? How about if he wants a 99.99 % chance of getting a match?

For the first question, Bob uses the reasonably accurate lower bound of formula (5.29) to set

$$\Pr(\text{match}) \approx 1 - e^{-n^2/N} = \frac{1}{2}.$$

It is easy to solve this for n:

$$e^{-n^2/N} = \frac{1}{2} \implies -\frac{n^2}{N} = \ln\left(\frac{1}{2}\right) \implies n = \sqrt{N \cdot \ln 2} \approx 0.83\sqrt{N}.$$

Thus it is enough to create lists that are a bit shorter than $\sqrt{N}$ in length.

The second question is similar, but now Bob solves

$$\Pr(\text{match}) \approx 1 - e^{-n^2/N} = 0.9999 = 1 - 10^{-4}.$$

The solution is

$$n = \sqrt{N \cdot \ln 10^4} \approx 3.035 \cdot \sqrt{N}.$$

Remark 5.42. Algorithms that rely on finding matching elements from within one or more lists go by a variety of names, including *collision algorithm*, *meet-in-the-middle algorithm*, *birthday paradox algorithm*, and *square root algorithm*. The last refers to the fact that the running time of a collision algorithm is generally a small multiple of the square root of the running time required by an exhaustive search. The connection with birthdays was briefly discussed in Sect. 5.4.1; see also Exercise 5.36. When one of these algorithms is used to break a cryptosystem, the word "algorithm" is often replaced by the word "attack," so cryptanalysts refer to *meet-in-the-middle attacks*, *square root attacks*, etc.

Remark 5.43. Collision algorithms tend to take approximately $\sqrt{N}$ steps in order to find a collision among N objects. A drawback of these algorithms is that they require creation of one or more lists of size approximately $\sqrt{N}$. When N is large, providing storage for $\sqrt{N}$ numbers may be more of an obstacle than doing the computation. In Sect. 5.5 we describe a collision method due to Pollard that, at the cost of a small amount of extra computation, requires essentially no storage.

5.4.3 A Discrete Logarithm Collision Algorithm

There are many applications of collision algorithms to cryptography. These may involve searching a space of keys or plaintexts or ciphertexts, or for public key cryptosystems, they may be aimed at solving the underlying hard mathematical problem. In this section we illustrate the general theory by formulating an abstract randomized collision algorithm to solve the discrete logarithm problem. For the finite field $\mathbb{F}_p$, it solves the discrete logarithm problem (DLP) in approximately $\sqrt{p}$ steps.

One may well ask why the probabilistic collision algorithm described in Proposition 5.44 with expected running time $\mathcal{O}(\sqrt{N})$ is interesting, since the baby step–giant step algorithm from Sect. 2.7 is deterministic and solves the same problem in the same amount of time. One answer is that both algorithms also require $\mathcal{O}(\sqrt{N})$ storage, which is a serious constraint if N is large. So the collision algorithm in Proposition 5.44 may be viewed as a warm-up for Pollard's ρ algorithm, which is a collision algorithm taking $\mathcal{O}(\sqrt{N})$ time, but using only $\mathcal{O}(1)$ storage. We will discuss Pollard's algorithm in Sect. 5.5.

One might also inquire why any of these $\mathcal{O}(\sqrt{N})$ collision algorithms are interesting, since, the index calculus described in Sect. 3.8 solves the DLP in $\mathbb{F}_p$ much more rapidly. But there are other groups, such as elliptic curve groups, for which collision algorithms are the fastest known way to solve the DLP. This explains why elliptic curve groups are used in cryptography; at present, the DLP in an elliptic curve group is much harder than the DLP in $\mathbb{F}_p^*$ for groups of about the same size. Elliptic curves and their use in cryptography is the subject of Chap. 6.

Proposition 5.44. *Let G be a group, and let $g \in G$ be an element of order N, i.e., $g^N = e$ and no smaller power of g is equal to e. Then, assuming that the discrete logarithm problem*

$$g^x = h \tag{5.30}$$

has a solution, a solution can be found in $\mathcal{O}(\sqrt{N})$ steps, where each step is an exponentiation in the group G. (Note that since $g^N = e$, the powering algorithm from Sect. 1.3.2 lets us raise g to any power using fewer than $2\log_2 N$ group multiplications.)

Proof. The idea is to write x as $x = y - z$ and look for a solution to

$$g^y = h \cdot g^z.$$

We do this by making a list of g^y values and a list of $h \cdot g^z$ values and looking for a match between the two lists.

We begin by choosing random exponents $y_1, y_2, \ldots, y_n$ between 1 and N and computing the values

$$g^{y_1},\ g^{y_2},\ g^{y_3},\ \ldots,\ g^{y_n} \quad \text{in } G. \tag{5.31}$$

Note that all of the values (5.31) are in the set

$$S = \{1, g, g^2, g^3, \ldots, g^{N-1}\},$$

so (5.31) is a selection of (approximately) n elements of S. In terms of the collision theorem (Theorem 5.38), we view S as an urn containing N balls and the list (5.31) as a way of coloring n of those balls red.

Next we choose additional random exponents $z_1, z_2, \ldots, z_n$ between 1 and k and compute the quantities

$$h \cdot g^{z_1},\ h \cdot g^{z_2},\ h \cdot g^{z_3},\ \ldots,\ h \cdot g^{z_n} \quad \text{in } G. \tag{5.32}$$

Since we are assuming that (5.30) has a solution, i.e., h is equal to some power of g, it follows that each of the values $h \cdot g^{z_i}$ is also in the set S. Thus the list (5.32) may be viewed as selecting n elements from the urn, and we would like to know the probability of selecting at least one red ball, i.e., the probability that at least one element in the list (5.32) matches an element in the list (5.31). The collision theorem (Theorem 5.38) says that

$$\Pr\begin{pmatrix}\text{at least one match} \\ \text{between (5.31) and (5.32)}\end{pmatrix} \approx 1 - \left(1 - \frac{n}{N}\right)^n \approx 1 - e^{-n^2/N}.$$

Thus if we choose (say) $n \approx 3\sqrt{N}$, then our probability of getting a match is greater than 99.98 %, so we are almost guaranteed a match. Or if that is not good enough, take $n \approx 5\sqrt{N}$ to get a probability of success greater than $1 - 10^{-10}$. Notice that as soon as we find a match between the two lists, say $g^y = h \cdot g^z$, then we have solved the discrete logarithm problem (5.30) by setting $x = y - z$.[16]

How long does it take us to find this solution? Each of the lists (5.31) and (5.32) has n elements, so it takes approximately $2n$ steps to assemble each list. More precisely, each element in each list requires us to compute g^i for some value of i between 1 and N, and it takes approximately $2\log_2(i)$ group multiplications to compute g^i using the fast exponentiation algorithm described in Sect. 1.3.2. (Here $\log_2$ is the logarithm to the base 2.) Thus it takes approximately $4n\log_2(N)$ multiplications to assemble the two lists. In addition, it takes about $\log_2(n)$ steps to check whether an element of the second list is in the first list (e.g., sort the first list), so $n\log_2(n)$ comparisons altogether. Hence the total computation time is approximately

$$4n\log_2(N) + n\log_2(n) = n\log_2(N^4 n) \text{ steps.}$$

Taking $n \approx 3\sqrt{N}$, which as we have seen gives us a 99.98 % chance of success, we find that

$$\text{Computation Time} \approx 13.5 \cdot \sqrt{N} \cdot \log_2(1.3 \cdot N). \qquad \square$$

[16]If this value of x happens to be negative and we want a positive solution, we can always use the fact that $g^N = 1$ to replace it with $x = y - z + N$.

t	g^t	$h \cdot g^t$
564	410	**422**
469	357	181
276	593	620
601	416	126
9	512	3
350	445	233

t	g^t	$h \cdot g^t$
53	10	605
332	651	175
178	121	401
477	450	206
503	116	428
198	426	72

t	g^t	$h \cdot g^t$
513	164	37
71	597	203
314	554	567
581	47	537
371	334	437
83	**422**	489

Table 5.9: Solving $2^x = 390$ in $\mathbb{F}_{659}$ with random exponent collisions

Example 5.45. We do an example with small numbers to illustrate the use of collisions. We solve the discrete logarithm problem

$$2^x = 390 \quad \text{in the finite field } \mathbb{F}_{659}.$$

The number 2 has order 658 modulo 659, so it is a primitive root. In this example $g = 2$ and $h = 390$. We choose random exponents t and compute the values of g^t and $h \cdot g^t$ until we get a match. The results are compiled in Table 5.9. We see that

$$2^{83} = 390 \cdot 2^{564} = 422 \quad \text{in } \mathbb{F}_{659}.$$

Hence using two lists of length 18, we have solved a discrete logarithm problem in $\mathbb{F}_{659}$. (We had a 39 % chance of getting a match with lists of length 18, so we were a little bit lucky.) The solution is

$$2^{83} \cdot 2^{-564} = 2^{-481} = 2^{177} = 390 \quad \text{in } \mathbb{F}_{659}.$$

Remark 5.46. The algorithms described in Propositions 2.21 and 5.44 solve the DLP in $\mathcal{O}(\sqrt{N})$ steps. It is thus interesting that, in a certain sense, Victor Shoup [130] has shown that there cannot exist a general algorithm to solve the DLP in an arbitrary finite group in fewer than $\mathcal{O}(\sqrt{p})$ steps, where p is the largest prime dividing the order of the group. This is the so-called *black box DLP*, in which you are given a box that instantaneously performs the group operations, but you're not allowed to look inside the box to see how it is doing the computations.

5.5 Pollard's ρ Method

As we noted in Remark 5.43, collision algorithms tend to require a considerable amount of storage. A beautiful idea of Pollard often allows one to use almost no storage, at the cost of a small amount of extra computation. We explain

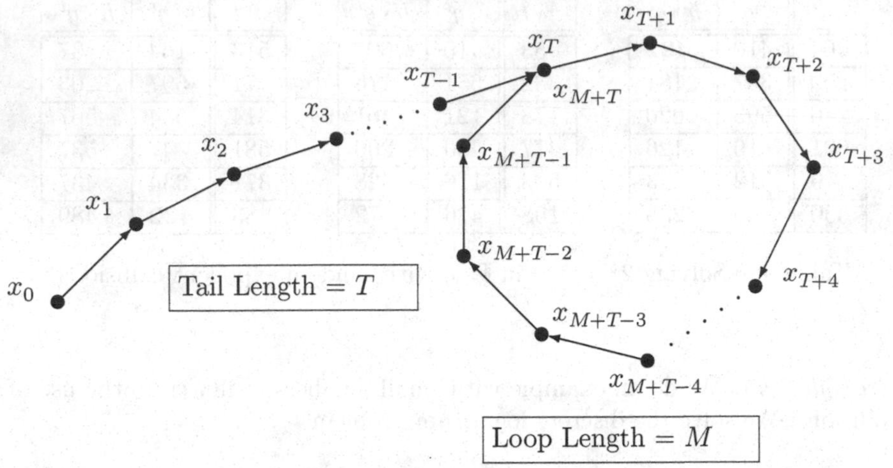

Figure 5.1: Pollard's ρ method

the basic idea behind Pollard's method and then illustrate it by yet again solving a small instance of the discrete logarithm problem in $\mathbb{F}_p$. See also Exercise 5.44 for a factorization algorithm based on the same ideas.

5.5.1 Abstract Formulation of Pollard's ρ Method

We begin in an abstract setting. Let S be a finite set and let

$$f : S \longrightarrow S$$

be a function that does a good job at mixing up the elements of S. Suppose that we start with some element $x \in S$ and we repeatedly apply f to create a sequence of elements

$$x_0 = x, \quad x_1 = f(x_0), \quad x_2 = f(x_1), \quad x_3 = f(x_2), \quad x_4 = f(x_3), \quad \ldots.$$

In other words,

$$x_i = \underbrace{(f \circ f \circ f \circ \cdots \circ f)}_{i \text{ iterations of } f}(x).$$

The map f from S to itself is an example of a *discrete dynamical system.* The sequence

$$x_0, x_1, x_2, x_3, x_4, \ldots \tag{5.33}$$

is called the (*forward*) *orbit of* x by the map f and is denoted by $O_f^+(x)$.

The set S is finite, so eventually there must be some element of S that appears twice in the orbit $O_f^+(x)$. We can illustrate the orbit as shown in Fig. 5.1. For a while the points $x_0, x_1, x_2, x_3, \ldots$ travel along a "path" without repeating until eventually they loop around to give a repeated element. Then they continue moving around the loop. As illustrated, we let T be the number of elements in the "tail" before getting to the loop, and we let M be the number of elements in the loop. Mathematically, T and M are defined by the conditions

$$T = \begin{pmatrix}\text{largest integer such that } x_{T-1} \\ \text{appears only once in } O_f^+(x)\end{pmatrix} \qquad M = \begin{pmatrix}\text{smallest integer such} \\ \text{that } x_{T+M} = x_T\end{pmatrix}$$

Remark 5.47. Look again at the illustration in Fig. 5.1. It may remind you of a certain Greek letter. For this reason, collision algorithms based on following the orbit of an element in a discrete dynamical system are called *ρ algorithms.* The first ρ algorithm was invented by Pollard in 1974.

Suppose that S contains N elements. Later, in Theorem 5.48, we will sketch a proof that the quantity $T + M$ is usually no more than a small multiple of $\sqrt{N}$. Since $x_T = x_{T+M}$ by definition, this means that we obtain a collision in $\mathcal{O}(\sqrt{N}\,)$ steps. However, since we don't know the values of T and M, it appears that we need to make a list of $x_0, x_1, x_2, x_3, \ldots, x_{T+M}$ in order to detect the collision.

Pollard's clever idea is that it is possible to detect a collision in $\mathcal{O}(\sqrt{N}\,)$ steps without storing all of the values. There are various ways to accomplish this. We describe one such method. Although not of optimal efficiency, it has the advantage of being easy to understand. (For more efficient methods, see [23, 28, §8.5], or [90].) The idea is to compute not only the sequence x_i, but also a second sequence y_i defined by

$$y_0 = x_0 \qquad \text{and} \qquad y_{i+1} = f\bigl(f(y_i)\bigr) \quad \text{for } i = 0, 1, 2, 3, \ldots.$$

In other words, every time that we apply f to generate the next element of the x_i sequence, we apply f twice to generate the next element of the y_i sequence. It is clear that

$$y_i = x_{2i}.$$

How long will it take to find an index i with $x_{2i} = x_i$? In general, for $j > i$ we have

$$x_j = x_i \qquad \text{if and only if} \qquad i \geq T \quad \text{and} \quad j \equiv i \pmod{M}.$$

This is clear from the ρ-shaped picture in Fig. 5.1, since we get $x_j = x_i$ precisely when we are past x_T, i.e., when $i \geq T$, and x_j has gone around the loop past x_i an integral number of times, i.e., when $j - i$ is a multiple of M.

Thus $x_{2i} = x_i$ if and only if $i \geq T$ and $2i \equiv i \pmod{M}$. The latter condition is equivalent to $M \mid i$, so we get $x_{2i} = x_i$ exactly when i is equal to the first multiple of M that is larger than T. Since one of the numbers T, $T+1, \ldots, T+M-1$ is divisible by M, this proves that

$$x_{2i} = x_i \quad \text{for some } 1 \leq i < T + M.$$

We show in the next theorem that the average value of $T + M$ is approximately $1.25 \cdot \sqrt{N}$, so we have a very good chance of getting a collision in a small multiple of $\sqrt{N}$ steps. This is more or less the same running time as the collision algorithm described in Sect. 5.4.3, but notice that we need to store only two numbers, namely the current values of the x_i sequence and the y_i sequence.

Theorem 5.48 (Pollard's ρ Method: abstract version)**.** *Let S be a finite set containing N elements, let $f : S \to S$ be a map, and let $x \in S$ be an initial point.*

(a) *Suppose that the forward orbit $O_f^+(x) = \{x_0, x_1, x_2, \ldots\}$ of x has a tail of length T and a loop of length M, as illustrated in Fig. 5.1. Then*

$$x_{2i} = x_i \quad \text{for some } 1 \le i < T + M. \tag{5.34}$$

(b) *If the map f is sufficiently random, then the expected value of $T + M$ is*

$$E(T + M) \approx 1.2533 \cdot \sqrt{N}.$$

Hence if N is large, then we are likely to find a collision as described by (5.34) in $O(\sqrt{N})$ steps, where a "step" is one evaluation of the function f.

Proof. (a) We proved this earlier in this section.
(b) We sketch the proof of (b) because it is an instructive blend of probability theory and analysis of algorithms. However, the reader desiring a rigorous proof will need to fill in some details. Suppose that we compute the first k values $x_0, x_1, x_2, \ldots, x_{k-1}$. What is the probability that we do not get any matches? If we assume that the successive x_i's are randomly chosen from the set S, then we can compute this probability as

$$\begin{aligned}
\Pr\begin{pmatrix} x_0, x_1, \ldots, x_{k-1} \\ \text{are all different} \end{pmatrix} &= \prod_{i=1}^{k-1} \Pr\left(\begin{matrix} x_i \ne x_j \text{ for} \\ \text{all } 0 \le j < i \end{matrix} \;\middle|\; \begin{matrix} x_0, x_1, \ldots, x_{i-1} \\ \text{are all different} \end{matrix}\right) \\
&= \prod_{i=1}^{k-1} \left(\frac{N-i}{N}\right) &\quad (5.35) \\
&= \prod_{i=1}^{k-1} \left(1 - \frac{i}{N}\right). &\quad (5.36)
\end{aligned}$$

Note that the probability formula (5.35) comes from the fact that if the first i choices $x_0, x_1, \ldots, x_{i-1}$ are distinct, then among the N possible choices for x_i, exactly $N - i$ of them are different from the previously chosen values. Hence the probability of getting a new value, assuming that the earlier values were distinct, is $\frac{N-i}{N}$.

We can approximate the product (5.36) using the estimate

$$1 - t \approx e^{-t}, \qquad \text{valid for small values of } t.$$

(Compare with the proof of Theorem 5.38(b), and see also Exercise 5.38.) In practice, k will be approximately $\sqrt{N}$ and N will be large, so $\frac{i}{N}$ will indeed be small for $1 \le i < k$. Hence

$$\Pr\begin{pmatrix} x_0, x_1, \ldots, x_{k-1} \\ \text{are all different} \end{pmatrix} \approx \prod_{i=1}^{k-1} e^{-i/N} = e^{-(1+2+\cdots+(k-1))/N} \approx e^{-k^2/2N}. \quad (5.37)$$

For the last approximation we are using the fact that

$$1 + 2 + \cdots + (k-1) = \frac{k^2 - k}{2} \approx \frac{k^2}{2} \quad \text{when } k \text{ is large.}$$

We now know the probability that $x_0, x_1, \ldots, x_{k-1}$ are all distinct. Assuming that they are distinct, what is the probability that the next choice x_k gives a match? There are k elements for it to match among the N possible elements, so this conditional probability is

$$\Pr(x_k \text{ is a match} \mid x_0, \ldots, x_{k-1} \text{ are distinct}) = \frac{k}{N}. \qquad (5.38)$$

Hence

$$\begin{aligned}
&\Pr(x_k \text{ is the first match}) \\
&\qquad = \Pr(x_k \text{ is a match AND } x_0, \ldots, x_{k-1} \text{ are distinct}) \\
&\qquad = \Pr(x_k \text{ is a match} \mid x_0, \ldots, x_{k-1} \text{ are distinct}) \\
&\qquad\qquad\qquad\qquad \cdot \Pr(x_0, \ldots, x_{k-1} \text{ are distinct}) \\
&\qquad \approx \frac{k}{N} \cdot e^{-k^2/2N} \qquad\qquad \text{from (5.37) and (5.38).}
\end{aligned}$$

The expected number of steps before finding the first match is then given by the formula

$$E(\text{first match}) = \sum_{k \ge 1} k \cdot \Pr(x_k \text{ is the first match}) \approx \sum_{k \ge 1} \frac{k^2}{N} \cdot e^{-k^2/2N}. \quad (5.39)$$

We want to know what this series looks like as a function of N. The following estimate, whose derivation uses elementary calculus, is helpful in estimating series of this sort.

Lemma 5.49. *Let $F(t)$ be a "nicely behaved" real valued function*[17] *with the property that $\int_0^\infty F(t)\,dt$ converges. Then for large values of n we have*

$$\sum_{k=1}^{\infty} F\left(\frac{k}{n}\right) \approx n \cdot \int_0^\infty F(t)\,dt. \qquad (5.40)$$

[17] For example, it would suffice that F have a continuous derivative.

Proof. We start with the definite integral of $F(t)$ over an interval $0 \le t \le A$. By definition, this integral is equal to a limit of Riemann sums,

$$\int_0^A F(t)\,dt = \lim_{n\to\infty} \sum_{k=1}^{An} F\left(\frac{k}{n}\right) \cdot \frac{1}{n},$$

where in the sum we have broken the interval $[0, A]$ into An pieces. In particular, if n is large, then

$$n \cdot \int_0^A F(t)\,dt \approx \sum_{k=1}^{An} F\left(\frac{k}{n}\right).$$

Now letting $A \to \infty$ yields (5.40). (We do not claim that this is a rigorous argument. Our aim is merely to convey the underlying idea. The interested reader may supply the details needed to complete the argument and to obtain explicit upper and lower bounds.) □

We use Lemma 5.49 to estimate

$$\begin{aligned} E(\text{first match}) &\approx \sum_{k\ge 1} \frac{k^2}{N} \cdot e^{-k^2/2N} && \text{from (5.39),} \\ &= \sum_{k\ge 1} F\left(\frac{k}{\sqrt{N}}\right) && \text{letting } F(t) = t^2 e^{-t^2/2}, \\ &\approx \sqrt{N} \cdot \int_0^\infty t^2 e^{-t^2/2}\,dt && \text{from (5.40) with } n = \sqrt{N}, \\ &\approx 1.2533 \cdot \sqrt{N} && \text{by numerical integration.} \end{aligned}$$

For the last line, we used a numerical method to estimate the definite integral, although in fact the integral can be evaluated exactly. (Its value turns out to be $\sqrt{\pi/2}$; see Exercise 5.43.) This completes the proof of (b), and combining (a) and (b) gives the final statement of Theorem 5.48. □

Remark 5.50. It is instructive to check numerically the accuracy of the estimates used in the proof of Theorem 5.48. In that proof we claimed that for large values of N, the expected number of steps before finding a match is given by each of the following three formulas:

$E_1 = \sum_{k\ge 1} \frac{k^2}{N} \prod_{i=1}^{k-1} \left(1 - \frac{i}{N}\right)$	$E_2 = \sum_{k\ge 1} \frac{k^2}{N} e^{-k^2/2N}$	$E_3 = \sqrt{N} \int_0^\infty t^2 e^{-t^2/2}\,dt$

More precisely, E_1 is the exact formula, but hard to compute exactly if N is very large, while E_2 and E_3 are approximations. We have computed the values of E_1, E_2, and E_3 for some moderate sized values of N and compiled the results in Table 5.10. As you can see, E_2 and E_3 are quite close to one another, and once N gets reasonably large, they also provide a good approximation for E_1. Hence for very large values of N, say $2^{80} < N < 2^{160}$, it is quite reasonable to estimate E_1 using E_3.

5.5.2 Discrete Logarithms via Pollard's ρ Method

In this section we describe how to use Pollard's ρ method to solve the discrete logarithm problem

$$g^t = h \quad \text{in } \mathbb{F}_p^*$$

N	E_1	E_2	E_3	E_1/E_3
100	12.210	12.533	12.533	0.97421
500	27.696	28.025	28.025	0.98827
1000	39.303	39.633	39.633	0.99167
5000	88.291	88.623	88.623	0.99626
10000	124.999	125.331	125.331	0.99735
20000	176.913	177.245	177.245	0.99812
50000	279.917	280.250	280.250	0.99881

Table 5.10: Expected number of steps until a ρ collision

when g is a primitive root modulo p. The idea is to find a collision between $g^i h^j$ and $g^k h^\ell$ for some known exponents i, j, k, ℓ. Then $g^{i-k} = h^{\ell-j}$, and taking roots in $\mathbb{F}_p$ will more or less solve the problem of expressing h as a power of g.

The difficulty is finding a function $f : \mathbb{F}_p \to \mathbb{F}_p$ that is complicated enough to mix up the elements of $\mathbb{F}_p$, yet simple enough to keep track of its orbits. Pollard [104] suggests using the function

$$f(x) = \begin{cases} gx & \text{if } 0 \le x < p/3, \\ x^2 & \text{if } p/3 \le x < 2p/3, \\ hx & \text{if } 2p/3 \le x < p. \end{cases} \tag{5.41}$$

Note that x must be reduced modulo p into the range $0 \le x < p$ before (5.41) is used to determine the value of $f(x)$.

Remark 5.51. No one has proven that the function $f(x)$ given by (5.41) is sufficiently random to guarantee that Theorem 5.48 is true for f, but experimentally, the function f works fairly well. However, Teske [144, 145] has shown that f is not sufficiently random to give optimal results, and she gives examples of somewhat more complicated functions that work better in practice.

Consider what happens when we repeatedly apply the function f given by (5.41) to the starting point $x_0 = 1$. At each step, we either multiply by g, multiply by h, or square the previous value. So after each step, we end up with a power of g multiplied by a power of h, say after i steps we have

$$x_i = (\underbrace{f \circ f \circ f \circ \cdots \circ f}_{i \text{ iterations of } f})(1) = g^{\alpha_i} \cdot h^{\beta_i}.$$

We cannot predict the values of α_i and β_i, but we can compute them at the same time that we are computing the x_i's using the definition (5.41) of f. Clearly $\alpha_0 = \beta_0 = 0$, and then subsequent values are given by

$$\alpha_{i+1} = \begin{cases} \alpha_i + 1 & \text{if } 0 \le x < p/3, \\ 2\alpha_i & \text{if } p/3 \le x < 2p/3, \\ \alpha_i & \text{if } 2p/3 \le x < p, \end{cases}$$

$$\beta_{i+1} = \begin{cases} \beta_i & \text{if } 0 \le x < p/3, \\ 2\beta_i & \text{if } p/3 \le x < 2p/3, \\ \beta_i + 1 & \text{if } 2p/3 \le x < p. \end{cases}$$

In computing α_i and β_i, it suffices to keep track of their values modulo $p-1$, since $g^{p-1} = 1$ and $h^{p-1} = 1$. This is important, since otherwise the values of α_i and β_i would become prohibitively large.

In a similar fashion we compute the sequence given by

$$y_0 = 1 \qquad \text{and} \qquad y_{i+1} = f\big(f(y_i)\big).$$

Then

$$y_i = x_{2i} = g^{\gamma_i} \cdot h^{\delta_i},$$

where the exponents γ_i and δ_i can be computed by two repetitions of the recursions used for α_i and β_i. Of course, the first time we use y_i to determine which case of (5.41) to apply, and the second time we use $f(y_i)$ to decide.

Applying the above procedure, we eventually find a collision in the x and the y sequences, say $y_i = x_i$. This means that

$$g^{\alpha_i} \cdot h^{\beta_i} = g^{\gamma_i} \cdot h^{\delta_i}.$$

So if we let

$$u \equiv \alpha_i - \gamma_i \pmod{p-1} \qquad \text{and} \qquad v \equiv \delta_i - \beta_i \pmod{p-1},$$

then $g^u = h^v$ in $\mathbb{F}_p$. Equivalently,

$$v \cdot \log_g(h) \equiv u \pmod{p-1}. \tag{5.42}$$

If $\gcd(v, p-1) = 1$, then we can multiply both sides of (5.42) by the inverse of v modulo $p-1$ to solve the discrete logarithm problem.

More generally, if $d = \gcd(v, p-1) \ge 2$, we use the extended Euclidean algorithm (Theorem 1.11) to find an integer s such that

$$s \cdot v \equiv d \pmod{p-1}.$$

Multiplying both sides of (5.42) by s yields

$$d \cdot \log_g(h) \equiv w \pmod{p-1}, \tag{5.43}$$

where $w \equiv s \cdot u \pmod{p-1}$. In this congruence we know all of the quantities except for $\log_g(h)$. The fact that d divides $p-1$ will force d to divide w, so w/d is one solution to (5.43), but there are others. The full set of solutions to (5.43) is obtained by starting with w/d and adding multiples of $(p-1)/d$,

$$\log_g(h) \in \left\{ \frac{w}{d} + k \cdot \frac{p-1}{d} : k = 0, 1, 2, \ldots, d-1 \right\}.$$

In practice, d will tend to be fairly small,[18] so it suffices to check each of the d possibilities for $\log_g(a)$ until the correct value is found.

Example 5.52. We illustrate Pollard's ρ method by solving the discrete logarithm problem

$$19^t \equiv 24717 \pmod{48611}.$$

The first step is to compute the x and y sequences until we find a match $y_i = x_i$, while also computing the exponent sequences $\alpha, \beta, \gamma, \delta$. The initial stages of this process and the final few steps before a collision has been found are given in Table 5.11.

i	x_i	$y_i = x_{2i}$	α_i	β_i	γ_i	δ_i
0	1	1	0	0	0	0
1	19	361	1	0	2	0
2	361	33099	2	0	4	0
3	6859	13523	3	0	4	2
4	33099	20703	4	0	6	2
5	33464	14974	4	1	13	4
6	13523	18931	4	2	14	5
7	13882	30726	5	2	56	20
8	20703	1000	6	2	113	40
9	11022	14714	12	4	228	80
			⋮			
542	21034	46993	13669	2519	27258	30257
543	20445	37138	27338	5038	27259	30258
544	40647	33210	6066	10076	5908	11908
545	28362	21034	6066	10077	5909	11909
546	36827	40647	12132	20154	23636	47636
547	11984	36827	12132	20155	47272	46664
548	33252	33252	12133	20155	47273	46665

Table 5.11: Pollard ρ computations to solve $19^t = 24717$ in $\mathbb{F}_{48611}$

[18]For most cryptographic applications, the prime p is chosen such that $p-1$ has precisely one large prime factor, since otherwise, the Pohlig–Hellman algorithm (Theorem 2.31) may be applicable. And it is unlikely that d will be divisible by the large prime factor of $p-1$.

From the table we see that $x_{1096} = x_{548} = 33252$ in $\mathbb{F}_{48611}$. The associated exponent values are

$$\alpha_{548} = 12133, \qquad \beta_{548} = 20155, \qquad \gamma_{548} = 47273, \qquad \delta_{548} = 46665,$$

so we know that

$$19^{12133} \cdot 24717^{20155} = 19^{47273} \cdot 24717^{46665} \quad \text{in } \mathbb{F}_{48611}.$$

(Before proceeding, we should probably check this equality to make sure that we didn't made an arithmetic error.) Moving the powers of 19 to one side and the powers of 24717 to the other side yields $19^{-35140} = 24717^{26510}$, and adding $48610 = p - 1$ to the exponent of 19 gives

$$19^{13470} = 24717^{26510} \quad \text{in } \mathbb{F}_{48611}. \tag{5.44}$$

We next observe that

$$\gcd(26510, 48610) = 10 \qquad \text{and} \qquad 970 \cdot 26510 \equiv 10 \pmod{48610}.$$

Raising both sides of (5.44) to the 970th power yields

$$19^{13470 \cdot 970} = 19^{13065900} = 19^{38420} = 24717^{10} \quad \text{in } \mathbb{F}_{48611}.$$

Hence

$$10 \cdot \log_{19}(24717) \equiv 38420 \pmod{48610},$$

which means that

$$\log_{19}(24717) \equiv 3842 \pmod{4861}.$$

The possible values for the discrete logarithm are obtained by adding multiples of 4861 to 3842, so $\log_{19}(24717)$ is one of the numbers in the set

$$\{3842, 8703, 13564, 18425, 23286, 28147, 33008, 37869, 42730, 47591\}.$$

To complete the solution, we compute 19 raised to each of these 10 values until we find the one that is equal to 24717:

$$\begin{gathered} 19^{3842} = 16580, \quad 19^{8703} = 29850, \quad 19^{13564} = 23894, \quad 19^{18425} = 20794, \\ 19^{23286} = 10170, \quad 19^{28147} = 32031, \quad 19^{33008} = 18761, \quad 19^{37869} = \boxed{24717}. \end{gathered}$$

This gives the solution $\log_{19}(24717) = 37869$. We check our answer

$$19^{37869} = 24717 \quad \text{in } \mathbb{F}_{48611}. \quad \checkmark$$

5.6 Information Theory

In 1948 and 1949, Claude Shannon published two papers [126, 127] that form the mathematical foundation of modern cryptography. In these papers he defines the concept of perfect (or unconditional) secrecy, introduces the idea of entropy of natural language and statistical analysis, provides the first proofs of security using probability theory, and gives precise connections between provable security and the size of the key, plaintext, and ciphertext spaces.

In public key cryptography, one is interested in how computationally difficult it is to break the system. The issue of security is thus a relative one—a given cryptosystem is hard to break if one assumes that some underlying problem is hard to solve. It requires some care to formulate these concepts properly. In this section we briefly introduce Shannon's ideas and explain their relevance to symmetric key systems. In [127], Shannon develops a theory of security for cryptosystems that assumes that no bounds are placed on the computational resources that may be brought to bear against them. For example, symmetric ciphers such as the simple substitution cipher (Sect. 1.1) and the Vigènere cipher (Sect. 5.2) are not computationally secure. With unlimited resources—indeed with very limited resources—an adversary can easily break these ciphers. If we seek unconditional security, we must either seek new algorithms or modify the implementation of known algorithms. In fact, Shannon shows that perfectly secure cryptosystems must have at least as many keys as plaintexts and that every key must be used with equal probability. This means that most practical cryptosystems are not unconditionally secure. We discuss the notion of perfect security in Sect. 5.6.1.

In [126] Shannon develops a mathematical theory that measures the amount of information that is revealed by a random variable. When the random variable represents the possible plaintexts or ciphertexts or keys of a cipher that is used to encrypt a natural language such as English, we obtain a framework for the rigorous mathematical study of cryptographic security. Shannon adopted the word *entropy* for this measure because of its formal similarity to Boltzmann's definition of entropy in statistical mechanics, and also because Shannon viewed language as a stochastic process, i.e., as a system governed by probabilities that produces a sequence of symbols. Later, the physicist E.T. Jaynes [60] argued that thermodynamic entropy could be interpreted as an application of a certain information-theoretic entropy. As a measure of "uncertainty" of a system, the logarithmic formula for entropy is determined, up to a constant, by requiring that it be continuous, monotonic, and satisfy a certain additive property. We discuss information-theoretic entropy and its application to cryptography in Sect. 5.6.2.

5.6.1 Perfect Secrecy

A cryptosystem has *perfect secrecy* if the interception of a ciphertext gives the cryptanalyst no information about the underlying plaintext and no information about any future encrypted messages. To formalize this concept, we introduce random variables M, C, and K representing the finite number of possible messages, ciphertexts, and keys. In other words, M is a random variable whose values are the possible messages (plaintexts), C is a random variable whose values are the possible ciphertexts, and K is a random variable whose values are the possible keys used for encryption and decryption. We let f_M, f_C, and f_K be the associated density functions.[19] The density functions f_M, f_K, and f_C are related to one another via the encryption/decryption formula $d_k(e_k(m)) = m$, which we will exploit shortly to prove (5.47).

We also have the joint densities and the conditional densities of all pairs of these random variables, such as $f_{(C,M)}(c,m)$ and $f_{C|M}(c \mid m)$, and so forth. We will let the variable names simplify the notation. For example, we write $f(c \mid m)$ for $f_{C|M}(c \mid m)$, the conditional probability density of the random variables C and M, i.e.,

$$f(c \mid m) = \Pr(C = c \text{ given that } M = m).$$

Similarly, we write $f(m)$ for $f_M(m)$, the probability that $M = m$.

Definition. A cryptosystem has *perfect secrecy* if

$$f(m \mid c) = f(m) \qquad \text{for all } m \in \mathcal{M} \text{ and all } c \in \mathcal{C}. \tag{5.45}$$

What does (5.45) mean? It says that the probability of any particular plaintext, $\Pr(M = m)$, is independent of the ciphertext. Intuitively, this means that the ciphertext reveals no knowledge of the plaintext.

Bayes's formula (Theorem 5.33) says that

$$f(m \mid c)f(c) = f(c \mid m)f(m),$$

which implies that perfect secrecy is equivalent to the condition

$$f(c \mid m) = f(c) \qquad \text{for all } c \in \mathcal{C} \text{ and all } m \in \mathcal{M} \text{ with } f(m) \neq 0. \tag{5.46}$$

Formula (5.46) says that the appearance of any particular ciphertext is equally likely, independent of the plaintext.

If we know f_K and f_M, then f_C is determined. To see this, we note that for a given key k, the probability that the ciphertext equals c is the same as the probability that the decryption of c is the plaintext, assuming of course

---

[19] As is typical, we have omitted reference to the underlying sample spaces. To be completely explicit, we have three probability spaces with sample spaces Ω_M, Ω_C, and Ω_K and probability functions $\Pr_M$, $\Pr_C$, and $\Pr_K$. Then M, C and K are random variables

$$M : \Omega_M \to \mathcal{M}, \qquad K : \Omega_K \to \mathcal{K}, \qquad C : \Omega_C \to \mathcal{C}.$$

Then by definition, the density function f_M is

$$f_M(m) = \Pr(M = m) = \Pr_M\bigl(\{\omega \in \Omega_M : M(\omega) = m\}\bigr),$$

and similarly for K and C.

that c is the encryption of some plaintext for key k. This allows us to compute the total probability $f_C(c)$ by summing over all possible keys and using the decomposition formula (5.20) of Proposition 5.24, or more precisely, its generalization described in Exercise 5.23. As usual, we let $\mathcal{K}$ denote the set of all possible keys and $e_k : \mathcal{M} \to \mathcal{C}$ and $d_k : \mathcal{C} \to \mathcal{M}$ be the encryption and decryption functions for the key $k \in \mathcal{K}$. Then the probability that the ciphertext is equal to c is given by the formula

	m_1	m_2	m_3
k_1	c_2	c_1	c_3
k_2	c_1	c_3	c_2

Table 5.12: Encryption of messages with keys k_1 and k_2

$$f_C(c) = \sum_{\substack{k \in \mathcal{K} \text{ such} \\ \text{that } c = e_k(m) \\ \text{for some } m \in \mathcal{M}}} f_K(k) f_M\big(d_k(c)\big); \tag{5.47}$$

see also Exercise 5.47. We note that if the encryption map $e_k : \mathcal{M} \to \mathcal{C}$ is onto for all keys k, which is often true in practice, then the sum in (5.47) is over all $k \in \mathcal{K}$.

Example 5.53. Consider the Shift Cipher described in Sect. 1.1. Suppose that each of the 26 possible keys (shift amounts) is chosen with equal probability and that each plaintext character is encrypted using a new, randomly chosen, shift amount. Then it is not hard to check that the resulting cryptosystem has perfect secrecy; see Exercise 5.46.

Recall that an encryption function is one-to-one, meaning that each message gives rise to a unique ciphertext. This implies that there are at least as many ciphertexts as plaintexts (messages). Perfect secrecy gives additional restrictions on the relative size of the key, message, and ciphertext spaces. We first investigate an example of a (tiny) cryptosystem that does not have perfect secrecy.

Example 5.54. Suppose that a cryptosystem has two keys k_1 and k_2, three messages m_1, m_2, and m_3, and three ciphertexts c_1, c_2, and c_3. Assume that the density function for the message random variable satisfies

$$f_M(m_1) = f_M(m_2) = \frac{1}{4} \quad \text{and} \quad f_M(m_3) = \frac{1}{2}. \tag{5.48}$$

Suppose further that Table 5.12 describes how the different keys act on the messages to produce ciphertexts.

For example, the encryption of the plaintext m_1 with the key k_1 is the ciphertext c_2. Under the assumption that the keys are used with equal probability, we can use (5.47) to compute the probability that the ciphertext is equal to c_1:

$$\begin{aligned} f(c_1) &= f(k_1)f_M(d_{k_1}(c_1)) + f(k_2)f_M(d_{k_2}(c_1)) \\ &= f(k_1)f(m_2) + f(k_2)f(m_1) \\ &= \frac{1}{2}\cdot\frac{1}{4} + \frac{1}{2}\cdot\frac{1}{4} = \frac{1}{4}. \end{aligned}$$

On the other hand, we see from the table that $f(c_1 \mid m_3) = 0$. Hence this cryptosystem does not have perfect secrecy.

This matches our intuition, since it is clear that seeing a ciphertext leaks some information about the plaintext. For example, if we see the ciphertext c_1, then we know that the message was either m_1 or m_2, it cannot be m_3.

As noted earlier, the number of ciphertexts must be at least as large as the number of plaintexts, since otherwise, decryption is not possible. It turns out that one consequence of perfect secrecy is that the number of keys must also be at least as large as the number of plaintexts.

Proposition 5.55. *If a cryptosystem has perfect secrecy, then $\#\mathcal{K} \ge \#\mathcal{C}^+$, where $\mathcal{C}^+ = \{m \in \mathcal{M} : f(m) > 0\}$ is the set of plaintexts that have a positive probability of being selected.*

Proof. We start by fixing some ciphertext $c \in \mathcal{C}$ with $f(c) > 0$. Perfect secrecy in the form of (5.46) tells us that

$$f(c \mid m) = f(c) > 0 \qquad \text{for all } m \in \mathcal{C}^+.$$

This says that there is a positive probability that $m \in \mathcal{C}^+$ encrypts to c, so in particular there is at least one key k satisfying $e_k(m) = c$. Further, if we start with a different plaintext $m' \in \mathcal{C}^+$, then we get a different key k', since otherwise $e_k(m) = c = e_k(m')$, which would contradict the one-to-one property of e_k.

To recapitulate, we have shown that for every $m \in \mathcal{C}^+$, the set

$$\{k \in \mathcal{K} : e_k(m) = c\}$$

is nonempty, and further, these sets are disjoint for different m's. Thus each plaintext $m \in \mathcal{C}^+$ is matched with one or more keys, and different m's are matched with different keys, which shows that the number of keys is at least as large as the number of plaintexts in $\mathcal{C}^+$. □

Given the restriction on the relative sizes of the key, ciphertext, and plaintext spaces in systems with perfect secrecy, namely

$$\#\mathcal{K} \ge \#\mathcal{M} \qquad \text{and} \qquad \#\mathcal{C} \ge \#\mathcal{M},$$

it is most efficient to assume that the key space, the plaintext space, and the ciphertext space are all of equal size. Assuming this, Shannon proves a theorem characterizing perfect secrecy.

Theorem 5.56. *Suppose that a cryptosystem satisfies*

$$\#\mathcal{K} = \#\mathcal{M} = \#\mathcal{C},$$

i.e., the numbers of keys, plaintexts, and ciphertexts are all equal. Then the system has perfect secrecy if and only if the following two conditions hold:
(a) *Each key $k \in \mathcal{K}$ is used with equal probability.*
(b) *For a given message $m \in \mathcal{M}$ and ciphertext $c \in \mathcal{C}$, there is exactly one key $k \in \mathcal{K}$ that encrypts m to c.*

Proof. Suppose first that a cryptosystem has perfect secrecy. We start by verifying (b). For any plaintext $m \in \mathcal{M}$ and ciphertext $c \in \mathcal{C}$, consider the (possibly empty) set of keys that encrypt m to c,

$$\mathcal{S}_{m,c} = \{k \in \mathcal{K} : e_k(m) = c\}.$$

We are going to prove that if the cryptosystem has perfect secrecy, then in fact $\#\mathcal{S}_{m,c} = 1$ for every $m \in \mathcal{M}$ and every $c \in \mathcal{C}$, which is equivalent to statement (b) of the theorem. We do this in three steps.

Claim 1. If $m \neq m'$, then $\mathcal{S}_{m,c} \cap \mathcal{S}_{m',c} = \emptyset$.

Suppose that $k \in \mathcal{S}_{m,c} \cap \mathcal{S}_{m',c}$. Then $e_k(m) = c = e_k(m')$, which implies that $m = m'$, since the encryption function e_k is injective. This proves Claim 1.

Claim 2. If the cryptosystem has perfect secrecy, then $\mathcal{S}_{m,c}$ is nonempty for every m and c.

We use the perfect secrecy assumption in the form $f(m,c) = f(m)f(c)$. We know that every $m \in \mathcal{M}$ is a valid plaintext for at least one key, so $f(m) > 0$. Similarly, every $c \in \mathcal{C}$ appears as the encryption of at least one plaintext using some key, so $f(c) > 0$. Hence perfect secrecy implies that

$$f(m,c) > 0 \quad \text{for all } m \in \mathcal{M} \text{ and all } c \in \mathcal{C}. \tag{5.49}$$

But the formula $f(m,c) > 0$ is simply another way of saying that c is a possible encryption of m. Hence there must be at least one key $k \in \mathcal{K}$ satisfying $e_k(m) = c$, i.e., there is some key $k \in \mathcal{S}_{m,c}$. This completes the proof of Claim 2.

Claim 3. If the cryptosystem has perfect secrecy, then $\#\mathcal{S}_{m,c} = 1$.

Fix a ciphertext $c \in \mathcal{C}$. Then

$$\begin{aligned} \#\mathcal{K} &\geq \#\Big(\bigcup_{m\in\mathcal{M}} \mathcal{S}_{m,c}\Big) && \text{since } \mathcal{K} \text{ contains every } \mathcal{S}_{m,c}, \\ &= \sum_{m\in\mathcal{M}} \#\mathcal{S}_{m,c} && \text{since the } \mathcal{S}_{m,c} \text{ are disjoint from Claim 1,} \end{aligned}$$

$$\begin{aligned} &\geq \#\mathcal{M} && \text{since } \#\mathcal{S}_{m,c} \geq 1 \text{ from Claim 2,} \\ &= \#\mathcal{K} && \text{since } \#\mathcal{K} = \#\mathcal{M} \text{ by assumption.} \end{aligned}$$

Thus all of these inequalities are equalities, so in particular,

$$\sum_{m \in \mathcal{M}} \#\mathcal{S}_{m,c} = \#\mathcal{M}.$$

Then the fact (Claim 2) that every $\#\mathcal{S}_{m,c}$ is greater than or equal to 1 implies that every $\#\mathcal{S}_{m,c}$ must equal 1. This completes the proof of Claim 3.

As noted above, Claim 3 is equivalent to statement (b) of the theorem. We turn now to statement (a). Consider the set of triples

$$(k, m, c) \in \mathcal{K} \times \mathcal{M} \times \mathcal{C} \quad \text{satisfying} \quad e_k(m) = c.$$

Clearly k and m determine a unique value for c, and (b) says that m and c determine a unique value for k. It is also not hard, using a similar argument and the assumption that $\#\mathcal{M} = \#\mathcal{C}$, to show that c and k determine a unique value for m; see Exercise 5.48.

For any triple (k, m, c) satisfying $e_k(m) = c$, we compute

$$\begin{aligned} f(m) = f(m \mid c) & \qquad \text{by perfect secrecy,} \\ = \frac{f(m,c)}{f(c)} & \qquad \text{definition of conditional probability,} \\ = \frac{f(m,k)}{f(c)} & \qquad \text{since any two of } m,\ k,\ c \text{ determine the third,} \\ = \frac{f(m)f(k)}{f(c)} & \qquad \text{since } M \text{ and } K \text{ are independent.} \end{aligned}$$

(There are cryptosystems in which the message forms part of the key; see for example Exercise 5.19, in which case M and K would not be independent.)

Canceling $f(m)$ from both sides, we have shown that

$$f(k) = f(c) \qquad \text{for every } k \in \mathcal{K} \text{ and every } c \in \mathcal{C}. \tag{5.50}$$

Note that our proof shows that (5.50) is true for every k and every c, because Exercise 5.48 tells us that for every (k, c) there is a (unique) m satisfying $e_k(m) = c$.

We sum (5.50) over all $c \in \mathcal{C}$ and divide by $\#\mathcal{C}$ to obtain

$$f(k) = \frac{1}{\#\mathcal{C}} \sum_{c \in \mathcal{C}} f(c) = \frac{1}{\#\mathcal{C}}.$$

This shows that $f(k)$ is constant, independent of the choice of $k \in \mathcal{K}$, which is precisely the assertion of (a). At the same time we have proven the useful fact that $f(c)$ is constant, i.e., every ciphertext is used with equal probability.

In the other direction, if a cryptosystem has properties (a) and (b), then the steps outlined to prove perfect secrecy of the shift cipher in Exercise 5.46 can be applied in this more general setting. We leave the details to the reader. □

Example 5.57 (The one-time pad). Vernam's one-time pad, patented in 1917, is an extremely simple, perfectly secret, albeit very inefficient, cryptosystem. The key k consists of a string of binary digits $k_0 k_1 \dots k_N$. It is used to encrypt a binary plaintext string $m = m_0 m_1 \dots m_N$ by XOR'ing the two strings together bit by bit. See (1.12) on page 44 for a description of the XOR operation, which for convenience we will denote by $\oplus$. Then the ciphertext $c = c_0 c_1 \dots c_N$ is given by

$$c_i = k_i \oplus m_i \quad \text{for } i = 0, 1, \dots, N.$$

Each key is *used only once* and then discarded, whence the name of the system. Since every key is used with equal probability, and since there is exactly one key that encrypts a given m to a given c, namely the key $m \oplus c$, Theorem 5.56 shows that Vernam's one-time pad has perfect secrecy.

Unfortunately, if Bob and Alice want to use a Vernam one-time pad to exchange N bits of information, they must already know N bits of shared secret information to use as the key. This makes one-time pads much too inefficient for large-scale communication networks. However, there are situations in which they have been used, such as top secret communications between diplomatic offices or for short messages between spies and their home bases.

It is also worth noting that a one-time pad remains completely secure only as long as its keys are never reused. When a key pad is used more than once, either due to error or to the difficulty of providing enough key material, then the cryptosystem may be vulnerable to cryptanalysis. This occurred in the real world when the Soviet Union reused some one-time pads during World War II. The United States mounted a massive cryptanalytic effort called the VENONA project that successfully decrypted a number of documents.

5.6.2 Entropy

In efficient cryptosystems, a single key must be used to encrypt many different plaintexts, so perfect secrecy is not possible. At best we can hope to build cryptosystems that are computationally secure. Unfortunately, anything less than perfect secrecy leaves open the possibility that a list of ciphertexts will reveal significant information about the key. To study this phenomenon, Shannon introduced the concept of *entropy*, which is a measure of the uncertainty in a system. Thus if we view $f_X(x) = \Pr(X = x)$ as being the probability that the outcome of a certain experiment is equal to x, then the entropy of X will be small if the outcome of a single experiment reveals a significant amount of information about the random variable X.

Let X be a random variable taking on finitely many values $x_1, x_2, \dots, x_n$, and let $p_1, p_2, \dots, p_n$ be the associated probabilities,

$$p_i = f_X(x_i) = \Pr(X = x_i).$$

The *entropy* $H(X)$ *of* X is a number that depends only on the probabilities $p_1, \ldots, p_n$ of the possible outcomes of X, so we write[20]

$$H(X) = H(p_1, \ldots, p_n).$$

We would like to capture the idea that H is the expected value of a random variable that measures the uncertainty that the outcome x_i has occurred. Thus the larger the value of $H(X)$, the less information about X that is revealed by the outcome of an experiment.

What properties should H possess?

Property $\mathbf{H_1}$ The function H should be continuous in the variables p_i. This reflects the intuition that a small change in p_i should produce a small change in the amount of information revealed by X.

Property $\mathbf{H_2}$ Let X_n be the random variable that is uniformly distributed on a set $\{x_1, \ldots, x_n\}$, i.e., the random variable X_n has n possible outcomes, each occurring with probability $\frac{1}{n}$. Then

$$H(X_{n+1}) > H(X_n) \quad \text{for all } n \geq 1.$$

This reflects the intuition that if all outcomes are equally likely, then the uncertainty should increase as the number of outcomes increases.

Property $\mathbf{H_3}$ The third property is subtler. It says that if an outcome of X is thought of as a choice, and if that choice can be broken down into two successive choices, then the original value of H is a weighted sum of the values of H for the successive choices. In order to quantify this intuition, we consider random variables X, Y, and $Z_1, \ldots, Z_n$ taking values in the sets

$$\begin{aligned} X &: \Omega \longrightarrow \{x_{ij} : 1 \leq i \leq n \text{ and } 1 \leq j \leq m_i\}, \\ Y &: \Omega \longrightarrow \{Z_1, \ldots, Z_n\}, \\ Z_i &: \Omega \longrightarrow \{x_{ij} : 1 \leq j \leq m_i\}, \end{aligned}$$

and satisfying

$$\Pr(X = x_{ij}) = \Pr(Y = Z_i \text{ and } Z_i = x_{ij}).$$

This reflects the intuition that the outcome $X = x_{ij}$ is being broken down into the successive choices $Y = Z_i$ followed by $Z_i = x_{ij}$. Then Property $\mathbf{H_3}$ is the formula

$$H(X) = H(Y) + \sum_{i=1}^{n} \Pr(Y = Z_i) H(Z_i).$$

[20] Although this notation is useful, it is important to remember that the domain of H is the set of random variables, not the set of n-tuples for some fixed value of n. Thus the domain of H is itself a set of functions.

Example 5.58. Let X_n be a uniformly distributed random variable on n objects. Then we claim that

$$H(X_{n^2}) = 2H(X_n).$$

To see this, we view X_{n^2} as choosing an element from $\{x_{ij} : 1 \le i, j \le n\}$, and we break this choice into two choices by first choosing an index i, and then choosing an index j. Property $\mathbf{H}_3$ says that

$$H(X_{n^2}) = H(X_n) + \sum_{i=1}^{n} \frac{1}{n} H(X_n) = 2H(X_n).$$

Example 5.59. We illustrate Property $\mathbf{H}_3$ with a more elaborate example. Suppose that X has five possible outcomes $\{x_1, x_2, x_3, x_4, x_5\}$ with probabilities

$$f_X(x_1) = \frac{1}{2}, \quad f_X(x_2) = \frac{1}{4}, \quad f_X(x_3) = \frac{1}{12}, \quad f_X(x_4) = \frac{1}{8}, \quad f_X(x_5) = \frac{1}{24}.$$

The five outcomes for X are illustrated by the branched tree in Fig. 5.2a.

Now suppose that X is written as two successive choices,the first deciding between the subsets $\{x_1, x_2, x_3\}$ and $\{x_4, x_5\}$, and the second choosing an element of the designated subset. So we have random variables Y, Z_1, Z_2, where

$$f_Y(Z_1) = \frac{5}{6} \quad \text{and} \quad f_Y(Z_2) = \frac{1}{6},$$

and

$$f_{Z_1}(x_1)=\frac{3}{5}, \quad f_{Z_1}(x_2) = \frac{3}{10}, \quad f_{Z_1}(x_3) = \frac{1}{10}, \quad f_{Z_2}(x_4) = \frac{3}{4}, \quad f_{Z_2}(x_5)= \frac{1}{4},$$

as illustrated in Fig. 5.2b. Then Property $\mathbf{H}_3$ for this example says that

$$\underset{\substack{\uparrow \\ H(X)}}{H\left(\frac{1}{2}, \frac{1}{4}, \frac{1}{12}, \frac{1}{8}, \frac{1}{24}\right)} = \underset{\substack{\uparrow \\ H(Y)}}{H\left(\frac{5}{6}, \frac{1}{6}\right)} + \frac{5}{6} \underset{\substack{\uparrow \\ H(Z_1)}}{H\left(\frac{3}{5}, \frac{3}{10}, \frac{1}{10}\right)} + \frac{1}{6} \underset{\substack{\uparrow \\ H(Z_2)}}{H\left(\frac{3}{4}, \frac{1}{4}\right)}$$

Theorem 5.60. *Every function having Properties* $\mathbf{H}_1$, $\mathbf{H}_2$, *and* $\mathbf{H}_3$ *is a constant multiple of the function*

$$H(p_1, \ldots, p_n) = -\sum_{i=1}^{n} p_i \log_2 p_i, \tag{5.51}$$

where $\log_2$ *denotes the logarithm to the base* 2, *and if* $p = 0$, *then we set* $p \log_2 p = 0$.[21]

[21] This convention makes sense, since we want H to be continuous in the p_i's, and it is true that $\lim_{p \to 0} p \log_2 p = 0$.

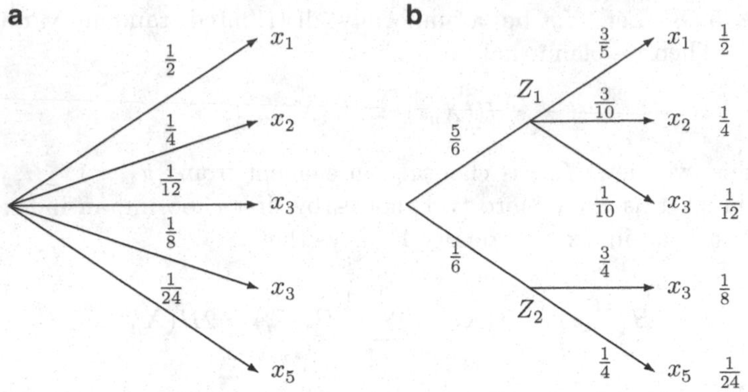

Figure 5.2: Splitting X into Y followed by Z_1 or Z_2. (**a**) Five outcomes of a choice. (**b**) Splitting into two choices

Proof. See Shannon's paper [126]. □

To illustrate the notion of uncertainty, consider what happens when one of the probabilities p_i is one and the other probabilities are zero. In this case, the formula (5.51) for entropy gives $H(p_1, \ldots, p_n) = 0$, which makes sense, since there is no uncertainty about the outcome of an experiment having only one possible outcome.

It turns out that the other extreme, namely maximal uncertainly, occurs when all of the probabilities p_i are equal. In order to prove this, we use an important inequality from real analysis known as Jensen's inequality. Before stating Jensen's inequality, we first need a definition.

Definition. A function F on the real line is called *concave* (*down*) on an interval I if the following inequality is true for all $0 \le \alpha \le 1$ and all s and t in I:

$$(1-\alpha)F(s) + \alpha F(t) \le F\big((1-\alpha)s + \alpha t\big). \tag{5.52}$$

This definition may seem mysterious, but it has a simple geometric interpretation. Notice that if we fix s and t and let a vary from 0 to 1, then the points $(1-\alpha)s + \alpha t$ trace out the interval from s to t on the real line. So inequality (5.52) is the geometric statement that the line segment connecting any two points on the graph of F lies below the graph of F. For example, the function $F(t) = 1 - t^2$ is concave. Illustrations of concave and noncave functions, with representative line segments, are given in Fig. 5.3. If the function F has a second derivative, then the second derivative test that you learned in calculus can be used to test for concavity (see Exercise 5.54).

Theorem 5.61 (Jensen's Inequality). *Suppose that F is concave on an interval I, and let $\alpha_1, \alpha_2, \ldots, \alpha_n$ be nonnegative numbers satisfying*

$$\alpha_1 + \alpha_2 + \cdots + \alpha_n = 1.$$

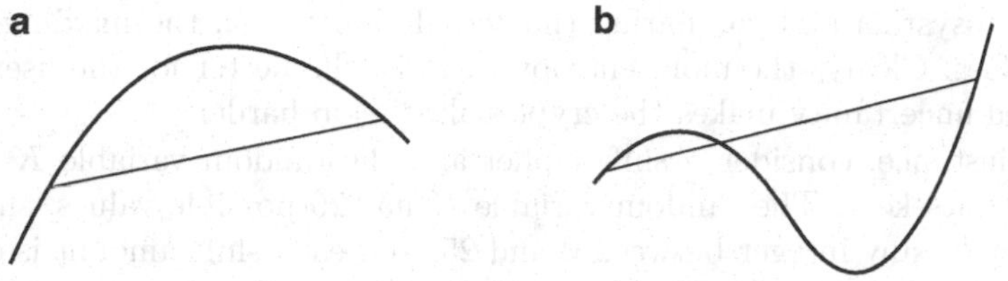

Figure 5.3: An illustration of concavity. (**a**) A concave function. (**b**) A non-concave function

Then

$$\sum_{i=1}^{n} \alpha_i F(t_i) \le F\Big(\sum_{i=1}^{n} \alpha_i t_i\Big) \qquad \textit{for all } t_1, t_2, \ldots, t_n \in I. \tag{5.53}$$

Further, equality holds in (5.53) *if and only if either F is a linear function or $t_1 = t_2 = \cdots = t_n$.*

Proof. Notice that for $n = 2$, the desired inequality (5.53) is exactly the definition of concavity (5.52). The general case is then proven by induction; see Exercise 5.55. □

Corollary 5.62. *Let X be a random variable that takes on finitely many possible values $x_1, \ldots, x_n$.*
(a) $H(X) \le \log_2 n$.
(b) $H(X) = \log_2 n$ *if and only if every event $X = x_i$ occurs with the same probability $1/n$.*

Proof. Let $p_i = \Pr(X = x_i)$ for $i = 1, 2, \ldots, n$. Then $p_1 + \cdots + p_n = 1$, so we may apply Jensen's inequality to the function $F(t) = \log_2 t$ with $\alpha_i = p_i$ and $t_i = 1/p_i$. (See Exercise 5.54 for a proof that $\log_2 t$ is a concave function.) The left-hand side of (5.53) is exactly the formula for entropy (5.51), so we find that

$$H(X) = -\sum_{i=1}^{n} p_i \log_2 p_i = \sum_{i=1}^{n} p_i \log_2 \frac{1}{p_i} \le \log_2 \left(\sum_{i=1}^{n} p_i \frac{1}{p_i}\right) = \log_2 n.$$

This proves (a). Further, the function $\log_2 t$ is not linear, so equality occurs if and only if $p_1 = p_2 = \cdots = p_n$, i.e., if all of the probabilities satisfy $p_i = 1/n$. This proves (b). □

Notice that Corollary 5.62 says that entropy is maximized when all of the probabilities are equal. This conforms to our intuitive understanding that uncertainty is maximized when every outcome is equally likely.

The theory of entropy is applied to cryptography by computing the entropy of random variables such as K, M, and C that are associated with the cryptosystem and comparing the actual values with the maximum possible values. Clearly the more entropy there is, the better for the user, since increased uncertainty makes the cryptanalyst's job harder.

For instance, consider a shift cipher and the random variable K associated with its keys. The random variable K has 26 possible values, since the shift may be any integer between 0 and 25, and each shift amount is equally probable, so K has maximal entropy $H(K) = \log_2(26)$.

Example 5.63. We consider the system with two keys described in Example 5.54 on page 265. Each key is equally likely, so $H(K) = \log_2(2) = 1$. Similarly, we can use the plaintext probabilities for this system as given by (5.48) to compute the entropy of the random variable M associated to the plaintexts.

$$H(M) = -\frac{1}{4}\log_2\left(\frac{1}{4}\right) - \frac{1}{4}\log_2\left(\frac{1}{4}\right) - \frac{1}{2}\log_2\left(\frac{1}{2}\right) = \frac{3}{2} = 1.5.$$

Notice that $H(M)$ is slightly smaller than $\log_2(3) \approx 1.585$, which would be the maximal possible entropy for M in a cryptosystem with three plaintexts.

We now introduce the concept of conditional entropy and its application to secrecy systems. Suppose that a signal is sent over a noisy channel, which means that the signal may be distorted during transmission. Shannon [126] defines the *equivocation* to be the conditional entropy of the original signal, given the received signal. He uses this quantity to measure the amount of uncertainty in transmissions across a noisy channel. Shannon [127] later observed that a noisy communication channel is also a model for a secrecy system. The original signal (the plaintext) is "distorted" by applying the encryption process, and the received signal (the ciphertext) is thus a noisy version of the original signal. In this way, the notion of equivocation can be applied to cryptography, where a large equivocation says that the ciphertext conceals most information about the plaintext.

Definition. Let X and Y be random variables, and let $x_1, \ldots, x_n$ be the possible values of X and $y_1, \ldots, y_m$ the possible values of Y. The *equivocation*, or *conditional entropy, of X on Y* is the quantity $H(X \mid Y)$ defined by

$$H(X \mid Y) = -\sum_{i=1}^{n}\sum_{j=1}^{m} f_Y(y_j) f_{X|Y}(x_i \mid y_j) \log_2 f_{X|Y}(x_i \mid y_j).$$

When $X = K$ is the key random variable and $Y = C$ is the ciphertext random variable, the quantity $H(K \mid C)$ is called the *key equivocation*. It measures the total amount of information about the key revealed by the ciphertext, or more precisely, it is the expected value of the conditional entropy $H(K \mid c)$ of K given a single observation c of C. The key equivocation can be determined by computing all of the conditional probabilities $f(k \mid c)$ of the cryptosystem. Alternatively, one can use the following result.

Proposition 5.64. *The key equivocation of a cryptosystem* $(\mathcal{K}, Mcal, \mathcal{C})$ *is related to the individual entropies of* K*,* M*, and* C *by the formula*

$$H(K \mid C) = H(K) + H(M) - H(C). \tag{5.54}$$

Proof. We leave the proof as an exercise; see Exercise 5.57 □

Example 5.65. We compute the key equivocation of the cryptosystem described in Examples 5.54 and 5.63. We already computed $H(K) = 1$ and $H(M) = \frac{3}{2}$, so it remains to compute $H(C)$. To do this, we need the values of $f(c)$ for each ciphertext $c \in \mathcal{C}$. We already computed $f(c_1) = \frac{1}{4}$, and a similar computation using (5.48) and Table 5.12 yields

$$f(c_2) = f(k_1)f(m_1) + f(k_2)f(m_3) = \left(\frac{1}{2}\right)\left(\frac{1}{4}\right) + \left(\frac{1}{2}\right)\left(\frac{1}{2}\right) = \frac{3}{8},$$

$$f(c_3) = f(k_1)f(m_3) + f(k_2)f(m_2) = \left(\frac{1}{2}\right)\left(\frac{1}{2}\right) + \left(\frac{1}{2}\right)\left(\frac{1}{4}\right) = \frac{3}{8}.$$

Therefore,

$$H(C) = -\frac{1}{4}\log_2\left(\frac{1}{4}\right) - 2 \cdot \frac{3}{8}\log_2\left(\frac{3}{8}\right) \approx 1.56,$$

and using (5.54), we find that

$$H(K \mid C) = H(K) + H(M) - H(C) \approx 1 + 1.5 - 1.56 \approx 0.94.$$

5.6.3 Redundancy and the Entropy of Natural Language

Suppose that the plaintext is written in a natural language such as English.[22] Then nearby letters, or nearby bits if the letters are converted to ASCII, are heavily dependent on one another, rather than looking random. For example, correlations between successive letters (bigrams or trigrams) can aid the cryptanalyst, as we saw when we cryptanalyzed a simple substitution cipher in Sect. 1.1. In this section we use the notion of entropy to quantify the redundancy inherent in a natural language.

We start by approximating the entropy of a single letter in English text. Let L denote the random variable whose values are the letters of the English language E with their associated probabilities as given in Table 1.3 on page 6. For example, the table says that

$$f_L(\mathtt{A}) = 0.0815, \quad f_L(\mathtt{B}) = 0.0144, \quad f_L(\mathtt{C}) = 0.0276, \quad \ldots, \quad f_L(\mathtt{Z}) = 0.0008.$$

[22]It should be noted that when implementing a modern public key cipher, one generally combines the plaintext with some random bits and then performs some sort of invertible transformation so that the resulting secondary plaintext looks more like a string of random bits. See Sect. 8.6.

We can use the values in Table 1.3 to compute the entropy of a single letter in English text,

$$H(L) = 0.0815 \log_2(0.0815) + \cdots + 0.0008 \log_2(0.0008) \approx 4.132.$$

If every letter were equally likely, the entropy would be $\log_2(26) \approx 4.7$. The fact that the entropy is only 4.132 shows that some letters in English are more prevalent than others.

The concept of entropy can be used to measure the amount of information conveyed by a language. Shannon [126] shows that $H(L)$ can be interpreted as the average number of bits of information conveyed by a single letter of a language. The value of $H(L)$ that we computed does reveal some redundancy: it says that a letter conveys only 4.132 bits of information on average, although it takes 4.7 bits on average to specify a letter in the English alphabet.

The fact that natural languages contain redundancy is obvious. For example, you will probably be able to read the following sentence, despite our having removed almost 40 % of the letters:

Th prblms o crptgry nd scrcy sysms frnsh n ntrstng aplcatn o comm thry.

However, the entropy $H(L)$ of a single letter does not take into account correlations between nearby letters, so it alone does not give a good value for the redundancy of the English language E. As a first step, we should take into account the correlations between pairs of letters (bigrams). Let L^2 denote the random variable whose values are pairs of English letters as they appear in typical English text. Some bigrams appear fairly frequently, for example

$$f_{L^2}(\texttt{TH}) = 0.00315 \quad \text{and} \quad f_{L^2}(\texttt{AN}) = 0.00172.$$

Others, such as `JX` and `ZQ`, never occur. Just as Table 1.3 was created experimentally by counting the letters in a long sample text, we can create a frequency table of bigrams and use it to obtain an experimental value for L^2. This leads to a value of $H(L^2) \approx 7.12$, so on average, each letter of E has entropy equal to half this value, namely 3.56. Continuing, we could experimentally compute the entropy of L^3, which is the random variable whose values are trigrams (triples of letters), and then $\frac{1}{3}H(L^3)$ would be an even better approximation to the entropy of E. Of course, we need to analyze a great deal of text in order to obtain a reliable estimate for trigram frequencies, and the problem becomes even harder as we look at L^4, L^5, L^6, and so on. However, this idea leads to the following important concept.

Definition. Let L be a language (e.g., English or French or C++), and for each $n \geq 1$, let L^n denote the random variables whose values are strings of n consecutive characters of L. The *entropy* of L is defined to be the quantity[23]

[23] To be rigorous, one should really define upper and lower densities using liminf and limsup, since it is not clear that limit defining $H(\mathsf{L})$ exists. We will not worry about such niceties here.

$$H(\mathsf{L}) = \lim_{n\to\infty} \frac{H(L^n)}{n}.$$

Although it is not possible to precisely determine the entropy of the English language E, experimentally it appears that

$$1.0 \le H(\mathsf{E}) \le 1.5.$$

This means that despite the fact that it requires almost five bits to represent each of the 26 letters used in English, each letter conveys less than one and a half bits of information. Thus English is approximately 70 % redundant![24]

5.6.4 The Algebra of Secrecy Systems

We make only a few brief remarks about the algebra of cryptosystems. In [127], Shannon considers ways of building new cryptosystems by taking algebraic combinations of old ones. The new systems are described in terms of linear combinations and products of the original encryption transformations.

Example 5.66 (Summation Systems). If R and T are two secrecy systems, then Shannon defines the *weighted sum* of R and T to be

$$S = pR + qT, \quad \text{where } p + q = 1.$$

The meaning of this notation is as follows. First one chooses either R or T, where the probability of choosing R is p and the probability of choosing T is q. Imagine that the choice is made by flipping an unbalanced coin, but note that both Bob and Alice need to have a copy of the output of the coin tosses. In other words, the list of choices, or a method for generating the list of choices, forms part of their private key.

The notion of summation extends to the sum of any number of secrecy systems. The systems R and T need to have the same message space, but they need not act on messages in a similar way. For example, the system R could be a substitution cipher and the system T could be a shift cipher. As another example, suppose that T_i is the shift cipher that encrypts a letter of the alphabet by shifting it i places. Then the system that encrypts by choosing a shift at random and encrypting according to the chosen shift is the summation cipher

$$\sum_{i=0}^{25} \frac{1}{26} T_i.$$

Example 5.67 (Product Systems). In order to define the product of two cryptosystems, it is necessary that the ciphertexts of the first system be plaintexts for the second system. Thus let

[24]This does not mean that one can remove 70 % of the letters and still have an intelligible message. What it means is that in principle, it is possible to take a long message that requires 4.7 bits to specify each letter and to compress it into a form that takes only 30 % as many bits.

$$e : \mathcal{M} \to \mathcal{C} \quad \text{and} \quad e' : \mathcal{M}' \to \mathcal{C}'$$

by two encryption functions, and suppose that $\mathcal{C} = \mathcal{M}'$, or more generally, that $\mathcal{C} \subseteq \mathcal{M}'$. Then the product system $e' \cdot e$ is defined to be the composition of e and e',

$$e' \cdot e : \mathcal{M} \xrightarrow{\ e\ } \mathcal{C} \subseteq \mathcal{M}' \xrightarrow{\ e'\ } \mathcal{C}'.$$

Product ciphers provide a means to strengthen security. They were used in the development of DES, the Digital Encryption Standard [97], the first national standard for symmetric encryption. DES features several rounds of a cipher called S-box encryption, so it is a multiple product of a cipher with itself. Further, each round consists of the composition of several different transformations. The use of product ciphers continues to be of importance in the development of new symmetric ciphers, including AES, the Advanced Encryption Standard. See Sect. 8.12 for a brief discussion of DES and AES.

5.7 Complexity Theory and $\mathcal{P}$ Versus $\mathcal{NP}$

A *decision problem* is a problem in a formal system that has a yes or no answer. For example, PRIME is the decision problem of determining whether a given integer is a prime. We discussed this problem in Sect. 3.4. Another example is the decision Diffie–Hellman problem (Exercise 2.7): given $g^a \bmod p$ and $g^b \bmod p$, determine whether a given number C is equal to $g^{ab} \bmod p$. Complexity theory attempts to understand and quantify the difficulty of solving particular decision problems.

The early history of this field is fascinating, as mathematicians tried to come to grips with the limitations on provability within formal systems. In 1936, Alan Turing proved that there is no algorithm that solves the *halting problem*. That is, there is no algorithm to determine whether an arbitrary computer program, given an arbitrary input, eventually halts execution. Such a problem is called *undecidable*. Earlier in that same year, Alonzo Church had published a proof of undecidability of a problem in the *lambda calculus*. He and Turing then showed that the lambda calculus and the notion of Turing machine are essentially equivalent. The breakthroughs on the theory of undecidability that appeared in the 1930s and 1940s began as a response to Hilbert's questions about the completeness of axiomatic systems and whether there exist unsolvable mathematical problems. Indeed, both Church and Turing were influenced by Gödel's discovery in 1930 that all sufficiently strong and precise axiomatic systems are incomplete, i.e., they contain true statements that are unprovable within the system.

There are uncountably many undecidable problems in mathematics, some of which have simple and interesting formulations. Here is an example of an easy to state undecidable problem called *Post's correspondence problem* [106]. Suppose that you are given a sequence of pairs of strings,

$$(s_1, t_1),\ (s_2, t_2),\ (s_3, t_3),\ \ldots,\ (s_k, t_k),$$

where a string is simply a list of characters from some alphabet containing at least two letters. The correspondence problem asks you to decide whether there is an integer $r \geq 1$ and a list of indices

$$i_1, i_2, \ldots, i_r \quad \text{between 1 and } k \tag{5.55}$$

such that the concatenations

$$s_{i_1} \parallel s_{i_2} \parallel \cdots \parallel s_{i_r} \quad \text{and} \quad t_{i_1} \parallel t_{i_2} \parallel \cdots \parallel t_{i_r} \quad \text{are equal.} \tag{5.56}$$

Note that if we bound the value of r, say $r \leq r_0$, then the problem becomes decidable, since there are only a finite number of concatentations to check. The problem with r restricted in this way is called the *bounded Post correspondence problem.*

On the other end of the spectrum are decision problems for which there exist quick algorithms leading to their solutions. We have already talked about algorithms being fast if they run in polynomial time and slow if they take exponential time; see the discussion in Sects. 2.6 and 3.4.2.

Definition. A decision problem belongs to the class $\mathcal{P}$ if there exists a polynomial-time algorithm that solves it. That is, given an input of length n, the answer will be produced in a polynomial (in n) number of steps. One says that the decision problems in $\mathcal{P}$ are those that can be solved in *polynomial time.*

The concept of verification in polynomial time has some subtlety that can be captured only by a more precise definition, which we do not give. The class $\mathcal{NP}$ is defined by the concept of a polynomial-time algorithm on a "nondeterministic" machine. This means, roughly speaking, that we are allowed to guess a solution, but the verification time to check that the guessed solution is correct must be polynomial in the length of the input.

An example of a decision problem in $\mathcal{P}$ is that of determining whether two integers have a nontrivial common factor. This problem is in $\mathcal{P}$ because the Euclidean algorithm takes fewer than $\mathcal{O}(n^3)$ steps. (Note that in this setting, the Euclidean algorithm takes more than $\mathcal{O}(n)$ steps, since we need to take account of the time it takes to add, subtract, multiply, and divide n-bit numbers.) Another decision problem in $\mathcal{P}$ is that of determining whether a given integer is prime. The famous AKS algorithm, Theorem 3.26, takes fewer than $\mathcal{O}(n^7)$ steps to check primality.

Definition. A decision problem belongs to the class $\mathcal{NP}$ if a yes-instance of the problem can be verified in polynomial time.

For example, the bounded Post correspondence problem is in $\mathcal{NP}$. It is clear that if you are given a list of indices (5.55) of bounded length such that the concatenations (5.56) are alleged to be the same, it takes a polynomial number of steps to verify that the concatenations are indeed the same. On the other hand, exhaustively checking all possible concatenations of length up to r_0 takes an exponential (in r_0) number of steps. It is less clear, but can be proven, that one cannot find a solution in a polynomial number of steps.

This brings us to one of the most famous open questions in all of mathematics and computer science[25]:

$$\boxed{\text{Does } \mathcal{P} = \mathcal{NP}?}$$

Since the status of $\mathcal{P}$ versus $\mathcal{NP}$ is currently unresolved, it is useful to characterize problems in terms of their relative difficulty. We say that problem A can be (*polynomial-time*) *reduced* to problem B if there is a constructive polynomial-time transformation that takes any instance of A and maps it to an instance of B. Thus any algorithm for solving B can be transformed into an algorithm for solving A. Hence if problem B belongs to $\mathcal{P}$, and if A is reducible to B, then A also belongs to $\mathcal{P}$. The intuition is that if A can be reduced to B, then solving A is no harder than solving B (up to a polynomial amount of computation).

Stephen Cook's 1971 paper [30] entitled "The Complexity of Theorem Proving Procedures" laid the foundations of the theory of $\mathcal{NP}$-completeness. In this paper, Cook works with a certain $\mathcal{NP}$ problem called "Satisfiability" (abbreviated SAT). The SAT problem asks, given a Boolean expression involving only variables, parentheses, OR, AND and NOT, whether there exists an assignment of truth values that makes the expression true. Cook proves that SAT has the following properties:

1. Every $\mathcal{NP}$ problem is polynomial-time reducible to SAT.

2. If there exists any problem in $\mathcal{NP}$ that fails to be in $\mathcal{P}$, then SAT is not in $\mathcal{P}$.

A problem that has these two properties is said to be $\mathcal{NP}$*-complete.* Since the publication of Cook's paper, many other problems have been shown to be $\mathcal{NP}$-complete.

A related notion is that of $\mathcal{NP}$-hardness. We say that a problem is $\mathcal{NP}$-*hard* if it has the reducibility property (1), although the problem itself need not belong to $\mathcal{NP}$. All $\mathcal{NP}$-complete problems are $\mathcal{NP}$-hard, but not conversely. For example, the halting problem is $\mathcal{NP}$-hard, but not $\mathcal{NP}$-complete.

In order to put our informal discussion onto a firm mathematical footing, it is necessary to introduce some formalism. We start with a finite set of symbols Σ, and we denote by Σ^* the set of all (finite) strings of these symbols. A subset of Σ^* is called a *language.* A decision problem is defined to be the problem of deciding whether an input string belongs to a language. The precise definitions of $\mathcal{P}$ and $\mathcal{NP}$ are then given within this formal framework, which

[25] As mentioned in Sect. 2.1, the question of whether $\mathcal{P} = \mathcal{NP}$ is one of the \$1,000,000 Millennium Prize problems.

we shall not develop further here. For an excellent introduction to the theory of complexity, see [46], and for additional material on complexity theory as it relates to cryptography, see for example [143, Chapters 2 and 3].

Up to now we have been discussing the complexity theory of decision problems, but not every problem has a yes/no answer. For example, the problem of integer factorization (given a composite number, find a nontrivial factor) has a solution that is an integer, as does the discrete logarithm problem (given g and h in a $\mathbb{F}_p^*$, find an x such that $g^x = h$). It is possible to formulate a theory of complexity for general computational problems, but we are content to give two examples. First, the integer factorization problem is in $\mathcal{NP}$, since given an integer N and a putative factor m, it can be verified in polynomial-time that m divides N. Second, the discrete logarithm problem is in $\mathcal{NP}$, since given a supposed solution x, one can verify in polynomial time (using the fast powering algorithm) that $g^x = h$. It is not known whether either of these computational problems is in $\mathcal{P}$, i.e., there are no known polynomial-time algorithms for either integer factorization or for discrete logarithms. The current general consensus seems to be that they are probably not in $\mathcal{P}$.

We turn now to the role of complexity theory in some of the problems that arise in cryptography. The problems of factoring integers and finding discrete logarithms are presumed to be difficult, since no one has yet discovered polynomial-time algorithms to produce solutions. However, the problem of producing a solution (this is called the function problem) may be different from the decision problem of determining whether a solution exists. Here is a version of the factoring problem phrased as a decision problem:

Does there exist a nontrivial factor of N that is less than k?

As we can see, a yes instance of this problem (i.e., N is composite) has a (trivial) polynomial-time verification algorithm, and so this decision problem belongs to $\mathcal{NP}$. It can also be shown that the complementary problem belongs to $\mathcal{NP}$. That is, if N is a no instance (i.e., N is prime), then the primality of N can be verified in polynomial time on a nondeterministic Turing machine. When both the yes and no instances of a problem can be verified in polynomial time, the decision problem is said to belong to the class co-$\mathcal{NP}$. Since it is widely believed that $\mathcal{NP}$ is not the same as co-$\mathcal{NP}$, it was also believed that factoring is not an $\mathcal{NP}$-complete problem. In 2004, Agrawal, Kayal and Saxena [1] showed that the decision problem of determining whether a number is prime does indeed belong to $\mathcal{P}$, settling the long-standing question whether this decision problem could be $\mathcal{NP}$-complete.

A cryptosystem is only as secure as its underlying hard problem, so it would be desirable to construct cryptosystems based on $\mathcal{NP}$-hard problems. There has been a great deal of interest in building efficient public key cryptosystems of this sort. A major difficulty is that one needs not only an $\mathcal{NP}$-hard problem, but also a trapdoor to the problem to use for decryption. This has led to a number of cryptosystems that are based special cases of $\mathcal{NP}$-hard problems, but it is not known whether these special cases are themselves $\mathcal{NP}$-hard.

The first example of a public key cryptosystem built around an $\mathcal{NP}$-complete problem was the knapsack cryptosystem of Merkle and Hellman. More precisely, they based their cryptosystem on the *subset-sum problem*, which asks the following:

Given n positive integers $a_1, \ldots, a_n$ and a target sum S, find a subset of the a_i such that

$$a_{i_1} + a_{i_2} + \cdots + a_{i_t} = S.$$

The subset-sum problem is $\mathcal{NP}$-complete, since one can show that any instance of SAT can be reduced to an instance of the subset-sum problem, and vice versa. In order to build a public key cryptosystem based on the (hard) subset-sum problem, Merkle and Hellman needed to build a trapdoor into the problem. They did this by using only certain special cases of the subset-sum problem, but unfortunately it turned out that these special cases are significantly easier than the general case and their cryptosystem was broken. And despite further work by a number of cryptographers, no one has been able to build a subset-sum cryptosystem that is both efficient and secure. See Sect. 7.2 for a detailed discussion of how subset-sum cryptosystems work and how they are broken.

Another cautionary note in going from theory to practice comes from the fact that even if a certain collection of problems is $\mathcal{NP}$-hard, that does not mean that every problem in the collection is hard. In some sense, $\mathcal{NP}$-hardness measures the difficulty of the hardest problem in the collection, not the average problem. It would not be good to base a cryptosystem on a problem for which a few instances are very hard, but most instances are very easy. Ideally, we want to use a collection of problems with the property that most instances are $\mathcal{NP}$-hard. An interesting example is the closest vector problem (CVP), which involves finding a vector in lattice that is close to a given vector. We discuss lattices and CVP in Chap. 7, but for now we note that CVP is $\mathcal{NP}$-hard. Our interest in CVP stems from a famous result of Ajtai and Dwork [4] in which they construct a cryptosystem based on CVP in a certain set of lattices. They show that the *average* difficulty of solving CVP for their lattices can be polynomially reduced to solving the *hardest* instance of CVP in a similar set of lattices (of somewhat smaller dimension). Although not practical, their public key cryptosystem was the first construction exhibiting worst-case/average-case equivalence.

Exercises

Section 5.1. Basic Principles of Counting

5.1. The Rhind papyrus is an ancient Egyptian mathematical manuscript that is more than 3500 years old. Problem 79 of the Rhind papyrus poses a problem that can be paraphrased as follows: there are seven houses; in each house lives seven cats;

each cat kills seven mice; each mouse has eaten seven spelt seeds[26]; each spelt seed would have produced seven hekat[27] of spelt. What is the sum of all of the named items? Solve this 3500 year old problem.

5.2. (a) How many n-tuples $(x_1, x_2, \ldots, x_n)$ are there if the coordinates are required to be integers satisfying $0 \le x_i < q$?
(b) Same question as (a), except now there are separate bounds $0 \le x_i < q_i$ for each coordinate.
(c) How many n-by-n matrices are there if the entries $x_{i,j}$ of the matrix are integers satisfying $0 \le x_{i,j} < q$?
(d) Same question as (a), except now the order of the coordinates does not matter. So for example, $(0, 0, 1, 3)$ and $(1, 0, 3, 0)$ are considered the same. (This one is rather tricky.)
(e) Twelve students are each taking four classes, for each class they need two loose-leaf notebooks, for each notebook they need 100 sheets of paper, and each sheet of paper has 32 lines on it. Altogether, how many students, classes, notebooks, sheets, and lines are there? (Bonus. Make this or a similar problem of your own devising into a rhyme like the St. Ives riddle.)

5.3. (a) List all of the permutations of the set $\{A, B, C\}$.
(b) List all of the permutations of the set $\{1, 2, 3, 4\}$.
(c) How many permutations are there of the set $\{1, 2, \ldots, 20\}$?
(d) Seven students are to be assigned to seven dormitory rooms, each student receiving his or her own room. In how many ways can this be done?
(e) How many different words can be formed with the four symbols A, A, B, C?

5.4. (a) List the 24 possible permutations of the letters A_1, A_2, B_1, B_2. If A_1 is indistinguishable from A_2, and B_1 is indistinguishable from B_2, show how the permutations become grouped into 6 distinct letter arrangements, each containing 4 of the original 24 permutations.
(b) Using the seven symbols A, A, A, A, B, B, B, how many different seven letter words can be formed?
(c) Using the nine symbols $A, A, A, A, B, B, B, C, C$, how many different nine letter words can be formed?
(d) Using the seven symbols A, A, A, A, B, B, B, how many different five letter words can be formed?

5.5. (a) There are 100 students eligible for an award, and the winner gets to choose from among 5 different possible prizes. How many possible outcomes are there?
(b) Same as in (a), but this time there is a first place winner, a second place winner, and a third place winner, each of whom gets to select a prize. However, there is only one of each prize. How many possible outcomes are there?
(c) Same as in (b), except that there are multiple copies of each prize, so each of the three winners may choose any of the prizes. Now how many possible outcomes are there? Is this larger or smaller than your answer from (b)?

[26] *Spelt* is an ancient type of wheat.
[27] A *hekat* is $\frac{1}{30}$ of a cubic cubit, which is approximately 4.8 l.

(d) Same as in (c), except that rather than specifying a first, second, and third place winner, we just choose three winning students without differentiating between them. Now how many possible outcomes are there? Compare the size of your answers to (b), (c), and (d).

5.6. Use the binomial theorem (Theorem 5.10) to compute each of the following quantities.
(a) $(5z + 2)^3$ (b) $(2a - 3b)^4$ (c) $(x - 2)^5$

5.7. The binomial coefficients satisfy many interesting identities. Give three proofs of the identity

$$\binom{n}{j} = \binom{n-1}{j-1} + \binom{n-1}{j}.$$

(a) For Proof #1, use the definition of $\binom{n}{j}$ as $\frac{n!}{(n-j)!j!}$.
(b) For Proof #2, use the binomial theorem (Theorem 5.10) and compare the coefficients of $x^j y^{n-j}$ on the two sides of the identity

$$(x + y)^n = (x + y)(x + y)^{n-1}.$$

(c) For Proof #3, argue directly that choosing j objects from a set of n objects can be decomposed into either choosing $j - 1$ objects from $n - 1$ objects or choosing j objects from $n - 1$ objects.

5.8. Let p be a prime number. This exercise sketches another proof of Fermat's little theorem (Theorem 1.24).
(a) If $1 \le j \le p - 1$, prove that the binomial coefficient $\binom{p}{j}$ is divisible by p.
(b) Use (a) and the binomial theorem (Theorem 5.10) to prove that

$$(a + b)^p \equiv a^p + b^p \pmod{p} \qquad \text{for all } a, b \in \mathbb{Z}.$$

(c) Use (b) with $b = 1$ and induction on a to prove that $a^p \equiv a \pmod{p}$ for all $a \ge 0$.
(d) Use (c) to deduce that $a^{p-1} \equiv 1 \pmod{p}$ for all a with $\gcd(p, a) = 1$.

5.9. We know that there are $n!$ different permutations of the set $\{1, 2, \ldots, n\}$.
(a) How many of these permutations leave no number fixed?
(b) How many of these permutations leave at least one number fixed?
(c) How many of these permutations leave exactly one number fixed?
(d) How many of these permutations leave at least two numbers fixed?
For each part of this problem, give a formula or algorithm that can be used to compute the answer for an arbitrary value of n, and then compute the value for $n = 10$ and $n = 26$. (This exercise generalizes Exercise 1.5.)

Section 5.2. The Vigenère Cipher

5.10. Encrypt each of the following Vigenère plaintexts using the given keyword and the Vigenère tableau (Table 5.1).
(a) Keyword: `hamlet`
Plaintext: `To be, or not to be, that is the question.`

(b) Keyword: `fortune`
Plaintext: `The treasure is buried under the big W.`

5.11. Decrypt each of the following Vigenère ciphertexts using the given keyword and the Vigenère tableau (Table 5.1).

(a) Keyword: `condiment`
Ciphertext:
```
rsghz bmcxt dvfsq hnigq xrnbm
pdnsq smbtr ku
```

(b) Keyword: `rabbithole`
Ciphertext:
```
khfeq ymsci etcsi gjvpw ffbsq
moapx zcsfx epsox yenpk daicx
cebsm ttptx zooeq laflg kipoc
zswqm taujw ghboh vrjtq hu
```

5.12. Explain how a cipher wheel with rotating inner wheel (see Fig. 1.1 on page 3) can be used in place of a Vigenère tableau (Table 5.1) to perform Vigenère encryption and decryption. Illustrate by describing the sequence of rotations used to perform a Vigenère encryption with the keyword `mouse`.

5.13. Let

$$\boldsymbol{s} = \text{“I am the very model of a modern major general.”}$$
$$\boldsymbol{t} = \text{“I have information vegetable, animal, and mineral.”}$$

(a) Make frequency tables for $\boldsymbol{s}$ and $\boldsymbol{t}$.
(b) Compute $\text{IndCo}(\boldsymbol{s})$ and $\text{IndCo}(\boldsymbol{t})$.
(c) Compute $\text{MutIndCo}(\boldsymbol{s}, \boldsymbol{t})$.

5.14. The following strings are blocks from a Vigenère encryption. It turns out that the keyword contains a repeated letter, so two of these blocks were encrypted with the same shift. Compute $\text{MutIndCo}(\boldsymbol{s}_i, \boldsymbol{s}_j)$ for $1 \le i < j \le 3$ and use these values to deduce which two strings were encrypted using the same shift.

$$\boldsymbol{s}_1 = \texttt{iwseesetftuonhdptbunnybioeatneghictdnsevi}$$
$$\boldsymbol{s}_2 = \texttt{qibfhroeqeickxmirbqlflgkrqkejbejpepldfjbk}$$
$$\boldsymbol{s}_3 = \texttt{iesnnciiheptevaireittuevmhooottrtaaflnatg}$$

5.15. (a) One of the following two strings was encrypted using a simple substitution cipher, while the other is a random string of letters. Compute the index of coincidence of each string and use the results to guess which is which.

$$s_1 = \texttt{RCZBWBFHSLPSCPILHBGZJTGBIBJGLYIJIBFHCQQFZBYFP},$$
$$s_2 = \texttt{KHQWGIZMGKPOYRKHUITDUXLXCWZOTWPAHFOHMGFEVUEJJ}.$$

(b) One of the following two strings was encrypted using a simple substitution cipher, while the other is a random permutation of the same set of letters.

$$s_1 = \texttt{NTDCFVDHCTHKGUNGKEPGXKEWNECKEGWEWETWKUEVHDKK}$$
$$\texttt{CDGCWXKDEEAMNHGNDIWUVWSSCTUNIGDSWKE}$$

nhqrk	vvvfe	fwgjo	mzjgc	kocgk	lejrj	wossy	wgvkk	hnesg	kwebi
bkkcj	vqazx	wnvll	zetjc	zwgqz	zwhah	kwdxj	fgnyw	gdfgh	bitig
mrkwn	nsuhy	iecru	ljjvs	qlvvw	zzxyv	woenx	ujgyr	kqbfj	lvjzx
dxjfg	nywus	rwoar	xhvvx	ssmja	vkrwt	uhktm	malcz	ygrsz	xwnvl
lzavs	hyigh	rvwpn	ljazl	nispv	jahym	ntewj	jvrzg	qvzcr	estul
fkwis	tfylk	ysnir	rddpb	svsux	zjgqk	xouhs	zzrjj	kyiwc	zckov
qyhdv	rhhny	wqhyi	rjdqm	iwutf	nkzgd	vvibg	oenwb	kolca	mskle
cuwwz	rgusl	zgfhy	etfre	ijjvy	ghfau	wvwtn	xlljv	vywyj	apgzw
trggr	dxfgs	ceyts	tiiih	vjjvt	tcxfj	hciiv	voaro	lrxij	vjnok
mvrgw	kmirt	twfer	oimsb	qgrgc					

Table 5.13: A Vigenère ciphertext for Exercise 5.16

togmg	gbymk	kcqiv	dmlxk	kbyif	vcuek	cuuis	vvxqs	pwwej	koqgg
phumt	whlsf	yovww	knhhm	rcqfq	vvhkw	psued	ugrsf	ctwij	khvfa
thkef	fwptj	ggviv	cgdra	pgwvm	osqxg	hkdvt	whuev	kcwyj	psgsn
gfwsl	jsfse	ooqhw	tofsh	aciin	gfbif	gabgj	adwsy	topml	ecqzw
asgvs	fwrqs	fsfvq	rhdrs	nmvmk	cbhrv	kblxk	gzi		

Table 5.14: A Vigenère ciphertext for Exercise 5.17

$$s_2 = \texttt{IGWSKGEHEXNGECKVWNKVWNKSUTEHTWHEKDNCDXWSIEKD}$$
$$\texttt{AECKFGNDCPUCKDNCUVWEMGEKWGEUTDGTWHD}$$

Thus their Indices of Coincidence are identical. Develop a method to compute a bigram index of coincidence, i.e., the frequency of pairs of letters, and use it to determine which string is most likely the encrypted text.

(Bonus: Decrypt the encrypted texts in (a) and (b), but be forewarned that the plaintexts are in Latin.)

5.16. Table 5.13 is a Vigenère ciphertext in which we have marked some of the repeated trigrams for you. How long do you think the keyword is? Why?

Bonus: Complete the cryptanalysis and recover the plaintext.

5.17. We applied a Kasiski test to the Vigenère ciphertext listed in Table 5.14 and found that the key length is probably 5. We then performed a mutual index of coincidence test to each shift of each pair of blocks and listed the results for you in Table 5.15. (This is the same type of table as Table 5.5 in the text, except that we haven't underlined the large values.) Use Table 5.15 to guess the relative rotations of the blocks, as we did in Table 5.6. This will give you a rotated version of the keyword. Try rotating it, as we did in Table 5.7, to find the correct keyword and decrypt the text.

5.18. Table 5.16 gives a Vigenère ciphertext for you to analyze from scratch. It is probably easiest to do so by writing a computer program, but you are welcome to try to decrypt it with just paper and pencil.

(a) Make a list of matching trigrams as we did in Table 5.3. Use the Kasiski test on matching trigrams to find the likely key length.

Blocks		Shift amount												
i	j	0	1	2	3	4	5	6	7	8	9	10	11	12
1	2	0.044	0.047	0.021	0.054	0.046	0.038	0.022	0.034	0.057	0.035	0.040	0.023	0.038
1	3	0.038	0.031	0.027	0.037	0.045	0.036	0.034	0.032	0.039	0.039	0.047	0.038	0.050
1	4	0.025	0.039	0.053	0.043	0.023	0.035	0.032	0.043	0.029	0.040	0.041	0.050	0.027
1	5	0.050	0.050	0.025	0.031	0.038	0.045	0.037	0.028	0.032	0.038	0.063	0.033	0.034
2	3	0.035	0.037	0.039	0.031	0.031	0.035	0.047	0.048	0.034	0.031	0.031	0.067	0.053
2	4	0.040	0.033	0.046	0.031	0.033	0.023	0.052	0.027	0.031	0.039	0.078	0.034	0.029
2	5	0.042	0.040	0.042	0.029	0.033	0.035	0.035	0.038	0.037	0.057	0.039	0.038	0.040
3	4	0.032	0.033	0.035	0.049	0.053	0.027	0.030	0.022	0.047	0.036	0.040	0.036	0.052
3	5	0.043	0.043	0.040	0.034	0.033	0.034	0.043	0.035	0.026	0.030	0.050	0.068	0.044
4	5	0.045	0.033	0.044	0.046	0.021	0.032	0.030	0.038	0.047	0.040	0.025	0.037	0.068

Blocks		Shift amount												
i	j	13	14	15	16	17	18	19	20	21	22	23	24	25
1	2	0.040	0.063	0.033	0.025	0.032	0.055	0.038	0.030	0.032	0.045	0.035	0.030	0.044
1	3	0.026	0.046	0.042	0.053	0.027	0.024	0.040	0.047	0.048	0.018	0.037	0.034	0.066
1	4	0.042	0.050	0.042	0.031	0.024	0.052	0.027	0.051	0.020	0.037	0.042	0.069	0.031
1	5	0.030	0.048	0.039	0.030	0.034	0.038	0.042	0.035	0.036	0.043	0.055	0.030	0.035
2	3	0.039	0.015	0.030	0.045	0.049	0.037	0.023	0.036	0.030	0.049	0.039	0.050	0.037
2	4	0.027	0.048	0.050	0.037	0.032	0.021	0.035	0.043	0.047	0.041	0.047	0.042	0.035
2	5	0.033	0.035	0.039	0.033	0.037	0.047	0.037	0.028	0.034	0.066	0.054	0.032	0.022
3	4	0.040	0.048	0.041	0.044	0.033	0.028	0.039	0.027	0.036	0.017	0.038	0.051	0.065
3	5	0.039	0.029	0.045	0.040	0.033	0.028	0.031	0.037	0.038	0.036	0.033	0.051	0.036
4	5	0.049	0.033	0.029	0.043	0.028	0.033	0.020	0.040	0.040	0.041	0.039	0.039	0.059

Table 5.15: Mutual indices of coincidence for Exercise 5.17

```
mgodt beida psgls akowu hxukc iawlr csoyh prtrt udrqh cengx
uuqtu habxw dgkie ktsnp sekld zlvnh wefss glzrn peaoy lbyig
uaafv eqgjo ewabz saawl rzjpv feyky gylwu btlyd kroec bpfvt
psgki puxfb uxfuq cvymy okagl sactt uwlrx psgiy ytpsf rjfuw
igxhr oyazd rakce dxeyr pdobr buehr uwcue ekfic zehrq ijezr
xsyor tcylf egcy
```

Table 5.16: A Vigenère ciphertext for Exercise 5.18

(b) Make a table of indices of coincidence for various key lengths, as we did in Table 5.4. Use your results to guess the probable key length.

(c) Using the probable key length from (a) or (b), make a table of mutual indices of coincidence between rotated blocks, as we did in Table 5.5. Pick the largest indices from your table and use them to guess the relative rotations of the blocks, as we did in Table 5.6.

(d) Use your results from (c) to guess a rotated version of the keyword, and then try the different rotations as we did in Table 5.7 to find the correct keyword and decrypt the text.

5.19. The *autokey cipher* is similar to the Vigenère cipher, except that rather than repeating the key, it simply uses the key to encrypt the first few letters and then uses the plaintext itself (shifted over) to continue the encryption. For example, in order to encrypt the message "`The autokey cipher is cool`" using the keyword `random`, we proceed as follows:

Plaintext	t h e a u t o k e y c i p h e r i s c o o l
Key	r a n d o m t h e a u t o k e y c i p h e r
Ciphertext	k h r d i f h r i y w b d r i p k a r v s c

The autokey cipher has the advantage that different messages are encrypted using different keys (except for the first few letters). Further, since the key does not repeat, there is no key length, so the autokey is not directly susceptible to a Kasiski or index of coincidence analysis. A disadvantage of the autokey is that a single mistake in encryption renders the remainder of the message unintelligible. According to [63], Vigenère invented the autokey cipher in 1586, but his invention was ignored and forgotten before being reinvented in the 1800s.

(a) Encrypt the following message using the autokey cipher:

Keyword: `LEAR`
Plaintext: `Come not between the dragon and his wrath.`

(b) Decrypt the following message using the autokey cipher:

Keyword: `CORDELIA`
Ciphertext: `pckkm yowvz ejwzk knyzv vurux cstri tgac`

(c) Eve intercepts an autokey ciphertext and manages to steal the accompanying plaintext:

Plaintext `ifmusicbethefoodofloveplayon`
Ciphertext `azdzwqvjjfbwnqphhmptjsszfjci`

Help Eve to figure out the keyword that was used for encryption. Describe your method in sufficient generality to show that the autokey cipher is susceptible to known plaintext attacks.

(d) Bonus Problem: Try to formulate a statistical or algebraic attack on the autokey cipher, assuming that you are given a large amount of ciphertext to analyze.

Section 5.3. Probability Theory

5.20. Use the definition (5.15) of the probability of an event to prove the following basic facts about probability theory.

(a) Let E and F be disjoint events. Then

$$\Pr(E \cup F) = \Pr(E) + \Pr(F).$$

(b) Let E and F be events that need not be disjoint. Then

$$\Pr(E \cup F) = \Pr(E) + \Pr(F) - \Pr(E \cap F).$$

(c) Let E be an event. Then $\Pr(E^c) = 1 - \Pr(E)$.

(d) Let E_1, E_2, E_3 be events. Prove that

$$\begin{aligned}\Pr(E_1 \cup E_2 \cup E_3) = {} & \Pr(E_1) + \Pr(E_2) + \Pr(E_3) - \Pr(E_1 \cap E_2) \\ & - \Pr(E_1 \cap E_3) - \Pr(E_2 \cap E_3) + \Pr(E_1 \cap E_2 \cap E_3).\end{aligned}$$

The formulas in (b) and (d) and their generalization to n events are known as the *inclusion–exclusion principle*.

5.21. We continue with the coin tossing scenario from Example 5.23, so our experiment consists in tossing a fair coin ten times. Compute the probabilities of the following events.

(a) The first and last tosses are both heads.
(b) Either the first toss or the last toss (or both) are heads.
(c) Either the first toss or the last toss (but not both) are heads.
(d) There are exactly k heads and $10 - k$ tails. Compute the probability for each value of k between 0 and 10. (*Hint.* To save time, note that the probability of exactly k heads is the same as the probability of exactly k tails.)
(e) There is an even number of heads.
(f) There is an odd number of heads.

5.22. Alice offers to make the following bet with you. She will toss a fair coin 14 times. If exactly 7 heads come up, she will give you \$4; otherwise you must give her \$1. Would you take this bet? If so, and if you repeated the bet 10000 times, how much money would you expect to win or lose?

5.23. Let E and F be events.
(a) Prove that $\Pr(E \mid E) = 1$. Explain in words why this is reasonable.
(b) If E and F are disjoint, prove that $\Pr(F \mid E) = 0$. Explain in words why this is reasonable.
(c) Let $F_1, \ldots, F_n$ be events satisfying $F_i \cap F_j = \emptyset$ for all $i \neq j$. We say that $F_1, \ldots, F_n$ are *pairwise disjoint*. Prove then that

$$\Pr\left(\bigcup_{i=1}^{n} F_i\right) = \sum_{i=1}^{n} \Pr(F_i).$$

(d) Let $F_1, \ldots, F_n$ be pairwise disjoint as in (c), and assume further that

$$F_1 \cup \cdots \cup F_n = \Omega,$$

where recall that Ω is the entire sample space. Prove the following general version of the decomposition formula (5.20) in Proposition 5.24(a):

$$\Pr(E) = \sum_{i=1}^{n} \Pr(E \mid F_i)\Pr(F_i).$$

(e) Prove a general version of Bayes's formula:

$$\Pr(F_i \mid E) = \frac{\Pr(E \mid F_i)\Pr(F_i)}{\Pr(E \mid F_1)\Pr(F_1) + \Pr(E \mid F_2)\Pr(F_2) + \cdots + \Pr(E \mid F_n)\Pr(F_n)}.$$

5.24. There are two urns containing pens and pencils. Urn #1 contains three pens and seven pencils and Urn #2 contains eight pens and four pencils.
(a) An urn is chosen at random and an object is drawn. What is the probability that it is a pencil?
(b) An urn is chosen at random and an object is drawn. If the object drawn is a pencil, what is the probability that it came from Urn #1?
(c) If an urn is chosen at random and two objects are drawn simultaneously, what is the probability that both are pencils?

5.25. An urn contains 20 silver coins and 10 gold coins. You are the sixth person in line to randomly draw and keep a coin from the urn.
(a) What is the probability that you draw a gold coin?

(b) If you draw a gold coin, what is the probability that the five people ahead of you all drew silver coins?

5.26. Consider the three prisoners scenario described in Example 5.26. Let A, B, and C denote respectively the events that Alice is to be released, Bob is to be released, and Carl is to be released, which we assume to be equally likely, so $\Pr(A) = \Pr(B) = \Pr(C) = \frac{1}{3}$. Also let J be the event that the jailer tells Aice that Bob is to stay in jail.

(a) Compute the values of $\Pr(B \mid J)$, $\Pr(J \mid B)$, and $\Pr(J \mid C)$.

(b) Compute the values of $\Pr(J \mid A^c)$ and $\Pr(J^c \mid A^c)$, where the event A^c is the event that Alice stays in jail.

(c) Suppose that if Alice is the one who is to be released, then the jailer flips a fair coin to decide whether to tell Alice that Bob stays in jail or that Carl stays in jail. What is the value of $\Pr(A \mid J)$?

(d) Suppose instead that if Alice is the one who is to be released, then the jailer always tells her that Bob will stay in jail. Now what is the value of $\Pr(A \mid J)$?

Other similar problems with counterintuitive conclusions include the Monty Hall problem (Exercise 5.27), Bertrand's box paradox, and the principle of restricted choice in contract bridge.

5.27. (*The Monty Hall Problem*) Monty Hall gives Dan the choice of three curtains. Behind one curtain is a car, while behind the other two curtains are goats. Dan chooses a curtain, but before it is opened, Monty Hall opens one of the other curtains and reveals a goat. He then offers Dan the option of keeping his original curtain or switching to the remaining closed curtain. The Monty Hall problem is to figure out Dan's best strategy: "To stick or to switch?"

(a) What is the probability that Dan wins the car if he always sticks to his first choice of curtain? What is the probability that Dan wins the car if he always switches curtains? Which is his best strategy? (If the answer seems counter-intuitive, suppose instead that there are 1000 curtains and that Monty Hall opens 998 goat curtains. Now what are the winning probabilities for the two strategies?)

(b) Suppose that we give Monty Hall another option, namely he's allowed to force Dan to stick with his first choice of curtain. Assuming that Monty Hall dislikes giving away cars, now what is Dan's best strategy, and what is his probability of winning a car?

(c) More generally, suppose that there are N curtains and M cars, and suppose that Monty Hall opens K curtains that have goats behind them. Compute the probabilities

$$\Pr(\text{Dan wins a car} \mid \text{Dan sticks}), \quad \Pr(\text{Dan wins a car} \mid \text{Dan switches}).$$

Which is the better strategy?

5.28. Let $\mathcal{S}$ be a set, let A be a property of interest, and suppose that for $m \in \mathcal{S}$, we have

$$\Pr(m \text{ does not have property } A) = \delta.$$

Suppose further that a Monte Carlo algorithm applied to m and a random number r satisfy:

(1) If the algorithm returns Yes, then m definitely has property A.
(2) If m has property A, then the probability that the algorithm returns Yes is at least p.

Notice that we can restate (1) and (2) as conditional probabilities:

(1) $\Pr(m \text{ has property } A \mid \text{algorithm returns Yes}) = 1$,
(2) $\Pr(\text{algorithm returns Yes} \mid m \text{ has property } A) \geq p$.

Suppose that we run the algorithm N times on the number m, and suppose that the algorithm returns No every single time. Derive a lower bound, in terms of δ, p, and N, for the probability that m does not have property A. (This generalizes the version of the Monte Carlo method that we studied in Sect. 5.3.3 with $\delta = 0.01$ and $p = \frac{1}{2}$. Be careful to distinguish p from $1 - p$ in your calculations.)

5.29. We continue with the setup described in Exercise 5.28.
(a) Suppose that $\delta = \frac{9}{10}$ and $p = \frac{3}{4}$. If we run the algorithm 25 times on the input m and always get back No, what is the probability that m does not have property A?
(b) Same question as (a), but this time we run the algorithm 100 times.
(c) Suppose that $\delta = \frac{99}{100}$ and $p = \frac{1}{2}$. How many times should we run the algorithm on m to be 99 % confident that m does not have property A, assuming that every output is No?
(d) Same question as (c), except now we want to be 99.9999 % confident.

5.30. If an integer n is composite, then the Miller–Rabin test has at least a 75 % chance of succeeding in proving that n is composite, while it never misidentifies a prime as being composite. (See Table 3.2 in Sect. 3.4 for a description of the Miller–Rabin test.) Suppose that we run the Miller–Rabin test N times on the integer n and that it fails to prove that n is composite. Show that the probability that n is prime satisfies (approximately)

$$\Pr(n \text{ is prime} \mid \text{the Miller–Rabin test fails } N \text{ times}) \geq 1 - \frac{\ln(n)}{4^N}.$$

(*Hint.* Use Exercise 5.28 with appropriate choices of A, $\mathcal{S}$, δ, and p. You may also use the estimate from Sect. 3.4.1 that the probability that n is prime is approximately $1/\ln(n)$.)

5.31. It is natural to assume that if $\Pr(E \mid F)$ is significantly larger than $\Pr(E)$, then somehow F is causing E. Baye's formula illustrates the fallacy of this sort of reasoning, since it says that

$$\frac{\Pr(E \mid F)}{\Pr(E)} = \frac{\Pr(F \mid E)}{\Pr(F)}.$$

So if F is "causing" E, then the same reasoning shows that E is "causing" F. All that one can really say is that E and F are correlated with one another, in the sense that either one of them being true makes it more likely that the other one is true. It is incorrect to deduce a cause-and-effect relation.

Here is a concrete example. Testing shows that first graders are more likely to be good spellers if their shoe sizes are larger than average. This is an experimental fact. Hence if we stretch a child's foot, it will make them a better speller! Alternatively, by Baye's formula, if we give them extra spelling lessons, then their feet will grow faster! Explain why these last two assertions are nonsense, and describe what's really going on.

5.32. Let $f_X(k)$ be the binomial density function (5.23). Prove directly, using the binomial theorem, that $\sum_{k=0}^{n} f_X(k) = 1$.

5.33. In Example 5.37 we used a differentiation trick to compute the value of the infinite series $\sum_{n=1}^{\infty} np(1-p)^{n-1}$. This exercise further develops this useful technique. The starting point is the formula for the geometric series

$$\sum_{n=0}^{\infty} x^n = \frac{1}{1-x} \qquad \text{for } |x| < 1 \tag{5.57}$$

and the differential operator

$$\mathcal{D} = x\frac{d}{dx}.$$

(a) Using the fact that $\mathcal{D}(x^n) = nx^n$, prove that

$$\sum_{n=1}^{\infty} nx^n = \frac{x}{(1-x)^2} \tag{5.58}$$

by applying $\mathcal{D}$ to both sides of (5.57). For which x does the left-hand side of (5.58) converge? (*Hint.* Use the ratio test.)

(b) Applying $\mathcal{D}$ again, prove that

$$\sum_{n=0}^{\infty} n^2 x^n = \frac{x+x^2}{(1-x)^3}. \tag{5.59}$$

(c) More generally, prove that for every value of k there is a polynomial $F_k(x)$ such that

$$\sum_{n=0}^{\infty} n^k x^n = \frac{F_k(x)}{(1-x)^{k+1}}. \tag{5.60}$$

(*Hint.* Use induction on k.)

(d) The first few polynomials $F_k(x)$ in (c) are $F_0(x) = 1$, $F_1(x) = x$, and $F_2(x) = x + x^2$. These follow from (5.57), (5.58), and (5.59). Compute $F_3(x)$ and $F_4(x)$.

(e) Prove that the polynomial $F_k(x)$ in (c) has degree k.

5.34. In each case, compute the expectation of the random variable X.

(a) The values of X are uniformly distributed on the set $\{0, 1, 2, \ldots, N-1\}$. (See Example 5.28.)

(b) The values of X are uniformly distributed on the set $\{1, 2, \ldots, N\}$.

(c) The values of X are uniformly distributed on the set $\{1, 3, 7, 11, 19, 23\}$.

(d) X is a random variable with a binomial density function; see formula (5.23) in Example 5.29 on page 240.

5.35. Let X be a random variable on the probability space Ω. It might seem more natural to define the expected value of X by the formula

$$\sum_{\omega \in \Omega} X(\omega) \cdot \Pr(\omega). \tag{5.61}$$

Prove that the formula (5.61) gives the same value as Eq. (5.27) on page 244, which we used in the text to define $E(X)$.

Section 5.4. Collision Algorithms and the Birthday Paradox

5.36. (a) In a group of 23 strangers, what is the probability that at least two of them have the same birthday? How about if there are 40 strangers? In a group of 200 strangers, what is the probability that one of them has the same birthday as your birthday? (*Hint.* See the discussion in Sect. 5.4.1.)

(b) Suppose that there are N days in a year (where N could be any number) and that there are n people. Develop a general formula, analogous to (5.28), for the probability that at least two of them have the same birthday. (*Hint.* Do a calculation similar to the proof of (5.28) in the collision theorem (Theorem 5.38), but note that the formula is a bit different because the birthdays are being selected from a single list of N days.)

(c) Find a lower bound of the form

$$\Pr(\text{at least one match}) \geq 1 - e^{-(\text{some function of } n \text{ and } N)}$$

for the probability in (b), analogous to the estimate (5.29).

5.37. A deck of cards is shuffled and the top eight cards are turned over.

(a) What is the probability that the king of hearts is visible?

(b) A second deck is shuffled and its top eight cards are turned over. What is the probability that a visible card from the first deck matches a visible card from the second deck? (Note that this is slightly different from Example 5.39 because the cards in the second deck are not being replaced.)

5.38. (a) Prove that

$$e^{-x} \geq 1 - x \quad \text{for all values of } x.$$

(*Hint.* Look at the graphs of e^{-x} and $1 - x$, or use calculus to compute the minimum of the function $f(x) = e^{-x} - (1 - x)$.)

(b) Prove that for all $a > 1$, the inequality

$$e^{-ax} \leq (1 - x)^a + \frac{1}{2}ax^2 \quad \text{is valid for all } 0 \leq x \leq 1.$$

(This is a challenging problem.)

(c) We used the inequality in (a) during the proof of the lower bound (5.29) in the collision theorem (Theorem 5.38). Use (b) to prove that

$$\Pr(\text{at least one red}) \leq 1 - e^{-mn/N} + \frac{mn^2}{2N^2}.$$

Thus if N is large and m and n are not much larger than $\sqrt{N}$, then the estimate

$$\Pr(\text{at least one red}) \approx 1 - e^{-mn/N}$$

is quite accurate. (*Hint.* Use (b) with $a = m$ and $x = n/N$.)

5.39. Solve the discrete logarithm problem $10^x = 106$ in the finite field $\mathbb{F}_{811}$ by finding a collision among the random powers 10^i and $106 \cdot 10^i$ that are listed in Table 5.17.

i	g^i	$h \cdot g^i$
116	96	444
497	326	494
225	757	764
233	517	465
677	787	700
622	523	290

i	g^i	$h \cdot g^i$
519	291	28
286	239	193
298	358	642
500	789	101
272	24	111
307	748	621

i	g^i	$h \cdot g^i$
791	496	672
385	437	95
178	527	714
471	117	237
42	448	450
258	413	795

i	g^i	$h \cdot g^i$
406	801	562
745	194	289
234	304	595
556	252	760
326	649	670
399	263	304

Table 5.17: Data for Exercise 5.39, $g = 10$, $h = 106$, $p = 811$

Section 5.5. Pollard's ρ Method

5.40. Table 5.18 gives some of the computations for the solution of the discrete logarithm problem

$$11^t = 41387 \quad \text{in } \mathbb{F}_{81799} \tag{5.62}$$

using Pollard's ρ method. (It is similar to Table 5.11 in Example 5.52.) Use the data in Table 5.18 to solve (5.62).

i	x_i	y_i	α_i	β_i	γ_i	δ_i
0	1	1	0	0	0	0
1	11	121	1	0	2	0
2	121	14641	2	0	4	0
3	1331	42876	3	0	12	2
4	14641	7150	4	0	25	4
			⋮			
151	4862	33573	40876	45662	29798	73363
152	23112	53431	81754	9527	37394	48058
153	8835	23112	81755	9527	67780	28637
154	15386	15386	81756	9527	67782	28637

Table 5.18: Computations to solve $11^t = 41387$ in $\mathbb{F}_{81799}$ for Exercise 5.40

5.41. Table 5.19 gives some of the computations for the solution of the discrete logarithm problem

$$7^t = 3018 \quad \text{in } \mathbb{F}_{7963} \tag{5.63}$$

using Pollard's ρ method. (It is similar to Table 5.11 in Example 5.52.) Extend Table 5.19 until you find a collision (we promise that it won't take too long) and then solve (5.63).

5.42. Write a computer program implementing Pollard's ρ method for solving the discrete logarithm problem and use it to solve each of the following:

(a) $2^t = 2495$ in $\mathbb{F}_{5011}$.
(b) $17^t = 14226$ in $\mathbb{F}_{17959}$.
(c) $29^t = 5953042$ in $\mathbb{F}_{15239131}$.

i	x_i	y_i	α_i	β_i	γ_i	δ_i
0	1	1	0	0	0	0
1	7	49	1	0	2	0
2	49	2401	2	0	4	0
3	343	6167	3	0	6	0
4	2401	1399	4	0	7	1
			$\vdots$			
87	1329	1494	6736	7647	3148	3904
88	1340	1539	6737	7647	3150	3904
89	1417	4767	6738	7647	6302	7808
90	1956	1329	6739	7647	4642	7655

Table 5.19: Computations to solve $7^t = 3018$ in $\mathbb{F}_{7963}$ for Exercise 5.41

5.43. Evaluate the integral $I = \int_0^\infty t^2 e^{-t^2/2}\,dt$ appearing in the proof of Theorem 5.48. (*Hint.* Write I^2 as an iterated integral,

$$I^2 = \int_0^\infty \int_0^\infty x^2 e^{-x^2/2} \cdot y^2 e^{-y^2/2}\,dx\,dy,$$

and switch to polar coordinates.)

5.44. This exercise describes Pollard's ρ factorization algorithm. It is particularly good at factoring numbers N that have a prime factor p with the property that p is considerably smaller than N/p. Later we will study an even faster, albeit more complicated, factorization algorithm with this property that is based on the theory of elliptic curves; see Sect. 6.6.

Let N be an integer that is not prime, and let

$$f : \mathbb{Z}/N\mathbb{Z} \longrightarrow \mathbb{Z}/N\mathbb{Z}$$

be a mixing function, for example $f(x) = x^2 + 1 \bmod N$. As in the abstract version of Pollard's ρ method (Theorem 5.48), let $x_0 = y_0$ be an initial value, and generate sequences by setting $x_{i+1} = f(x_i)$ and $y_{i+1} = f(f(y_i))$. At each step, also compute the greatest common divisor

$$g_i = \gcd\bigl(|x_i - y_i|, N\bigr).$$

(a) Let p be the smallest prime divisor of N. If the function f is sufficiently random, show that with high probability we have

$$g_k = p \quad \text{for some } k = \mathcal{O}(\sqrt{p}).$$

Hence the algorithm factors N in $\mathcal{O}(\sqrt{p})$ steps.

(b) Program Pollard's ρ algorithm with $f(x) = x^2 + 1$ and $x_0 = y_0 = 0$, and use it to factor the following numbers. In each case, give the smallest value of k such that g_k is a nontrivial factor of N and print the ratio $k/\sqrt{N}$.

(i) $N = 2201$. (ii) $N = 9409613$. (iii) $N = 1782886219$.

(c) Repeat your computations in (b) using the function $f(x) = x^2 + 2$. Do the running times change?
(d) Explain what happens if you run Pollard's ρ algorithm and N is prime.
(e) Explain what happens if you run Pollard's ρ algorithm with $f(x) = x^2$ and any initial values for x_0.
(f) Try running Pollard's ρ algorithm with the function $f(x) = x^2 - 2$. Explain what is happening. (*Hint.* This part is more challenging. It may help to use the identity $f^n(u + u^{-1}) = u^{2^n} + u^{-2^n}$, which you can prove by induction.)

Section 5.6. Information Theory

5.45. Consider the cipher that has three keys, three plaintexts, and four ciphertexts that are combined using the following encryption table (which is similar to Table 5.12 used in Example 5.54 on page 265).

	m_1	m_2	m_3
k_1	c_2	c_4	c_1
k_2	c_1	c_3	c_2
k_3	c_3	c_1	c_2

Suppose further that the plaintexts and keys are used with the following probabilities:

$$f(m_1) = f(m_2) = \frac{2}{5}, \qquad f(m_3) = \frac{1}{5}, \qquad f(k_1) = f(k_2) = f(k_3) = \frac{1}{3}.$$

(a) Compute $f(c_1)$, $f(c_2)$, $f(c_3)$, and $f(c_4)$.
(b) Compute $f(c_1 \mid m_1)$, $f(c_1 \mid m_2)$, and $f(c_1 \mid m_3)$. Does this cryptosystem have perfect secrecy?
(c) Compute $f(c_2 \mid m_1)$ and $f(c_3 \mid m_1)$.
(d) Compute $f(k_1 \mid c_3)$ and $f(k_2 \mid c_3)$.

5.46. Suppose that a shift cipher is employed such that each key, i.e., each shift amount from 0 to 25, is used with equal probability and such that a new key is chosen to encrypt each successive letter. Show that this cryptosystem has perfect secrecy by filling in the details of the following steps.
(a) Show that $\sum_{k \in \mathcal{K}} f_M(d_k(c)) = 1$ for every ciphertext $c \in \mathcal{C}$.
(b) Compute the ciphertext density function f_C using (5.47), which in this case says that

$$f_C(c) = \sum_{k \in \mathcal{K}} f_K(k) f_M(d_k(c)).$$

(c) Compare $f_C(c)$ to $f_{C|M}(c \mid m)$.

5.47. Give the details of the proof of (5.47), which says that

$$f_C(c) = \sum_{\substack{k \in \mathcal{K} \text{ such} \\ \text{that } c \in e_k(\mathcal{M})}} f_K(k) f_M(d_k(c)).$$

(*Hint.* Use the decomposition formula from Exercise 5.23(d)).

5.48. Suppose that a cryptosystem has the same number of plaintexts as it does ciphertexts ($\#\mathcal{M} = \#\mathcal{C}$). Prove that for any given key $k \in \mathcal{K}$ and any given ciphertext $c \in \mathcal{C}$, there is a unique plaintext $m \in \mathcal{M}$ that encrypts to c using the key k. (We used this fact during the proof of Theorem 5.56. Notice that the proof does not require the cryptosystem to have perfect secrecy; all that is needed is that $\#\mathcal{M} = \#\mathcal{C}$.)

5.49. Let $\mathcal{S}_{m,c} = \{k \in \mathcal{K} : e_k(m) = c\}$ be the set used during the proof of Theorem 5.56. Prove that if $c \neq c'$, then $\mathcal{S}_{m,c} \cap \mathcal{S}_{m,c'} = \emptyset$. (Prove this for any cryptosystem; it is not necessary to assume perfect secrecy.)

5.50. Suppose that a cryptosystem satisfies $\#\mathcal{K} = \#\mathcal{M} = \#\mathcal{C}$ and that it has perfect secrecy. Prove that every ciphertext is used with equal probability and that every plaintext is used with equal probability. (*Hint.* We proved one of these during the course of proving Theorem 5.56. The proof of the other is similar.)

5.51. Prove the "only if" part of Theorem 5.56, i.e., prove that if a cryptosystem with an equal number of keys, plaintexts, and ciphertexts satisfies conditions (a) and (b) of Theorem 5.56, then it has perfect secrecy.

5.52. Let X_n be a uniformly distributed random variable on n objects, and let $r \geq 1$. Prove directly from Property $\mathbf{H}_3$ of entropy that

$$H(X_{n^r}) = rH(X_n).$$

This generalizes Example 5.58.

5.53. Let X, Y, and $Z_1, \ldots, Z_m$ be random variables as described in Property $\mathbf{H}_3$ on page 270. Let

$$p_i = \Pr(Y = Z_i) \quad \text{and} \quad q_{ij} = \Pr(Z_i = x_{ij}), \quad \text{so} \quad \Pr(X = x_{ij}) = p_i q_{ij}.$$

With this notation, Property $\mathbf{H}_3$ says that

$$H\Bigl((p_i q_{ij})_{\substack{1\le i\le n\\ 1\le j\le m_i}}\Bigr) = H\bigl((p_i)_{1\le i\le n}\bigr) + \sum_{i=1}^{n} p_i H\bigl((q_{ij})_{1\le j\le m_i}\bigr).$$

(See Example 5.59.) Then the formula (5.51) for entropy given in Theorem 5.60 implies that

$$\sum_{i=1}^{n}\sum_{j=1}^{m_i} p_i q_{ij}\log_2(p_i q_{ij}) = \sum_{i=1}^{n} p_i \log_2(p_i) + \sum_{i=1}^{n} p_i \sum_{j=1}^{m_i} q_{ij}\log_2(q_{ij}). \tag{5.64}$$

Prove directly that (5.64) is true. (*Hint.* Remember that the probabilities satisfy $\sum_i p_i = 1$ and $\sum_j q_{ij} = 1$.)

5.54. Let $F(x)$ be a twice differentiable function with the property that $F''(x) < 0$ for all x in its domain. Prove that F is concave in the sense of (5.52). Conclude in particular that the function $F(x) = \log x$ is concave for all $x > 0$.

5.55. Use induction to prove Jensen's inequality (Theorem 5.61).

5.56. Let X and Y be independent random variables.

(a) Prove that the equivocation $H(X \mid Y)$ is equal to the entropy $H(X)$.
(b) If $H(X \mid Y) = H(X)$, is it necessarily true that X and Y are independent?

5.57. Prove that key equivocation satisfies the formula

$$H(K \mid C) = H(K) + H(M) - H(C)$$

as described in Proposition 5.64.

5.58. We continue with the cipher described in Exercise 5.45.
(a) Compute the entropies $H(K)$, $H(M)$, and $H(C)$.
(b) Compute the key equivocation $H(K \mid C)$.

5.59. Suppose that the key equivocation of a certain cryptosystem vanishes, i.e., suppose that $H(K \mid C) = 0$. Prove that even a single observed ciphertext uniquely determines which key was used.

5.60. Write a computer program that reads a text file and performs the following tasks:

[1] Convert all alphabetic characters to lowercase and convert all strings of consecutive nonalphabetic characters to a single space. (The reason for leaving in a space is that when you count bigrams and trigrams, you will want to know where words begin and end.)

[2] Count the frequency of each letter `a`-to-`z`, print a frequency table, and use your frequency table to estimate the entropy of a single letter in English, as we did in Sect. 5.6.3 using Table 1.3.

[3] Count the frequency of each bigram `aa`, `ab`,...,`zz`, being careful to include only bigrams that appear within words. (As an alternative, also allow bigrams that either start or end with a space, in which case there are $27^2 - 1 = 728$ possible bigrams.) Print a frequency table of the 25 most common bigrams and their probabilities, and use your full frequency table to estimate the entropy of bigrams in English. In the notation of Sect. 5.6.3, this is the quantity $H(L^2)$. Compare $\frac{1}{2}H(L^2)$ with the value of $H(L)$ from step [1].

[4] Repeat [3], but this time with trigrams. Compare $\frac{1}{3}H(L^3)$ with the values of $H(L)$ and $\frac{1}{2}H(L^2)$ from [2] and [3]. (Note that for this part, you will need a large quantity of text in order to get some reasonable frequencies.)

Try running your program on some long blocks of text. For example, the following noncopyrighted material is available in the form of ordinary text files from Project Gutenberg at `http://www.gutenberg.org/`. To what extent are the letter frequencies similar and to what extent do they differ in these different texts?

(a) *Alice's Adventures in Wonderland* by Lewis Carroll,
`http://www.gutenberg.org/etext/11`
(b) *Relativity: the Special and General Theory* by Albert Einstein,
`http://www.gutenberg.org/etext/5001`
(c) The Old Testament (translated from the original Hebrew, of course!),
`http://www.gutenberg.org/etext/1609`
(d) *20000 Lieues Sous Les Mers* (20000 Leagues Under the Sea) by Jules Verne,
`http://www.gutenberg.org/etext/5097`. Note that this one is a little trickier, since first you will need to convert all of the letters to their unaccented forms.

Chapter 6

Elliptic Curves and Cryptography

The subject of elliptic curves encompasses a vast amount of mathematics.[1] Our aim in this section is to summarize just enough of the basic theory for cryptographic applications. For additional reading, there are a number of survey articles and books devoted to elliptic curve cryptography [14, 68, 81, 135], and many others that describe the number theoretic aspects of the theory of elliptic curves, including [25, 65, 73, 74, 136, 134, 138].

6.1 Elliptic Curves

An *elliptic curve*[2] is the set of solutions to an equation of the form

$$Y^2 = X^3 + AX + B.$$

Equations of this type are called *Weierstrass equations* after the mathematician who studied them extensively during the nineteenth century. Two examples of elliptic curves,

$$E_1 : Y^2 = X^3 - 3X + 3 \qquad \text{and} \qquad E_2 : Y^2 = X^3 - 6X + 5,$$

are illustrated in Fig. 6.1.

[1]Indeed, even before elliptic curves burst into cryptographic prominence, a well-known mathematician [73] opined that "it is possible to write endlessly on elliptic curves!"

[2]A word of warning. You may recall from high school geometry that an ellipse is a geometric object that looks like a squashed circle. Elliptic curves are *not* ellipses, and indeed, despite their somewhat unfortunate name, elliptic curves and ellipses have only the most tenuous connection with one another.

J. Hoffstein et al., *An Introduction to Mathematical Cryptography*,
Undergraduate Texts in Mathematics, DOI 10.1007/978-1-4939-1711-2_6

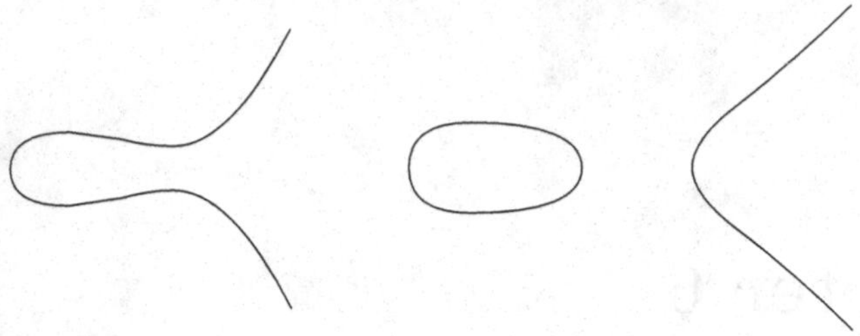

$E_1 : Y^2 = X^3 - 3X + 3$ $E_2 : Y^2 = X^3 - 6X + 5$

Figure 6.1: Two examples of elliptic curves

An amazing feature of elliptic curves is that there is a natural way to take two points on an elliptic curve and "add" them to produce a third point. We put quotation marks around "add" because we are referring to an operation that combines two points in a manner analogous to addition in some respects (it is commutative and associative, and there is an identity), but very unlike addition in other ways. The most natural way to describe the "addition law" on elliptic curves is to use geometry.

Let P and Q be two points on an elliptic curve E, as illustrated in Fig. 6.2. We start by drawing the line L through P and Q. This line L intersects E at three points, namely P, Q, and one other point R. We take that point R and reflect it across the x-axis (i.e., we multiply its Y-coordinate by -1) to get a new point R'. The point R' is called the "sum of P and Q," although as you can see, this process is nothing like ordinary addition. For now, we denote this strange addition law by the symbol $\oplus$. Thus we write[3]

$$P \oplus Q = R'.$$

Example 6.1. Let E be the elliptic curve

$$Y^2 = X^3 - 15X + 18. \tag{6.1}$$

The points $P = (7, 16)$ and $Q = (1, 2)$ are on the curve E. The line L connecting them is given by the equation[4]

$$L : Y = \frac{7}{3}X - \frac{1}{3}. \tag{6.2}$$

In order to find the points where E and L intersect, we substitute (6.2) into (6.1) and solve for X. Thus

[3]Not to be confused with the identical symbol $\oplus$ that we used to denote the XOR operation in a different context!

[4]Recall that the equation of the line through two points (x_1, y_1) and (x_2, y_2) is given by the point–slope formula $Y - y_1 = \lambda \cdot (X - x_1)$, where the slope λ is equal to $\frac{y_2 - y_1}{x_2 - x_1}$.

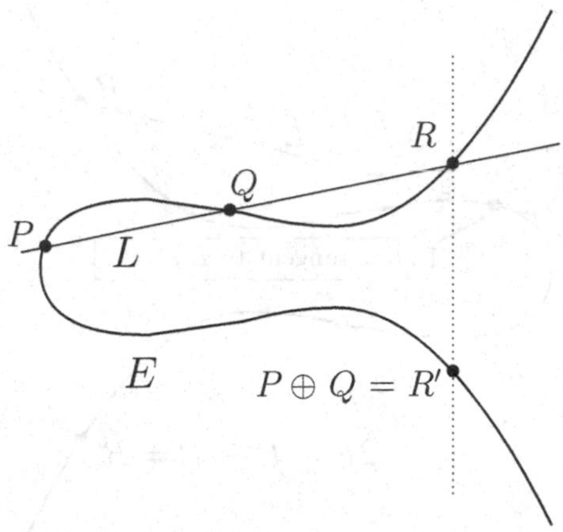

Figure 6.2: The addition law on an elliptic curve

$$\begin{aligned}\left(\frac{7}{3}X - \frac{1}{3}\right)^2 &= X^3 - 15X + 18,\\ \frac{49}{9}X^2 - \frac{14}{9}X + \frac{1}{9} &= X^3 - 15X + 18,\\ 0 &= X^3 - \frac{49}{9}X^2 - \frac{121}{9}X + \frac{161}{9}.\end{aligned}$$

We need to find the roots of this cubic polynomial. In general, finding the roots of a cubic is difficult. However, in this case we already know two of the roots, namely $X = 7$ and $X = 1$, since we know that P and Q are in the intersection $E \cap L$. It is then easy to find the other factor,

$$X^3 - \frac{49}{9}X^2 - \frac{121}{9}X + \frac{161}{9} = (X-7)\cdot(X-1)\cdot\left(X + \frac{23}{9}\right),$$

so the third point of intersection of L and E has X-coordinate equal to $-\frac{23}{9}$. Next we find the Y-coordinate by substituting $X = -\frac{23}{9}$ into Eq. (6.2). This gives $R = \left(-\frac{23}{9}, -\frac{170}{27}\right)$. Finally, we reflect across the X-axis to obtain

$$P \oplus Q = \left(-\frac{23}{9}, \frac{170}{27}\right).$$

There are a few subtleties to elliptic curve addition that need to be addressed. First, what happens if we want to add a point P to itself? Imagine what happens to the line L connecting P and Q if the point Q slides along the curve and gets closer and closer to P. In the limit, as Q approaches P, the line L becomes the tangent line to E at P. Thus in order to add P to itself, we simply take L to be the tangent line to E at P, as illustrated in Fig. 6.3. Then L intersects E at P and at one other point R, so we can proceed as

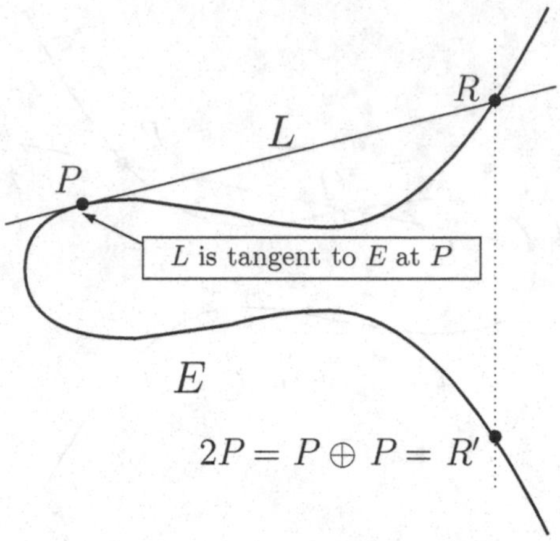

Figure 6.3: Adding a point P to itself

before. In some sense, L still intersects E at three points, but P counts as two of them.

Example 6.2. Continuing with the curve E and point P from Example 6.1, we compute $P \oplus P$. The slope of E at P is computed by implicitly differentiating equation (6.1). Thus

$$2Y\frac{dY}{dX} = 3X^2 - 15, \qquad \text{so} \qquad \frac{dY}{dX} = \frac{3X^2 - 15}{2Y}.$$

Substituting the coordinates of $P = (7, 16)$ gives slope $\lambda = \frac{33}{8}$, so the tangent line to E at P is given by the equation

$$L : Y = \frac{33}{8}X - \frac{103}{8}. \tag{6.3}$$

Now we substitute (6.3) into Eq. (6.1) for E, simplify, and factor:

$$\left(\frac{33}{8}X - \frac{103}{8}\right)^2 = X^3 - 15X + 18,$$
$$X^3 - \frac{1089}{64}X^2 + \frac{2919}{32}X - \frac{9457}{64} = 0,$$
$$(X-7)^2 \cdot \left(X - \frac{193}{64}\right) = 0.$$

Notice that the X-coordinate of P, which is $X = 7$, appears as a double root of the cubic polynomial, so it was easy for us to factor the cubic. Finally, we substitute $X = \frac{193}{64}$ into Eq. (6.3) for L to get $Y = -\frac{223}{512}$, and then we switch the sign on Y to get

$$P \oplus P = \left(\frac{193}{64}, \frac{223}{512}\right).$$

A second potential problem with our "addition law" arises if we try to add a point $P = (a, b)$ to its reflection about the X-axis $P' = (a, -b)$. The line L through P and P' is the vertical line $x = a$, and this line intersects E in only the two points P and P'. (See Fig. 6.4.) There is no third point of intersection, so it appears that we are stuck! But there is a way out. The solution is to create an extra point $\mathcal{O}$ that lives "at infinity." More precisely, the point $\mathcal{O}$ does not exist in the XY-plane, but we pretend that it lies on every vertical line. We then set

$$P \oplus P' = \mathcal{O}.$$

We also need to figure out how to add $\mathcal{O}$ to an ordinary point $P = (a, b)$ on E. The line L connecting P to $\mathcal{O}$ is the vertical line through P, since $\mathcal{O}$ lies on vertical lines, and that vertical line intersects E at the points P, $\mathcal{O}$, and $P' = (a, -b)$. To add P to $\mathcal{O}$, we reflect P' across the X-axis, which gets us back to P. In other words, $P \oplus \mathcal{O} = P$, so $\mathcal{O}$ acts like zero for elliptic curve addition.

Example 6.3. Continuing with the curve E from Example 6.1, notice that the point $T = (3, 0)$ is on the curve E and that the tangent line to E at T is the vertical line $X = 3$. Thus if we add T to itself, we get $T \oplus T = \mathcal{O}$.

Definition. An *elliptic curve* E is the set of solutions to a Weierstrass equation

$$E : Y^2 = X^3 + AX + B,$$

together with an extra point $\mathcal{O}$, where the constants A and B must satisfy

$$4A^3 + 27B^2 \neq 0.$$

The *addition law on* E is defined as follows. Let P and Q be two points on E. Let L be the line connecting P and Q, or the tangent line to E at P if $P = Q$. Then the intersection of E and L consists of three points P, Q, and R, counted with appropriate multiplicities and with the understanding that $\mathcal{O}$ lies on every vertical line. Writing $R = (a, b)$, the sum of P and Q is defined to be the reflection $R' = (a, -b)$ of R across the X-axis. This sum is denoted by $P \oplus Q$, or simply by $P + Q$.

Further, if $P = (a, b)$, we denote the reflected point by $\ominus P = (a, -b)$, or simply by $-P$; and we define $P \ominus Q$ (or $P - Q$) to be $P \oplus (\ominus Q)$. Similarly, repeated addition is represented as multiplication of a point by an integer,

$$nP = \underbrace{P + P + P + \cdots + P}_{n \text{ copies}}.$$

Remark 6.4. What is this extra condition $4A^3 + 27B^2 \neq 0$? The quantity $\Delta_E = 4A^3 + 27B^2$ is called the *discriminant of* E. The condition $\Delta_E \neq 0$ is equivalent to the condition that the cubic polynomial $X^3 + AX + B$ have no repeated roots, i.e., if we factor $X^3 + AX + B$ completely as

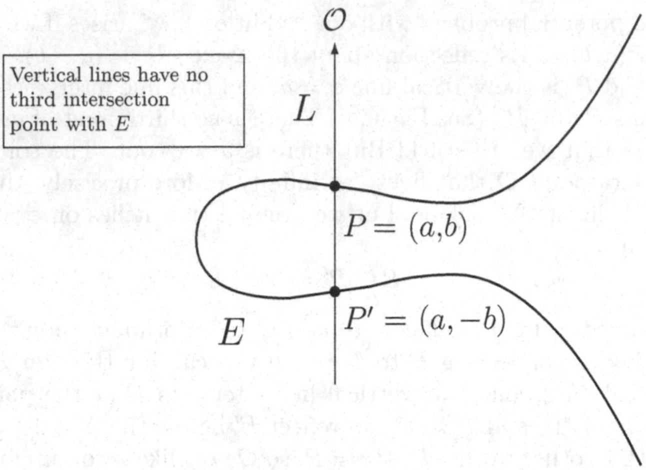

Figure 6.4: The vertical line L through $P = (a, b)$ and $P' = (a, -b)$

$$X^3 + AX + B = (X - e_1)(X - e_2)(X - e_3),$$

where e_1, e_2, e_3 are allowed to be complex numbers, then

$$4A^3 + 27B^2 \neq 0 \qquad \text{if and only if} \qquad e_1, e_2, e_3 \text{ are distinct.}$$

(See Exercise 6.3.) Curves with $\Delta_E = 0$ have singular points (see Exercise 6.4). The addition law does not work well on these curves. That is why we include the requirement that $\Delta_E \neq 0$ in our definition of an elliptic curve.

Theorem 6.5. *Let E be an elliptic curve. Then the addition law on E has the following properties:*

(a) $P + \mathcal{O} = \mathcal{O} + P = P$ *for all* $P \in E$. [Identity]

(b) $P + (-P) = \mathcal{O}$ *for all* $P \in E$. [Inverse]

(c) $(P + Q) + R = P + (Q + R)$ *for all* $P, Q, R \in E$. [Associative]

(d) $P + Q = Q + P$ *for all* $P, Q \in E$. [Commutative]

In other words, the addition law makes the points of E into an abelian group. (See Sect. 2.5 for a general discussion of groups and their axioms.)

Proof. As we explained earlier, the identity law (a) and inverse law (b) are true because $\mathcal{O}$ lies on all vertical lines. The commutative law (d) is easy to verify, since the line that goes through P and Q is the same as the line that goes through Q and P, so the order of the points does not matter.

The remaining piece of Theorem 6.5 is the associative law (c). One might not think that this would be hard to prove, but if you draw a picture and start to put in all of the lines needed to verify (c), you will see that it is quite complicated. There are many ways to prove the associative law, but none of the proofs are easy. After we develop explicit formulas for the addition law on E (Theorem 6.6), you can use those formulas to check the associative law by a direct (but painful) calculation. More perspicacious, but less elementary, proofs may be found in [74, 136, 138] and other books on elliptic curves. □

Our next task is to find explicit formulas to enable us to easily add and subtract points on an elliptic curve. The derivation of these formulas uses elementary analytic geometry, a little bit of differential calculus to find a tangent line, and a certain amount of algebraic manipulation. We state the results in the form of an algorithm, and then briefly indicate the proof.

Theorem 6.6 (Elliptic Curve Addition Algorithm). *Let*

$$E : Y^2 = X^3 + AX + B$$

be an elliptic curve and let P_1 and P_2 be points on E.
(a) *If $P_1 = \mathcal{O}$, then $P_1 + P_2 = P_2$.*
(b) *Otherwise, if $P_2 = \mathcal{O}$, then $P_1 + P_2 = P_1$.*
(c) *Otherwise, write $P_1 = (x_1, y_1)$ and $P_2 = (x_2, y_2)$.*
(d) *If $x_1 = x_2$ and $y_1 = -y_2$, then $P_1 + P_2 = \mathcal{O}$.*
(e) *Otherwise, define λ by*

$$\lambda = \begin{cases} \dfrac{y_2 - y_1}{x_2 - x_1} & \text{if } P_1 \neq P_2, \\[2ex] \dfrac{3x_1^2 + A}{2y_1} & \text{if } P_1 = P_2, \end{cases}$$

and let

$$x_3 = \lambda^2 - x_1 - x_2 \qquad \text{and} \qquad y_3 = \lambda(x_1 - x_3) - y_1.$$

Then $P_1 + P_2 = (x_3, y_3)$.

Proof. Parts (a) and (b) are clear, and (d) is the case that the line through P_1 and P_2 is vertical, so $P_1 + P_2 = \mathcal{O}$. (Note that if $y_1 = y_2 = 0$, then the tangent line is vertical, so that case works, too.) For (e), we note that if $P_1 \neq P_2$, then λ is the slope of the line through P_1 and P_2, and if $P_1 = P_2$, then λ is the slope of the tangent line at $P_1 = P_2$. In either case the line L is given by the equation $Y = \lambda X + \nu$ with $\nu = y_1 - \lambda x_1$. Substituting the equation for L into the equation for E gives

$$(\lambda X + \nu)^2 = X^3 + AX + B,$$

so

$$X^3 - \lambda^2 X^2 + (A - 2\lambda\nu)X + (B - \nu^2) = 0.$$

We know that this cubic has x_1 and x_2 as two of its roots. If we call the third root x_3, then it factors as

$$X^3 - \lambda^2 X^2 + (A - 2\lambda\nu)X + (B - \nu^2) = (X - x_1)(X - x_2)(X - x_3).$$

Now multiply out the right-hand side and look at the coefficient of X^2 on each side. The coefficient of X^2 on the right-hand side is $-x_1 - x_2 - x_3$, which must equal $-\lambda^2$, the coefficient of X^2 on the left-hand side. This allows us to solve for $x_3 = \lambda^2 - x_1 - x_2$, and then the Y-coordinate of the third intersection point of E and L is given by $\lambda x_3 + \nu$. Finally, in order to get $P_1 + P_2$, we must reflect across the X-axis, which means replacing the Y-coordinate with its negative. □

6.2 Elliptic Curves over Finite Fields

In the previous section we developed the theory of elliptic curves geometrically. For example, the sum of two distinct points P and Q on an elliptic curve E is defined by drawing the line L connecting P to Q and then finding the third point where L and E intersect, as illustrated in Fig. 6.2. However, in order to apply the theory of elliptic curves to cryptography, we need to look at elliptic curves whose points have coordinates in a finite field $\mathbb{F}_p$. This is easy to do.

Definition. Let $p \geq 3$ be a prime. An *elliptic curve over* $\mathbb{F}_p$ is an equation of the form

$$E : Y^2 = X^3 + AX + B \qquad \text{with } A, B \in \mathbb{F}_p \text{ satisfying } 4A^3 + 27B^2 \neq 0.$$

The *set of points on E with coordinates in* $\mathbb{F}_p$ is the set

$$E(\mathbb{F}_p) = \big\{(x, y) : x, y \in \mathbb{F}_p \text{ satisfy } y^2 = x^3 + Ax + B\big\} \cup \{\mathcal{O}\}.$$

Remark 6.7. Elliptic curves over $\mathbb{F}_2$ are actually quite important in cryptography, but they require more complicated equations, so we delay our discussion of them until Sect. 6.7.

Example 6.8. Consider the elliptic curve

$$E : Y^2 = X^3 + 3X + 8 \quad \text{over the field } \mathbb{F}_{13}.$$

We can find the points of $E(\mathbb{F}_{13})$ by substituting in all possible values $X = 0, 1, 2, \ldots, 12$ and checking for which X values the quantity $X^3 + 3X + 8$ is a square modulo 13. For example, putting $X = 0$ gives 8, and 8 is not a square modulo 13. Next we try $X = 1$, which gives $1+3+8 = 12$. It turns out that 12 is a square modulo 13; in fact, it has two square roots,

$$5^2 \equiv 12 \pmod{13} \qquad \text{and} \qquad 8^2 \equiv 12 \pmod{13}.$$

This gives two points $(1, 5)$ and $(1, 8)$ in $E(\mathbb{F}_{13})$. Continuing in this fashion, we end up with a complete list,

$$E(\mathbb{F}_{13}) = \{\mathcal{O}, (1,5), (1,8), (2,3), (2,10), (9,6), (9,7), (12,2), (12,11)\}.$$

Thus $E(\mathbb{F}_{13})$ consists of nine points.

Suppose now that P and Q are two points in $E(\mathbb{F}_p)$ and that we want to "add" the points P and Q. One possibility is to develop a theory of geometry using the field $\mathbb{F}_p$ instead of $\mathbb{R}$. Then we could mimic our earlier constructions to define $P + Q$. This can be done, and it leads to a fascinating field of mathematics called algebraic geometry. However, in the interests of brevity of exposition, we instead use the explicit formulas given in Theorem 6.6 to add points in $E(\mathbb{F}_p)$. But we note that if one wants to gain a deeper understanding of the theory of elliptic curves, then it is necessary to use some of the machinery and some of the formalism of algebraic geometry.

Let $P = (x_1, y_1)$ and $Q = (x_2, y_2)$ be points in $E(\mathbb{F}_p)$. We define the sum $P + Q$ to be the point (x_3, y_3) obtained by applying the elliptic curve addition algorithm (Theorem 6.6). Notice that in this algorithm, the only operations used are addition, subtraction, multiplication, and division involving the coefficients of E and the coordinates of P and Q. Since those coefficients and coordinates are in the field $\mathbb{F}_p$, we end up with a point (x_3, y_3) whose coordinates are in $\mathbb{F}_p$. Of course, it is not completely clear that (x_3, y_3) is a point in $E(\mathbb{F}_p)$.

Theorem 6.9. *Let E be an elliptic curve over $\mathbb{F}_p$ and let P and Q be points in $E(\mathbb{F}_p)$.*

(a) *The elliptic curve addition algorithm (Theorem 6.6) applied to P and Q yields a point in $E(\mathbb{F}_p)$. We denote this point by $P + Q$.*

(b) *This addition law on $E(\mathbb{F}_p)$ satisfies all of the properties listed in Theorem 6.5. In other words, this addition law makes $E(\mathbb{F}_p)$ into a finite group.*

Proof. The formulas in Theorem 6.6(e) are derived by substituting the equation of a line into the equation for E and solving for X, so the resulting point is automatically a point on E, i.e., it is a solution to the equation defining E. This shows why (a) is true, although when $P = Q$, a small additional argument is needed to indicate why the resulting cubic polynomial has a double root. For (b), the identity law follows from the addition algorithm steps (a) and (b), the inverse law is clear from the addition algorithm Step (d), and the commutative law is easy, since a brief examination of the addition algorithm shows that switching the two points leads to the same result. Unfortunately, the associative law is not so clear. It is possible to verify the associative law directly using the addition algorithm formulas, although there are many special cases to consider. The alternative is to develop more of the general theory of elliptic curves, as is done in the references cited in the proof of Theorem 6.5. □

Example 6.10. We continue with the elliptic curve

$$E : Y^2 = X^3 + 3X + 8 \quad \text{over } \mathbb{F}_{13}$$

from Example 6.8, and we use the addition algorithm (Theorem 6.6) to add the points $P = (9,7)$ and $Q = (1,8)$ in $E(\mathbb{F}_{13})$. Step (e) of that algorithm tells us to first compute

$$\lambda = \frac{y_2 - y_1}{x_2 - x_1} = \frac{8-7}{1-9} = \frac{1}{-8} = \frac{1}{5} = 8,$$

where recall that all computations[5] are being performed in the field $\mathbb{F}_{13}$, so $-8 = 5$ and $\frac{1}{5} = 5^{-1} = 8$. Next we compute

$$\nu = y_1 - \lambda x_1 = 7 - 8 \cdot 9 = -65 = 0.$$

Finally, the addition algorithm tells us to compute

$$x_3 = \lambda^2 - x_1 - x_2 = 64 - 9 - 1 = 54 = 2,$$
$$y_3 = -(\lambda x_3 + \nu) = -8 \cdot 2 = -16 = 10.$$

This completes the computation of

$$P + Q = (1,8) + (9,7) = (2,10) \qquad \text{in } E(\mathbb{F}_{13}).$$

Similarly, we can use the addition algorithm to add $P = (9,7)$ to itself. Keeping in mind that all calculations are in $\mathbb{F}_{13}$, we find that

$$\lambda = \frac{3x_1^2 + A}{2y_1} = \frac{3 \cdot 9^2 + 3}{2 \cdot 7} = \frac{246}{14} = 12 \quad \text{and} \quad \nu = y_1 - \lambda x_1 = 7 - 12 \cdot 9 = 3.$$

Then

$$x_3 = \lambda^2 - x_1 - x_2 = (12)^2 - 9 - 9 = 9 \quad \text{and} \quad y_3 = -(\lambda x_3 + \nu) = -(12 \cdot 9 + 3) = 6,$$

so $P + P = (9,7) + (9,7) = (9,6)$ in $E(\mathbb{F}_{13})$. In a similar fashion, we can compute the sum of every pair of points in $E(\mathbb{F}_{13})$. The results are listed in Table 6.1.

It is clear that the set of points $E(\mathbb{F}_p)$ is a finite set, since there are only finitely many possibilities for the X- and Y-coordinates. More precisely, there are p possibilities for X, and then for each X, the equation

$$Y^2 = X^3 + AX + B$$

shows that there are at most two possibilities for Y. (See Exercise 1.36.) Adding in the extra point $\mathcal{O}$, this shows that $\#E(\mathbb{F}_p)$ has at most $2p+1$ points. However, this estimate is considerably larger than the true size.

[5]This is a good time to learn that $\frac{1}{5}$ is a *symbol* for a solution to the equation $5x = 1$. In order to assign a value to the symbol $\frac{1}{5}$, you must know where that value lives. In $\mathbb{Q}$, the value of $\frac{1}{5}$ is the usual number with which you are familiar, but in $\mathbb{F}_{13}$ the value of $\frac{1}{5}$ is 8, while in $\mathbb{F}_{11}$ the value of $\frac{1}{5}$ is 9. And in $\mathbb{F}_5$ the symbol $\frac{1}{5}$ is not assigned a value.

	$\mathcal{O}$	(1,5)	(1,8)	(2,3)	(2,10)	(9,6)	(9,7)	(12,2)	(12,11)
$\mathcal{O}$	$\mathcal{O}$	(1,5)	(1,8)	(2,3)	(2,10)	(9,6)	(9,7)	(12,2)	(12,11)
(1,5)	(1,5)	(2,10)	$\mathcal{O}$	(1,8)	(9,7)	(2,3)	(12,2)	(12,11)	(9,6)
(1,8)	(1,8)	$\mathcal{O}$	(2,3)	(9,6)	(1,5)	(12,11)	(2,10)	(9,7)	(12,2)
(2,3)	(2,3)	(1,8)	(9,6)	(12,11)	$\mathcal{O}$	(12,2)	(1,5)	(2,10)	(9,7)
(2,10)	(2,10)	(9,7)	(1,5)	$\mathcal{O}$	(12,2)	(1,8)	(12,11)	(9,6)	(2,3)
(9,6)	(9,6)	(2,3)	(12,11)	(12,2)	(1,8)	(9,7)	$\mathcal{O}$	(1,5)	(2,10)
(9,7)	(9,7)	(12,2)	(2,10)	(1,5)	(12,11)	$\mathcal{O}$	(9,6)	(2,3)	(1,8)
(12,2)	(12,2)	(12,11)	(9,7)	(2,10)	(9,6)	(1,5)	(2,3)	(1,8)	$\mathcal{O}$
(12,11)	(12,11)	(9,6)	(12,2)	(9,7)	(2,3)	(2,10)	(1,8)	$\mathcal{O}$	(1,5)

Table 6.1: Addition table for $E : Y^2 = X^3 + 3X + 8$ over $\mathbb{F}_{13}$

When we plug in a value for X, there are three possibilities for the value of the quantity

$$X^3 + AX + B.$$

First, it may be a quadratic residue modulo p, in which case it has two square roots and we get two points in $E(\mathbb{F}_p)$. This happens about 50 % of the time. Second, it may be a nonresidue modulo p, in which case we discard X. This also happens about 50 % of the time. Third, it might equal 0, in which case we get one point in $E(\mathbb{F}_p)$, but this case happens very rarely.[6] Thus we might expect that the number of points in $E(\mathbb{F}_p)$ is approximately

$$\#E(\mathbb{F}_p) \approx 50\,\% \cdot 2 \cdot p + 1 = p + 1.$$

A famous theorem of Hasse, later vastly generalized by Weil and Deligne, says that this is true up to random fluctuations.

Theorem 6.11 (Hasse). *Let E be an elliptic curve over $\mathbb{F}_p$. Then*

$$\#E(\mathbb{F}_p) = p + 1 - t_p \qquad \textit{with } t_p \textit{ satisfying } \quad |t_p| \le 2\sqrt{p}.$$

Definition. The quantity

$$t_p = p + 1 - \#E(\mathbb{F}_p)$$

appearing in Theorem 6.11 is called the *trace of Frobenius* for $E/\mathbb{F}_p$. We will not explain the somewhat technical reasons for this name, other than to say that t_p appears as the trace of a certain 2-by-2 matrix that acts as a linear transformation on a certain two-dimensional vector space associated to $E/\mathbb{F}_p$.

Example 6.12. Let E be given by the equation

$$E : Y^2 = X^3 + 4X + 6.$$

We can think of E as an elliptic curve over $\mathbb{F}_p$ for different finite fields $\mathbb{F}_p$ and count the number of points in $E(\mathbb{F}_p)$. Table 6.2 lists the results for the first few primes, together with the value of t_p and, for comparison purposes, the value of $2\sqrt{p}$.

[6]The congruence $X^3 + AX + B \equiv 0 \pmod{p}$ has at most three solutions, and if p is large, the chance of randomly choosing one of them is very small.

p	$\#E(\mathbb{F}_p)$	t_p	$2\sqrt{p}$
3	4	0	3.46
5	8	−2	4.47
7	11	−3	5.29
11	16	−4	6.63
13	14	0	7.21
17	15	3	8.25

Table 6.2: Number of points and trace of Frobenius for $E : Y^2 = X^3 + 4X + 6$

Remark 6.13. Hasse's theorem (Theorem 6.11) gives a bound for $\#E(\mathbb{F}_p)$, but it does not provide a method for calculating this quantity. In principle, one can substitute in each value for X and check the value of $X^3 + AX + B$ against a table of squares modulo p, but this takes time $\mathcal{O}(p)$, so is very inefficient. Schoof [120] found an algorithm to compute $\#E(\mathbb{F}_p)$ in time $O\big((\log p)^6\big)$, i.e., he found a polynomial-time algorithm. Schoof's algorithm was improved and made practical by Elkies and Atkin, so it is now known as the *SEA algorithm*. We will not describe SEA, which uses advanced techniques from the theory of elliptic curves, but see [121]. Also see Remark 6.32 in Sect. 6.7 for another counting algorithm due to Satoh that is designed for a different type of finite field.

6.3 The Elliptic Curve Discrete Logarithm Problem (ECDLP)

In Chap. 2 we talked about the discrete logarithm problem (DLP) in the finite field $\mathbb{F}_p^*$. In order to create a cryptosystem based on the DLP for $\mathbb{F}_p^*$, Alice publishes two numbers g and h, and her secret is the exponent x that solves the congruence

$$h \equiv g^x \pmod{p}.$$

Let's consider how Alice can do something similar with an elliptic curve E over $\mathbb{F}_p$. If Alice views g and h as being elements of the group $\mathbb{F}_p^*$, then the discrete logarithm problem requires Alice's adversary Eve to find an x such that

$$h \equiv \underbrace{g \cdot g \cdot g \cdots g}_{x \text{ multiplications}} \pmod{p}.$$

In other words, Eve needs to determine how many times g must be multiplied by itself in order to get to h.

With this formulation, it is clear that Alice can do the same thing with the group of points $E(\mathbb{F}_p)$ of an elliptic curve E over a finite field $\mathbb{F}_p$. She chooses and publishes two points P and Q in $E(\mathbb{F}_p)$, and her secret is an integer n that makes

$$Q = \underbrace{P + P + P + \cdots + P}_{n \text{ additions on } E} = nP.$$

Then Eve needs to find out how many times P must be added to itself in order to get Q. Keep in mind that although the "addition law" on an elliptic curve is conventionally written with a plus sign, addition on E is actually a very complicated operation, so this elliptic analogue of the discrete logarithm problem may be quite difficult to solve.

Definition. Let E be an elliptic curve over the finite field $\mathbb{F}_p$ and let P and Q be points in $E(\mathbb{F}_p)$. The *Elliptic Curve Discrete Logarithm Problem* (ECDLP) is the problem of finding an integer n such that $Q = nP$. By analogy with the discrete logarithm problem for $\mathbb{F}_p^*$, we denote this integer n by

$$n = \log_P(Q)$$

and we call n the *elliptic discrete logarithm of Q with respect to P.*

Remark 6.14. Our definition of $\log_P(Q)$ is not quite precise. The first difficulty is that there may be points $P, Q \in E(\mathbb{F}_p)$ such that Q is not a multiple of P. In this case, $\log_P(Q)$ is not defined. However, for cryptographic purposes, Alice starts out with a public point P and a private integer n and she computes and publishes the value of $Q = nP$. So in practical applications, $\log_P(Q)$ exists and its value is Alice's secret.

The second difficulty is that if there is one value of n satisfying $Q = nP$, then there are many such values. To see this, we first note that there exists a positive integer s such that $sP = \mathcal{O}$. We recall the easy proof of this fact (cf. Proposition 2.12). Since $E(\mathbb{F}_p)$ is finite, the points in the list $P, 2P, 3P, 4P, \ldots$ cannot all be distinct. Hence there are integers $k > j$ such that $kP = jP$, and we can take $s = k - j$. The smallest such $s \geq 1$ is called the *order of* P. (Proposition 2.13 tells us that the order of P divides $\#E(\mathbb{F}_p)$.) Thus if s is the order of P and if n_0 is any integer such that $Q = n_0 P$, then the solutions to $Q = nP$ are the integers $n = n_0 + is$ with $i \in \mathbb{Z}$. (See Exercise 6.9.)

This means that the value of $\log_P(Q)$ is really an element of $\mathbb{Z}/s\mathbb{Z}$, i.e., $\log_P(Q)$ is an integer modulo s, where s is the order of P. For concreteness we could set $\log_P(Q)$ equal to n_0. However the advantage of defining the values to be in $\mathbb{Z}/s\mathbb{Z}$ is that the elliptic discrete logarithm then satisfies

$$\log_P(Q_1 + Q_2) = \log_P(Q_1) + \log_P(Q_2) \qquad \text{for all } Q_1, Q_2 \in E(\mathbb{F}_p). \quad (6.4)$$

Notice the analogy with the ordinary logarithm $\log(\alpha\beta) = \log(\alpha) + \log(\beta)$ and the discrete logarithm for $\mathbb{F}_p^*$ (cf. Remark 2.2). The fact that the discrete logarithm for $E(\mathbb{F}_p)$ satisfies (6.4) means that it respects the addition law when the group $E(\mathbb{F}_p)$ is mapped to the group $\mathbb{Z}/s\mathbb{Z}$. We say that the map $\log_P$ defines a *group homomorphism* (cf. Exercise 2.13)

$$\log_P : E(\mathbb{F}_p) \longrightarrow \mathbb{Z}/s\mathbb{Z}.$$

Example 6.15. Consider the elliptic curve

$$E : Y^2 = X^3 + 8X + 7 \quad \text{over } \mathbb{F}_{73}.$$

The points $P = (32, 53)$ and $Q = (39, 17)$ are both in $E(\mathbb{F}_{73})$, and it is easy to verify (by hand if you're patient and with a computer if not) that

$$Q = 11P, \qquad \text{so} \qquad \log_P(Q) = 11.$$

Similarly, $R = (35, 47) \in E(\mathbb{F}_{73})$ and $S = (58, 4) \in E(\mathbb{F}_{73})$, and after some computation we find that they satisfy $R = 37P$ and $S = 28P$, so

$$\log_P(R) = 37 \qquad \text{and} \qquad \log_P(S) = 28.$$

Finally, we mention that $\#E(\mathbb{F}_{73}) = 82$, but P satisfies $41P = \mathcal{O}$. Thus P has order $41 = 82/2$, so only half of the points in $E(\mathbb{F}_{73})$ are multiples of P. For example, $(20, 65)$ is in $E(\mathbb{F}_{73})$, but it does not equal a multiple of P.

6.3.1 The Double-and-Add Algorithm

It appears to be quite difficult to recover the value of n from the two points P and $Q = nP$ in $E(\mathbb{F}_p)$, i.e., it is difficult to solve the ECDLP. We will say more about the difficulty of the ECDLP in later sections. However, in order to use the function

$$\mathbb{Z} \longrightarrow E(\mathbb{F}_p), \qquad n \longmapsto nP,$$

for cryptography, we need to efficiently compute nP from the known values n and P. If n is large, we certainly do not want to compute nP by computing $P, 2P, 3P, 4P, \ldots$.

The most efficient way to compute nP is very similar to the method that we described in Sect. 1.3.2 for computing powers $a^n \pmod{N}$, which we needed for Diffie–Hellman key exchange (Sect. 2.3) and for the Elgamal and RSA public key cryptosystems (Sects. 2.4 and 3.2). However, since the operation on an elliptic curve is written as addition instead of as multiplication, we call it "double-and-add" instead of "square-and-multiply."

The underlying idea is the same as before. We first write n in binary form as

$$n = n_0 + n_1 \cdot 2 + n_2 \cdot 4 + n_3 \cdot 8 + \cdots + n_r \cdot 2^r \qquad \text{with } n_0, n_1, \ldots, n_r \in \{0, 1\}.$$

(We also assume that $n_r = 1$.) Next we compute the following quantities:

$$Q_0 = P, \quad Q_1 = 2Q_0, \quad Q_2 = 2Q_1, \quad \ldots \quad , Q_r = 2Q_{r-1}.$$

Notice that Q_i is simply twice the previous Q_{i-1}, so

$$Q_i = 2^i P.$$

	Input. Point $P \in E(\mathbb{F}_p)$ and integer $n \geq 1$.
1.	Set $Q = P$ and $R = \mathcal{O}$.
2.	Loop while $n > 0$.
	3. If $n \equiv 1 \pmod 2$, set $R = R + Q$.
	4. Set $Q = 2Q$ and $n = \lfloor n/2 \rfloor$.
	5. If $n > 0$, continue with loop at Step **2**.
6.	Return the point R, which equals nP.

Table 6.3: The double-and-add algorithm for elliptic curves

These points are referred to as 2-power multiples of P, and computing them requires r doublings. Finally, we compute nP using at most r additional additions,

$$nP = n_0Q_0 + n_1Q_1 + n_2Q_2 + \cdots + n_rQ_r.$$

We'll refer to the addition of two points in $E(\mathbb{F}_p)$ as a *point operation.* Thus the total time to compute nP is at most $2r$ point operations in $E(\mathbb{F}_p)$. Notice that $n \geq 2^r$, so it takes no more than $2\log_2(n)$ point operations to compute nP. This makes it feasible to compute nP even for very large values of n. We have summarized the double-and-add algorithm in Table 6.3.

Example 6.16. We use the Double-and-Add Algorithm as described in Table 6.3 to compute nP in $E(\mathbb{F}_p)$ for

$$n = 947, \qquad E : Y^2 = X^3 + 14X + 19, \qquad p = 3623, \qquad P = (6, 730).$$

The binary expansion of n is

$$n = 947 = 1 + 2 + 2^4 + 2^5 + 2^7 + 2^8 + 2^9.$$

The step by step calculation, which requires nine doublings and six additions, is given in Table 6.4. The final result is $947P = (3492, 60)$. (The n column in Table 6.4 refers to the n used in the algorithm described in Table 6.3.)

Remark 6.17. There is an additional technique that can be used to further reduce the time required to compute nP. The idea is to write n using sums and differences of powers of 2. The reason that this is advantageous is because there are generally fewer terms, so fewer point additions are needed to compute nP. It is important to observe that subtracting two points on an elliptic curve is as easy as adding them, since $-(x, y) = (x, -y)$. This is rather different from $\mathbb{F}_p^*$, where computing a^{-1} takes significantly more time than it takes to multiply two elements.

An example will help to illustrate the idea. We saw in Example 6.16 that $947 = 1+2+2^4+2^5+2^7+2^8+2^9$, so it takes 15 point operations (9 doublings and 6 additions) to compute $947P$. But if we instead write

$$947 = 1 + 2 - 2^4 - 2^6 + 2^{10},$$

Step i	n	$Q = 2^i P$	R
0	947	$(6, 730)$	$\mathcal{O}$
1	473	$(2521, 3601)$	$(6, 730)$
2	236	$(2277, 502)$	$(2149, 196)$
3	118	$(3375, 535)$	$(2149, 196)$
4	59	$(1610, 1851)$	$(2149, 196)$
5	29	$(1753, 2436)$	$(2838, 2175)$
6	14	$(2005, 1764)$	$(600, 2449)$
7	7	$(2425, 1791)$	$(600, 2449)$
8	3	$(3529, 2158)$	$(3247, 2849)$
9	1	$(2742, 3254)$	$(932, 1204)$
10	0	$(1814, 3480)$	$(3492, 60)$

Table 6.4: Computing $947 \cdot (6, 730)$ on $Y^2 = X^3 + 14X + 19$ modulo 3623

then we can compute

$$947P = P + 2P - 2^4 P - 2^6 P + 2^{10} P$$

using 10 doublings and 4 additions, for a total of 14 point operations. Writing a number n as a sum of positive and negative powers of 2 is called a *ternary expansion of* n.

How much savings can we expect? Suppose that n is a large number and let $k = \lfloor \log n \rfloor + 1$. In the worst case, if n has the form $2^k - 1$, then computing nP using a binary expansion of n requires $2k$ point operations (k doublings and k additions), since

$$2^k - 1 = 1 + 2 + 2^2 + \cdots + 2^{k-1}.$$

But if we allow ternary expansions, then we prove below (Proposition 6.18) that computing nP never requires more than $\frac{3}{2}k + 1$ point operations ($k + 1$ doublings and $\frac{1}{2}k$ additions).

This is the worst case scenario, but it's also important to know what happens on average. The binary expansion of a random number has approximately the same number of 1's and 0's, so for most n, computing nP using the binary expansion of n takes about $\frac{3}{2}k$ steps (k doublings and $\frac{1}{2}k$ additions). But if we allow sums and differences of powers of 2, then one can show that most n have an expansion with $\frac{2}{3}$ of the terms being 0. So for most n, we can compute nP in about $\frac{4}{3}k + 1$ steps ($k + 1$ doublings and $\frac{1}{3}k$ additions).

Proposition 6.18. *Let n be a positive integer and let $k = \lfloor \log n \rfloor + 1$, which means that $2^k > n$. Then we can always write*

$$n = u_0 + u_1 \cdot 2 + u_2 \cdot 4 + u_3 \cdot 8 + \cdots + u_k \cdot 2^k \tag{6.5}$$

with $u_0, u_1, \ldots, u_k \in \{-1, 0, 1\}$ and at most $\frac{1}{2}k$ of the u_i nonzero.

Proof. The proof is essentially an algorithm for writing n in the desired form. We start by writing n in binary,

$$n = n_0 + n_1 \cdot 2 + n_2 \cdot 4 + \cdots + n_{k-1} \cdot 2^{k-1} \qquad \text{with } n_0, \ldots, n_{k-1} \in \{0, 1\}.$$

Working from left to right, we look for the first occurrence of two or more consecutive nonzero n_i coefficients. For example, suppose that

$$n_s = n_{s+1} = \cdots = n_{s+t-1} = 1 \qquad \text{and} \qquad n_{s+t} = 0$$

for some $t \geq 2$. In other words, the quantity

$$2^s + 2^{s+1} + \cdots + 2^{s+t-1} + 0 \cdot 2^{s+t} \tag{6.6}$$

appears in the binary expansion of n. We observe that

$$2^s + 2^{s+1} + \cdots + 2^{s+t-1} + 0 \cdot 2^{s+t} = 2^s(1 + 2 + 4 + \cdots + 2^{t-1}) = 2^s(2^t - 1),$$

so we can replace (6.6) with

$$-2^s + 2^{s+t}.$$

Repeating this procedure, we end up with an expansion of n of the form (6.5) in which no two consecutive u_i are nonzero. (Note that although the original binary expansion went up to only 2^{k-1}, the new expansion might go up to 2^k.) □

6.3.2 How Hard Is the ECDLP?

The collision algorithms described in Sect. 5.4 are easily adapted to any group, for example to the group of points $E(\mathbb{F}_p)$ on an elliptic curve. In order to solve $Q = nP$, Eve chooses random integers $j_1, \ldots, j_r$ and $k_1, \ldots, k_r$ between 1 and p and makes two lists of points:

List #1. $j_1P,\ j_2P,\ j_3P, \ldots,\ j_rP,$

List #2. $k_1P + Q,\ k_2P + Q,\ k_3P + Q, \ldots,\ k_rP + Q.$

As soon as she finds a match (collision) between the two lists, she is done, since if she finds $j_uP = k_vP + Q$, then $Q = (j_u - k_v)P$ provides the solution. As we saw in Sect. 5.4, if r is somewhat larger than $\sqrt{p}$, say $r \approx 3\sqrt{p}$, then there is a very good chance that there will be a collision.

This naive collision algorithm requires quite a lot of storage for the two lists. However, it is not hard to adapt Pollard's ρ method from Sect. 5.5 to devise a storage-free collision algorithm with a similar running time. (See Exercise 6.13.) In any case, there are certainly algorithms that solve the ECDLP for $E(\mathbb{F}_p)$ in $\mathcal{O}(\sqrt{p})$ steps.

We have seen that there are much faster ways to solve the discrete logarithm problem for $\mathbb{F}_p^*$. In particular, the index calculus described in Sect. 3.8

has a subexponential running time, i.e., the running time is $\mathcal{O}(p^\epsilon)$ for every $\epsilon > 0$. The principal reason that elliptic curves are used in cryptography is the fact that there are no index calculus algorithms known for the ECDLP, and indeed, there are no general algorithms known that solve the ECDLP in fewer than $O(\sqrt{p})$ steps. In other words, despite the highly structured nature of the group $E(\mathbb{F}_p)$, the fastest known algorithms to solve the ECDLP are no better than the generic algorithm that works equally well to solve the discrete logarithm problem in any group. This fact is sufficiently important that it bears highlighting.

The fastest known algorithm to solve ECDLP in $E(\mathbb{F}_p)$ takes approximately $\sqrt{p}$ steps.

Thus the ECDLP appears to be much more difficult than the DLP. Recall, however, there are some primes p for which the DLP in $\mathbb{F}_p^*$ is comparatively easy. For example, if $p-1$ is a product of small primes, then the Pohlig–Hellman algorithm (Theorem 2.31) gives a quick solution to the DLP in $\mathbb{F}_p^*$. In a similar fashion, there are some elliptic curves and some primes for which the ECDLP in $E(\mathbb{F}_p)$ is comparatively easy. We discuss some of these special cases, which must be avoided in the construction of secure cryptosystems, in Sect. 6.9.1.

6.4 Elliptic Curve Cryptography

It is finally time to apply elliptic curves to cryptography. We start with the easiest application, Diffie–Hellman key exchange, which involves little more than replacing the discrete logarithm problem for the finite field $\mathbb{F}_p$ with the discrete logarithm problem for an elliptic curve $E(\mathbb{F}_p)$. We then describe elliptic analogues of the Elgamal public key cryptosystem and the digital signature algorithm (DSA).

6.4.1 Elliptic Diffie–Hellman Key Exchange

Alice and Bob agree to use a particular elliptic curve $E(\mathbb{F}_p)$ and a particular point $P \in E(\mathbb{F}_p)$. Alice chooses a secret integer n_A and Bob chooses a secret integer n_B. They compute the associated multiples

$$\overbrace{Q_A = n_A P}^{\text{Alice computes this}} \qquad \text{and} \qquad \overbrace{Q_B = n_B P}^{\text{Bob computes this}},$$

and they exchange the values of Q_A and Q_B. Alice then uses her secret multiplier to compute $n_A Q_B$, and Bob similarly computes $n_B Q_A$. They now have the shared secret value

Public parameter creation	
A trusted party chooses and publishes a (large) prime p, an elliptic curve E over $\mathbb{F}_p$, and a point P in $E(\mathbb{F}_p)$.	
Private computations	
Alice	**Bob**
Chooses a secret integer n_A. Computes the point $Q_A = n_A P$.	Chooses a secret integer n_B. Computes the point $Q_B = n_B P$.
Public exchange of values	
Alice sends Q_A to Bob $\longrightarrow$	Q_A
Q_B $\longleftarrow$	Bob sends Q_B to Alice
Further private computations	
Alice	**Bob**
Computes the point $n_A Q_B$.	Computes the point $n_B Q_A$.
The shared secret value is $n_A Q_B = n_A(n_B P) = n_B(n_A P) = n_B Q_A$.	

Table 6.5: Diffie–Hellman key exchange using elliptic curves

$$n_A Q_B = (n_A n_B)P = n_B Q_A,$$

which they can use as a key to communicate privately via a symmetric cipher. Table 6.5 summarizes elliptic Diffie–Hellman key exchange.

Example 6.19. Alice and Bob decide to use elliptic Diffie–Hellman with the following prime, curve, and point:

$$p = 3851, \qquad E : Y^2 = X^3 + 324X + 1287, \qquad P = (920, 303) \in E(\mathbb{F}_{3851}).$$

Alice and Bob choose respective secret values $n_A = 1194$ and $n_B = 1759$, and then

$$\begin{aligned} &\text{Alice computes} \quad Q_A = 1194P = (2067, 2178) \in E(\mathbb{F}_{3851}),\\ &\text{Bob computes} \quad Q_B = 1759P = (3684, 3125) \in E(\mathbb{F}_{3851}). \end{aligned}$$

Alice sends Q_A to Bob and Bob sends Q_B to Alice. Finally,

$$\begin{aligned} &\text{Alice computes} \quad n_A Q_B = 1194(3684, 3125) = (3347, 1242) \in E(\mathbb{F}_{3851}),\\ &\text{Bob computes} \quad n_B Q_A = 1759(2067, 2178) = (3347, 1242) \in E(\mathbb{F}_{3851}). \end{aligned}$$

Bob and Alice have exchanged the secret point $(3347, 1242)$. As will be explained in Remark 6.20, they should discard the y-coordinate and treat only the value $x = 3347$ as a secret shared value.

One way for Eve to discover Alice and Bob's secret is to solve the ECDLP

$$nP = Q_A,$$

since if Eve can solve this problem, then she knows n_A and can use it to compute $n_A Q_B$. Of course, there might be some other way for Eve to compute their secret without actually solving the ECDLP. The precise problem that Eve needs to solve is the elliptic analogue of the Diffie–Hellman problem described on page 69.

Definition. Let $E(\mathbb{F}_p)$ be an elliptic curve over a finite field and let $P \in E(\mathbb{F}_p)$. The *Elliptic Curve Diffie–Hellman Problem* is the problem of computing the value of $n_1 n_2 P$ from the known values of $n_1 P$ and $n_2 P$.

Remark 6.20. Elliptic Diffie–Hellman key exchange requires Alice and Bob to exchange points on an elliptic curve. A point Q in $E(\mathbb{F}_p)$ consists of two coordinates $Q = (x_Q, y_Q)$, where x_Q and y_Q are elements of the finite field $\mathbb{F}_p$, so it appears that Alice must send Bob two numbers in $\mathbb{F}_p$. However, those two numbers modulo p do not contain as much information as two arbitrary numbers, since they are related by the formula

$$y_Q^2 = x_Q^3 + Ax_Q + B \qquad \text{in } \mathbb{F}_p.$$

Note that Eve knows A and B, so if she can guess the correct value of x_Q, then there are only two possible values for y_Q, and in practice it is not too hard for her to actually compute the two values of y_Q.

There is thus little reason for Alice to send both coordinates of Q_A to Bob, since the y-coordinate contains so little additional information. Instead, she sends Bob only the x-coordinate of Q_A. Bob then computes and uses one of the two possible y-coordinates. If he happens to choose the "correct" y, then he is using Q_A, and if he chooses the "incorrect" y (which is the negative of the correct y), then he is using $-Q_A$. In any case, Bob ends up computing one of

$$\pm n_B Q_A = \pm(n_A n_B)P.$$

Similarly, Alice ends up computing one of $\pm(n_A n_B)P$. Then Alice and Bob use the x-coordinate as their shared secret value, since that x-coordinate is the same regardless of which y they use.

Example 6.21. Alice and Bob decide to exchange another secret value using the same public parameters as in Example 6.19:

$$p = 3851, \qquad E : Y^2 = X^3 + 324X + 1287, \qquad P = (920, 303) \in E(\mathbb{F}_{3851}).$$

However, this time they want to send fewer bits to one another. Alice and Bob respectively choose new secret values $n_A = 2489$ and $n_B = 2286$, and as before,

$$\begin{aligned} &\text{Alice computes} \quad Q_A = n_A P = 2489(920, 303) = (593, 719) \in E(\mathbb{F}_{3851}), \\ &\text{Bob computes} \quad Q_B = n_B P = 2286(920, 303) = (3681, 612) \in E(\mathbb{F}_{3851}). \end{aligned}$$

However, rather than sending both coordinates, Alice sends only $x_A = 593$ to Bob and Bob sends only $x_B = 3681$ to Alice.

Alice substitutes $x_B = 3681$ into the equation for E and finds that

$$y_B^2 = x_B^3 + 324x_B + 1287 = 3681^3 + 324 \cdot 3681 + 1287 = 997.$$

(Recall that all calculations are performed in $\mathbb{F}_{3851}$.) Alice needs to compute a square root of 997 modulo 3851. This is not hard to do, especially for primes satisfying $p \equiv 3 \pmod 4$, since Proposition 2.26 tells her that $b^{(p+1)/4}$ is a square root of b modulo p. So Alice sets

$$y_B = 997^{(3851+1)/4} = 997^{963} \equiv 612 \pmod{3851}.$$

It happens that she gets the same point $Q_B = (x_B, y_B) = (3681, 612)$ that Bob used, and she computes $n_A Q_B = 2489(3681, 612) = (509, 1108)$.

Similarly, Bob substitutes $x_A = 593$ into the equation for E and takes a square root,

$$\begin{aligned} y_A^2 &= x_A^3 + 324x_A + 1287 = 593^3 + 324 \cdot 593 + 1287 = 927, \\ y_A &= 927^{(3851+1)/4} = 927^{963} \equiv 3132 \pmod{3851}. \end{aligned}$$

Bob then uses the point $Q'_A = (593, 3132)$, which is not Alice's point Q_A, to compute $n_B Q'_A = 2286(593, 3132) = (509, 2743)$. Bob and Alice end up with points that are negatives of one another in $E(\mathbb{F}_p)$, but that is all right, since their shared secret value is the x-coordinate $x = 509$, which is the same for both points.

6.4.2 Elliptic Elgamal Public Key Cryptosystem

It is easy to create a direct analogue of the Elgamal public key cryptosystem described in Sect. 2.4. Briefly, Alice and Bob agree to use a particular prime p, elliptic curve E, and point $P \in E(\mathbb{F}_p)$. Alice chooses a secret multiplier n_A and publishes the point $Q_A = n_A P$ as her public key. Bob's plaintext is a point $M \in E(\mathbb{F}_p)$. He chooses an integer k to be his random element and computes

$$C_1 = kP \quad \text{and} \quad C_2 = M + kQ_A.$$

He sends the two points (C_1, C_2) to Alice, who computes

$$C_2 - n_A C_1 = (M + kQ_A) - n_A(kP) = M + k(n_A P) - n_A(kP) = M$$

Public parameter creation	
A trusted party chooses and publishes a (large) prime p, an elliptic curve E over $\mathbb{F}_p$, and a point P in $E(\mathbb{F}_p)$.	
Alice	**Bob**
Key creation	
Choose a private key n_A. Compute $Q_A = n_A P$ in $E(\mathbb{F}_p)$. Publish the public key Q_A.	
Encryption	
	Choose plaintext $M \in E(\mathbb{F}_p)$. Choose a random element k. Use Alice's public key Q_A to compute $C_1 = kP \in E(\mathbb{F}_p)$. and $C_2 = M + kQ_A \in E(\mathbb{F}_p)$. Send ciphertext (C_1, C_2) to Alice.
Decryption	
Compute $C_2 - n_A C_1 \in E(\mathbb{F}_p)$. This quantity is equal to M.	

Table 6.6: Elliptic Elgamal key creation, encryption, and decryption

to recover the plaintext. The elliptic Elgamal public key cryptosystem is summarized in Table 6.6.

In principle, the elliptic Elgamal cryptosystem works fine, but there are some practical difficulties.

1. There is no obvious way to attach plaintext messages to points in $E(\mathbb{F}_p)$.
2. The elliptic Elgamal cryptosystem has 4-to-1 message expansion, as compared to the 2-to-1 expansion ratio of Elgamal using $\mathbb{F}_p$. (See Remark 2.9.)

The reason that elliptic Elgamal has a 4-to-1 message expansion lies in the fact that the plaintext M is a single point in $E(\mathbb{F}_p)$. By Hasse's theorem (Theorem 6.11) there are approximately p different points in $E(\mathbb{F}_p)$, hence only about p different plaintexts. However, the ciphertext (C_1, C_2) consists of four numbers modulo p, since each point in $E(\mathbb{F}_p)$ has two coordinates.

Various methods have been proposed to solve these problems. The difficulty of associating plaintexts to points can be circumvented by choosing M randomly and using it as a mask for the actual plaintext. One such method, which also decreases message expansion, is described in Exercise 6.17.

Another natural way to improve message expansion is to send only the x-coordinates of C_1 and C_2, as was suggested for Diffie–Hellman key exchange in Remark 6.20. Unfortunately, since Alice must compute the difference $C_2 - n_A C_1$, she needs the correct values of both the x-and y-coordinates of C_1 and C_2. (Note that the points $C_2 - n_A C_1$ and $C_2 + n_A C_1$ are quite different!) However, the x-coordinate of a point determines the y-coordinate up to change of sign, so Bob can send one extra bit, for example

$$\text{Extra bit} = \begin{cases} 0 & \text{if } 0 \le y < \frac{1}{2}p, \\ 1 & \text{if } \frac{1}{2}p < y < p \end{cases}$$

(See Exercise 6.16.) In this way, Bob needs to send only the x-coordinates of C_1 and C_2, plus two extra bits. This idea is sometimes referred to as *point compression.*

6.4.3 Elliptic Curve Signatures

The *Elliptic Curve Digital Signature Algorithm* (ECDSA), which is described in Table 6.7, is a straightforward analogue of the digital signature algorithm (DSA) described in Table 4.3 of Sect. 4.3. ECDSA is in widespread use, especially, but not only, in situations where signature size is important. Official specifications for implementing ECDSA are described in [6, 142]. (See also Sect. 8.8 for an amusing real-world implementation of digital cash using ECDSA.)

In order to prove that ECDSA works, i.e., that the verification step succeeds in verifying a valid signature, we compute

$$\begin{aligned} v_1 G + v_2 V &= d s_2^{-1} G + s_1 s_2^{-1}(sG) \\ &= (d + s s_1) s_2^{-1} G \\ &= (e s_2) s_2^{-1} G \\ &= eG \in E(\mathbb{F}_p). \end{aligned}$$

Hence

$$x(v_1 G + v_2 V) \bmod q = x(eG) \pmod{q} = s_1,$$

so the signature is accepted as valid.

6.5 The Evolution of Public Key Cryptography

The invention of RSA in the late 1970s catapulted the problem of factoring large integers into prominence, leading to improved factorization methods such as the quadratic and number field sieves described in Sect. 3.7. In 1984, Hendrik Lenstra Jr. circulated a manuscript describing a new factorization method using elliptic curves. Lenstra's algorithm [75], which we describe in Sect. 6.6, is an elliptic analogue of Pollard's $p - 1$ factorization algorithm

Public parameter creation	
A trusted party chooses a finite field $\mathbb{F}_p$, an elliptic curve $E/\mathbb{F}_p$, and a point $G \in E(\mathbb{F}_p)$ of large prime order q.	
Samantha	**Victor**
Key creation	
Choose secret signing key $1 < s < q-1$. Compute $V = sG \in E(\mathbb{F}_p)$. Publish the verification key V.	
Signing	
Choose document $d \bmod q$. Choose random element $e \bmod q$. Compute $eG \in E(\mathbb{F}_p)$ and then, $s_1 = x(eG) \bmod q$ and $s_2 \equiv (d + s s_1)e^{-1} \pmod{q}$. Publish the signature (s_1, s_2).	
Verification	
	Compute $v_1 \equiv d s_2^{-1} \pmod{q}$ and $v_2 \equiv s_1 s_2^{-1} \pmod{q}$. Compute $v_1 G + v_2 V \in E(\mathbb{F}_p)$ and verify that $x(v_1 G + v_2 V) \bmod q = s_1$.

Table 6.7: The elliptic curve digital signature algorithm (ECDSA)

(Sect. 3.5) and exploits the fact that the number of points in $E(\mathbb{F}_p)$ varies as one chooses different elliptic curves. Although less efficient than sieve methods for the factorization problems that occur in cryptography, Lenstra's algorithm helped introduce elliptic curves to the cryptographic community.

The importance of factorization algorithms for cryptography is that they are used to break RSA and other similar cryptosystems. In 1985, Neal Koblitz and Victor Miller independently proposed using elliptic curves to create cryptosystems. They suggested that the elliptic curve discrete logarithm problem might be more difficult than the classical discrete logarithm problem modulo p. Thus Diffie–Hellman key exchange and the Elgamal public key cryptosystem, implemented using elliptic curves as described in Sect. 6.4, might require smaller keys and run more efficiently than RSA because one could use smaller numbers.

Koblitz [67] and Miller [88] each published their ideas as academic papers, but neither of them pursued the commercial aspects of elliptic curve cryptography. Indeed, at the time, there was virtually no research on the ECDLP, so it was difficult to say with any confidence that the ECDLP was indeed significantly more difficult than the classical DLP. However, the potential of what became known as elliptic curve cryptography (ECC) was noted by Scott

Vanstone and Ron Mullin, who had started a cryptographic company called Certicom in 1985. They joined with other researchers in both academia and the business world to promote ECC as an alternative to RSA and Elgamal.

All was not smooth sailing. For example, during the late 1980s, various cryptographers proposed using so-called supersingular elliptic curves for added efficiency, but in 1990, the MOV algorithm (see Sect. 6.9.1) showed that supersingular curves are vulnerable to attack. Some saw this as an indictment of ECC as a whole, while others pointed out that RSA also has weak instances that must be avoided, e.g., RSA must avoid using numbers that can be easily factored by Pollard's $p-1$ method.

The purely mathematical question of whether ECC provided a secure and efficient alternative to RSA was clouded by the fact that there were commercial and financial issues at stake. In order to be commercially successful, cryptographic methods must be standardized for use in areas such as communications and banking. RSA had the initial lead, since it was invented first, but RSA was patented, and some companies resisted the idea that standards approved by trade groups or government bodies should mandate the use of a patented technology. Elgamal, after it was invented in 1985, provided a royalty-free alternative, so many standards specified Elgamal as an alternative to RSA. In the meantime, ECC was growing in stature, but even as late as 1997, more than a decade after its introduction, leading experts indicated their doubts about the security of ECC.[7]

A major dilemma pervading the field of cryptography is that no one knows the actual difficulty of the supposedly hard problems on which it is based. Currently, the security of public key cryptosystems depends on the perception and consensus of experts as to the difficulty of problems such as integer factorization and discrete logarithms. All that can be said is that "such-and-such a problem has been extensively studied for N years, and here is the fastest known method for solving it." Proponents of factorization-based cryptosystems point to the fact that, in some sense, people have been trying to factor numbers since antiquity; but in truth, the modern theory of factorization requires high-speed computing devices and barely predates the invention of RSA. Serious study of the elliptic curve discrete logarithm problem started in the late 1980s, so modern factorization methods have a 10–15 year head start on ECDLP. In Chap. 7 we will describe public key cryptosystems (NTRU, GGH) whose security is based on certain hard problems in the theory of lattices. Lattices have been extensively investigated since the nineteenth century, but again the invention and analysis of modern computational algorithms is much more recent, having been initiated by fundamental work of

[7]In 1997, the RSA corporation posted the following quote by RSA co-inventor Ron Rivest on its website: "But the security of cryptosystems based on elliptic curves is not well understood, due in large part to the abstruse nature of elliptic curves. ...

Over time, this may change, but for now trying to get an evaluation of the security of an elliptic-curve cryptosystem is a bit like trying to get an evaluation of some recently discovered Chaldean poetry. Until elliptic curves have been further studied and evaluated, I would advise against fielding any large-scale applications based on them."

Lenstra, Lenstra, and Lovász in the early 1980s. Lattices appeared as a tool for cryptanalysis during the 1980s and as a means of creating cryptosystems in the 1990s.

RSA, the first public key cryptosystem, was patented by its inventors. The issue of patents in cryptography is fraught with controversy. One might argue that the RSA patent, which ran from 1983 to 2000, set back the use of cryptography by requiring users to pay licensing fees. However, it is also true that in order to build a company, an inventor needs investors willing to risk their money, and it is much easier to raise funds if there is an exclusive product to offer. Further, the fact that RSA was originally the "only game in town" meant that it automatically received extensive scrutiny from the academic community, which helped to validate its security.

The invention and eventual commercial implementation of ECC followed a different path. Since neither Koblitz nor Miller applied for a patent, the basic underlying idea of ECC became freely available for all to use. This led Certicom and other companies to apply for patents giving improvements to the basic ECC idea. Some of these improvements were based on significant new research ideas, while others were less innovative and might almost be characterized as routine homework problems.[8] Unfortunately, the United States Patents and Trademark Office (USPTO) does not have the expertise to effectively evaluate the flood of cryptographic patent applications that it receives. The result has been a significant amount of uncertainty in the marketplace as to which versions of ECC are free and which require licenses, even assuming that all of the issued patents can withstand a legal challenge.

6.6 Lenstra's Elliptic Curve Factorization Algorithm

Pollard's $p-1$ factorization method, which we discussed in Sect. 3.5, finds factors of $N = pq$ by searching for a power a^L with the property that

$$a^L \equiv 1 \pmod{p} \qquad \text{and} \qquad a^L \not\equiv 1 \pmod{q}.$$

Fermat's little theorem tells us that this is likely to work if $p-1$ divides L and $q-1$ does not divide L. So what we do is to take $L = n!$ for some moderate value of n. Then we hope that $p-1$ or $q-1$, but not both, is a product of small primes, hence divides $n!$. Clearly Pollard's method works well for some numbers, but not for all numbers. The determining factor is whether $p-1$ or $q-1$ is a product of small primes.

What is it about the quantity $p-1$ that makes it so important for Pollard's method? The answer lies in Fermat's little theorem. Intrinsically, $p-1$ is

[8]For example, at the end of Sect. 6.4.2 we described how to save bandwidth in elliptic Elgamal by sending the x-coordinate and one additional bit to specify the y-coordinate. This idea is called "point compression" and is covered by US Patent 6,141,420.

important because there are $p-1$ elements in $\mathbb{F}_p^*$, so every element α of $\mathbb{F}_p^*$ satisfies $\alpha^{p-1}=1$. Now consider that last statement as it relates to the theme of this chapter, which is that the points and the addition law for an elliptic curve $E(\mathbb{F}_p)$ are very much analogous to the elements and the multiplication law for $\mathbb{F}_p^*$. Hendrik Lenstra [75] made this analogy precise by devising a factorization algorithm that uses the group law on an elliptic curve E in place of multiplication modulo N.

In order to describe Lenstra's algorithm, we need to work with an elliptic curve modulo N, where the integer N is not prime, so the ring $\mathbb{Z}/N\mathbb{Z}$ is not a field. However, suppose that we start with an equation

$$E : Y^2 = X^3 + AX + B$$

and suppose that $P=(a,b)$ is a point on E modulo N, by which we mean that

$$b^2 \equiv a^3 + A \cdot a + B \pmod{N}.$$

Then we can apply the elliptic curve addition algorithm (Theorem 6.6) to compute $2P, 3P, 4P, \ldots$, since the only operations required by that algorithm are addition, subtraction, multiplication, and division (by numbers relatively prime to N).

Example 6.22. Let $N=187$ and consider the elliptic curve

$$E : Y^2 = X^3 + 3X + 7$$

modulo 187 and the point $P=(38,112)$, that is on E modulo 187. In order to compute $2P \bmod 187$, we follow the elliptic curve addition algorithm and compute

$$\begin{aligned}
\frac{1}{2y(P)} &= \frac{1}{224} \equiv 91 \pmod{187},\\
\lambda = \frac{3x(P)^2 + A}{2y(P)} &= \frac{4335}{224} \equiv 34 \cdot 91 \equiv 102 \pmod{187},\\
x(2P) &= \lambda^2 - 2x(P) = 10328 \equiv 43 \pmod{187},\\
y(2P) &= \lambda\big(x(P) - x(2P)\big) - y(P) = 102(38-43) - 112 \equiv 126 \pmod{187}.
\end{aligned}$$

Thus $2P=(43,126)$ as a point on the curve E modulo 187.

For clarity, we have written $x(P)$ and $y(P)$ for the x-and y-coordinates of P, and similarly for $2P$. Also, during the calculation we needed to find the reciprocal of 224 modulo 187, i.e., we needed to solve the congruence

$$224d \equiv 1 \pmod{187}.$$

This was easily accomplished using the extended Euclidean algorithm (Theorem 1.11; see also Remark 1.15 and Exercise 1.12), since it turns out that $\gcd(224,187)=1$.

We next compute $3P = 2P + P$ in a similar fashion. In this case, we are adding distinct points, so the formula for λ is different, but the computation is virtually the same:

$$\begin{aligned}
\frac{1}{x(2P) - x(P)} &= \frac{1}{5} \equiv 75 \pmod{187}, \\
\lambda = \frac{y(2P) - y(P)}{x(2P) - x(P)} &= \frac{14}{5} \equiv 14 \cdot 75 \equiv 115 \pmod{187}, \\
x(3P) = \lambda^2 - x(2P) - x(P) &= 13144 \equiv 54 \pmod{187}, \\
y(3P) = \lambda\big(x(P) - x(3P)\big) - y(P) &= 115(38 - 54) - 112 \equiv 105 \pmod{187}.
\end{aligned}$$

Thus $3P = (54, 105)$ on the curve E modulo 187. Again we needed to compute a reciprocal, in this case, the reciprocal of 5 modulo 187. We leave it to you to continue the calculations. For example, it is instructive to check that $P + 3P$ and $2P + 2P$ give the same answer, namely $4P = (93, 64)$.

Example 6.23. Continuing with Example 6.22, we attempt to compute $5P$ for the point $P = (38, 112)$ on the elliptic curve

$$E : Y^2 = X^3 + 3X + 7 \quad \text{modulo } 187.$$

We already computed $2P = (43, 126)$ and $3P = (54, 105)$. The first step in computing $5P = 3P + 2P$ is to compute the reciprocal of

$$x(3P) - x(2P) = 54 - 43 = 11 \quad \text{modulo } 187.$$

However, when we apply the extended Euclidean algorithm to 11 and 187, we find that $\gcd(11, 187) = 11$, so 11 does not have a reciprocal modulo 187.

It seems that we have hit a dead end, but in fact, we have struck it rich! Notice that since the quantity $\gcd(11, 187)$ is greater than 1, it gives us a divisor of 187. So our failure to compute $5P$ also tells us that 11 divides 187, which allows us to factor 187 as $187 = 11 \cdot 17$. This idea underlies Lenstra's elliptic curve factorization algorithm.

We examine more closely why we were not able to compute $5P$ modulo 187. If we instead look at the elliptic curve E modulo 11, then a quick computation shows that the point

$$P = (38, 112) \equiv (5, 2) \pmod{11} \quad \text{satisfies } 5P = \mathcal{O} \text{ in } E(\mathbb{F}_{11}).$$

This means that when we attempt to compute $5P$ modulo 11, we end up with the point $\mathcal{O}$ at infinity, so at some stage of the calculation we have tried to divide by zero. But here "zero" means zero in $\mathbb{F}_{11}$, so we actually end up trying to find the reciprocal modulo 11 of some integer that is divisible by 11.

Following the lead from Examples 6.22 and 6.23, we replace multiplication modulo N in Pollard's factorization method with addition modulo N on an elliptic curve. We start with an elliptic curve E and a point P on E modulo N and we compute

$$2! \cdot P,\ 3! \cdot P,\ 4! \cdot P,\ 5! \cdot P, \ldots \pmod{N}.$$

Notice that once we have computed $Q = (n-1)! \cdot P$, it is easy to compute $n! \cdot P$, since it equals nQ. At each stage, there are three things that may happen. First, we may be able to compute $n! \cdot P$. Second, during the computation we may need to find the reciprocal of a number d that is a multiple of N, which would not be helpful, but luckily this situation is quite unlikely to occur. Third, we may need to find the reciprocal of a number d that satisfies $1 < \gcd(d, N) < N$, in which case the computation of $n! \cdot P$ fails, but $\gcd(d, N)$ is a nontrivial factor of N, so we are happy.

Input. Integer N to be factored.

1. Choose random values A, a, and b modulo N.

2. Set $P = (a, b)$ and $B \equiv b^2 - a^3 - A \cdot a \pmod{N}$.
Let E be the elliptic curve $E : Y^2 = X^3 + AX + B$.

3. Loop $j = 2, 3, 4, \ldots$ up to a specified bound.

- **4.** Compute $Q \equiv jP \pmod{N}$ and set $P = Q$.
- **5.** If computation in Step **4** fails, then we have found a $d > 1$ with $d \mid N$.
- **6.** If $d < N$, then **success**, return d.
- **7.** If $d = N$, go to Step **1** and choose a new curve and point.

8. Increment j and loop again at Step **2**.

Table 6.8: Lenstra's elliptic curve factorization algorithm

This completes the description of Lenstra's elliptic curve factorization algorithm, other than the minor problem of finding an initial point P on an elliptic curve E modulo N. The obvious method is to fix an equation for the curve E, plug in values of X, and check whether the quantity $X^3 + AX + B$ is a square modulo N. Unfortunately, this is difficult to do unless we know how to factor N. The solution to this dilemma is to first choose the point $P = (a, b)$ at random, second choose a random value for A, and third set

$$B \equiv b^2 - a^3 - A \cdot a \pmod{N}.$$

Then the point P is automatically on the curve $E : Y^2 = X^3 + AX + B$ modulo N. Lenstra's algorithm is summarized in Table 6.8.

Example 6.24. We illustrate Lenstra's algorithm by factoring $N = 6887$. We begin by randomly selecting a point $P = (1512, 3166)$ and a number $A = 14$ and computing

$$B \equiv 3166^2 - 1512^3 - 14 \cdot 1512 \equiv 19 \pmod{6887}.$$

We let E be the elliptic curve

$$E : Y^2 = X^3 + 14X + 19,$$

so by construction, the point P is automatically on E modulo 6887. Now we start computing multiples of P modulo 6887. First we find that

$$2P \equiv (3466, 2996) \pmod{6887}.$$

Next we compute

$$3! \cdot P = 3 \cdot (2P) = 3 \cdot (3466, 2996) \equiv (3067, 396) \pmod{6887}.$$

n	$n! \cdot P \bmod 6887$		
1	P	$=$	$(1512, 3166)$
2	$2! \cdot P$	$=$	$(3466, 2996)$
3	$3! \cdot P$	$=$	$(3067, 396)$
4	$4! \cdot P$	$=$	$(6507, 2654)$
5	$5! \cdot P$	$=$	$(2783, 6278)$
6	$6! \cdot P$	$=$	$(6141, 5581)$

Table 6.9: Multiples of $P = (1512, 3166)$ on $Y^2 \equiv X^3 + 14X + 19 \pmod{6887}$

And so on. The values up to $6! \cdot P$ are listed in Table 6.9. These values are not, in and of themselves, interesting. It is only when we try, and fail, to compute $7! \cdot P$, that something interesting happens.

From Table 6.9 we read off the value of $Q = 6! \cdot P = (6141, 5581)$, and we want to compute $7Q$. First we compute

$$\begin{aligned} 2Q &\equiv (5380, 174) && \pmod{6887}, \\ 4Q &\equiv 2 \cdot 2Q \equiv (203, 2038) && \pmod{6887}. \end{aligned}$$

Then we compute $7Q$ as

$$\begin{aligned} Q &\equiv (Q + 2Q) + 4Q && \pmod{6887} \\ &\equiv \big((6141, 5581) + (5380, 174)\big) + (203, 2038) && \pmod{6887} \\ &\equiv (984, 589) + (203, 2038) && \pmod{6887}. \end{aligned}$$

When we attempt to perform the final step, we need to compute the reciprocal of $203 - 984$ modulo 6887, but we find that

$$\gcd(203 - 984, 6887) = \gcd(-781, 6887) = 71.$$

Thus we have discovered a nontrivial divisor of 6887, namely 71, which gives the factorization $6887 = 71 \cdot 97$.

It turns out that in $E(\mathbb{F}_{71})$, the point P satisfies $63P \equiv \mathcal{O} \pmod{71}$, while in $E(\mathbb{F}_{97})$, the point P satisfies $107P \equiv \mathcal{O} \pmod{97}$. The reason that we succeeded in factoring 6887 using $7! \cdot P$, but not with a smaller multiple of P, is precisely because $7!$ is the smallest factorial that is divisible by 63.

Remark 6.25. In Sect. 3.7 we discussed the speed of sieve factorization methods and saw that the average running time of the quadratic sieve to factor a composite number N is approximately

$$O\left(e^{\sqrt{(\log N)(\log\log N)}}\right) \text{ steps.} \tag{6.7}$$

Notice that the running time depends on the size of the integer N.

On the other hand, the most naive possible factorization method, namely trying each possible divisor $2, 3, 4, 5, \ldots$, has a running time that depends on the smallest prime factor of N. More precisely, this trial division algorithm takes exactly p steps, where p is the smallest prime factor of N. If it happens that $N = pq$ with p and q approximately the same size, then the running time is approximately $\sqrt{N}$, which is much slower than sieve methods; but if N happens to have a very small prime factor, trial division may be helpful in finding it.

It is an interesting and useful property of the elliptic curve factorization algorithm that its expected running time depends on the smallest prime factor of N, rather than on N itself. (See Exercise 5.44 for another, albeit slower, factorization algorithm with this property.) More precisely, if p is the smallest factor of N, then the elliptic curve factorization algorithm has average running time approximately

$$O\left(e^{\sqrt{2(\log p)(\log\log p)}}\right) \text{ steps.} \tag{6.8}$$

If $N = pq$ is a product of two primes with $p \approx q$, the running times in (6.7) and (6.8) are approximately equal, and then the fact that a sieve step is much faster than an elliptic curve step makes sieve methods faster in practice. However, the elliptic curve method is quite useful for finding moderately large factors of extremely large numbers, because its running time depends on the smallest prime factor.

6.7 Elliptic Curves over $\mathbb{F}_2$ and over $\mathbb{F}_{2^k}$

Computers speak binary, so they are especially well suited to doing calculations modulo 2. This suggests that it might be more efficient to use elliptic curves modulo 2. Unfortunately, if E is an elliptic curve defined over $\mathbb{F}_2$, then $E(\mathbb{F}_2)$ contains at most 5 points, so $E(\mathbb{F}_2)$ is not useful for cryptographic purposes.

However, there are other finite fields in which $2 = 0$. These are the fields $\mathbb{F}_{2^k}$ containing 2^k elements. Recall from Sect. 2.10.4 that for every prime power p^k there exists a field $\mathbb{F}_{p^k}$ with p^k elements; and further, up to relabeling the elements, there is exactly one such field. So we can take an elliptic curve whose Weierstrass equation has coefficients in a field $\mathbb{F}_{p^k}$ and look at the group of points on that curve having coordinates in $\mathbb{F}_{p^k}$. Hasse's theorem (Theorem 6.11) is true in this more general setting.

Theorem 6.26 (Hasse). *Let E be an elliptic curve over $\mathbb{F}_{p^k}$. Then*

$$\#E(\mathbb{F}_{p^k}) = p^k + 1 - t_{p^k} \qquad \text{with } t_{p^k} \text{ satisfying } |t_{p^k}| \le 2p^{k/2}.$$

Example 6.27. We work with the field

$$\mathbb{F}_9 = \{a + bi : a, b \in \mathbb{F}_3\}, \qquad \text{where } i^2 = -1.$$

(See Example 2.58 for a discussion of $\mathbb{F}_{p^2}$ for primes $p \equiv 3 \pmod 4$.) Let E be the elliptic curve over $\mathbb{F}_9$ defined by the equation

$$E : Y^2 = X^3 + (1+i)X + (2+i).$$

By trial and error we find that there are 10 points in $E(\mathbb{F}_9)$,

$$(2i, 1+2i), \quad (2i, 2+i), \quad (1+i, 1+i), \quad (1+i, 2+2i), \quad (2,0),$$
$$(2+i, i), \quad (2+i, 2i), \quad (2+2i, 1), \quad (2+2i, 2), \quad \mathcal{O}.$$

Points can be doubled or added to one another using the formulas for the addition of points, always keeping in mind that $i^2 = -1$ and that we are working modulo 3. For example, you can check that

$$(2,0) + (2+i, 2i) = (2i, 1+2i) \qquad \text{and} \qquad 2(1+i, 2+2i) = (2+i, i).$$

Our goal is to use elliptic curves over $\mathbb{F}_{2^k}$ for cryptography, but there is one difficulty that we must first address. The problem is that we cheated a little bit when we defined an elliptic curve as a curve given by a Weierstrass equation $Y^2 = X^3 + AX + B$ satisfying $\Delta = 4A^3 + 27B^2 \neq 0$. In fact, the correct definition of the discriminant Δ is

$$\Delta = -16(4A^3 + 27B^2).$$

As long as we work in a field where $2 \neq 0$, then the condition $\Delta \neq 0$ is the same with either definition, but for fields such as $\mathbb{F}_{2^k}$ where $2 = 0$, we have $\Delta = 0$ for every standard Weierstrass equation. The solution is to enlarge the collection of allowable Weierstrass equations.

Definition. An *elliptic curve* E is the set of solutions to a *generalized Weierstrass equation*

$$E : Y^2 + a_1XY + a_3Y = X^3 + a_2X^2 + a_4X + a_6,$$

together with an extra point $\mathcal{O}$. The coefficients $a_1, \ldots, a_6$ are required to satisfy $\Delta \neq 0$, where the *discriminant* Δ is defined in terms of certain quantities b_2, b_4, b_6, b_8 as follows:

$$b_2 = a_1^2 + 4a_2, \qquad b_4 = 2a_4 + a_1a_3, \qquad b_6 = a_3^2 + 4a_6,$$
$$b_8 = a_1^2a_6 + 4a_2a_6 - a_1a_3a_4 + a_2a_3^2 - a_4^2,$$

$$\Delta = -b_2^2 b_8 - 8b_4^3 - 27b_6^2 + 9b_2 b_4 b_6.$$

(Although these formulas look complicated, they are easy enough to compute, and the condition $\Delta \neq 0$ is exactly what is required to ensure that the curve E is nonsingular.)

The geometric definition of the addition law on E is similar to our earlier definition, the only change being that the old reflection step $(x, y) \mapsto (x, -y)$ is replaced by the slightly more complicated reflection step

$$(x, y) \longmapsto (x, -y - a_1 x - a_3).$$

This is also the formula for the negative of a point.

Working with generalized Weierstrass equations, it is not hard to derive an addition algorithm similar to the algorithm described in Theorem 6.6; see Exercise 6.22 for details. For example, if $P_1 = (x_1, y_1)$ and $P_2 = (x_2, y_2)$ are points with $P_1 \neq \pm P_2$, then the x-coordinate of their sum is given by

$$x(P_1 + P_2) = \lambda^2 + a_1\lambda - a_2 - x_1 - x_2 \qquad \text{with} \quad \lambda = \frac{y_2 - y_1}{x_2 - x_1}.$$

Similarly, the x-coordinate of twice a point $P = (x, y)$ is given by the duplication formula

$$x(2P) = \frac{x^4 - b_4 x^2 - 2b_6 x - b_8}{4x^3 + b_2 x^2 + 4b_4 x + b_6}.$$

Example 6.28. The polynomial $T^3 + T + 1$ is irreducible in $\mathbb{F}_2[T]$, so as explained in Sect. 2.10.4, the quotient ring $\mathbb{F}_2[T]/(T^3 + T + 1)$ is a field $\mathbb{F}_8$ with eight elements. Every element in $\mathbb{F}_8$ can be represented by an expression of the form

$$a + bT + cT^2 \qquad \text{with } a, b, c \in \mathbb{F}_2,$$

with the understanding that when we multiply two elements, we divide the product by $T^3 + T + 1$ and take the remainder.

Now consider the elliptic curve E defined over the field $\mathbb{F}_8$ by the generalized Weierstrass equation

$$E : Y^2 + (1 + T)Y = X^3 + (1 + T^2)X + T.$$

The discriminant of E is $\Delta = 1 + T + T^2$. There are nine points in $E(\mathbb{F}_8)$,

$$\begin{gathered}(0, T), \quad (0, 1), \quad (T, 0), \quad (T, 1 + T), \quad (1 + T, T), \\ (1 + T, 1), \quad (1 + T^2, T + T^2), \quad (1 + T^2, 1 + T^2), \quad \mathcal{O}.\end{gathered}$$

Using the group law described in Exercise 6.22, we can add and double points, for example

$$(1 + T^2, T + T^2) + (1 + T, T) = (1 + T^2, 1 + T^2) \quad \text{and} \quad 2(T, 1 + T) = (T, 0).$$

There are some computational advantages to working with elliptic curves defined over $\mathbb{F}_{2^k}$, rather than over $\mathbb{F}_p$. We already mentioned the first, the binary nature of computers tends to make them operate more efficiently in situations in which $2 = 0$. A second advantage is the option to take k composite, in which case $\mathbb{F}_{2^k}$ contains other finite fields intermediate between $\mathbb{F}_2$ and $\mathbb{F}_{2^k}$. (The precise statement is that $\mathbb{F}_{2^j}$ is a subfield of $\mathbb{F}_{2^k}$ if and only if $j \mid k$.) These intermediate fields can sometimes be used to speed up computations, but there are also situations in which they cause security problems. So as is often the case, increased efficiency may come at the cost of decreased security; to avoid potential problems, it is often safest to use fields $\mathbb{F}_{2^k}$ with k prime.

The third, and most important, advantage of working over $\mathbb{F}_{2^k}$ lies in a suggestion of Neal Koblitz to use an elliptic curve E over $\mathbb{F}_2$, while taking points on E with coordinates in $\mathbb{F}_{2^k}$. As we now explain, this allows the use of the Frobenius map instead of the doubling map and leads to a significant gain in efficiency.

Definition. The *(p-power) Frobenius map* τ is the map from the field $\mathbb{F}_{p^k}$ to itself defined by the simple rule

$$\tau : \mathbb{F}_{p^k} \longrightarrow \mathbb{F}_{p^k}, \qquad \alpha \longmapsto \alpha^p.$$

The Frobenius map has the surprising property that it preserves addition and multiplication,[9]

$$\tau(\alpha + \beta) = \tau(\alpha) + \tau(\beta) \qquad \text{and} \qquad \tau(\alpha \cdot \beta) = \tau(\alpha) \cdot \tau(\beta).$$

The multiplication rule is obvious, since

$$\tau(\alpha \cdot \beta) = (\alpha \cdot \beta)^p = \alpha^p \cdot \beta^p = \tau(\alpha) \cdot \tau(\beta).$$

In general, the addition rule is a consequence of the binomial theorem (see Exercise 6.24). For $p = 2$, which is what we will need, the proof is easy,

$$\tau(\alpha + \beta) = (\alpha + \beta)^2 = \alpha^2 + 2\alpha \cdot \beta + \beta^2 = \alpha^2 + \beta^2 = \tau(\alpha) + \tau(\beta),$$

where we have used the fact that $2 = 0$ in $\mathbb{F}_{2^k}$. We also note that $\tau(\alpha) = \alpha$ for every $\alpha \in \mathbb{F}_2$, which is clear, since $\mathbb{F}_2 = \{0, 1\}$.

Now let E be an elliptic curve defined over $\mathbb{F}_2$, i.e., given by a generalized Weierstrass equation with coefficients in $\mathbb{F}_2$, and let $P = (x, y) \in E(\mathbb{F}_{2^k})$ be a point on E with coordinates in some larger field $\mathbb{F}_{2^k}$. We define a Frobenius map on points in $E(\mathbb{F}_{2^k})$ by applying τ to each coordinate,

$$\tau(P) = \bigl(\tau(x), \tau(y)\bigr). \tag{6.9}$$

We are going to show that the map τ has some nice properties. For example, we claim that

$$\tau(P) \in E(\mathbb{F}_{2^k}). \tag{6.10}$$

[9]In mathematical terminology, the Frobenius map τ is a field automorphism of $\mathbb{F}_{p^k}$. It also fixes $\mathbb{F}_p$. One can show that the Galois group of $\mathbb{F}_{p^k}/\mathbb{F}_p$ is cyclic of order k and is generated by τ.

Further, if $P, Q \in E(\mathbb{F}_{2^k})$, then we claim that

$$\tau(P+Q) = \tau(P) + \tau(Q). \tag{6.11}$$

In other words, τ maps $E(\mathbb{F}_{2^k})$ to itself, and it respects the addition law. (In mathematical terminology, the Frobenius map is a group homomorphism of $E(\mathbb{F}_{2^k})$ to itself.)

It is easy to check (6.10). We are given that $P = (x, y) \in E(\mathbb{F}_{2^k})$, so

$$y^2 + a_1xy + a_3y - x^3 - a_2x^2 - a_4x - a_6 = 0.$$

Applying τ to both sides and using the fact that τ respects addition and multiplication in $\mathbb{F}_{2^k}$, we find that

$$\tau(y)^2 + \tau(a_1)\tau(x)\tau(y) + \tau(a_3)\tau(y) - \tau(x)^3 - \tau(a_2)\tau(x)^2 - \tau(a_4)\tau(x) - \tau(a_6) = 0.$$

By assumption, the Weierstrass equation has coefficients in $\mathbb{F}_2$, and we know that τ fixes elements of $\mathbb{F}_2$, so

$$\tau(y)^2 + a_1\tau(x)\tau(y) + a_3\tau(y) - \tau(x)^3 - a_2\tau(x)^2 - a_4\tau(x) - a_6 = 0.$$

Hence $\tau(P) = \big(\tau(x), \tau(y)\big)$ is a point of $E(\mathbb{F}_{2^k})$.

A similar computation, which we omit, shows that (6.11) is true. The key fact is that the addition law on E requires only addition, subtraction, multiplication, and division of the coordinates of points and the coefficients of the Weierstrass equation.

Our next result shows that the Frobenius map is closely related to the number of points in $E(\mathbb{F}_p)$.

Theorem 6.29. *Let E be an elliptic curve over $\mathbb{F}_p$ and let*

$$t = p + 1 - \#E(\mathbb{F}_p).$$

Notice that Hasse's theorem (Theorem 6.11) says that $|t| \le 2\sqrt{p}$.

(a) *Let α and β be the complex roots of the quadratic polynomial $Z^2 - tZ + p$. Then $|\alpha| = |\beta| = \sqrt{p}$, and for every $k \ge 1$ we have*

$$\#E(\mathbb{F}_{p^k}) = p^k + 1 - \alpha^k - \beta^k.$$

(b) *Let*

$$\tau : E(\mathbb{F}_{p^k}) \longrightarrow E(\mathbb{F}_{p^k}), \qquad (x, y) \longmapsto (x^p, y^p),$$

be the Frobenius map. Then for every point $Q \in E(\mathbb{F}_{p^k})$ we have

$$\tau^2(Q) - t \cdot \tau(Q) + p \cdot Q = \mathcal{O},$$

where $\tau^2(Q)$ denotes the composition $\tau(\tau(Q))$.

Proof. The proof requires more tools than we have at our disposal; see for example [136, V §2] or [147]. □

Recall from Sect. 6.3.1 that to compute a multiple nP of a point P, we first expressed n as a sum of powers of 2 and then used a double-and-add method to compute nP. For random values of n, this required approximately $\log n$ doublings and $\frac{1}{2}\log n$ additions. A refinement of this method using both positive and negative powers of 2 reduces the time to approximately $\log n$ doublings and $\frac{1}{3}\log n$ additions. Notice that the number of doublings remains at $\log n$. Koblitz's idea is to replace the doubling map with the Frobenius map. This leads to a large savings, because it takes much less time to compute $\tau(P)$ than it does to compute $2P$. The key to the approach is Theorem 6.29, which tells us that the action of the Frobenius map on $E(\mathbb{F}_{2^k})$ satisfies a quadratic equation.

Definition. A *Koblitz curve* is an elliptic curve defined over $\mathbb{F}_2$ by an equation of the form

$$E_a : Y^2 + XY = X^3 + aX^2 + 1$$

with $a \in \{0, 1\}$. The discriminant of E_a is $\Delta = 1$.

For concreteness we restrict attention to the curve

$$E_0 : Y^2 + XY = X^3 + 1.$$

It is easy to check that

$$E_0(\mathbb{F}_2) = \{(0,1), (1,0), (1,1), \mathcal{O}\},$$

so $\#E_0(\mathbb{F}_2) = 4$ and

$$t = 2 + 1 - \#E_0(\mathbb{F}_2) = -1.$$

To apply Theorem 6.29, we use the quadratic formula to find the roots of the polynomial $Z^2 + Z + 2$. The roots are

$$\frac{-1 + \sqrt{-7}}{2} \quad \text{and} \quad \frac{-1 - \sqrt{-7}}{2}.$$

Then Theorem 6.29(a) tells us that

$$\#E_0(\mathbb{F}_{2^k}) = 2^k + 1 - \left(\frac{-1 + \sqrt{-7}}{2}\right)^k - \left(\frac{-1 - \sqrt{-7}}{2}\right)^k. \tag{6.12}$$

This formula easily allows us to compute the number of points in $\#E_0(\mathbb{F}_{2^k})$, even for very large values of k. For example,

$$\#E_0(\mathbb{F}_{2^{97}}) = 158456325028528296935114828764.$$

(See also Exercise 6.25.)

Further, Theorem 6.29(b) says that the Frobenius map τ satisfies the equation $\tau^2 + \tau + 2 = 0$ when it acts on points of $E(\mathbb{F}_{2^k})$, i.e.,

$$\tau^2(P) + \tau(P) + 2P = \mathcal{O} \qquad \text{for all } P \in E(\mathbb{F}_{2^k}).$$

The idea now is to write an arbitrary integer n as a sum of powers of τ, subject to the assumption that $\tau^2 = -2 - \tau$. Say we have written n as

$$n = v_0 + v_1\tau + v_2\tau^2 + \cdots + v_\ell\tau^\ell \qquad \text{with } v_i \in \{-1, 0, 1\}.$$

Then we can compute nP efficiently using the formula

$$\begin{aligned} nP &= (v_0 + v_1\tau + v_2\tau^2 + \cdots + v_\ell\tau^\ell)P \\ &= v_0P + v_1\tau(P) + v_2\tau^2(P) + \cdots + v_\ell\tau^\ell(P). \end{aligned}$$

This takes less time than using the binary or ternary method because it is far easier to compute $\tau^i(P)$ than it is to compute 2^iP.

Proposition 6.30. *Let n be a positive integer. Then n can be written in the form*

$$n = v_0 + v_1\tau + v_2\tau^2 + \cdots + v_\ell\tau^\ell \quad \textit{with } v_i \in \{-1, 0, 1\}, \tag{6.13}$$

under the assumption that τ satisfies $\tau^2 = -2 - \tau$. Further, this can always be done with $\ell \approx 2\log n$ and with at most $\frac{1}{3}$ of the v_i nonzero.

Proof. The proof is similar to Proposition 6.18, the basic idea being that we write integers as $2a + b$ with $b \in \{0, 1, -1\}$ and replace 2 with $-\tau - \tau^2$; see Exercise 6.27. With more work, it is possible to find an expansion (6.13) with $\ell \approx \log n$ and approximately $\frac{1}{3}$ of the v_i nonzero; see [29, §15.1]. □

Example 6.31. We illustrate Proposition 6.30 with a numerical example. Let $n = 7$. Then

$$\begin{aligned} 7 &= 1 + 3 \cdot 2 = 1 + 3 \cdot (-\tau - \tau^2) = 1 - 3\tau - 3\tau^2 = 1 - \tau - \tau^2 - 2\tau - 2\tau^2 \\ &= 1 - \tau - \tau^2 - (-\tau - \tau^2)\tau - (-\tau - \tau^2)\tau^2 = 1 - \tau + 2\tau^3 + \tau^4 \\ &= 1 - \tau + (-\tau - \tau^2)\tau^3 + \tau^4 = 1 - \tau - \tau^5. \end{aligned}$$

Thus $7 = 1 - \tau - \tau^5$.

Remark 6.32. As we have seen, computing $\#E(\mathbb{F}_{2^k})$ for Koblitz curves is very easy. However, for general elliptic curves over $\mathbb{F}_{2^k}$, this is a more difficult task. The SEA algorithm and its variants [120, 121] that we mentioned in Remark 6.13 are reasonably efficient at counting the number of points in $E(\mathbb{F}_q)$ for any fields with a large number of elements. Satoh [113] devised an alternative method that is often faster than SEA when $q = p^e$ for a small prime p and (moderately) large exponent e. Satoh's original paper dealt only with the case $p \geq 3$, but subsequent work [44, 140] covers also the cryptographically important case of $p = 2$.

6.8 Bilinear Pairings on Elliptic Curves

You have probably seen examples of bilinear pairings in a linear algebra class. For example, the dot product is a bilinear pairing on the vector space $\mathbb{R}^n$,

$$\beta(\boldsymbol{v}, \boldsymbol{w}) = \boldsymbol{v} \cdot \boldsymbol{w} = v_1 w_1 + v_2 w_2 + \cdots + v_n w_n.$$

It is a pairing in the sense that it takes a pair of vectors and returns a number, and it is bilinear in the sense that it is a linear transformation in each of its variables. In other words, for any vectors $\boldsymbol{v}_1, \boldsymbol{v}_2, \boldsymbol{w}_1, \boldsymbol{w}_2$ and any real numbers a_1, a_2, b_1, b_2, we have

$$\begin{aligned} \beta(a_1\boldsymbol{v}_1 + a_2\boldsymbol{v}_2, \boldsymbol{w}) &= a_1\beta(\boldsymbol{v}_1, \boldsymbol{w}) + a_2\beta(\boldsymbol{v}_2, \boldsymbol{w}), \\ \beta(\boldsymbol{v}, b_1\boldsymbol{w}_1 + b_2\boldsymbol{w}_2) &= b_1\beta(\boldsymbol{v}, \boldsymbol{w}_1) + b_2\beta(\boldsymbol{v}, \boldsymbol{w}_2). \end{aligned} \tag{6.14}$$

More generally, if A is any n-by-n matrix, then the function $\beta(\boldsymbol{v}, \boldsymbol{w}) = \boldsymbol{v}A\boldsymbol{w}^t$ is a bilinear pairing on $\mathbb{R}^n$, where we write $\boldsymbol{v}$ as a row vector and we write $\boldsymbol{w}^t$, the transpose of $\boldsymbol{w}$, as a column vector.

Another bilinear pairing that you have seen is the determinant map on $\mathbb{R}^2$. Thus if $\boldsymbol{v} = (v_1, v_2)$ and $\boldsymbol{w} = (w_1, w_2)$, then

$$\delta(\boldsymbol{v}, \boldsymbol{w}) = \det \begin{pmatrix} v_1 & v_2 \\ w_1 & w_2 \end{pmatrix} = v_1 w_2 - v_2 w_1$$

is a bilinear map. The determinant map has the further property that it is alternating, which means that if we switch the vectors, the value changes sign,

$$\delta(\boldsymbol{v}, \boldsymbol{w}) = -\delta(\boldsymbol{w}, \boldsymbol{v}).$$

Notice that the alternating property implies that $\delta(\boldsymbol{v}, \boldsymbol{v}) = 0$ for every vector $\boldsymbol{v}$.

The bilinear pairings that we discuss in this section are similar in that they take as input two points on an elliptic curve and give as output a number. However, the bilinearity condition is slightly different, because the output value is a nonzero element of a finite field, so the sum on the right-hand side of (6.14) is replaced by a product.

Bilinear pairings on elliptic curves have a number of important cryptographic applications. For most of these applications it is necessary to work with finite fields $\mathbb{F}_{p^k}$ of prime power order. Fields of prime power order are discussed in Sect. 2.10.4, but even if you have not covered that material, you can just imagine a field that is similar to $\mathbb{F}_p$, but that has p^k elements. (N.B. The field $\mathbb{F}_{p^k}$ is very different from the ring $\mathbb{Z}/p^k\mathbb{Z}$; see Exercise 2.40.) Standard references for the material used in this section are [136] and [147].

6.8.1 Points of Finite Order on Elliptic Curves

We begin by briefly describing the points of finite order on an elliptic curve.

Definition. Let $m \geq 1$ be an integer. A point $P \in E$ satisfying $mP = \mathcal{O}$ is called a *point of order* m in the group E. We denote the set of points of order m by

$$E[m] = \{P \in E : mP = \mathcal{O}\}.$$

Such points are called *points of finite order* or *torsion points.*

It is easy to see that if P and Q are in $E[m]$, then $P+Q$ and $-P$ are also in $E[m]$, so $E[m]$ is a subgroup of E. If we want the coordinates of P to lie in a particular field K, for example in $\mathbb{Q}$ or $\mathbb{R}$ or $\mathbb{C}$ or $\mathbb{F}_p$, then we write $E(K)[m]$. (See Exercise 2.12.)

The group of points of order m has a fairly simple structure, at least if we allow the coordinates of the points to be in a sufficiently large field.

Proposition 6.33. *Let $m \geq 1$ be an integer.*

(a) *Let E be an elliptic curve over $\mathbb{Q}$ or $\mathbb{R}$ or $\mathbb{C}$. Then*

$$E(\mathbb{C})[m] \cong \mathbb{Z}/m\mathbb{Z} \times \mathbb{Z}/m\mathbb{Z}$$

is a product of two cyclic groups of order m.

(b) *Let E be an elliptic curve over $\mathbb{F}_p$ and assume that p does not divide m. Then there exists a value of k such that*

$$E(\mathbb{F}_{p^{jk}})[m] \cong \mathbb{Z}/m\mathbb{Z} \times \mathbb{Z}/m\mathbb{Z} \quad \textit{for all } j \geq 1.$$

Proof. For the proof, which is beyond the scope of this book, see any standard text on elliptic curves, for example [136, Corollary III.6.4]. □

Remark 6.34. Notice that if ℓ is prime and if K is a field such that

$$E(K)[\ell] = \mathbb{Z}/\ell\mathbb{Z} \times \mathbb{Z}/\ell\mathbb{Z},$$

then we may view $E[\ell]$ as a 2-dimensional vector space over the field $\mathbb{Z}/\ell\mathbb{Z}$. And even if m is not prime,

$$E(K)[m] = \mathbb{Z}/m\mathbb{Z} \times \mathbb{Z}/m\mathbb{Z}$$

still has a "basis" $\{P_1, P_2\}$ in the sense that every point $P = E[m]$ can be written as a linear combination

$$P = aP_1 + bP_2$$

for a unique choice of coefficients $a, b \in \mathbb{Z}/m\mathbb{Z}$. Of course, if m is large, it may be very difficult to find a and b. Indeed, if P is a multiple of P_1, then finding the value of a is the same as solving the ECDLP for P and P_1.

6.8.2 Rational Functions and Divisors on Elliptic Curves

In order to define the Weil and Tate pairings, we need to explain how a rational function on an elliptic curve is related to its zeros and poles. We start with the simpler case of a rational function of one variable. A *rational function* is a ratio of polynomials

$$f(X) = \frac{a_0 + a_1X + a_2X^2 + \cdots + a_nX^n}{b_0 + b_1X + b_2X^2 + \cdots + b_mX^m}.$$

Any nonzero polynomial can be factored completely if we allow complex numbers, so a nonzero rational function can be factored as

$$f(X) = \frac{a(X-\alpha_1)^{e_1}(X-\alpha_2)^{e_2}\cdots(X-\alpha_r)^{e_r}}{b(X-\beta_1)^{d_1}(X-\beta_2)^{d_2}\cdots(X-\beta_s)^{d_s}}.$$

We may assume that $\alpha_1,\ldots,\alpha_r,\beta_1,\ldots,\beta_s$ are distinct numbers, since otherwise we can cancel some of the terms in the numerator with some of the terms in the denominator. The numbers $\alpha_1,\ldots,\alpha_r$ are called the *zeros of* $f(X)$ and the numbers $\beta_1,\ldots,\beta_s$ are called the *poles of* $f(X)$. The exponents $e_1,\ldots,e_r,d_1,\ldots,d_s$ are the associated *multiplicities*. We keep track of the zeros and poles of $f(X)$ and their multiplicities by defining the *divisor of* $f(X)$ to be the formal sum

$$\operatorname{div}(f(X)) = e_1[\alpha_1] + e_2[\alpha_2] + \cdots + e_r[\alpha_r] - d_1[\beta_1] - d_2[\beta_2] - \cdots - d_r[\beta_r].$$

Note that this is simply a convenient shorthand way of saying that $f(X)$ has a zero of multiplicity e_1 at α_1, a zero of multiplicity e_2 at α_2, etc.

If E is an elliptic curve,

$$E : Y^2 = X^3 + AX + B,$$

and if $f(X,Y)$ is a nonzero rational function of two variables, we may view f as defining a function on E by writing points as $P = (x,y)$ and setting $f(P) = f(x,y)$. Then just as for rational functions of one variable, there are points of E where the numerator of f vanishes and there are points of E where the denominator of f vanishes, so f has zeros and poles on E. Further, one can assign multiplicities to the zeros and poles, so f has an associated divisor

$$\operatorname{div}(f) = \sum_{P\in E} n_P[P].$$

In this formal sum, the coefficients n_P are integers, and only finitely many of the n_P are nonzero, so $\operatorname{div}(f)$ is a finite sum. Of course, the coordinates of the zeros and poles of f may require moving to a larger field. For example, if E is defined over $\mathbb{F}_p$, then the poles and zeros of f have coordinates in $\mathbb{F}_{p^k}$ for some k, but the value of k will, in general, depend on the function f.

Example 6.35. Suppose that the cubic polynomial used to define E factors as

$$X^3 + AX + B = (X - \alpha_1)(X - \alpha_2)(X - \alpha_3).$$

Then the points $P_1 = (\alpha_1, 0)$, $P_2 = (\alpha_2, 0)$, and $P_3 = (\alpha_3, 0)$ are distinct (see Remark 6.4) and satisfy $2P_1 = 2P_2 = 2P_3 = \mathcal{O}$, i.e., they are points of order 2. The function Y, which remember is defined by

$$Y(P) = (\text{the } y\text{-coordinate of the point } P),$$

vanishes at these three points and at no other points $P = (x, y)$. The divisor of Y has the form $[P_1] + [P_2] + [P_3] - n[\mathcal{O}]$ for some integer n, and it follows from Theorem 6.36 that $n = 3$, so

$$\operatorname{div}(Y) = [P_1] + [P_2] + [P_3] - 3[\mathcal{O}].$$

More generally, we define a *divisor on* E to be any formal sum

$$D = \sum_{P \in E} n_P[P] \quad \text{with } n_P \in \mathbb{Z} \text{ and } n_P = 0 \text{ for all but finitely many } P.$$

The *degree of a divisor* is the sum of its coefficients,

$$\deg(D) = \deg\Bigl(\sum_{P \in E} n_P[P]\Bigr) = \sum_{P \in E} n_P.$$

We define the *sum of a divisor* by dropping the square brackets; thus

$$\operatorname{Sum}(D) = \operatorname{Sum}\Bigl(\sum_{P \in E} n_P[P]\Bigr) = \sum_{P \in E} n_P P.$$

Note that $n_P P$ means to add P to itself n_P times using the addition law on E. It is natural to ask which divisors are divisors of functions, and to what extent the divisor of a function determines the function. These questions are answered by the following theorem.

Theorem 6.36. *Let E be an elliptic curve.*

(a) *Let f and g be nonzero rational functions on E. If $\operatorname{div}(f) = \operatorname{div}(g)$, then there is a nonzero constant c such that $f = cg$.*

(b) *Let $D = \sum_{P \in E} n_P[P]$ be a divisor on E. Then D is the divisor of a rational function on E if and only if*

$$\deg(D) = 0 \quad \textit{and} \quad \operatorname{Sum}(D) = \mathcal{O}.$$

In particular, if a rational function on E has no zeros or no poles, then it is constant.

Proof. Again we refer the reader to any elliptic curve textbook such as [136, Propositions II.3.1 and III.3.4]. □

Example 6.37. Suppose that $P \in E[m]$ is a point of order m. By definition, $mP = \mathcal{O}$, so the divisor

$$m[P] - m[\mathcal{O}]$$

satisfies the conditions of Theorem 6.36(b). Hence there is a rational function $f_P(X,Y)$ on E satisfying

$$\operatorname{div}(f_P) = m[P] - m[\mathcal{O}].$$

The case $m = 2$ is particularly simple. A point $P \in E$ has order 2 if and only if its Y-coordinate vanishes. If we let $P = (\alpha, 0) \in E[2]$, then the function $f_P = X - \alpha$ satisfies

$$\operatorname{div}(X - \alpha) = 2[P] - 2[\mathcal{O}];$$

see Exercise 6.30.

6.8.3 The Weil Pairing

The Weil pairing, which is denoted by e_m, takes as input a pair of points $P, Q \in E[m]$ and gives as output an mth root of unity $e_m(P,Q)$. The bilinearity of the Weil pairing is expressed by the equations

$$\begin{aligned} e_m(P_1 + P_2, Q) &= e_m(P_1, Q)e_m(P_2, Q), \\ e_m(P, Q_1 + Q_2) &= e_m(P, Q_1)e_m(P, Q_2). \end{aligned} \tag{6.15}$$

This is similar to the vector space bilinearity described in (6.14), but note that the bilinearity in (6.15) is multiplicative, in the sense that the quantities on the right-hand side are multiplied, while the bilinearity in (6.14) is additive, in the sense that the quantities on the right-hand side are added.

Definition. Let $P, Q \in E[m]$, i.e., P and Q are points of order m in the group E. Let f_P and f_Q be rational functions on E satisfying

$$\operatorname{div}(f_P) = m[P] - m[\mathcal{O}] \qquad \text{and} \qquad \operatorname{div}(f_Q) = m[Q] - m[\mathcal{O}].$$

(See Example 6.37.) The *Weil pairing of P and Q* is the quantity

$$e_m(P,Q) = \frac{f_P(Q+S)}{f_P(S)} \bigg/ \frac{f_Q(P-S)}{f_Q(-S)}, \tag{6.16}$$

where $S \in E$ is any point satisfying $S \notin \{\mathcal{O}, P, -Q, P-Q\}$. (This ensures that all of the quantities on the right-hand side of (6.16) are defined and nonzero.) One can check that the value of $e_m(P,Q)$ does not depend on the choice of f_P, f_Q, and S; see Exercise 6.32.

Despite its somewhat arcane definition, the Weil pairing e_m has many useful properties.

Theorem 6.38. (a) *The values of the Weil pairing satisfy*

$$e_m(P,Q)^m = 1 \quad \textit{for all } P,Q \in E[m].$$

In other words, $e_m(P,Q)$ is an mth root of unity.

(b) *The Weil pairing is* bilinear, *which means that*

$$e_m(P_1+P_2,Q) = e_m(P_1,Q)e_m(P_2,Q) \quad \textit{for all } P_1,P_2,Q \in E[m],$$

and

$$e_m(P,Q_1+Q_2) = e_m(P,Q_1)e_m(P,Q_2) \quad \textit{for all } P,Q_1,Q_2 \in E[m].$$

(c) *The Weil pairing is* alternating, *which means that*

$$e_m(P,P) = 1 \quad \textit{for all } P \in E[m].$$

This implies that $e_m(P,Q) = e_m(Q,P)^{-1}$ for all $P,Q, \in E[m]$, see Exercise 6.31.

(d) *The Weil pairing is* nondegenerate, *which means that*

$$\textit{if } e_m(P,Q) = 1 \textit{ for all } Q \in E[m], \textit{ then } P = \mathcal{O}.$$

Proof. Some parts of Theorem 6.38 are easy to prove, while other parts are not so easy. For a complete proof, see for example [136, Section III.8]. □

Remark 6.39. Where does the Weil pairing come from? According to Proposition 6.33 (see also Remark 6.34), if we allow points with coordinates in a sufficiently large field, then $E[m]$ looks like a 2-dimensional "vector space" over the "field" $\mathbb{Z}/m\mathbb{Z}$. So if we choose a basis $P_1, P_2 \in E[m]$, then any element $P \in E[m]$ can be written in terms of this basis as

$$P = a_P P_1 + b_P P_2 \quad \text{for unique } a_P, b_P \in \mathbb{Z}/m\mathbb{Z},$$

and then we can define an alternating bilinear pairing by using the determinant,

$$E[m] \times E[m] \longrightarrow \mathbb{Z}/m\mathbb{Z}, \qquad (P,Q) \longmapsto \det\begin{pmatrix} a_P & a_Q \\ b_P & b_Q \end{pmatrix} = a_P b_Q - a_Q b_P.$$

But there are two problems with this pairing. First, it depends on choosing a basis, and second, there's no easy way to compute it other than writing P and Q in terms of the basis. However, it should come as no surprise that the determinant and the Weil pairing are closely related to one another. To be precise, if we let $\zeta = e_m(P_1,P_2)$, then it is easy to check that (see Exercise 6.33)

$$e_m(P,Q) = \zeta^{\det\begin{pmatrix} a_P & a_Q \\ b_P & b_Q \end{pmatrix}} = \zeta^{a_P b_Q - a_Q b_P}.$$

The glory[10] of the Weil pairing is that it can be computed quite efficiently without first expressing P and Q in terms of any particular basis of $E[m]$. (See Sect. 6.8.4 for a double-and-add algorithm to compute $e_m(P, Q)$.) This is good, since expressing a point in terms of the basis P_1 and P_2 is at least as difficult as solving the ECDLP; see Exercise 6.10.

Example 6.40. We are going to compute e_2 directly from the definition. Let E be given by the equation

$$Y^2 = X^3 + Ax + B = (X - \alpha_1)(X - \alpha_2)(X - \alpha_3).$$

Note that $\alpha_1 + \alpha_2 + \alpha_3 = 0$, since the left-hand side has no X^2 term. The points

$$P_1 = (\alpha_1, 0), \qquad P_2 = (\alpha_2, 0), \qquad P_3 = (\alpha_3, 0),$$

are points of order 2, and as noted in Example 6.37 (see also Exercise 6.30),

$$\operatorname{div}(X - \alpha_i) = 2[P_i] - 2[\mathcal{O}].$$

In order to compute $e_2(P_1, P_2)$, we can take an arbitrary point $S = (x, y) \in E$. Using the addition formula, we find that the x-coordinate of $P_1 - S$ is equal to

$$\begin{aligned} X(P_1 - S) &= \left(\frac{-y}{x - \alpha_1}\right)^2 - x - \alpha_1 \\ &= \frac{y^2 - (x - \alpha_1)^2(x + \alpha_1)}{(x - \alpha_1)^2} \\ &= \frac{(x - \alpha_1)(x - \alpha_2)(x - \alpha_3) - (x - \alpha_1)^2(x + \alpha_1)}{(x - \alpha_1)^2} \\ &\qquad\qquad \text{since } y^2 = (x - \alpha_1)(x - \alpha_2)(x - \alpha_3), \\ &= \frac{(x - \alpha_2)(x - \alpha_3) - (x - \alpha_1)(x + \alpha_1)}{x - \alpha_1} \\ &= \frac{(-\alpha_2 - \alpha_3)x + \alpha_2\alpha_3 + \alpha_1^2}{x - \alpha_1} \\ &= \frac{\alpha_1 x + \alpha_2\alpha_3 + \alpha_1^2}{x - \alpha_1} \qquad \text{since } \alpha_1 + \alpha_2 + \alpha_3 = 0. \end{aligned}$$

Similarly,

$$X(P_2 + S) = \frac{\alpha_2 x + \alpha_1\alpha_3 + \alpha_2^2}{x - \alpha_2}.$$

[10]For those who have taken a course in abstract algebra, we mention that the other glorious property of the Weil pairing is that it interacts well with Galois theory. Thus let E be an elliptic curve over a field K, let L/K be a Galois extension, and let $P, Q \in E(L)[m]$. Then for every element $g \in \mathrm{Gal}(L/K)$, the Weil pairing obeys the rule $e_m\big(g(P), g(Q)\big) = g\big(e_m(P, Q)\big)$.

Using the rational functions $f_{P_i} = X - \alpha_i$ and assuming that P_1 and P_2 are distinct nonzero points in $E[2]$, we find directly from the definition of e_m that

$$\begin{aligned}
e_2(P_1, P_2) &= \frac{f_{P_1}(P_2+S)}{f_{P_1}(S)} \Big/ \frac{f_{P_2}(P_1-S)}{f_{P_2}(-S)} \\
&= \frac{X(P_2+S)-\alpha_1}{X(S)-\alpha_1} \Big/ \frac{X(P_1-S)-\alpha_2}{X(-S)-\alpha_2} \\
&= \frac{\frac{\alpha_2 x+\alpha_1\alpha_3+\alpha_2^2}{x-\alpha_2} - \alpha_1}{x-\alpha_1} \Big/ \frac{\frac{\alpha_1 x+\alpha_2\alpha_3+\alpha_1^2}{x-\alpha_1} - \alpha_2}{x-\alpha_2} \\
&= \frac{(\alpha_2-\alpha_1)x+\alpha_1\alpha_3+\alpha_2^2+\alpha_1\alpha_2}{(\alpha_1-\alpha_2)x+\alpha_2\alpha_3+\alpha_1^2+\alpha_1\alpha_2} \\
&= \frac{(\alpha_2-\alpha_1)x+\alpha_2^2-\alpha_1^2}{(\alpha_1-\alpha_2)x+\alpha_1^2-\alpha_2^2} \qquad \text{since } \alpha_1+\alpha_2+\alpha_3=0, \\
&= -1.
\end{aligned}$$

6.8.4 An Efficient Algorithm to Compute the Weil Pairing

In this section we describe a double-and-add method that can be used to efficiently compute the Weil pairing. The key idea, which is due to Victor Miller [89], is an algorithm to rapidly evaluate certain functions with specified divisors, as explained in the next theorem. (For further material on Miller's algorithm, see [136, Section XI.8].)

Theorem 6.41. *Let E be an elliptic curve and let $P = (x_P, y_P)$ and $Q = (x_Q, y_Q)$ be nonzero points on E.*

(a) *Let λ be the slope of the line connecting P and Q, or the slope of the tangent line to E at P if $P = Q$. (If the line is vertical, we let $\lambda = \infty$.) Define a function $g_{P,Q}$ on E as follows:*

$$g_{P,Q} = \begin{cases} \dfrac{y - y_P - \lambda(x - x_P)}{x + x_P + x_Q - \lambda^2} & \text{if } \lambda \neq \infty, \\ x - x_P & \text{if } \lambda = \infty. \end{cases}$$

Then

$$\operatorname{div}(g_{P,Q}) = [P] + [Q] - [P+Q] - [\mathcal{O}]. \tag{6.17}$$

(b) (Miller's Algorithm) *Let $m \geq 1$ and write the binary expansion of m as*

$$m = m_0 + m_1 \cdot 2 + m_2 \cdot 2^2 + \cdots + m_{n-1} 2^{n-1}$$

with $m_i \in \{0, 1\}$ and $m_{n-1} \neq 0$. The following algorithm returns a function f_P whose divisor satisfies

$$\operatorname{div}(f_P) = m[P] - [mP] - (m-1)[\mathcal{O}],$$

where the functions $g_{T,T}$ and $g_{T,P}$ used by the algorithm are as defined in (a).

```
[1]  Set T = P and f = 1
[2]  Loop i = n − 2 down to 0
[3]      Set f = f^2 · g_{T,T}
[4]      Set T = 2T
[5]      If m_i = 1
[6]          Set f = f · g_{T,P}
[7]          Set T = T + P
[8]      End If
[9]  End i Loop
[10] Return the value f
```

In particular, if $P \in E[m]$, *then* $\operatorname{div}(f_P) = m[P] - m[\mathcal{O}]$.

Proof. (a) Suppose first that $\lambda \neq \infty$ and let $y = \lambda x + \nu$ be the line through P and Q or the tangent line at P if $P = Q$. This line intersects E at the three points P, Q, and $-P-Q$, so

$$\operatorname{div}(y - \lambda x - \nu) = [P] + [Q] + [-P-Q] - 3[\mathcal{O}].$$

Vertical lines intersect E at points and their negatives, so

$$\operatorname{div}(x - x_{P+Q}) = [P+Q] + [-P-Q] - 2[\mathcal{O}].$$

It follows that

$$g_{P,Q} = \frac{y - \lambda x - \nu}{x - x_{P+Q}}$$

has the desired divisor (6.17). Finally, the addition formula (Theorem 6.6) tells us that $x_{P+Q} = \lambda^2 - x_P - x_Q$, and we can eliminate ν from the numerator of $g_{P,Q}$ using $y_P = \lambda x_P + \nu$.

If $\lambda = \infty$, then $P+Q = \mathcal{O}$, so we want $g_{P,Q}$ to have divisor $[P]+[-P]-2[\mathcal{O}]$. The function $x - x_P$ has this divisor.

(b) This is a standard double-and-add algorithm, similar to others that we have seen in the past. The key to the algorithm comes from (a), which tells us that the functions $g_{T,T}$ and $g_{T,P}$ used in Steps 3 and 6 have divisors

$$\operatorname{div}(g_{T,T}) = 2[T] - [2T] - [\mathcal{O}] \quad \text{and} \quad \operatorname{div}(g_{T,P}) = [T] + [P] - [T+P] - [\mathcal{O}].$$

We leave to the reader the remainder of the proof, which is a simple induction using these relations. □

Let $P \in E[m]$. The algorithm described in Theorem 6.41 tells us how to compute a function f_P with divisor $m[P] - m[\mathcal{O}]$. Further, if R is any point of E, then we can compute $f_P(R)$ directly by evaluating the functions $g_{T,T}(R)$ and $g_{T,P}(R)$ each time we execute Steps 3 and 6 of the algorithm. Notice that quantities of the form $f_P(R)$ are exactly what are needed in order to evaluate the Weil pairing $e_m(P,Q)$. More precisely, given nonzero points $P,Q \in E[m]$, we choose a point $S \notin \{\mathcal{O}, P, -Q, P-Q\}$ and use Theorem 6.41 to evaluate

$$e_m(P,Q) = \frac{f_P(Q+S)}{f_P(S)} \Bigg/ \frac{f_Q(P-S)}{f_Q(-S)}$$

by computing each of the functions at the indicated point.

Remark 6.42. For added efficiency, one can compute $f_P(Q+S)$ and $f_P(S)$ simultaneously, and similarly for $f_Q(P-S)$ and $f_Q(-S)$. Further savings are available using the Tate pairing, which is a variant of the Weil pairing that we describe briefly in Sect. 6.8.5.

Example 6.43. We take the elliptic curve

$$y^2 = x^3 + 30x + 34 \quad \text{over the finite field } \mathbb{F}_{631}.$$

The curve has $\#E(\mathbb{F}_{631}) = 650 = 2 \cdot 5^2 \cdot 13$ points, and it turns out that it has 25 points of order 5. The points

$$P = (36, 60) \quad \text{and} \quad Q = (121, 387)$$

generate the points of order 5 in $E(\mathbb{F}_{631})$. In order to compute the Weil pairing using Miller's algorithm, we want a point S that is not in the subgroup spanned by P and Q. We take $S = (0, 36)$. The point S has order 130. Then Miller's algorithm gives

$$\frac{f_P(Q+S)}{f_P(S)} = \frac{103}{219} = 473 \in \mathbb{F}_{631}.$$

Reversing the roles of P and Q and replacing S by $-S$, Miller's algorithm also gives

$$\frac{f_Q(P-S)}{f_Q(-S)} = \frac{284}{204} = 88 \in \mathbb{F}_{631}.$$

Finally, taking the ratio of these two values yields

$$e_5(P,Q) = \frac{473}{88} = 242 \in \mathbb{F}_{631}.$$

We check that $(242)^5 = 1$, so $e_5(P,Q)$ is a fifth root of unity in $\mathbb{F}_{631}$.

Continuing to work on the same curve, we take $P' = (617, 5)$ and $Q' = (121, 244)$. Then a similar calculation gives

$$\frac{f_{P'}(Q'+S)}{f_{P'}(S)} = \frac{326}{523} = 219 \quad \text{and} \quad \frac{f_{Q'}(P'-S)}{f_{Q'}(-S)} = \frac{483}{576} = 83,$$

and taking the ratio of these two values yields

$$e_5(P',Q') = \frac{219}{83} = 512 \in \mathbb{F}_{631}.$$

It turns out that $P' = 3P$ and $Q' = 4Q$. We check that

$$e_5(P,Q)^{12} = 242^{12} = 512 = e_5(P',Q') = e_5(3P,4Q),$$

which illustrates the bilinearity property of the Weil pairing.

6.8.5 The Tate Pairing

The Weil pairing is a nondegenerate bilinear form on elliptic curves defined over any field. For elliptic curves over finite fields there is another pairing, called the *Tate pairing* (or sometimes the *Tate–Lichtenbaum pairing*), that is often used in cryptography because it is computationally somewhat more efficient than the Weil pairing. In this section we briefly describe the Tate pairing. (For further material on the Tate pairing, see [136, Section XI.9].)

Definition. Let E be an elliptic curve over $\mathbb{F}_q$, let ℓ be a prime, let $P \in E(\mathbb{F}_q)[\ell]$, and let $Q \in E(\mathbb{F}_q)$. Choose a rational function f_P on E with

$$\operatorname{div}(f_P) = \ell[P] - \ell[\mathcal{O}].$$

The *Tate pairing of P and Q* is the quantity

$$\tau(P,Q) = \frac{f_P(Q+S)}{f_P(S)} \in \mathbb{F}_q^*,$$

where S is any point in $E(\mathbb{F}_q)$ such that $f_P(Q+S)$ and $f_P(S)$ are defined and nonzero. It turns out that the value of the Tate pairing is well-defined only up to multiplying it by the ℓth power of an element of $\mathbb{F}_q^*$. If $q \equiv 1 \pmod{\ell}$, we define the *(modified) Tate pairing of P and Q* to be

$$\hat{\tau}(P,Q) = \tau(P,Q)^{(q-1)/\ell} = \left(\frac{f_P(Q+S)}{f_P(S)}\right)^{(q-1)/\ell} \in \mathbb{F}_q^*.$$

Theorem 6.44. *Let E be an elliptic curve over $\mathbb{F}_q$ and let ℓ be a prime with*

$$q \equiv 1 \pmod{\ell} \qquad \text{and} \qquad E(\mathbb{F}_q)[\ell] \cong \mathbb{Z}/\ell\mathbb{Z}.$$

Then the modified Tate pairing gives a well-defined map

$$\hat{\tau} : E(\mathbb{F}_q)[\ell] \times E(\mathbb{F}_q)[\ell] \longrightarrow \mathbb{F}_q^*$$

having the following properties:
(a) Bilinearity:

$$\hat{\tau}(P_1+P_2,Q) = \hat{\tau}(P_1,Q)\hat{\tau}(P_2,Q) \quad \text{and} \quad \hat{\tau}(P,Q_1+Q_2) = \hat{\tau}(P,Q_1)\hat{\tau}(P,Q_2).$$

(b) Nondegeneracy:

$\hat{\tau}(P,P)$ is a primitive ℓth root of unity for all nonzero $P \in E(\mathbb{F}_q)[\ell]$.

(A primitive ℓth root of unity is a number $\zeta \neq 1$ such that $\zeta^\ell = 1$.)

In applications such as tripartite Diffie–Hellman (Sect. 6.10.1) and ID-based cryptography (Sect. 6.10.2), one may use the Tate pairing in place of the Weil pairing. Note that Miller's algorithm gives an efficient way to compute the Tate pairing, since Theorem 6.41(b) explains how to rapidly compute the value of f_P.

6.9 The Weil Pairing over Fields of Prime Power Order

There are many applications of the Weil pairing in which it is necessary to work in fields $\mathbb{F}_{p^k}$ of prime power order. In this section we discuss the m-embedding degree, which is the smallest value of k such that $E(\mathbb{F}_{p^k})[m]$ is as large as possible, and we give an application called the MOV algorithm that reduces the ECDLP in $E(\mathbb{F}_p)$ to the DLP in $\mathbb{F}_{p^k}^*$. We then describe distortion maps on E and use them to define a modified Weil pairing $\hat{e}_m$ for which $\hat{e}_m(P, P)$ is nontrivial.

6.9.1 Embedding Degree and the MOV Algorithm

Let E be an elliptic curve over $\mathbb{F}_p$ and let $m \geq 1$ be an integer with $p \nmid m$. In order to obtain nontrivial values of the Weil pairing e_m, we need to use independent points of order m on E. According to Proposition 6.33(b), the curve E has m^2 points of order m, but their coordinates may lie in a larger finite field.

Definition. Let E be an elliptic curve over $\mathbb{F}_p$ and let $m \geq 1$ be an integer with $p \nmid m$. The *embedding degree of E with respect to m* is the smallest value of k such that

$$E(\mathbb{F}_{p^k})[m] \cong \mathbb{Z}/m\mathbb{Z} \times \mathbb{Z}/m\mathbb{Z}.$$

For cryptographic applications, the most interesting case occurs when m is a (large) prime, in which case there are alternative characterizations of the embedding degree, as in the following result.

Proposition 6.45. *Let E be an elliptic curve over $\mathbb{F}_p$ and let $\ell \neq p$ be a prime. Assume that $E(\mathbb{F}_p)$ contains a point of order ℓ. Then the embedding degree of E with respect to ℓ is given by one of the following cases:*

(i) *The embedding degree of E is 1. (This cannot happen if $\ell > \sqrt{p} + 1$; see Exercise 6.39.)*

(ii) $p \equiv 1 \pmod{\ell}$ *and the embedding degree is ℓ.*

(iii) $p \not\equiv 1 \pmod{\ell}$ *and the embedding degree is the smallest value of $k \geq 2$ such that*

$$p^k \equiv 1 \pmod{\ell}.$$

Proof. The proof uses more advanced methods than we have at our disposal. See [147, Proposition 5.9] for a proof of case (iii), which is the case that most often occurs in practice. □

The significance of the embedding degree k is that the Weil pairing embeds the ECDLP on the elliptic curve $E(\mathbb{F}_p)$ into the DLP in the field $\mathbb{F}_{p^k}$. The basic setup is as follows. Let E be an elliptic curve over $\mathbb{F}_p$ and let $P \in E(\mathbb{F}_p)$ be a point of order ℓ, where ℓ is a large prime, say $\ell > \sqrt{p} + 1$. Let k be the embedding degree with respect to ℓ and suppose that we know how to solve the discrete logarithm problem in the field $\mathbb{F}_{p^k}$. Let $Q \in E(\mathbb{F}_p)$ be a point that is a multiple of P. Then the following algorithm of Menezes, Okamoto, and Vanstone [82] solves the elliptic curve discrete logarithm problem for P and Q.

The MOV Algorithm

1. Compute the number of points $N = \#E(\mathbb{F}_{p^k})$. This is feasible if k is not too large, since there are polynomial-time algorithms to count the number of points on an elliptic curve; see Remarks 6.13 and 6.32. Note that $\ell \mid N$, since by assumption $E(\mathbb{F}_p)$ has a point of order ℓ.
2. Choose a random point $T \in E(\mathbb{F}_{p^k})$ with $T \notin E(\mathbb{F}_p)$.
3. Compute $T' = (N/\ell)T$. If $T' = \mathcal{O}$, go back to Step 2. Otherwise, T' is a point of order ℓ, so proceed to Step 4.
4. Compute the Weil pairing values
$$\alpha = e_\ell(P,T') \in \mathbb{F}_{p^k}^* \qquad \text{and} \qquad \beta = e_\ell(Q,T') \in \mathbb{F}_{p^k}^*.$$
This can be done quite efficiently, in time proportional to $\log(p^k)$, see Sect. 6.8.4. If $\alpha = 1$, return to Step 2.
5. Solve the DLP for α and β in $\mathbb{F}_{p^k}^*$, i.e., find an exponent n such that $\beta = \alpha^n$. If p^k is not too large, this can be done using the index calculus. Note that the index calculus (Sect. 3.8) is a subexponential algorithm, so it is considerably faster than collision algorithms such as Pollard's ρ method (Sects. 5.4 and 5.5).
6. Then also $Q = nP$, so the ECDLP has been solved.

The MOV algorithm is summarized in Table 6.10. A few comments are in order.

Remark 6.46. How does one generate a random point $T \in E(\mathbb{F}_{p^k})$ with $T \notin E(\mathbb{F}_p)$ in Step 2? One method is to choose random values $x \in \mathbb{F}_{p^k}$ and check whether $x^3 + Ax + B$ is a square in $\mathbb{F}_{p^k}$, which is easy to do, since z is a square in $\mathbb{F}_{p^k}$ if and only if $z^{(p^k-1)/2} = 1$. (We are assuming that p is an odd prime.) There then exist practical (i.e., polynomial time) algorithms to compute square roots in finite fields, but to describe them would take us too far afield; see [28, §§1.5.1, 1.5.2].

Remark 6.47. Why does the MOV algorithm solve the ECDLP? The point T' constructed by the algorithm is generally independent of P, so the pair of points $\{P, T'\}$ forms a basis for the 2-dimensional vector space
$$E[\ell] = \mathbb{Z}/\ell\mathbb{Z} \times \mathbb{Z}/\ell\mathbb{Z}.$$

It follows from the nondegeneracy of the Weil pairing that $e_\ell(P, T')$ is a nontrivial ℓth root of unity in $\mathbb{F}_{p^k}^*$. In other words,

$$e_\ell(P, T')^r = 1 \quad \text{if and only if} \quad \ell \mid r.$$

Suppose now that $Q = jP$ and that our goal is to find the value of j modulo ℓ. The MOV algorithm finds an integer n satisfying $e_\ell(Q, T') = e_\ell(P, T')^n$. The linearity of the Weil pairing implies that

$$e_\ell(P, T')^n = e_\ell(Q, T') = e_\ell(jP, T') = e_\ell(P, T')^j,$$

so $e_\ell(P, T')^{n-j} = 1$. Hence $n \equiv j \pmod{\ell}$, which shows that n solves the ECDLP for P and Q.

Remark 6.48. How practical is the MOV algorithm? The answer, obviously, depends on the size of k. If k is large, say $k > (\ln p)^2$, then the MOV algorithm is completely infeasible. For example, if $p \approx 2^{160}$, then we would have to solve the DLP in $\mathbb{F}_{p^k}$ with $k > 4000$. Since a randomly chosen elliptic curves over $\mathbb{F}_p$ almost always has embedding degree that is much larger than $(\ln p)^2$, it would seem that the MOV algorithm is not useful. However, there are certain special sorts of curves whose embedding degree is small. An important class of such curves consists of those satisfying

$$\#E(\mathbb{F}_p) = p + 1.$$

These *supersingular elliptic curves* generally have embedding degree $k = 2$, and in any case $k \le 6$. For example,

$$E : y^2 = x^3 + x$$

is supersingular for any prime $p \equiv 3 \pmod{4}$, and it has embedding degree 2 for any $\ell > \sqrt{p} + 1$. This means that solving ECLDP in $E(\mathbb{F}_p)$ is no harder than solving DLP in $\mathbb{F}_{p^2}^*$, which makes E a very poor choice for use in cryptography.[11]

Remark 6.49. An elliptic curve E over a finite field $\mathbb{F}_p$ is called *anomalous* if $\#E(\mathbb{F}_p) = p$. A number of people [114, 122, 141] more or less simultaneously observed that there is a very fast (linear time) algorithm to solve the ECDLP on anomalous elliptic curves, so such curves must be avoided in cryptographic constructions.

There are also some cases in which the ECDLP is easier than expected for elliptic curves E over finite fields $\mathbb{F}_{2^m}$ when m is composite. (A reason to use such fields is that field operations can sometimes be done more efficiently.) This attack uses a tool called Weil descent and was originally suggested by Gerhard Frey. The idea is to transfer an ECDLP in $E(\mathbb{F}_{2^m})$ to a discrete logarithm problem on a hyperelliptic curve (see Sect. 8.10) over a smaller field $\mathbb{F}_{2^k}$, where k divides m. The details are complicated and beyond the scope of this book. See [29, §22.3] for details.

[11] Or so it would seem, but we will see in Sect. 6.9.3 that the ECDLP on E does have its uses in cryptography!

1. Compute the number of points $N = \#E(\mathbb{F}_{p^k})$.
2. Choose a random point $T \in E(\mathbb{F}_{p^k})$ with $T \notin E(\mathbb{F}_p)$.
3. Let $T' = (N/\ell)T$. If $T' = \mathcal{O}$, go back to Step 2. Otherwise T' is a point of order ℓ, so proceed to Step 4.
4. Compute the Weil pairing values
$$\alpha = e_\ell(P, T') \in \mathbb{F}_{p^k}^* \quad \text{and} \quad \beta = e_\ell(Q, T') \in \mathbb{F}_{p^k}^*.$$
If $\alpha = 1$, go to Step 2.
5. Solve the DLP for α and β in $\mathbb{F}_{p^k}^*$, i.e., find an exponent n such that $\beta = \alpha^n$.
6. Then also $Q = nP$, so the ECDLP has been solved.

Table 6.10: The MOV algorithm to solve the ECDLP

6.9.2 Distortion Maps and a Modified Weil Pairing

The Weil pairing is alternating, which means that $e_m(P, P) = 1$ for all P. In cryptographic applications we generally want to evaluate the pairing at points $P_1 = aP$ and $P_2 = bP$, but using the Weil pairing directly is not helpful, since
$$e_m(P_1, P_2) = e_m(aP, bP) = e_m(P, P)^{ab} = 1^{ab} = 1.$$
One way around this dilemma is to choose an elliptic curve that has a "nice" map $\phi : E \to E$ with the property that P and $\phi(P)$ are "independent" in $E[m]$. Then we can evaluate
$$e_m\bigl(P_1, \phi(P_2)\bigr) = e_m\bigl(aP, \phi(bP)\bigr) = e_m\bigl(aP, b\phi(P)\bigr) = e_m\bigl(P, \phi(P)\bigr)^{ab}.$$
For cryptographic applications one generally takes m to be prime, so we restrict our attention to this case.

Definition. Let $\ell \geq 3$ be a prime, let E be an elliptic curve, let $P \in E[\ell]$ be a point of order ℓ, and let $\phi : E \to E$ be a map from E to itself. We say that ϕ is an *ℓ-distortion map for P* if it has the following two properties[12]:

(i) $\phi(nP) = n\phi(P)$ for all $n \geq 1$.

(ii) The number $e_\ell\bigl(P, \phi(P)\bigr)$ is a *primitive ℓth root of unity*. This means that
$$e_\ell\bigl(P, \phi(P)\bigr)^r = 1 \qquad \text{if and only if} \qquad r \text{ is a multiple of } \ell.$$

The next proposition gives various ways to check condition (ii).

Proposition 6.50. *Let E be an elliptic curve, let $\ell \geq 3$ be a prime, and view $E[\ell] = \mathbb{Z}/\ell\mathbb{Z} \times \mathbb{Z}/\ell\mathbb{Z}$ as a 2-dimensional vector space over the field $\mathbb{Z}/\ell\mathbb{Z}$. Let $P, Q \in E[\ell]$. Then the following are equivalent:*

[12] There are various definitions of distortion maps in the literature. The one that we give distills the essential properties needed for most cryptographic applications. In practice, one also requires an efficient algorithm to compute ϕ.

(a) *P and Q form a basis for the vector space $E[\ell]$.*
(b) *$P \neq \mathcal{O}$ and Q is not a multiple of P.*
(c) *$e_\ell(P,Q)$ is a primitive ℓth root of unity.*
(d) *$e_\ell(P,Q) \neq 1$.*

Proof. It is clear that (a) implies (b), since a basis consists of independent vectors. Conversely, suppose that (a) is false. This means that there is a linear relation

$$uP + vQ = \mathcal{O} \quad \text{with } u,v \in \mathbb{Z}/\ell\mathbb{Z} \text{ not both } 0.$$

If $v = 0$, then $P = \mathcal{O}$, so (b) is false. And if $v \neq 0$, then v has an inverse in $\mathbb{Z}/\ell\mathbb{Z}$, so $Q = -v^{-1}uP$ is a multiple of P, again showing that (b) is false. This completes the proof that (a) and (b) are equivalent.

To ease notation, we let

$$\zeta = e_\ell(P,Q).$$

From the definition of the Weil pairing, we know that $\zeta^\ell = 1$. Let $r \geq 1$ be the smallest integer such that $\zeta^r = 1$. Use the extended Euclidean algorithm (Theorem 1.11) to write the greatest common divisor of r and ℓ as

$$sr + t\ell = \gcd(r,\ell) \quad \text{for some } s,t \in \mathbb{Z}.$$

Then

$$\zeta^{\gcd(r,\ell)} = \zeta^{sr+t\ell} = (\zeta^r)^s(\zeta^\ell)^t = 1.$$

The minimality of r tells us that $r = \gcd(r,\ell)$, so $r \mid \ell$. Since ℓ is prime, it follows that either $r = 1$, so $\zeta = 1$, or else $r = \ell$. This proves that (c) and (d) are equivalent.

We next verify that (a) implies (d). So we are given that P and Q are a basis for $E[\ell]$. In particular, $P \neq \mathcal{O}$, so the nondegeneracy of the Weil pairing tells us that there is a point $R \in E[\ell]$ with $e_\ell(P,R) \neq 1$. Since P and Q are a basis for $E[\ell]$, we can write R as a linear combination of P and Q, say

$$R = uP + vQ.$$

Then the bilinearity and alternating properties of the Weil pairing yield

$$1 \neq e_\ell(P,R) = e_\ell(P, uP + vQ) = e_\ell(P,P)^u e_\ell(P,Q)^v = e_\ell(P,Q)^v.$$

Hence $e_\ell(P,Q) \neq 1$, which shows that (d) is true.

Finally, we show that (d) implies (b) by assuming that (b) is false and deducing that (d) is false. The assumption that (b) is false means that either $P = \mathcal{O}$ or $Q = uP$ for some $u \in \mathbb{Z}/\ell\mathbb{Z}$. But if $P = \mathcal{O}$, then $e_\ell(P,Q) = e_\ell(\mathcal{O},Q) = 1$ by bilinearity, while if $Q = uP$, then

$$e_\ell(P,Q) = e_\ell(P,uP) = e_\ell(P,P)^u = 1^u = 1$$

by the alternating property of e_ℓ. Thus in both cases we find that $e_\ell(P,Q) = 1$, so (d) is false. □

Definition. Let E be an elliptic curve, let $P \in E[\ell]$, and let ϕ be an ℓ-distortion map for P. The *modified Weil pairing* $\hat{e}_\ell$ *on* $E[\ell]$ *(relative to* ϕ*)* is defined by

$$\hat{e}_\ell(Q, Q') = e_\ell\bigl(Q, \phi(Q')\bigr).$$

In cryptographic applications, the modified Weil pairing is evaluated at points that are multiples of P. The crucial property of the modified Weil pairing is its nondegeneracy, as described in the next result.

Proposition 6.51. *Let E be an elliptic curve, let $P \in E[\ell]$, let ϕ be an ℓ-distortion map for P, and let $\hat{e}_\ell$ be the modified Weil pairing relative to ϕ. Let Q and Q' be multiples of P. Then*

$$\hat{e}_\ell(Q, Q') = 1 \quad \text{if and only if} \quad Q = \mathcal{O} \quad \text{or} \quad Q' = \mathcal{O}.$$

Proof. We are given that Q and Q' are multiples of P, so we can write them as $Q = sP$ and $Q' = tP$. The definition of distortion map and the linearity of the Weil pairing imply that

$$\hat{e}_\ell(Q, Q') = \hat{e}_\ell(sP, tP) = e_\ell\bigl(sP, \phi(tP)\bigr) = e_\ell\bigl(sP, t\phi(P)\bigr) = e_\ell\bigl(P, \phi(P)\bigr)^{st}.$$

The quantity $e_\ell\bigl(P, \phi(P)\bigr)$ is a primitive ℓth root of unity, so

$$\begin{aligned} \hat{e}_\ell(Q, Q') = 1 \quad &\iff \quad \ell \mid st \\ &\iff \quad \ell \mid s \quad \text{or} \quad \ell \mid t \\ &\iff \quad Q = \mathcal{O} \quad \text{or} \quad Q' = \mathcal{O}. \end{aligned}$$

□

6.9.3 A Distortion Map on $y^2 = x^3 + x$

In order to use the modified Weil pairing for cryptographic purposes, we need to give at least one example of an elliptic curve with a distortion map. In this section we give such an example for the elliptic curve $y^2 = x^3 + x$ over the field $\mathbb{F}_p$ with $p \equiv 3 \pmod 4$. (See Exercise 6.43 for another example.) We start by describing the map ϕ.

Proposition 6.52. *Let E be the elliptic curve*

$$E : y^2 = x^3 + x$$

over a field K and suppose that K has an element $\alpha \in K$ satisfying $\alpha^2 = -1$. Define a map ϕ by

$$\phi(x, y) = (-x, \alpha y) \quad \text{and} \quad \phi(\mathcal{O}) = \mathcal{O}.$$

(a) *Let $P \in E(K)$. Then $\phi(P) \in E(K)$, so ϕ is a map from $E(K)$ to itself.*

(b) *The map* ϕ *respects the addition law on* E,[13]

$$\phi(P_1 + P_2) = \phi(P_1) + \phi(P_2) \quad \text{for all } P_1, P_2 \in E(K).$$

In particular, $\phi(nP) = n\phi(P)$ *for all* $P \in E(K)$ *and all* $n \geq 1$.

Proof. (a) Let $P = (x, y) \in E(K)$. Then

$$(\alpha y)^2 = -y^2 = -(x^3 + x) = (-x)^3 + (-x),$$

so $\phi(P) = (-x, \alpha y) \in E(K)$.

(b) Suppose that $P_1 = (x_1, y_1)$ and $P_2 = (x_2, y_2)$ are distinct points. Then using the elliptic curve addition algorithm (Theorem 6.6), we find that the x-coordinate of $\phi(P_1) + \phi(P_2)$ is

$$\begin{aligned} x(\phi(P_1) + \phi(P_2)) &= \left(\frac{\alpha y_2 - \alpha y_1}{(-x_2) - (-x_1)}\right)^2 - (-x_1) - (-x_2) \\ &= \alpha^2 \left(\frac{y_2 - y_1}{x_2 - x_1}\right)^2 + x_1 + x_2 \\ &= -\left(\left(\frac{y_2 - y_1}{x_2 - x_1}\right)^2 - x_1 - x_2\right) \\ &= -x(P_1 + P_2). \end{aligned}$$

Similarly, the y-coordinate of $\phi(P_1) + \phi(P_2)$ is

$$\begin{aligned} y(\phi(P_1) + \phi(P_2)) &= \left(\frac{\alpha y_2 - \alpha y_1}{(-x_2) - (-x_1)}\right)\big(-x_1 - x(\phi(P_1) + \phi(P_2))\big) - \alpha y_1 \\ &= -\alpha \left(\frac{y_2 - y_1}{x_2 - x_1}\right)\big(-x_1 + x(P_1 + P_2)\big) - \alpha y_1 \\ &= \alpha \left(\left(\frac{y_2 - y_1}{x_2 - x_1}\right)\big(x_1 - x(P_1 + P_2)\big) + y_1\right) \\ &= \alpha y(P_1 + P_2). \end{aligned}$$

Hence

$$\phi(P_1) + \phi(P_2) = \big(-x(P_1 + P_2), \alpha y(P_1 + P_2)\big) = \phi(P_1 + P_2).$$

This handles the case that $P_1 \neq P_2$. We leave the case $P_1 = P_2$ for the reader; see Exercise 6.38. □

We now have the tools needed to construct a distortion map on the curve $y^2 = x^3 + x$ over certain finite fields.

---

[13] In the language of abstract algebra, the map ϕ is a *homomorphism* of the group $E(K)$ to itself; see Exercise 2.13. In the language of algebraic geometry, a homomorphism from an elliptic curve to itself is called an *isogeny*.

Proposition 6.53. *Fix the following quantities.*

- *A prime p satisfying $p \equiv 3 \pmod 4$.*
- *The elliptic curve $E : y^2 = x^3 + x$.*
- *An element $\alpha \in \mathbb{F}_{p^2}$ satisfying $\alpha^2 = -1$.*
- *The map $\phi(x, y) = (-x, \alpha y)$.*
- *A prime $\ell \geq 3$ such that there exists a nonzero point $P \in E(\mathbb{F}_p)[\ell]$.*

Then ϕ is an ℓ-distortion map for P, i.e., the quantity

$$\hat{e}_\ell(P, P) = e_\ell(P, \phi(P))$$

is a primitive ℓth root of unity.

Proof. We first note that $\mathbb{F}_p$ does *not* contain an element satisfying $\alpha^2 = -1$. This is part of quadratic reciprocity (Theorem 3.62), but it is also easy to prove directly from the fact that $\mathbb{F}_p^*$ is a group of order $p - 1$, so it cannot have any elements of order 4, since $p \equiv 3 \pmod 4$.

However, the field $\mathbb{F}_{p^2}$ of order p^2 does contain a square root of -1, since if g is a primitive root for $\mathbb{F}_{p^2}^*$ (see Theorem 2.62), then $\alpha = g^{(p^2-1)/4}$ satisfies $\alpha^4 = 1$ and $\alpha^2 \neq 1$, so $\alpha^2 = -1$.

Since $P \neq \mathcal{O}$, it is clear that $\phi(P) \neq \mathcal{O}$. Since P is a point of order ℓ, Proposition 6.52(b) says that

$$\ell\phi(P) = \phi(\ell P) = \phi(\mathcal{O}) = \mathcal{O},$$

so $\phi(P)$ is a point of order ℓ. We are going to prove that $\phi(P)$ is not a multiple of P, and then Proposition 6.50 tells us that $e_\ell(P, \phi(P))$ is a primitive ℓth root of unity.

Suppose to the contrary that $\phi(P)$ is a multiple of P. We write $P = (x, y) \in E(\mathbb{F}_p)$. The coordinates of P are in $\mathbb{F}_p$, so the coordinates of any multiple of P are also in $\mathbb{F}_p$. Thus the coordinates of $\phi(P) = (-x, \alpha y)$ would be in $\mathbb{F}_p$. But $\alpha \notin \mathbb{F}_p$, since $\mathbb{F}_p$ does not contain a square root of -1, so we must have $y = 0$. Then $P = (x, 0)$ is a point of order 2, which is not possible, since P is a point of order ℓ with $\ell \geq 3$. Hence $\phi(P)$ is not a multiple of P and we are done. □

Remark 6.54. We recall from Example 2.58 that if $p \equiv 3 \pmod 4$, then the field with p^2 elements looks like

$$\mathbb{F}_{p^2} = \{a + bi : a, b \in \mathbb{F}_p\},$$

where i satisfies $i^2 = -1$. This makes it quite easy to work with the field $\mathbb{F}_{p^2}$ in the context of Proposition 6.53.

Example 6.55. We take $E : y^2 = x^3 + x$ and the prime $p = 547$. Then

$$\#E(\mathbb{F}_{547}) = 548 = 2^2 \cdot 137.$$

By trial and error we find the point $P_0 = (2, 253) \in E(\mathbb{F}_{547})$, and then

$$P = (67, 481) = 4P_0 = 4(2, 253) \in E(\mathbb{F}_{547})$$

is a point of order 137.

In order to find more points of order 137, we go to the larger field

$$\mathbb{F}_{547^2} = \{a + bi : a, b \in \mathbb{F}_{547}\}, \quad \text{where } i^2 = -1.$$

The distortion map gives

$$\phi(P) = (-67, 481i) \in E(\mathbb{F}_{547^2}).$$

In order to compute the Weil pairing of P and $\phi(P)$, we randomly choose a point

$$S = (256 + 110i, 441 + 15i) \in E(\mathbb{F}_{547^2})$$

and use Miller's algorithm to compute

$$\frac{f_P(\phi(P) + S)}{f_P(S)} = \frac{376 + 138i}{384 + 76i} = 510 + 96i,$$

$$\frac{f_{\phi(P)}(P - S)}{f_{\phi(P)}(-S)} = \frac{498 + 286i}{393 + 120i} = 451 + 37i.$$

Then

$$\hat{e}_{137}(P, P) = e_{137}(P, \phi(P)) = \frac{510 + 96i}{451 + 37i} = 37 + 452i \in \mathbb{F}_{547^2}.$$

We check that $(37 + 452i)^{137} = 1$, so $\hat{e}_{137}(P, P)$ is indeed a primitive 137th root of unity in $\mathbb{F}_{547^2}$.

Example 6.56. Continuing with the curve E, prime $p = 547$, and point $P = (67, 481)$ from Example 6.55, we use the MOV method to solve the ECDLP for the point

$$Q = (167, 405) \in E(\mathbb{F}_{547}).$$

The distortion map gives $\phi(Q) = (380, 405i)$, and we use the randomly chosen point $S = (402 + 397i, 271 + 205i) \in E(\mathbb{F}_{547^2})$ to compute

$$\hat{e}_{547}(P, Q) = e_{547}(P, \phi(Q)) = \frac{\frac{368+305i}{348+66i}}{\frac{320+206i}{175+351i}} = 530 + 455i \in \mathbb{F}_{547^2}.$$

From the previous example we have $\hat{e}_{137}(P, P) = 37 + 452i$, so we need to solve the DLP

$$(37 + 452i)^n = 530 + 455i \quad \text{in } \mathbb{F}_{547^2}.$$

The solution to this DLP is $n = 83$, and the MOV algorithm tells us that $n = 83$ is also a solution to the ECDLP. We check by verifying that $Q = 83P$.

6.10 Applications of the Weil Pairing

In Sect. 6.9.1 we described a negative application of the Weil pairing to cryptography, namely the MOV algorithm to solve the ECDLP for an elliptic curve over $\mathbb{F}_p$ by reducing the problem to the DLP in $\mathbb{F}_q$, where q is a certain power of p. In this section we describe two positive applications of the Weil pairing to cryptography. The first is a version of Diffie–Hellman key exchange involving three people, and the second is an ID-based public key cryptosystem in which the public keys can be selected by their owners.

6.10.1 Tripartite Diffie–Hellman Key Exchange

We have seen in Sect. 6.4.1 how two people can perform a Diffie–Hellman key exchange using elliptic curves. Suppose that three people, Alice, Bob, and Carl, want to perform a triple exchange of keys with only one pass of information between each pair of people. This is possible using a clever pairing-based construction due to Antoine Joux [61, 62].

The first step is for Alice, Bob, and Carl to agree on an elliptic curve E and a point $P \in E(\mathbb{F}_q)[\ell]$ of prime order such that there is an ℓ-distortion map for P. Let $\hat{e}_\ell$ be the associated modified Weil pairing.

As in ordinary Diffie–Hellman, they each choose a secret integer, say Alice chooses n_A, Bob chooses n_B, and Carl chooses n_C. They compute the associated multiples

$$\overbrace{Q_A = n_A P,}^{\text{Alice computes this}} \qquad \overbrace{Q_B = n_B P,}^{\text{Bob computes this}} \qquad \text{and} \qquad \overbrace{Q_C = n_C P.}^{\text{Carl computes this}}$$

They now publish the values of Q_A, Q_B, and Q_C.

In order to compute the shared value, Alice computes the modified pairing of the public points Q_B and Q_C and then raises the result to the n_A power, where n_A is her secret integer. Thus Alice computes

$$\hat{e}_\ell(Q_B, Q_C)^{n_A}.$$

The points Q_B and Q_C are certain multiples of P, and although Alice doesn't know what multiples, the bilinearity of the modified Weil pairing implies that the value computed by Alice is equal to

$$\hat{e}_\ell(Q_B, Q_C)^{n_A} = \hat{e}_\ell(n_B P, n_C P)^{n_A} = \hat{e}_\ell(P, P)^{n_B n_C n_A}.$$

Bob and Carl use their secret integers and the public points to perform similar computations.

$$\begin{aligned} &\text{Bob computes:} \quad \hat{e}_\ell(Q_A, Q_C)^{n_B} = \hat{e}_\ell(n_A P, n_C P)^{n_B} = \hat{e}_\ell(P, P)^{n_A n_C n_B}, \\ &\text{Carl computes:} \quad \hat{e}_\ell(Q_A, Q_B)^{n_C} = \hat{e}_\ell(n_A P, n_B P)^{n_C} = \hat{e}_\ell(P, P)^{n_A n_B n_C}. \end{aligned}$$

Alice, Bob, and Carl have now shared the secret value $\hat{e}_\ell(P, P)^{n_A n_B n_C}$. Tripartite (three-person) Diffie–Hellman key exchange is summarized in Table 6.11.

Public parameter creation		
A trusted authority publishes a finite field $\mathbb{F}_q$, an elliptic curve $E/\mathbb{F}_q$, a point $P \in E(\mathbb{F}_q)$ of prime order ℓ, and an ℓ-distortion map ϕ for P.		
Private computations		
Alice	**Bob**	**Carl**
Choose secret n_A.	Choose secret n_B.	Choose secret n_C.
Compute $Q_A = n_A P$.	Compute $Q_B = n_B P$.	Compute $Q_C = n_C P$.
Publication of values		
Alice, Bob, and Carl publish their points Q_A, Q_B, and Q_C		
Further private computations		
Alice	**Bob**	**Carl**
Compute $\hat{e}_\ell(Q_B, Q_C)^{n_A}$.	Compute $\hat{e}_\ell(Q_A, Q_C)^{n_B}$.	Compute $\hat{e}_\ell(Q_A, Q_B)^{n_C}$.
The shared secret value is $\hat{e}_\ell(P, P)^{n_A n_B n_C}$.		

Table 6.11: Tripartite Diffie–Hellman key exchange using elliptic curves

If Eve can solve the ECDLP, then clearly she can break tripartite Diffie–Hellman key exchange, since she will be able to recover the secret integers n_A, n_B, and n_C. (Recovering any one of them would suffice.) But the security of tripartite DH does not rely solely on the difficulty of the ECDLP. Eve can use Alice's public point Q_A and the public point P to compute both

$$\hat{e}_\ell(P, P) \qquad \text{and} \qquad \hat{e}_\ell(Q_A, P) = \hat{e}_\ell(n_A P, P) = \hat{e}_\ell(P, P)^{n_A}.$$

Thus Eve can recover n_A if she can solve the equation $a^n = b$ in $\mathbb{F}_q$, where she knows the values of $a = \hat{e}_\ell(P, P)$ and $b = \hat{e}_\ell(Q_A, P)$. In other words, the security of tripartite Diffie–Hellman also rests on the difficulty of solving the classical discrete logarithm problem for a subgroup of $\mathbb{F}_q^*$ of order ℓ. (See also Exercise 6.48.)

Since there are subexponential algorithms to solve the DLP in $\mathbb{F}_q$ (see Sect. 3.8), using tripartite Diffie–Hellman securely requires a larger field than does two-person elliptic curve Diffie–Hellman. This is a drawback, to be sure, but since there are no other methods known to do tripartite Diffie–Hellman, one accepts half a loaf in preference to going hungry.

Example 6.57. We illustrate tripartite Diffie–Hellman with a numerical example using the curve

$$E : y^2 = x^3 + x \quad \text{over the field } \mathbb{F}_{1303}.$$

This curve has $\#E(\mathbb{F}_{1303}) = 1304 = 2^3 \cdot 163$ points. The point $P = (334, 920) \in E(\mathbb{F}_{1303})$ has order 163. Alice, Bob, and Carl choose the secret values

$$n_A = 71, \qquad n_B = 3, \qquad n_C = 126.$$

They use their secret values to compute and publish:

$$\begin{aligned} \text{Alice publishes the point} \quad & Q_A = n_A P = (1279, 1171), \\ \text{Bob publishes the point} \quad & Q_B = n_B P = (872, 515), \\ \text{Carl publishes the point} \quad & Q_C = n_C P = (196, 815). \end{aligned}$$

Finally, Alice, Bob, and Carl use their own secret integers and the public points to compute:

$$\begin{aligned} \text{Alice computes} \quad & \hat{e}_{163}(Q_B, Q_C)^{71} = (172 + 256i)^{71} &= 768 + 662i, \\ \text{Bob computes} \quad & \hat{e}_{163}(Q_A, Q_C)^{3} = (1227 + 206i)^{3} &= 768 + 662i, \\ \text{Carl computes} \quad & \hat{e}_{163}(Q_A, Q_B)^{126} = (282 + 173i)^{126} &= 768 + 662i. \end{aligned}$$

Their shared secret value is $768 + 662i$.

6.10.2 ID-Based Public Key Cryptosystems

The goal of *ID-based cryptography* is very simple. One would like a public key cryptosystem in which the user's public key can be chosen by the user. For example, Alice might use her email address `alice@liveshere.com` as her identity-based public key, and then anyone who knows how to send her email automatically knows her public key. Of course, this idea is too simplistic; Alice must have some secret information that is used for decryption, and somehow that secret information must be used during the encryption process.

Here is a more sophisticated version of the same idea. We assume that there is a trusted authority Tom who is available to perform computations and distribute information. Tom publishes a master public key $\mathsf{Tom}^{\mathrm{Pub}}$ and keeps secret an associated private key $\mathsf{Tom}^{\mathrm{Pri}}$. When Bob wants to send Alice a message, he uses the master public key $\mathsf{Tom}^{\mathrm{Pub}}$ and Alice's ID-based public key $\mathsf{Alice}^{\mathrm{Pub}}$ (which, recall, could simply be her email address) in some sort of cryptographic algorithm to encrypt his message.

In the meantime, Alice tells Tom that she wants to use $\mathsf{Alice}^{\mathrm{Pub}}$ as her ID-based public key. Tom uses the master private key $\mathsf{Tom}^{\mathrm{Pri}}$ and Alice's ID-based public key $\mathsf{Alice}^{\mathrm{Pub}}$ to create a private key $\mathsf{Alice}^{\mathrm{Pri}}$ for Alice. Alice then uses $\mathsf{Alice}^{\mathrm{Pri}}$ to decrypt and read Bob's message.

The principle of ID-based cryptography is clear, but it is not easy to see how one might create a practical and secure ID-based public key cryptosystem.

Remark 6.58. The trusted authority Tom needs to keep track of which public keys he has assigned, since otherwise Eve could send Alice's public key to Tom and ask him to create and send her the associated private key, which would be the same as Alice's private key. But there is another threat that must be countered. Eve is allowed to send Tom a large number of public keys of her choice (other than ones that have already been assigned to other people) and ask Tom to create the associated private keys. It is essential that knowledge of these additional private keys not allow Eve to recover Tom's master private key $\mathsf{Tom}^{\mathrm{Pri}}$, since otherwise Eve would be able to reconstitute everyone's private keys! Further, Eve's possession of a large number of public–private key pairs should not allow her to create any additional public–private key pairs.

The idea of ID-based cryptography was initially described by Shamir in 1984 [125], and a practical ID-based system was devised by Boneh and Franklin in 2001 [20, 21]. This system, which we now describe, uses pairings on elliptic curves.

The first step is for Tom, the trusted authority, to select a finite field $\mathbb{F}_q$, an elliptic curve E, and a point $P \in E(\mathbb{F}_q)[\ell]$ of prime order such that there is an ℓ-distortion map for P. Let $\hat{e}_\ell$ be the modified Weil pairing relative to the map. Tom also needs to publish two hash functions H_1 and H_2. (A *hash function* is a function that is easy to compute, but hard to invert. See Sect. 8.1 for a discussion of hash functions.) The first one assigns a point in $E(\mathbb{F}_q)[\ell]$ to each possible user ID,[14]

$$H_1 : \{\text{User IDs}\} \longrightarrow E(\mathbb{F}_q)[\ell].$$

The second hash function assigns to each element of $\mathbb{F}_q^*$ a binary string of length B,

$$H_2 : \mathbb{F}_q^* \longrightarrow \{\text{bit strings of length } B\},$$

where the set of plaintexts $\mathcal{M}$ is the set of all binary strings of length B.

Tom creates his master key by choosing a secret (nonzero) integer s modulo ℓ and computing the point

$$P^{\mathsf{Tom}} = sP \in E(\mathbb{F}_q)[\ell].$$

Tom's master private key is the integer s and his master public key is the point P^{Tom}.

Now suppose that Bob wants to send Alice a message $M \in \mathcal{M}$ using her ID-based public key $\mathsf{Alice}^{\mathrm{Pub}}$. He uses her public key and the hash function H_1 to compute the point

$$P^{\mathsf{Alice}} = H_1(\mathsf{Alice}^{\mathrm{Pub}}) \in E(\mathbb{F}_q)[\ell].$$

He also chooses a random number (a random element) $1 < r < q$ and computes the two quantities

$$C_1 = rP \qquad \text{and} \qquad C_2 = M \,\mathsf{xor}\, H_2\big(\hat{e}_\ell(P^{\mathsf{Alice}}, P^{\mathsf{Tom}})^r\big). \tag{6.18}$$

Here, to avoid confusion with addition of points on the elliptic curve, we write xor for the XOR operation on bit strings; see (1.12) on page 44. The ciphertext is the pair $C = (C_1, C_2)$.

[14]There are various ways define a hash function H_1 with values in $E(\mathbb{F}_q)[\ell]$. For example, take a given User ID I, convert it to a binary string β, apply a hash function to β that takes values uniformly in $\{1, 2, \ldots, \ell - 1\}$ to get an integer m, and set $H_1(I) = mP$.

Public parameter creation
A trusted authority Tom publishes a finite field $\mathbb{F}_q$, an elliptic curve $E/\mathbb{F}_q$, a point $P \in E(\mathbb{F}_q)[\ell]$ of prime order ℓ, and an ℓ-distortion map ϕ for P. Tom also chooses hash functions $$H_1 : \{\text{IDs}\} \to E(\mathbb{F}_q)[\ell] \quad \text{and} \quad H_2 : \mathbb{F}_q^* \to \{0,1\}^B.$$
Master key creation
Tom chooses a secret integer s modulo ℓ. Tom publishes the point $P^{\mathsf{Tom}} = sP \in E(\mathbb{F}_q)[\ell]$.
Private key extraction
Alice chooses an ID-based public key $\mathsf{Alice}^{\text{Pub}}$. Tom computes the point $P^{\mathsf{Alice}} = H_1(\mathsf{Alice}^{\text{Pub}}) \in E(\mathbb{F}_q)[\ell]$. Tom sends the point $Q^{\mathsf{Alice}} = sP^{\mathsf{Alice}} \in E(\mathbb{F}_q)[\ell]$ to Alice.
Encryption
Bob chooses a plaintext M and a random number r modulo $q-1$. Bob computes the point $P^{\mathsf{Alice}} = H_1(\mathsf{Alice}^{\text{Pub}}) \in E(\mathbb{F}_q)[\ell]$. Bob's ciphertext is the pair $$(C_1, C_2) = \bigl(rP, M \operatorname{\mathsf{xor}} H_2(\hat{e}_\ell(P^{\mathsf{Alice}}, P^{\mathsf{Tom}})^r)\bigr).$$
Decryption
Alice decrypts the ciphertext (C_1, C_2) by computing $$C_2 \operatorname{\mathsf{xor}} H_2\bigl(\hat{e}_\ell(Q^{\mathsf{Alice}}, C_1)\bigr).$$

Table 6.12: Identity-based encryption using pairings on elliptic curves

In order to decrypt Bob's message, Alice needs to request that Tom give her the private key $\mathsf{Alice}^{\text{Pri}}$ associated to her ID-based public key $\mathsf{Alice}^{\text{Pub}}$. She can do this ahead of time, or she can wait until she has received Bob's message. In any case, the private key that Tom gives to Alice is the point

$$Q^{\mathsf{Alice}} = sP^{\mathsf{Alice}} = sH_1(\mathsf{Alice}^{\text{Pub}}) \in E(\mathbb{F}_q)[\ell].$$

In other words, Tom feeds Alice's public key to the hash function H_1 to get a point in $E(\mathbb{F}_q)[\ell]$, and then he multiplies that point by his secret key s.

Alice is finally ready to decrypt Bob's message (C_1, C_2). She first computes $\hat{e}_\ell(Q^{\mathsf{Alice}}, C_1)$, which, by a chain of calculations using bilinearity, is equal to

$$\begin{aligned} \hat{e}_\ell(Q^{\mathsf{Alice}}, C_1) &= \hat{e}_\ell(sP^{\mathsf{Alice}}, rP) = \hat{e}_\ell(P^{\mathsf{Alice}}, P)^{rs} \\ &= \hat{e}_\ell(P^{\mathsf{Alice}}, sP)^r = \hat{e}_\ell(P^{\mathsf{Alice}}, P^{\mathsf{Tom}})^r. \end{aligned}$$

Notice that this is exactly the quantity that Bob used in (6.18) to create the second part of his ciphertext. Hence Alice can recover the plaintext by computing

$$C_2 \operatorname{\mathsf{xor}} H_2\bigl(\hat{e}_\ell(Q^{\mathsf{Alice}}, C_1)\bigr)$$

$$= \big(M \,\mathsf{xor}\, H_2\big(\hat{e}_\ell(P^{\mathsf{Alice}}, P^{\mathsf{Tom}})^r\big)\big) \,\mathsf{xor}\, H_2\big(\hat{e}_\ell(P^{\mathsf{Alice}}, P^{\mathsf{Tom}})^r\big) = M.$$

The last step follows because $M \,\mathsf{xor}\, N \,\mathsf{xor}\, N = M$ for any bit strings M and N.

The full process of ID-based encryption is summarized in Table 6.12.

Exercises

Section 6.1. Elliptic Curves

6.1. Let E be the elliptic curve $E : Y^2 = X^3 - 2X + 4$ and let $P = (0, 2)$ and $Q = (3, -5)$. (You should check that P and Q are on the curve E.)
(a) Compute $P \oplus Q$.
(b) Compute $P \oplus P$ and $Q \oplus Q$.
(c) Compute $P \oplus P \oplus P$ and $Q \oplus Q \oplus Q$.

6.2. Check that the points $P = (-1, 4)$ and $Q = (2, 5)$ are points on the elliptic curve $E : Y^2 = X^3 + 17$.
(a) Compute the points $P \oplus Q$ and $P \ominus Q$.
(b) Compute the points $2P$ and $2Q$.
(*Bonus.* How many points with integer coordinates can you find on E?)

6.3. Suppose that the cubic polynomial $X^3 + AX + B$ factors as

$$X^3 + AX + B = (X - e_1)(X - e_2)(X - e_3).$$

Prove that $4A^3 + 27B^2 = 0$ if and only if two (or more) of e_1, e_2, and e_3 are the same. (*Hint.* Multiply out the right-hand side and compare coefficients to relate A and B to e_1, e_2, and e_3.)

6.4. Sketch each of the following curves, as was done in Fig. 6.1 on page 300.
(a) $E : Y^2 = X^3 - 7X + 3$.
(b) $E : Y^2 = X^3 - 7X + 9$.
(c) $E : Y^2 = X^3 - 7X - 12$.
(d) $E : Y^2 = X^3 - 3X + 2$.
(e) $E : Y^2 = X^3$.
Notice that the curves in (d) and (e) have $\Delta_E = 0$, so they are not elliptic curves. How do their pictures differ from the pictures in (a), (b), and (c)? Each of the curves (d) and (e) has one point that is somewhat unusual. These unusual points are called *singular points.*

Section 6.2. Elliptic Curves over Finite Fields

6.5. For each of the following elliptic curves E and finite fields $\mathbb{F}_p$, make a list of the set of points $E(\mathbb{F}_p)$.
(a) $E : Y^2 = X^3 + 3X + 2$ over $\mathbb{F}_7$.
(b) $E : Y^2 = X^3 + 2X + 7$ over $\mathbb{F}_{11}$.

(c) $E : Y^2 = X^3 + 4X + 5$ over $\mathbb{F}_{11}$.
(d) $E : Y^2 = X^3 + 9X + 5$ over $\mathbb{F}_{11}$.
(e) $E : Y^2 = X^3 + 9X + 5$ over $\mathbb{F}_{13}$.

6.6. Make an addition table for E over $\mathbb{F}_p$, as we did in Table 6.1.
(a) $E : Y^2 = X^3 + X + 2$ over $\mathbb{F}_5$.
(b) $E : Y^2 = X^3 + 2X + 3$ over $\mathbb{F}_7$.
(c) $E : Y^2 = X^3 + 2X + 5$ over $\mathbb{F}_{11}$.
You may want to write a computer program for (c), since $E(\mathbb{F}_{11})$ has a lot of points!

6.7. Let E be the elliptic curve

$$E : y^2 = x^3 + x + 1.$$

Compute the number of points in the group $E(\mathbb{F}_p)$ for each of the following primes:
(a) $p = 3$. (b) $p = 5$. (c) $p = 7$. (d) $p = 11$.
In each case, also compute the trace of Frobenius

$$t_p = p + 1 - \#E(\mathbb{F}_p)$$

and verify that $|t_p|$ is smaller than $2\sqrt{p}$.

Section 6.3. The Elliptic Curve Discrete Logarithm Problem

6.8. Let E be the elliptic curve

$$E : y^2 = x^3 + x + 1$$

and let $P = (4, 2)$ and $Q = (0, 1)$ be points on E modulo 5. Solve the elliptic curve discrete logarithm problem for P and Q, that is, find a positive integer n such that $Q = nP$.

6.9. Let E be an elliptic curve over $\mathbb{F}_p$ and let P and Q be points in $E(\mathbb{F}_p)$. Assume that Q is a multiple of P and let $n_0 > 0$ be the smallest solution to $Q = nP$. Also let $s > 0$ be the smallest solution to $sP = \mathcal{O}$. Prove that every solution to $Q = nP$ looks like $n_0 + is$ for some $i \in \mathbb{Z}$. (*Hint.* Write n as $n = is + r$ for some $0 \le r < s$ and determine the value of r.)

6.10. Let $\{P_1, P_2\}$ be a basis for $E[m]$. The *Basis Problem* for $\{P_1, P_2\}$ is to express an arbitrary point $P \in E[m]$ as a linear combination of the basis vectors, i.e., to find n_1 and n_2 so that $P = n_1P_1 + n_2P_2$. Prove that an algorithm that solves the basis problem for $\{P_1, P_2\}$ can be used to solve the ECDLP for points in $E[m]$.

6.11. Use the double-and-add algorithm (Table 6.3) to compute nP in $E(\mathbb{F}_p)$ for each of the following curves and points, as we did in Fig. 6.4.

(a) $E : Y^2 = X^3 + 23X + 13$, $p = 83$, $P = (24, 14)$, $n = 19$;
(b) $E : Y^2 = X^3 + 143X + 367$, $p = 613$, $P = (195, 9)$, $n = 23$;
(c) $E : Y^2 = X^3 + 1828X + 1675$, $p = 1999$, $P = (1756, 348)$, $n = 11$;
(d) $E : Y^2 = X^3 + 1541X + 1335$, $p = 3221$, $P = (2898, 439)$, $n = 3211$.

6.12. Convert the proof of Proposition 6.18 into an algorithm and use it to write each of the following numbers n as a sum of positive and negative powers of 2 with at most $\frac{1}{2}\lfloor \log n \rfloor + 1$ nonzero terms. Compare the number of nonzero terms in the binary expansion of n with the number of nonzero terms in the ternary expansion of n.
(a) 349. (b) 9337. (c) 38728. (d) 8379483273489.

6.13. In Sect. 5.5 we gave an abstract description of Pollard's ρ method, and in Sect. 5.5.2 we gave an explicit version to solve the discrete logarithm problem in $\mathbb{F}_p$. Adapt this material to create a Pollard ρ algorithm to solve the ECDLP.

Section 6.4. Elliptic Curve Cryptography

6.14. Alice and Bob agree to use elliptic Diffie–Hellman key exchange with the prime, elliptic curve, and point

$$p = 2671, \qquad E : Y^2 = X^3 + 171X + 853, \qquad P = (1980, 431) \in E(\mathbb{F}_{2671}).$$

(a) Alice sends Bob the point $Q_A = (2110, 543)$. Bob decides to use the secret multiplier $n_B = 1943$. What point should Bob send to Alice?
(b) What is their secret shared value?
(c) How difficult is it for Eve to figure out Alice's secret multiplier n_A? If you know how to program, use a computer to find n_A.
(d) Alice and Bob decide to exchange a new piece of secret information using the same prime, curve, and point. This time Alice sends Bob only the x-coordinate $x_A = 2$ of her point Q_A. Bob decides to use the secret multiplier $n_B = 875$. What single number modulo p should Bob send to Alice, and what is their secret shared value?

6.15. Exercise 2.10 on page 109 describes a multistep public key cryptosystem based on the discrete logarithm problem for $\mathbb{F}_p$. Describe a version of this cryptosystem that uses the elliptic curve discrete logarithm problem. (You may assume that Alice and Bob know the order of the point P in the group $E(\mathbb{F}_p)$, i.e., they know the smallest integer $N \geq 1$ with the property that $NP = \mathcal{O}$.)

6.16. A shortcoming of using an elliptic curve $E(\mathbb{F}_p)$ for cryptography is the fact that it takes two coordinates to specify a point in $E(\mathbb{F}_p)$. However, as discussed briefly at the end of Sect. 6.4.2, the second coordinate actually conveys very little additional information.
(a) Suppose that Bob wants to send Alice the value of a point $R \in E(\mathbb{F}_p)$. Explain why it suffices for Bob to send Alice the x-coordinate of $R = (x_R, y_R)$ together with the single bit

$$\beta_R = \begin{cases} 0 & \text{if } 0 \leq y_R < \frac{1}{2}p, \\ 1 & \text{if } \frac{1}{2}p < y_R < p. \end{cases}$$

(You may assume that Alice is able to efficiently compute square roots modulo p. This is certainly true, for example, if $p \equiv 3 \pmod 4$; see Proposition 2.26.)
(b) Alice and Bob decide to use the prime $p = 1123$ and the elliptic curve

$$E : Y^2 = X^3 + 54X + 87.$$

Bob sends Alice the x-coordinate $x = 278$ and the bit $\beta = 0$. What point is Bob trying to convey to Alice? What about if instead Bob had sent $\beta = 1$?

6.17. The Menezes–Vanstone variant of the elliptic Elgamal public key cryptosystem improves message expansion while avoiding the difficulty of directly attaching plaintexts to points in $E(\mathbb{F}_p)$. The MV-Elgamal cryptosystem is described in Table 6.13 on page 365.

(a) The last line of Table 6.13 claims that $m_1' = m_1$ and $m_2' = m_2$. Prove that this is true, so the decryption process does work.

(b) What is the message expansion of MV-Elgamal?

(c) Alice and Bob agree to use

$$p = 1201, \qquad E : Y^2 = X^3 + 19X + 17, \qquad P = (278, 285) \in E(\mathbb{F}_p),$$

for MV-Elgamal. Alice's secret value is $n_A = 595$. What is her public key? Bob sends Alice the encrypted message $((1147, 640), 279, 1189)$. What is the plaintext?

6.18. This exercise continues the discussion of the MV-Elgamal cryptosystem described in Table 6.13 on page 365.

(a) Eve knows the elliptic curve E and the ciphertext values c_1 and c_2. Show how Eve can use this knowledge to write down a polynomial equation (modulo p) that relates the two pieces m_1 and m_2 of the plaintext. In particular, if Eve can figure out one piece of the plaintext, then she can recover the other piece by finding the roots of a certain polynomial modulo p.

(b) Alice and Bob exchange a message using MV-Elgamal with the prime, elliptic curve, and point in Exercise 6.17(c). Eve intercepts the ciphertext

$$((269, 339), 814, 1050),$$

and through other sources she discovers that the first part of the plaintext is $m_1 = 1050$. Use your algorithm in (a) to recover the second part of the plaintext.

6.19. Section 6.4.3 describes ECDSA, an elliptic analogue of DSA. Formulate an elliptic analogue of the simpler Elgamal digital signature algorithm described in Table 4.2 in Sect. 4.3.

6.20. This exercise asks you to compute some numerical instances of the elliptic curve digital signature algorithm described in Table 6.7 for the public parameters

$$E : y^2 = x^3 + 231x + 473, \quad p = 17389, \quad q = 1321, \quad G = (11259, 11278) \in E(\mathbb{F}_p).$$

You should begin by verifying that G is a point of order q in $E(\mathbb{F}_p)$.

(a) Samantha's private signing key is $s = 542$. What is her public verification key? What is her digital signature on the document $d = 644$ using the random element $e = 847$?

(b) Tabitha's public verification key is $V = (11017, 14637)$. Is $(s_1, s_2) = (907, 296)$ a valid signature on the document $d = 993$?

(c) Umberto's public verification key is $V = (14594, 308)$. Use any method that you want to find Umberto's private signing key, and then use the private key to forge his signature on the document $d = 516$ using the random element $e = 365$.

Public Parameter Creation	
A trusted party chooses and publishes a (large) prime p, an elliptic curve E over $\mathbb{F}_p$, and a point P in $E(\mathbb{F}_p)$.	
Alice	**Bob**
Key Creation	
Chooses a secret multiplier n_A. Computes $Q_A = n_A P$. Publishes the public key Q_A.	
Encryption	
	Chooses plaintext values m_1 and m_2 modulo p. Chooses a random number k. Computes $R = kP$. Computes $S = kQ_A$ and writes it as $S = (x_S, y_S)$. Sets $c_1 \equiv x_S m_1 \pmod{p}$ and $c_2 \equiv y_S m_2 \pmod{p}$. Sends ciphertext (R, c_1, c_2) to Alice.
Decryption	
Computes $T = n_A R$ and writes it as $T = (x_T, y_T)$. Sets $m_1' \equiv x_T^{-1} c_1 \pmod{p}$ and $m_2' \equiv y_T^{-1} c_2 \pmod{p}$. Then $m_1' = m_1$ and $m_2' = m_2$.	

Table 6.13: Menezes–Vanstone variant of Elgamal (Exercises 6.17, 6.18)

Section 6.6. Lenstra's Elliptic Curve Factorization Algorithm

6.21. Use the elliptic curve factorization algorithm to factor each of the numbers N using the given elliptic curve E and point P.

(a) $N = 589$, $E : Y^2 = X^3 + 4X + 9$, $P = (2, 5)$.

(b) $N = 26167$, $E : Y^2 = X^3 + 4X + 128$, $P = (2, 12)$.

(c) $N = 1386493$, $E : Y^2 = X^3 + 3X - 3$, $P = (1, 1)$.

(d) $N = 28102844557$, $E : Y^2 = X^3 + 18X - 453$, $P = (7, 4)$.

Section 6.7. Elliptic Curves over $\mathbb{F}_2$ and over $\mathbb{F}_{2^k}$

6.22. Let E be an elliptic curve given by a generalized Weierstrass equation

$$E : Y^2 + a_1 XY + a_3 Y = X^3 + a_2 X^2 + a_4 X + a_6.$$

Let $P_1 = (x_1, y_1)$ and $P_2 = (x_2, y_2)$ be points on E. Prove that the following algorithm computes their sum $P_3 = P_1 + P_2$.

First, if $x_1 = x_2$ and $y_1 + y_2 + a_1 x_2 + a_3 = 0$, then $P_1 + P_2 = \mathcal{O}$.
Otherwise define quantities λ and ν as follows:

$$\begin{array}{lll} [\text{If } x_1 \neq x_2] & \lambda = \dfrac{y_2 - y_1}{x_2 - x_1}, & \nu = \dfrac{y_1 x_2 - y_2 x_1}{x_2 - x_1}, \\ [\text{If } x_1 = x_2] & \lambda = \dfrac{3x_1^2 + 2a_2 x_1 + a_4 - a_1 y_1}{2y_1 + a_1 x_1 + a_3}, & \nu = \dfrac{-x_1^3 + a_4 x_1 + 2a_6 - a_3 y_1}{2y_1 + a_1 x_1 + a_3}. \end{array}$$

Then

$$P_3 = P_1 + P_2 = (\lambda^2 + a_1\lambda - a_2 - x_1 - x_2,\ -(\lambda + a_1)x_3 - \nu - a_3).$$

6.23. Let $\mathbb{F}_8 = \mathbb{F}_2[T]/(T^3 + T + 1)$ be as in Example 6.28, and let E be the elliptic curve

$$E : Y^2 + XY + Y = X^3 + TX + (T + 1).$$

(a) Calculate the discriminant of E.

(b) Verify that the points

$$P = (1 + T + T^2, 1 + T), \quad Q = (T^2, T), \quad R = (1 + T + T^2, 1 + T^2),$$

are in $E(\mathbb{F}_8)$ and compute the values of $P + Q$ and $2R$.

(c) Find all of the points in $E(\mathbb{F}_8)$.

(d) Find a point $P \in E(\mathbb{F}_8)$ such that every point in $E(\mathbb{F}_8)$ is a multiple of P.

6.24. Let $\tau(\alpha) = \alpha^p$ be the Frobenius map on $\mathbb{F}_{p^k}$.

(a) Prove that

$$\tau(\alpha + \beta) = \tau(\alpha) + \tau(\beta) \qquad \text{and} \qquad \tau(\alpha \cdot \beta) = \tau(\alpha) \cdot \tau(\beta) \qquad \text{for all } \alpha, \beta \in \mathbb{F}_{p^k}.$$

(*Hint.* For the addition formula, use the binomial theorem (Theorem 5.10).)

(b) Prove that $\tau(\alpha) = \alpha$ for all $\alpha \in \mathbb{F}_p$.

(c) Let E be an elliptic curve over $\mathbb{F}_p$ and let $\tau(x, y) = (x^p, y^p)$ be the Frobenius map from $E(\mathbb{F}_{p^k})$ to itself. Prove that

$$\tau(P + Q) = \tau(P) + \tau(Q) \quad \text{for all } P \in E(\mathbb{F}_{p^k}).$$

6.25. Let E_0 be the Koblitz curve $Y^2 + XY = X^3 + 1$ over the field $\mathbb{F}_2$, and for every $k \geq 1$, let

$$t_k = 2^k + 1 - \#E(\mathbb{F}_{2^k}).$$

(a) Prove that $t_1 = -1$ and $t_2 = -3$.

(b) Prove that t_k satisfies the recursion

$$t_k = t_1 t_{k-1} - 2t_{k-2} \qquad \text{for all } k \geq 3.$$

(You may use the formula (6.12) that we stated, but did not prove, on page 334.)

(c) Use the recursion in (b) to compute $\#E(\mathbb{F}_{16})$.

(d) Program a computer to calculate the recursion and use it to compute the values of $\#E(\mathbb{F}_{2^{11}})$, $\#E(\mathbb{F}_{2^{31}})$, and $\#E(\mathbb{F}_{2^{101}})$.

6.26. Let E be an elliptic curve over $\mathbb{F}_p$, and for $k \geq 1$, let

$$t_k = p^k + 1 - \#E(\mathbb{F}_{p^k}).$$

(a) Prove that

$$t_k = t_1 t_{k-1} - p t_{k-2} \qquad \text{for all } k \geq 2,$$

where by convention we set $t_0 = 2$.

(b) Use (a) to express t_2, t_3, and t_4 in terms of p and t_1.

(*Hint.* Use Theorem 6.29(a). This generalizes Exercise 6.25.)

6.27. Let τ satisfy $\tau^2 = -2 - \tau$. Prove that the following algorithm gives coefficients $v_i \in \{-1, 0, 1\}$ such that the positive integer n is equal to

$$n = v_0 + v_1\tau + v_2\tau^2 + \cdots + v_\ell\tau^\ell. \qquad (6.19)$$

Further prove that $\ell \leq 2\lceil \log(n) \rceil + 1$.

```
[1]    Set n₀ = n and n₁ = 0 and i = 0
[2]    Loop while n₀ ≠ 0 or n₁ ≠ 0
[3]        If n₀ is odd
[4]            Set vᵢ = 2 − ((n₀ − 2n₁) mod 4)
[5]            Set n₀ = n₀ − vᵢ
[6]        Else
[7]            Set vᵢ = 0
[8]        End If
[9]        Set i = i + 1
[10]       Set (n₀, n₁) = (n₁ − ½n₀, −½n₀)
[11]   End Loop
```

6.28. Implement the algorithm in Exercise 6.27 and use it to compute the τ-expansion (6.19) of the following integers. What is the highest power of τ that appears and how many nonzero terms are there?

(a) $n = 931$ (b) $n = 32755$ (c) $n = 82793729188$

Section 6.8. Bilinear Pairings on Elliptic Curves

6.29. Let $R(x)$ and $S(x)$ be rational functions. Prove that the divisor of a product is the sum of the divisors, i.e.,

$$\operatorname{div}\bigl(R(x)S(x)\bigr) = \operatorname{div}\bigl(R(x)\bigr) + \operatorname{div}\bigl(S(x)\bigr).$$

6.30. This exercise sketches a proof that if $P = (\alpha, 0) \in E$, then $\operatorname{div}(X - \alpha) = 2[P] - 2[\mathcal{O}]$.

(a) Prove that

$$\operatorname{div}(X - \alpha) = m[P] - m[O]$$

for some integer $m \geq 1$.

(b) Prove that the Weierstrass equation of E can be written in the form

$$E : Y^2 = (X - \alpha)(X^2 + aX + b),$$

and that the polynomials of $X - \alpha$ and $X^2 + aX + b$ have no common roots.

(c) Prove that

$$\operatorname{div}(X - \alpha) = 2n[P] - 2n[O]$$

for some integer $n \geq 1$. (*Hint.* Take the divisor of both sides of $Y^2 = (X - \alpha)(X^2 + aX + b)$ and use (b).)

(d) Prove that

$$\operatorname{div}(X - \alpha) = 2[P] - 2[O].$$

(*Warning.* This part requires some knowledge of discrete valuation rings that is not developed in this book.)

6.31. Prove that the Weil pairing satisfies

$$e_m(P, Q) = e_m(Q, P)^{-1} \qquad \text{for all } P, Q, \in E[m].$$

(*Hint.* Use the fact that $e_m(P + Q, P + Q) = 1$ and expand using bilinearity.)

6.32. This exercise asks you to verify that the Weil pairing e_m is well-defined.

(a) Prove that the value of $e_m(P, Q)$ is independent of the choice of rational functions f_P and f_Q.

(b) Prove that the value of $e_m(P, Q)$ is independent of the auxiliary point S. (*Hint.* Fix the points P and Q and consider the quantity

$$F(S) = \frac{f_P(Q + S)}{f_P(S)} \bigg/ \frac{f_Q(P - S)}{f_Q(-S)}$$

as a function of S. Compute the divisor of F and use the fact that every nonconstant function on E has at least one zero.)

You might also try to prove that the Weil pairing is bilinear, but do not be discouraged if you do not succeed, since the standard proofs use more tools than we have developed in the text.

6.33. Choose a basis $\{P_1, P_2\}$ for $E[m]$ and write each $P \in E[m]$ as a linear combination $P = a_P P_1 + b_P P_2$. (See Remark 6.39.) Use the basic properties of the Weil pairing described in Theorem 6.38 to prove that

$$e_m(P,Q) = e_m(P_1,P_2)^{\det\begin{pmatrix} a_P & a_Q \\ b_P & b_Q \end{pmatrix}} = e_m(P_1,P_2)^{a_P b_Q - a_Q b_P}.$$

6.34. Complete the proof of Proposition 6.52 by proving that $\phi(2P) = 2\phi(P)$.

6.35. For each of the following elliptic curves E, finite fields $\mathbb{F}_p$, points P and Q of order m, and auxiliary points S, use Miller's algorithm to compute the Weil pairing $e_m(P,Q)$. (See Example 6.43.)

	E	p	P	Q	m	S
(a)	$y^2 = x^3 + 23$	1051	(109 203)	(240 203)	5	(1,554)
(b)	$y^2 = x^3 - 35x - 9$	883	(5, 66)	(103, 602)	7	(1,197)
(c)	$y^2 = x^3 + 37x$	1009	(8, 703)	(49, 20)	7	(0,0)
(d)	$y^2 = x^3 + 37x$	1009	(417, 952)	(561, 153)	7	(0,0)

Notice that (c) and (d) use the same elliptic curve. Letting P' and Q' denote the points in (d), verify that

$$P' = 2P, \quad Q' = 3Q, \quad \text{and} \quad e_7(P', Q') = e_7(P,Q)^6.$$

6.36. Let E over $\mathbb{F}_q$ and ℓ be as described in Theorem 6.44. Prove that the modified Tate pairing is symmetric, in the sense that

$$\hat{\tau}(P,Q) = \hat{\tau}(Q,P) \qquad \text{for all } P, Q \in E(\mathbb{F}_q)[\ell].$$

6.37. Let E be an elliptic curve over $\mathbb{F}_q$ and let $P, Q \in E(\mathbb{F}_q)[\ell]$. Prove that the Weil pairing and the Tate pairing are related by the formula

$$e_\ell(P,Q) = \frac{\tau(P,Q)}{\tau(Q,P)},$$

provided that the Tate pairings on the right-hand side are computed consistently. Thus the Weil pairing requires approximately twice as much work to compute as does the Tate pairing.

Section 6.9. The Weil Pairing over Fields of Prime Power Order

6.38. Prove Proposition 6.52(b) in the case $P_1 = P_2$.

6.39. Let E be an elliptic curve over $\mathbb{F}_p$ and let ℓ be a prime. Suppose that $E(\mathbb{F}_p)$ contains a point of order ℓ and that $\ell > \sqrt{p} + 1$. Prove that $E(\mathbb{F}_p)[\ell] \cong \mathbb{Z}/\ell\mathbb{Z}$.

6.40. Let E be an elliptic curve over a finite field $\mathbb{F}_q$ and let ℓ be a prime. Suppose that we are given four points $P, aP, bP, cP \in E(\mathbb{F}_q)[\ell]$. The (*elliptic*) *decision Diffie–Hellman problem* is to determine whether cP is equal to abP. Of course, if we could solve the Diffie–Hellman problem itself, then we could compute abP and compare it with cP, but the Diffie–Hellman problem is often difficult to solve.

Suppose that there exists a distortion map ϕ for $E[\ell]$. Show how to use the modified Weil pairing to solve the elliptic decision Diffie–Hellman problem without actually having to compute abP.

6.41. Let E be the elliptic curve $E : y^2 = x^3 + x$ and let $\phi(x,y) = (-x, \alpha y)$ be the map described in Proposition 6.52. Prove that $\phi(\phi(P)) = -P$ for all $P \in E$. (Intuitively, ϕ behaves like multiplication by $\sqrt{-1}$ when it is applied to points of E.)

6.42. Let $p \equiv 3 \pmod 4$, let $E : y^2 = x^3 + x$, let $P \in E(\mathbb{F}_p)[\ell]$, and let $\phi(x,y) = (-x, \alpha y)$ be the ℓ-distortion map for P described in Proposition 6.53. Suppose further that $\ell \equiv 3 \pmod 4$. Prove that ϕ is an ℓ-distortion map for every point in $E[\ell]$. In other words, if $Q \in E$ is any point of order ℓ, prove that $e_\ell(Q, \phi(Q))$ is a primitive ℓth root of unity.

6.43. Let E be the elliptic curve

$$E : y^2 = x^3 + 1$$

over a field K, and suppose that K contains an element $\beta \neq 1$ satisfying $\beta^3 = 1$. (We say that β is a *primitive cube root of unity.*) Define a map ϕ by

$$\phi(x,y) = (\beta x, y) \quad \text{and} \quad \phi(\mathcal{O}) = \mathcal{O}.$$

(a) Let $P \in E(K)$. Prove that $\phi(P) \in E(K)$.
(b) Prove that ϕ respects the addition law on E, i.e., $\phi(P_1 + P_2) = \phi(P_1) + \phi(P_2)$ for all $P_1, P_2 \in E(K)$.

6.44. Let $E : y^2 = x^3 + 1$ be the elliptic curve in Exercise 6.43.
(a) Let $p \geq 3$ be a prime with $p \equiv 2 \pmod 3$. Prove that $\mathbb{F}_p$ does not contain a primitive cube root of unity, but that $\mathbb{F}_{p^2}$ does contain a primitive cube root of unity.
(b) Let $\beta \in \mathbb{F}_{p^2}$ be a primitive cube root of unity and define a map $\phi(x,y) = (\beta x, y)$ as in Exercise 6.43. Suppose that $E(\mathbb{F}_p)$ contains a point P of prime order $\ell \geq 5$. Prove that ϕ is an ℓ-distortion map for P.

6.45. Let E be the elliptic curve $E : y^2 = x^3 + x$ over the field $\mathbb{F}_{691}$. The point $P = (301, 14) \in E(\mathbb{F}_{691})$ has order 173. Use the distortion map on E from Exercise 6.42 to compute $\hat{e}_{173}(P,P)$ (cf. Example 6.55). Verify that the value is a primitive 173rd root of unity.

6.46. Continuing with the curve E, prime $p = 691$, and point $P = (301, 14)$ from Exercise 6.45, let

$$Q = (143, 27) \in E(\mathbb{F}_{691}).$$

Use the MOV method to solve the ECDLP for P and Q, i.e., compute $\hat{e}_{173}(P,Q)$ and express it as the nth power of $\hat{e}_{173}(P,P)$. Check your answer by verifying that nP is equal to Q.

Section 6.10. Applications of the Weil Pairing

6.47. Alice, Bob, and Carl use tripartite Diffie–Hellman with the curve

$$E : y^2 = x^3 + x \quad \text{over the field } \mathbb{F}_{1723}.$$

They use the point

$$P = (668, 995) \quad \text{of order 431.}$$

(a) Alice chooses the secret value $n_A = 278$. What is Alice's public point Q_A?

(b) Bob's public point is $Q_B = (1275, 1550)$ and Carl's public point is $Q_C = (897, 1323)$. What is the value of $\hat{e}_{431}(Q_B, Q_C)$?
(c) What is their shared value?
(d) Bob's secret value is $n_B = 224$. Verify that $\hat{e}_{431}(Q_A, Q_C)^{n_B}$ is the same as the value that you got in (c).
(e) Figure out Carl's secret value n_C. (Since P has order 431, you can do this on a computer by trying all possible values.)

6.48. Show that Eve can break tripartite Diffie–Hellman key exchange as described in Table 6.10.1 if she knows how to solve the Diffie–Hellman problem (page 69) for the field $\mathbb{F}_q$.

6.49. In this exercise we consider what is required to break the identity-based encryption scheme described in Table 6.12 on page 360.
(a) Show that if Eve can solve the discrete logarithm problem in either $E(\mathbb{F}_q)$ or in $\mathbb{F}_q^*$, then she can recover Tom's secret key s, which means that she can do anything that Tom can do, including decrypting everyone's ciphtertexts.
(b) Suppose that Eve only knows how to solve the elliptic curve Diffie–Hellman problem in $E(\mathbb{F}_q)$, as described on page 318. Show that she can decrypt all ciphertexts.
(c) What if Eve only knows how to solve the Diffie–Hellman problem in $\mathbb{F}_q^*$. Can she still decrypt all ciphertexts?

Chapter 7

Lattices and Cryptography

The security of all of the public key cryptosystems that we have previously studied has been based, either directly or indirectly, on either the difficulty of factoring large numbers or the difficulty of finding discrete logarithms in a finite group. In this chapter we investigate a new type of hard problem arising in the theory of lattices that can be used as the basis for a public key cryptosystem. Lattice-based cryptosystems offer several potential advantages over earlier systems, including faster encryption/decryption and so-called quantum resistance. The latter means that at present there are no known quantum algorithms to rapidly solve hard lattice problems; see Sect. 8.11. Further, we will see that the theory of lattices has applications in cryptography beyond simply providing a new source of hard problems.

Recall that a vector space V over the real numbers $\mathbb{R}$ is a set of vectors, where two vectors can be added together and a vector can be multiplied by a real number. A lattice is similar to a vector space, except that we are restricted to multiplying the vectors in a lattice by integers. This seemingly minor restriction leads to many interesting and subtle questions. Since the subject of lattices can appear somewhat abstruse and removed from the everyday reality of cryptography, we begin this chapter with two motivating examples in which lattices are not mentioned, but where they are lurking in the background, waiting to be used for cryptanalysis. We then review the theory of vector spaces in Sect. 7.3 and formally introduce lattices in Sect. 7.4.

7.1 A Congruential Public Key Cryptosystem

In this section we describe a toy model of a real public key cryptosystem. This version turns out to have an unexpected connection with lattices of dimension 2, and hence a fatal vulnerability, since the dimension is so low. However,

J. Hoffstein et al., *An Introduction to Mathematical Cryptography*, Undergraduate Texts in Mathematics, DOI 10.1007/978-1-4939-1711-2_7

it is instructive as an example of how lattices may appear in cryptanalysis even when the underlying hard problem appears to have nothing to do with lattices. Further, it provides a lowest-dimensional introduction to the NTRU public key cryptosystem, which will be described in Sect. 7.10.

Alice begins by choosing a large positive integer q, which is a public parameter, and two other secret positive integers f and g satisfying

$$f < \sqrt{q/2}, \qquad \sqrt{q/4} < g < \sqrt{q/2}, \qquad \text{and} \qquad \gcd(f, qg) = 1.$$

She then computes the quantity

$$h \equiv f^{-1}g \pmod{q} \qquad \text{with } 0 < h < q.$$

Notice that f and g are small compared to q, since they are $\mathcal{O}(\sqrt{q})$, while the quantity h will generally be $\mathcal{O}(q)$, which is considerably larger. Alice's private key is the pair of small integers f and g and her public key is the large integer h.

In order to send a message, Bob chooses a plaintext m and a random integer r (a random element) satisfying the inequalities

$$0 < m < \sqrt{q/4} \qquad \text{and} \qquad 0 < r < \sqrt{q/2}.$$

He computes the ciphertext

$$e \equiv rh + m \pmod{q} \qquad \text{with } 0 < e < q$$

and sends it to Alice.

Alice decrypts the message by first computing

$$a \equiv fe \pmod{q} \qquad \text{with } 0 < a < q,$$

and then computing

$$b \equiv f^{-1}a \pmod{g} \qquad \text{with } 0 < b < g. \tag{7.1}$$

Note that f^{-1} in (7.1) is the inverse of f modulo g.

We now verify that $b = m$, which will show that Alice has recovered Bob's plaintext. We first observe that the quantity a satisfies

$$a \equiv fe \equiv f(rh + m) \equiv frf^{-1}g + fm \equiv rg + fm \pmod{q}.$$

The size restrictions on f, g, r, m imply that the integer $rg + fm$ is small,

$$rg + fm < \sqrt{\frac{q}{2}}\sqrt{\frac{q}{2}} + \sqrt{\frac{q}{2}}\sqrt{\frac{q}{4}} < q.$$

Thus when Alice computes $a \equiv fe \pmod{q}$ with $0 < a < q$, she gets the exact value

Alice	Bob
Key Creation	
Choose a large integer modulus q. Choose secret integers f and g with $f < \sqrt{q/2}$, $\sqrt{q/4} < g < \sqrt{q/2}$, and $\gcd(f, qg) = 1$. Compute $h \equiv f^{-1}g \pmod{q}$. Publish the public key (q, h).	
Encryption	
	Choose plaintext m with $m < \sqrt{q/4}$. Use Alice's public key (q, h) to compute $e \equiv rh + m \pmod{q}$. Send ciphertext e to Alice.
Decryption	
Compute $a \equiv fe \pmod{q}$ with $0 < a < q$. Compute $b \equiv f^{-1}a \pmod{g}$ with $0 < b < g$. Then b is the plaintext m.	

Table 7.1: A congruential public key cryptosystem

$$a = rg + fm. \tag{7.2}$$

This is the key point: the formula (7.2) is an equality of integers and not merely a congruence modulo q. Finally Alice computes

$$b \equiv f^{-1}a \equiv f^{-1}(rg + fm) \equiv f^{-1}fm \equiv m \pmod{g} \qquad \text{with } 0 < b < g.$$

Since $m < \sqrt{q/4} < g$, it follows that $b = m$. The congruential cryptosystem is summarized in Table 7.1.

Example 7.1. Alice chooses

$$q = 122430513841, \quad f = 231231, \quad \text{and} \quad g = 195698.$$

Here $f \approx 0.66\sqrt{q}$ and $g \approx 0.56\sqrt{q}$ are allowable values. Alice computes

$$f^{-1} \equiv 49194372303 \pmod{q} \quad \text{and} \quad h \equiv f^{-1}g \equiv 39245579300 \pmod{q}.$$

Alice's public key is the pair $(q, h) = (122430513841, 39245579300)$.

Bob decides to send Alice the plaintext $m = 123456$ using the random value $r = 101010$. He uses Alice's public key to compute the ciphertext

$$e \equiv rh + m \equiv 18357558717 \pmod{q},$$

which he sends to Alice.

In order to decrypt e, Alice first uses her secret value f to compute

$$a \equiv fe \equiv 48314309316 \pmod{q}.$$

(Note that $a = 48314309316 < 122430513841 = q$.) She then uses the value $f^{-1} \equiv 193495 \pmod{g}$ to compute

$$f^{-1}a \equiv 193495 \cdot 48314309316 \equiv 123456 \pmod{g},$$

and, as predicted by the theory, this is Bob's plaintext m.

How might Eve attack this system? She might try doing a brute-force search through all possible private keys or through all possible plaintexts, but this takes $\mathcal{O}(q)$ operations. Let's consider in more detail Eve's task if she tries to find the private key (f, g) from the known public key (q, h). It is not hard to see that if Eve can find any pair of positive integers F and G satisfying

$$Fh \equiv G \pmod{q} \quad \text{and} \quad F = \mathcal{O}(\sqrt{q}) \quad \text{and} \quad G = \mathcal{O}(\sqrt{q}), \tag{7.3}$$

then (F, G) is likely to serve as a decryption key. Rewriting the congruence (7.3) as $Fh = G + qR$, we reformulate Eve's task as that of finding a pair of comparatively small integers (F, G) with the property that

$$\underbrace{F}_{\text{unknown integer}}\,\underbrace{(1,h)}_{\text{known vector}} - \underbrace{R}_{\text{unknown integer}}\,\underbrace{(0,q)}_{\text{known vector}} = \overbrace{(F,G)}^{\text{unknown small vector}}.$$

Thus Eve knows two vectors $\boldsymbol{v}_1 = (1, h)$ and $\boldsymbol{v}_2 = (0, q)$, each of which has length $\mathcal{O}(q)$, and she wants to find a linear combination $\boldsymbol{w} = a_1\boldsymbol{v}_1 + a_2\boldsymbol{v}_2$ such that $\boldsymbol{w}$ has length $\mathcal{O}(\sqrt{q})$, but keep in mind that the coefficients a_1 and a_2 are required to be integers. Thus Eve needs to find a short nonzero vector in the set of vectors

$$L = \{a_1\boldsymbol{v}_1 + a_2\boldsymbol{v}_2 : a_1, a_2 \in \mathbb{Z}\}.$$

This set L is an example of a two-dimensional lattice. Notice that it looks sort of like a two-dimensional vector space with basis $\{\boldsymbol{v}_1, \boldsymbol{v}_2\}$, except that we are allowed to take only integer linear combinations of $\boldsymbol{v}_1$ and $\boldsymbol{v}_2$.

Unfortunately for Bob and Alice, there is an extremely rapid method for finding short vectors in two-dimensional lattices. This method, which is due to Gauss, is described in Sect. 7.13.1 and used to break the congruential cryptosystem in Sect. 7.14.1.

7.2 Subset-Sum Problems and Knapsack Cryptosystems

The first attempt to base a cryptosystem on an $\mathcal{NP}$-complete problem[1] was made by Merkle and Hellman in the late 1970s [84]. They used a version of the following mathematical problem, which generalizes the classical knapsack problem.

> **The Subset-Sum Problem**
> Suppose that you are given a list of positive integers $(M_1, M_2, \ldots, M_n)$ and another integer S. Find a subset of the elements in the list whose sum is S. (You may assume that there is at least one such subset.)

Example 7.2. Let $\boldsymbol{M} = (2, 3, 4, 9, 14, 23)$ and $S = 21$. Then a bit of trial and error yields the subset $\{3, 4, 14\}$ whose sum is 21, and it is not hard to check that this is the only subset that sums to 21. Similarly, if we take $S = 29$, then we find that $\{2, 4, 23\}$ has the desired sum. But in this case there is a second solution, since $\{2, 4, 9, 14\}$ also sums to 29.

Here is another way to describe the subset-sum problem. The list

$$\boldsymbol{M} = (M_1, M_2, \ldots, M_n)$$

of positive integers is public knowledge. Bob chooses a secret binary vector $\boldsymbol{x} = (x_1, x_2, \ldots, x_n)$, i.e., each x_i may be either 0 or 1. Bob computes the sum

$$S = \sum_{i=1}^{n} x_i M_i$$

and sends S to Alice. The subset-sum problem asks Alice to find either the original vector $\boldsymbol{x}$ or another binary vector giving the same sum. Notice that the vector $\boldsymbol{x}$ tells Alice which M_i to include in S, since M_i is in the sum S if and only if $x_i = 1$. Thus specifying the binary vector $\boldsymbol{x}$ is the same as specifying a subset of $\boldsymbol{M}$.

It is clear that Alice can find $\boldsymbol{x}$ by checking all 2^n binary vectors of length n. A simple collision algorithm allows Alice to cut the exponent in half.

Proposition 7.3. *Let $\boldsymbol{M} = (M_1, M_2, \ldots, M_n)$ and let $(\boldsymbol{M}, S)$ be a subset-sum problem. For all sets of integers $\boldsymbol{I}$ and $\boldsymbol{J}$ satisfying*

$$\boldsymbol{I} \subset \{i : 1 \le i \le \tfrac{1}{2}n\} \qquad \textit{and} \qquad \boldsymbol{J} \subset \{j : \tfrac{1}{2}n < j \le n\},$$

compute and make a list of the values

[1] $\mathcal{NP}$-complete problems are discussed in Sect. 5.7. However, if you have not read that section, suffice it to say that $\mathcal{NP}$-complete problems are considered to be very hard to solve in a computational sense.

$$A_I = \sum_{i \in I} M_i \quad \textit{and} \quad B_J = S - \sum_{j \in J} M_j.$$

Then these lists include a pair of sets I_0 *and* J_0 *satisfying* $A_{I_0} = B_{J_0}$, *and the sets* I_0 *and* J_0 *give a solution to the subset-sum problem,*

$$S = \sum_{i \in I_0} M_i + \sum_{j \in J_0} M_j.$$

The number of entries in each list is at most $2^{n/2}$, *so the running time of the algorithm is* $\mathcal{O}(2^{n/2+\epsilon})$, *where* ϵ *is some small value that accounts for sorting and comparing the lists.*

Proof. It suffices to note that if $\boldsymbol{x}$ is a binary vector giving a solution to the given subset-sum problem, then we can write the solution as

$$\sum_{1 \le i \le \frac{1}{2}n} x_i M_i = S - \sum_{\frac{1}{2}n < i \le n} x_i M_i.$$

The number of subsets I and J is $\mathcal{O}(2^{n/2})$, since they are subsets of sets of order $n/2$. □

If n is large, then in general it is difficult to solve a random instance of a subset-sum problem. Suppose, however, that Alice possesses some secret knowledge or trapdoor information about $\boldsymbol{M}$ that enables her to guarantee that the solution $\boldsymbol{x}$ is unique and that allows her to easily find $\boldsymbol{x}$. Then Alice can use the subset sum problem as a public key cryptosystem. Bob's plaintext is the vector $\boldsymbol{x}$, his encrypted message is the sum $S = \sum x_i M_i$, and only Alice can easily recover $\boldsymbol{x}$ from knowledge of S.

But what sort of sneaky trick can Alice use to ensure that she can solve this particular subset-sum problem, but that nobody else can? One possibility is to use a subset-sum problem that is extremely easy to solve, but somehow to disguise the easy solution from other people.

Definition. A *superincreasing sequence* of integers is a list of positive integers $\boldsymbol{r} = (r_1, r_2, \ldots, r_n)$ with the property that

$$r_{i+1} \ge 2r_i \quad \text{for all } 1 \le i \le n-1.$$

The following estimate explains the name of such sequences.

Lemma 7.4. *Let* $\boldsymbol{r} = (r_1, r_2, \ldots, r_n)$ *be a superincreasing sequence. Then*

$$r_k > r_{k-1} + \cdots + r_2 + r_1 \quad \textit{for all } 2 \le k \le n.$$

Proof. We give a proof by induction on k. For $k = 2$ we have $r_2 \ge 2r_1 > r_1$, which gets the induction started. Now suppose that the lemma is true for some $2 \le k < n$. Then first using the superincreasing property and next the induction hypothesis, we find that

$$r_{k+1} \ge 2r_k = r_k + r_k > r_k + (r_{k-1} + \cdots + r_2 + r_1).$$

This shows that the lemma is also true for $k+1$. □

A subset-sum problem in which the integers in $\boldsymbol{M}$ form a superincreasing sequence is very easy to solve.

Proposition 7.5. *Let $(\boldsymbol{M}, S)$ be a subset-sum problem in which the integers in $\boldsymbol{M}$ form a superincreasing sequence. Assuming that a solution $\boldsymbol{x}$ exists, it is unique and may be computed by the following fast algorithm:*

```
Loop i from n down to 1
    If S ≥ M_i, set x_i = 1 and subtract M_i from S
    Else set x_i = 0
End Loop
```

Proof. The assumption that $\boldsymbol{M}$ is a superincreasing sequence means that $M_{i+1} \ge 2M_i$. We are given that a solution exists, so to distinguish it from the vector $\boldsymbol{x}$ produced by the algorithm, we call the actual solution $\boldsymbol{y}$. Thus we are assuming that $\boldsymbol{y} \cdot \boldsymbol{M} = S$ and we need to show that $\boldsymbol{x} = \boldsymbol{y}$.

We prove by downward induction that $x_k = y_k$ for all $1 \le k \le n$. Our inductive hypothesis is that $x_i = y_i$ for all $k < i \le n$ and we need to prove that $x_k = y_k$. (Note that we allow $k = n$, in which case our inductive hypothesis is vacuously true.) The hypothesis means that when we performed the algorithm from $i = n$ down to $i = k+1$, we had $x_i = y_i$ at each stage. So before executing the loop with $i = k$, the value of S has been reduced to

$$S_k = S - \sum_{i=k+1}^{n} x_i M_i = \sum_{i=1}^{n} y_i M_i - \sum_{i=k+1}^{n} x_i M_i = \sum_{i=1}^{k} y_i M_i.$$

Now consider what happens when we execute the loop with $i = k$. There are two possibilities:

$$\begin{array}{llll} (1) & y_k = 1 \implies & S_k \ge M_k & \implies x_k = 1, \ \checkmark \\ (2) & y_k = 0 \implies & S_k \le M_{k-1} + \cdots + M_1 < M_k & \implies x_k = 0. \ \checkmark \end{array}$$

(Note that in Case (2) we have used Lemma 7.4 to deduce that $M_{k-1} + \cdots + M_1$ is strictly smaller than M_k.) In both cases we get $x_k = y_k$, which completes the proof that $\boldsymbol{x} = \boldsymbol{y}$. Further, it shows that the solution is unique, since we have shown that any solution agrees with the output of the algorithm, which by its nature returns a unique vector $\boldsymbol{x}$ for any given input S. □

Example 7.6. The set $\boldsymbol{M} = (3, 11, 24, 50, 115)$ is superincreasing. We write $S = 142$ as a sum of elements in $\boldsymbol{M}$ by following the algorithm. First $S \geq 115$, so $x_5 = 1$ and we replace S with $S - 115 = 27$. Next $27 < 50$, so $x_4 = 0$. Continuing, $27 \geq 24$, so $x_3 = 1$ and S becomes $27 - 24 = 3$. Then $3 < 11$, so $x_2 = 0$, and finally $3 \geq 3$, so $x_1 = 1$. Notice that S is reduced to $3 - 3 = 0$, which tells us that $\boldsymbol{x} = (1, 0, 1, 0, 1)$ is a solution. We check our answer,

$$1 \cdot 3 + 0 \cdot 11 + 1 \cdot 24 + 0 \cdot 50 + 1 \cdot 115 = 142. \qquad \checkmark$$

Merkle and Hellman proposed a public key cryptosystem based on a superincreasing subset-sum problem that is disguised using congruences. In order to create the public/private key pair, Alice starts with a superincreasing sequence $\boldsymbol{r} = (r_1, \ldots, r_n)$. She also chooses two large secret integers A and B satisfying

$$B > 2r_n \quad \text{and} \quad \gcd(A, B) = 1.$$

Alice creates a new sequence $\boldsymbol{M}$ that is not superincreasing by setting

$$M_i \equiv Ar_i \pmod{B} \qquad \text{with } 0 \leq M_i < B.$$

The sequence $\boldsymbol{M}$ is Alice's public key.

In order to encrypt a message, Bob chooses a plaintext $\boldsymbol{x}$ that is a binary vector and computes and sends to Alice the ciphertext

$$S = \boldsymbol{x} \cdot \boldsymbol{M} = \sum_{i=1}^{n} x_i M_i.$$

Alice decrypts S by first computing

$$S' \equiv A^{-1} S \pmod{B} \qquad \text{with } 0 \leq S' < B.$$

Then Alice solves the subset-sum problem for S' using the superincreasing sequence $\boldsymbol{r}$ and the fast algorithm described in Proposition 7.5.

The reason that decryption works is because S' is congruent to

$$S' \equiv A^{-1} S \equiv A^{-1} \sum_{i=1}^{n} x_i M_i \equiv A^{-1} \sum_{i=1}^{n} x_i A r_i \equiv \sum_{i=1}^{n} x_i r_i \pmod{B}.$$

The assumption that $B > 2r_n$ and Lemma 7.4 tell Alice that

$$\sum_{i=1}^{n} x_i r_i \leq \sum_{i=1}^{n} r_i < 2r_n < B,$$

so by choosing S' in the range from 0 to $B - 1$, she ensures that she gets an exact equality $S' = \sum x_i r_i$, rather than just a congruence.

The Merkle–Hellman cryptosystem is summarized in Table 7.2.

<table>
<tr><th>Alice</th><th>Bob</th></tr>
<tr><td colspan="2" align="center">Key Creation</td></tr>
<tr><td>Choose superincreasing $\boldsymbol{r} = (r_1, \ldots, r_n)$.
Choose A and B with $B > 2r_n$ and $\gcd(A, B) = 1$.
Compute $M_i = Ar_i \pmod{B}$ for $1 \le i \le n$.
Publish the public key $\boldsymbol{M} = (M_1, \ldots, M_n)$.</td><td></td></tr>
<tr><td colspan="2" align="center">Encryption</td></tr>
<tr><td></td><td>Choose binary plaintext $\boldsymbol{x}$.
Use Alice's public key $\boldsymbol{M}$
to compute $S = \boldsymbol{x} \cdot \boldsymbol{M}$.
Send ciphertext S to Alice.</td></tr>
<tr><td colspan="2" align="center">Decryption</td></tr>
<tr><td>Compute $S' \equiv A^{-1}S \pmod{B}$.
Solve the subset-sum problem S'
using the superincreasing sequence $\boldsymbol{r}$.
The plaintext $\boldsymbol{x}$ satisfies $\boldsymbol{x} \cdot \boldsymbol{r} = S'$.</td><td></td></tr>
</table>

Table 7.2: The Merkle–Hellman subset-sum cryptosystem

Example 7.7. Let $\boldsymbol{r} = (3, 11, 24, 50, 115)$ be Alice's secret superincreasing sequence, and suppose that she chooses $A = 113$ and $B = 250$. Then her disguised sequence is

$$\begin{aligned} \boldsymbol{M} &\equiv (113 \cdot 3, 113 \cdot 11, 113 \cdot 24, 113 \cdot 50, 113 \cdot 115) \pmod{250} \\ &= (89, 243, 212, 150, 245). \end{aligned}$$

Notice that $\boldsymbol{M}$ is not even close to being superincreasing (even if she rearranges the terms so that they are increasing).

Bob decides to send Alice the secret message $\boldsymbol{x} = (1, 0, 1, 0, 1)$. He encrypts $\boldsymbol{x}$ by computing

$$S = \boldsymbol{x} \cdot \boldsymbol{M} = 1 \cdot 89 + 0 \cdot 243 + 1 \cdot 212 + 0 \cdot 150 + 1 \cdot 245 = 546.$$

Upon receiving S, Alice multiplies by 177, the inverse of 113 modulo 250, to obtain

$$S' \equiv 177 \cdot 546 = 142 \pmod{250}.$$

Then Alice uses the algorithm in Proposition 7.5 to solve $S' = \boldsymbol{x} \cdot \boldsymbol{r}$ for the superincreasing sequence $\boldsymbol{r}$. (See Example 7.6.) In this way she recovers the plaintext $\boldsymbol{x}$.

Cryptosystems based on disguised subset-sum problems are known as *subset-sum cryptosystems* or *knapsack cryptosystems*. The general idea is to start with a secret superincreasing sequence, disguise it using secret modular linear operations, and publish the disguised sequence as the public key. The original Merkle and Hellman system suggested applying a secret permutation to the entries of $A\boldsymbol{r} \pmod{B}$ as an additional layer of security. Later versions, proposed by a number of people, involved multiple multiplications and reductions modulo several different moduli. For an excellent survey of knapsack cryptosystems, see the article by Odlyzko [103].

Remark 7.8. An important question that must be considered concerning knapsack systems is the size of the various parameters required to obtain a desired level of security. There are 2^n binary vectors $\boldsymbol{x} = (x_1, \ldots, x_n)$, and we have seen in Proposition 7.3 that there is a collision algorithm, so it is possible to break a knapsack cryptosystem in $\mathcal{O}(2^{n/2})$ operations. Thus in order to obtain security on the order of 2^k, it is necessary to take $n > 2k$, so for example, 2^{80} security requires $n > 160$. But although this provides security against a collision attack, it does not preclude the existence of other, more efficient attacks, which, as we will see in Sect. 7.14.2, actually do exist. (See also Remark 7.10.)

Remark 7.9. Assuming that we have chosen a value for n, how large must we take the other parameters? It turns out that if r_1 is too small, then there are easy attacks, so we must insist that $r_1 > 2^n$. The superincreasing nature of the sequence implies that

$$r_n > 2r_{n-1} > 4r_{n-1} > \cdots > 2^n r_1 > 2^{2n}.$$

Then $B > 2r_n = 2^{2n+1}$, so we find that the entries M_i in the public key and the ciphertext S satisfy

$$M_i = \mathcal{O}(2^{2n}) \qquad \text{and} \qquad S = \mathcal{O}(2^{2n}).$$

Thus the public key $\boldsymbol{M}$ is a list of n integers, each approximately $2n$ bits long, while the plaintext $\boldsymbol{x}$ consists of n bits of information, and the ciphertext is approximately $2n$ bits. Notice that the message expansion ratio is 2-to-1.

For example, suppose that $n = 160$. Then the public key size is about $2n^2 = 51200$ bits. Compare this to RSA or Diffie–Hellman, where, for security on the order of 2^{80}, the public key size is only about 1000 bits. This larger key size might seem to be a major disadvantage, but it is compensated for by the tremendous speed of the knapsack systems. Indeed, a knapsack decryption requires only one (or a very few) modular multiplications, and a knapsack encryption requires none at all. This is far more efficient than the large number of computationally intensive modular exponentiations used by RSA and Diffie–Hellman. Historically, this made knapsack cryptosystems quite appealing.

Remark 7.10. The best known algorithms to solve a randomly chosen subset-sum problem are versions of the collision algorithm such as Proposition 7.3. Unfortunately, a randomly chosen subset-sum problem has no trapdoor, hence cannot be used to create a cryptosystem. And it turns out that the use of a disguised superincreasing subset-sum problem allows other, more efficient, algorithms. The first such attacks, by Shamir, Odlyzko, Lagarias and others, used various ad hoc methods, but after the publication of the famous LLL[2] lattice reduction paper [77] in 1985, it became clear that knapsack-based cryptosystems have a fundamental weakness. Roughly speaking, if n is smaller than around 300, then lattice reduction allows an attacker to recover the plaintext $\boldsymbol{x}$ from the ciphertext S in a disconcertingly short amount time. Hence a secure system requires $n > 300$, in which case the private key length is greater than $2n^2 = 180000$ bits ≈ 176 kB. This is so large as to make secure knapsack systems impractical.

We now briefly describe how Eve can reformulate the subset-sum problem using vectors. Suppose that she wants to write S as a subset-sum from the set $\boldsymbol{M} = (m_1, \ldots, m_n)$. Her first step is to form the matrix

$$\begin{pmatrix} 2 & 0 & 0 & \cdots & 0 & m_1 \\ 0 & 2 & 0 & \cdots & 0 & m_2 \\ 0 & 0 & 2 & \cdots & 0 & m_3 \\ \vdots & \vdots & \vdots & \ddots & \vdots & \vdots \\ 0 & 0 & 0 & \cdots & 2 & m_n \\ 1 & 1 & 1 & \cdots & 1 & S \end{pmatrix}. \tag{7.4}$$

The relevant vectors are the rows of the matrix (7.4), which we label as

$$\begin{aligned} \boldsymbol{v}_1 &= (2, 0, 0, \ldots, 0, m_1), \\ \boldsymbol{v}_2 &= (0, 2, 0, \ldots, 0, m_2), \\ &\ \vdots \qquad \vdots \\ \boldsymbol{v}_n &= (0, 0, 0, \ldots, 2, m_n), \\ \boldsymbol{v}_{n+1} &= (1, 1, 1, \ldots, 1, S). \end{aligned}$$

Just as in the 2-dimensional example described at the end of Sect. 7.1, Eve looks at the set of all *integer* linear combinations of $\boldsymbol{v}_1, \ldots, \boldsymbol{v}_{n+1}$,

$$L = \{a_1\boldsymbol{v}_1 + a_2\boldsymbol{v}_2 + \cdots + a_n\boldsymbol{v}_n + a_{n+1}\boldsymbol{v}_{n+1} : a_1, a_2, \ldots, a_{n+1} \in \mathbb{Z}\}.$$

The set L is another example of a *lattice.*

Suppose now that $\boldsymbol{x} = (x_1, \ldots, x_n)$ is a solution to the given subset-sum problem. Then the lattice L contains the vector

$$\boldsymbol{t} = \sum_{i=1}^{n} x_i\boldsymbol{v}_i - \boldsymbol{v}_{n+1} = (2x_1 - 1, 2x_2 - 1, \ldots, 2x_n - 1, 0),$$

where the last coordinate of $\boldsymbol{t}$ is 0 because $S = x_1m_1 + \cdots + x_nm_n$.

[2]The three L's are A.K. Lenstra, H.W. Lenstra, and L. Lovász.

We now come to the crux of the matter. Since the x_i are all 0 or 1, all of the $2x_i - 1$ values are ± 1, so the vector $\boldsymbol{t}$ is quite short, $\|\boldsymbol{t}\| = \sqrt{n}$. On the other hand, we have seen that $m_i = \mathcal{O}(2^{2n})$ and $S = \mathcal{O}(2^{2n})$, so the vectors generating L all have lengths $\|\boldsymbol{v}_i\| = \mathcal{O}(2^{2n})$. Thus it is unlikely that L contains any nonzero vectors, other than $\boldsymbol{t}$, whose length is as small as $\sqrt{n}$. If we postulate that Eve knows an algorithm that can find small nonzero vectors in lattices, then she will be able to find $\boldsymbol{t}$, and hence to recover the plaintext $\boldsymbol{x}$.

Algorithms that find short vectors in lattices are called *lattice reduction algorithms*. The most famous of these is the LLL algorithm, to which we alluded earlier, and its variants such as LLL-BKZ. The remainder of this chapter is devoted to describing lattices, cryptosystems based on lattices, the LLL algorithm, and cryptographic applications of LLL. A more detailed analysis of knapsack cryptosystems is given in Sect. 7.14.2; see also Example 7.33.

7.3 A Brief Review of Vector Spaces

Before starting our discussion of lattices, we pause to remind the reader of some important definitions and ideas from linear algebra. Vector spaces can be defined in vast generality,[3] but for our purposes in this chapter, it is enough to consider vector spaces that are contained in $\mathbb{R}^m$ for some positive integer m.

We start with the basic definitions that are essential for studying vector spaces.

Vector Spaces. A *vector space* V is a subset of $\mathbb{R}^m$ with the property that

$$\alpha_1 \boldsymbol{v}_1 + \alpha_2 \boldsymbol{v}_2 \in V \quad \text{for all } \boldsymbol{v}_1, \boldsymbol{v}_2 \in V \text{ and all } \alpha_1, \alpha_2 \in \mathbb{R}.$$

Equivalently, a vector space is a subset of $\mathbb{R}^m$ that is closed under addition and under scalar multiplication by elements of $\mathbb{R}$.

Linear Combinations. Let $\boldsymbol{v}_1, \boldsymbol{v}_2, \ldots, \boldsymbol{v}_k \in V$. A *linear combination* of $\boldsymbol{v}_1, \boldsymbol{v}_2, \ldots, \boldsymbol{v}_k \in V$ is any vector of the form

$$\boldsymbol{w} = \alpha_1 \boldsymbol{v}_1 + \alpha_2 \boldsymbol{v}_2 + \cdots + \alpha_k \boldsymbol{v}_k \quad \text{with } \alpha_1, \ldots, \alpha_k \in \mathbb{R}.$$

The collection of all such linear combinations,

$$\{\alpha_1 \boldsymbol{v}_1 + \cdots + \alpha_k \boldsymbol{v}_k : \alpha_1, \ldots, \alpha_k \in \mathbb{R}\},$$

is called the *span* of $\{\boldsymbol{v}_1, \ldots, \boldsymbol{v}_k\}$.

Independence. A set of vectors $\boldsymbol{v}_1, \boldsymbol{v}_2, \ldots, \boldsymbol{v}_k \in V$ is (*linearly*) *independent* if the only way to get

$$\alpha_1 \boldsymbol{v}_1 + \alpha_2 \boldsymbol{v}_2 + \cdots + \alpha_k \boldsymbol{v}_k = \boldsymbol{0} \tag{7.5}$$

is to have $\alpha_1 = \alpha_2 = \cdots = \alpha_k = 0$. The set is (*linearly*) *dependent* if we can make (7.5) true with at least one α_i nonzero.

[3]For example, we saw in Sect. 3.6 a nice application of vector spaces over the field $\mathbb{F}_2$.

Bases. A *basis* for V is a set of linearly independent vectors $\boldsymbol{v}_1, \ldots, \boldsymbol{v}_n$ that span V. This is equivalent to saying that every vector $\boldsymbol{w} \in V$ can be written in the form

$$\boldsymbol{w} = \alpha_1 \boldsymbol{v}_1 + \alpha_2 \boldsymbol{v}_2 + \cdots + \alpha_n \boldsymbol{v}_n$$

for a *unique* choice of $\alpha_1, \ldots, \alpha_n \in \mathbb{R}$.

We next describe the relationship between different bases and the important concept of dimension.

Proposition 7.11. *Let $V \subset \mathbb{R}^m$ be a vector space.*

(a) *There exists a basis for V.*

(b) *Any two bases for V have the same number of elements. The number of elements in a basis for V is called the* dimension *of V.*

(c) *Let $\boldsymbol{v}_1, \ldots, \boldsymbol{v}_n$ be a basis for V and let $\boldsymbol{w}_1, \ldots, \boldsymbol{w}_n$ be another set of n vectors in V. Write each $\boldsymbol{w}_j$ as a linear combination of the $\boldsymbol{v}_i$,*

$$\begin{aligned} \boldsymbol{w}_1 &= \alpha_{11}\boldsymbol{v}_1 + \alpha_{12}\boldsymbol{v}_2 + \cdots + \alpha_{1n}\boldsymbol{v}_n, \\ \boldsymbol{w}_2 &= \alpha_{21}\boldsymbol{v}_1 + \alpha_{22}\boldsymbol{v}_2 + \cdots + \alpha_{2n}\boldsymbol{v}_n, \\ \vdots \quad & \qquad \vdots \\ \boldsymbol{w}_n &= \alpha_{n1}\boldsymbol{v}_1 + \alpha_{n2}\boldsymbol{v}_2 + \cdots + \alpha_{nn}\boldsymbol{v}_n. \end{aligned}$$

Then $\boldsymbol{w}_1, \ldots, \boldsymbol{w}_n$ is also a basis for V if and only if the determinant of the matrix

$$\begin{pmatrix} \alpha_{11} & \alpha_{12} & \cdots & \alpha_{1n} \\ \alpha_{21} & \alpha_{22} & \cdots & \alpha_{2n} \\ \vdots & \vdots & \ddots & \vdots \\ \alpha_{n1} & \alpha_{n2} & \cdots & \alpha_{nn} \end{pmatrix}$$

is not equal to 0.

We next explain how to measure lengths of vectors in $\mathbb{R}^n$ and the angles between pairs of vectors. These important concepts are tied up with the notion of dot product and the Euclidean norm.

Definition. Let $\boldsymbol{v}, \boldsymbol{w} \in V \subset \mathbb{R}^m$ and write $\boldsymbol{v}$ and $\boldsymbol{w}$ using coordinates as

$$\boldsymbol{v} = (x_1, x_2, \ldots, x_m) \quad \text{and} \quad \boldsymbol{w} = (y_1, y_2, \ldots, y_m).$$

The *dot product of $\boldsymbol{v}$ and $\boldsymbol{w}$* is the quantity

$$\boldsymbol{v} \cdot \boldsymbol{w} = x_1 y_1 + x_2 y_2 + \cdots + x_m y_m.$$

We say that $\boldsymbol{v}$ and $\boldsymbol{w}$ are *orthogonal* to one another if $\boldsymbol{v} \cdot \boldsymbol{w} = 0$.

The *length*, or *Euclidean norm*, of $\boldsymbol{v}$ is the quantity

$$\|\boldsymbol{v}\| = \sqrt{x_1^2 + x_2^2 + \cdots + x_m^2}.$$

Notice that dot products and norms are related by the formula

$$\boldsymbol{v} \cdot \boldsymbol{v} = \|\boldsymbol{v}\|^2.$$

Proposition 7.12. *Let $\boldsymbol{v}, \boldsymbol{w} \in V \subset \mathbb{R}^m$.*

(a) *Let θ be the angle between the vectors $\boldsymbol{v}$ and $\boldsymbol{w}$, where we place the starting points of $\boldsymbol{v}$ and $\boldsymbol{w}$ at the origin $\boldsymbol{0}$. Then*

$$\boldsymbol{v} \cdot \boldsymbol{w} = \|\boldsymbol{v}\| \, \|\boldsymbol{w}\| \cos(\theta), \tag{7.6}$$

(b) **(Cauchy–Schwarz inequality)**

$$|\boldsymbol{v} \cdot \boldsymbol{w}| \le \|\boldsymbol{v}\| \, \|\boldsymbol{w}\|. \tag{7.7}$$

Proof. For (a), see any standard linear algebra textbook. We observe that the Cauchy–Schwarz inequality (b) follows immediately from (a), but we feel that it is of sufficient importance to warrant a direct proof. If $\boldsymbol{w} = \boldsymbol{0}$, there is nothing to prove, so we may assume that $\boldsymbol{w} \ne \boldsymbol{0}$. We consider the function

$$\begin{aligned} f(t) = \|\boldsymbol{v} - t\boldsymbol{w}\|^2 &= (\boldsymbol{v} - t\boldsymbol{w}) \cdot (\boldsymbol{v} - t\boldsymbol{w}) \\ &= \boldsymbol{v} \cdot \boldsymbol{v} - 2t\boldsymbol{v} \cdot \boldsymbol{w} + t^2 \boldsymbol{w} \cdot \boldsymbol{w} \\ &= \|\boldsymbol{v}\|^2 - 2t\boldsymbol{v} \cdot \boldsymbol{w} + t^2 \|\boldsymbol{w}\|^2. \end{aligned}$$

We know that $f(t) \ge 0$ for all $t \in \mathbb{R}$, so we choose the value of t that minimizes $f(t)$ and see what it gives. This minimizing value is $t = \boldsymbol{v} \cdot \boldsymbol{w} / \|\boldsymbol{w}\|^2$. Hence

$$0 \le f\left(\frac{\boldsymbol{v} \cdot \boldsymbol{w}}{\|\boldsymbol{w}\|^2}\right) = \|\boldsymbol{v}\|^2 - \frac{(\boldsymbol{v} \cdot \boldsymbol{w})^2}{\|\boldsymbol{w}\|^2}.$$

Simplifying this expression and taking square roots gives the desired result. □

Definition. An *orthogonal basis* for a vector space V is a basis $\boldsymbol{v}_1, \ldots, \boldsymbol{v}_n$ with the property that

$$\boldsymbol{v}_i \cdot \boldsymbol{v}_j = 0 \quad \text{for all } i \ne j.$$

The basis is *orthonormal* if in addition, $\|\boldsymbol{v}_i\| = 1$ for all i.

There are many formulas that become much simpler using an orthogonal or orthonormal basis. In particular, if $\boldsymbol{v}_1, \ldots, \boldsymbol{v}_n$ is an orthogonal basis and if $\boldsymbol{v} = a_1\boldsymbol{v}_1 + \cdots + a_n\boldsymbol{v}_n$ is a linear combination of the basis vectors, then

$$\begin{aligned} \|\boldsymbol{v}\|^2 &= \|a_1\boldsymbol{v}_1 + \cdots + a_n\boldsymbol{v}_n\|^2 \\ &= (a_1\boldsymbol{v}_1 + \cdots + a_n\boldsymbol{v}_n) \cdot (a_1\boldsymbol{v}_1 + \cdots + a_n\boldsymbol{v}_n) \end{aligned}$$

$$= \sum_{i=1}^{n} \sum_{j=1}^{n} a_i a_j (\boldsymbol{v}_i \cdot \boldsymbol{v}_j)$$

$$= \sum_{i=1}^{n} a_i^2 \|\boldsymbol{v}_i\|^2 \quad \text{since } \boldsymbol{v}_i \cdot \boldsymbol{v}_j = 0 \text{ for } i \neq j.$$

If the basis is orthonormal, then this further simplifies to $\|\boldsymbol{v}\|^2 = \sum a_i^2$.

There is a standard method, called the Gram–Schmidt algorithm, for creating an orthonormal basis. We describe a variant of the usual algorithm that gives an orthogonal basis, since it is this version that is most relevant for our later applications.

Theorem 7.13 (Gram–Schmidt Algorithm)**.** *Let $\boldsymbol{v}_1, \ldots, \boldsymbol{v}_n$ be a basis for a vector space $V \subset \mathbb{R}^m$. The following algorithm creates an orthogonal basis $\boldsymbol{v}_1^*, \ldots, \boldsymbol{v}_n^*$ for V:*

```
Set v*_1 = v_1.
Loop i = 2, 3, ..., n.
    Compute  μ_ij = v_i · v*_j / ||v*_j||^2   for 1 ≤ j < i.
    Set  v*_i = v_i − Σ_{j=1}^{i−1} μ_ij v*_j.
End Loop
```

The two bases have the property that

$$\operatorname{Span}\{\boldsymbol{v}_1, \ldots, \boldsymbol{v}_i\} = \operatorname{Span}\{\boldsymbol{v}_1^*, \ldots, \boldsymbol{v}_i^*\} \quad \textit{for all } i = 1, 2, \ldots, n.$$

Proof. The proof of orthogonality is by induction, so we suppose that the vectors $\boldsymbol{v}_1^*, \ldots, \boldsymbol{v}_{i-1}^*$ are pairwise orthogonal and we need to prove that $\boldsymbol{v}_i^*$ is orthogonal to all of the previous starred vectors. To do this, we take any $k < i$ and compute

$$\boldsymbol{v}_i^* \cdot \boldsymbol{v}_k^* = \left(\boldsymbol{v}_i - \sum_{j=1}^{i-1} \mu_{ij} \boldsymbol{v}_j^* \right) \cdot \boldsymbol{v}_k^*$$

$$= \boldsymbol{v}_i \cdot \boldsymbol{v}_k^* - \mu_{ik} \|\boldsymbol{v}_k^*\|^2 \quad \text{since } \boldsymbol{v}_k^* \cdot \boldsymbol{v}_j^* = 0 \text{ for } j \neq k,$$

$$= 0 \quad \text{from the definition of } \mu_{ik}.$$

To prove the final statement about the spans, we note first that it is clear from the definition of $\boldsymbol{v}_i^*$ that $\boldsymbol{v}_i$ is in the span of $\boldsymbol{v}_1^*, \ldots, \boldsymbol{v}_i^*$. We prove the other inclusion by induction, so we suppose that $\boldsymbol{v}_1^*, \ldots, \boldsymbol{v}_{i-1}^*$ are in the span of $\boldsymbol{v}_1, \ldots, \boldsymbol{v}_{i-1}$ and we need to prove that $\boldsymbol{v}_i^*$ is in the span of $\boldsymbol{v}_1, \ldots, \boldsymbol{v}_i$. But from the definition of $\boldsymbol{v}_i^*$, we see that it is in the span of $\boldsymbol{v}_1^*, \ldots, \boldsymbol{v}_{i-1}^*, \boldsymbol{v}_i$, so we are done by the induction hypothesis. □

7.4 Lattices: Basic Definitions and Properties

After seeing the examples in Sects. 7.1 and 7.2 and being reminded of the fundamental properties of vector spaces in Sect. 7.3, the reader will not be surprised by the formal definitions of a lattice and its properties.

Definition. Let $\boldsymbol{v}_1, \ldots, \boldsymbol{v}_n \in \mathbb{R}^m$ be a set of linearly independent vectors. The *lattice L generated by* $\boldsymbol{v}_1, \ldots, \boldsymbol{v}_n$ is the set of linear combinations of $\boldsymbol{v}_1, \ldots, \boldsymbol{v}_n$ with coefficients in $\mathbb{Z}$,

$$L = \{a_1\boldsymbol{v}_1 + a_2\boldsymbol{v}_2 + \cdots + a_n\boldsymbol{v}_n : a_1, a_2, \ldots, a_n \in \mathbb{Z}\}.$$

A *basis* for L is any set of independent vectors that generates L. Any two such sets have the same number of elements. The *dimension of* L is the number of vectors in a basis for L.

Suppose that $\boldsymbol{v}_1, \ldots, \boldsymbol{v}_n$ is a basis for a lattice L and that $\boldsymbol{w}_1, \ldots, \boldsymbol{w}_n \in L$ is another collection of vectors in L. Just as we did for vector spaces, we can write each $\boldsymbol{w}_j$ as a linear combination of the basis vectors,

$$\begin{aligned}
\boldsymbol{w}_1 &= a_{11}\boldsymbol{v}_1 + a_{12}\boldsymbol{v}_2 + \cdots + a_{1n}\boldsymbol{v}_n, \\
\boldsymbol{w}_2 &= a_{21}\boldsymbol{v}_1 + a_{22}\boldsymbol{v}_2 + \cdots + a_{2n}\boldsymbol{v}_n, \\
&\vdots \qquad\qquad\qquad \vdots \\
\boldsymbol{w}_n &= a_{n1}\boldsymbol{v}_1 + a_{n2}\boldsymbol{v}_2 + \cdots + a_{nn}\boldsymbol{v}_n,
\end{aligned}$$

but since now we are dealing with lattices, we know that all of the a_{ij} coefficients are integers.

Suppose that we try to express the $\boldsymbol{v}_i$ in terms of the $\boldsymbol{w}_j$. This involves inverting the matrix

$$A = \begin{pmatrix} a_{11} & a_{12} & \cdots & a_{1n} \\ a_{21} & a_{22} & \cdots & a_{2n} \\ \vdots & \vdots & \ddots & \vdots \\ a_{n1} & a_{n2} & \cdots & a_{nn} \end{pmatrix}.$$

Note that we need the $\boldsymbol{v}_i$ to be linear combinations of the $\boldsymbol{w}_j$ using *integer* coefficients, so we need the entries of A^{-1} to have integer entries. Hence

$$1 = \det(I) = \det(AA^{-1}) = \det(A)\det(A^{-1}),$$

where $\det(A)$ and $\det(A^{-1})$ are integers, so we must have $\det(A) = \pm 1$. Conversely, if $\det(A) = \pm 1$, then the theory of the adjoint matrix tells us that A^{-1} does indeed have integer entries. (See Exercise 7.10.) This proves the following useful result.

Proposition 7.14. *Any two bases for a lattice L are related by a matrix having integer coefficients and determinant equal to ± 1.*

For computational purposes, it is often convenient to work with lattices whose vectors have integer coordinates. For example,

$$\mathbb{Z}^n = \{(x_1, x_2, \ldots, x_n) : x_1, \ldots, x_n \in \mathbb{Z}\}$$

is the lattice consisting of all vectors with integer coordinates.

Definition. An *integral* (or *integer*) *lattice* is a lattice all of whose vectors have integer coordinates. Equivalently, an integral lattice is an additive subgroup of $\mathbb{Z}^m$ for some $m \ge 1$.

Example 7.15. Consider the three-dimensional lattice $L \subset \mathbb{R}^3$ generated by the three vectors

$$\boldsymbol{v}_1 = (2,1,3), \quad \boldsymbol{v}_2 = (1,2,0), \quad \boldsymbol{v}_3 = (2,-3,-5).$$

It is convenient to form a matrix using $\boldsymbol{v}_1, \boldsymbol{v}_2, \boldsymbol{v}_3$ as the rows of the matrix,

$$A = \begin{pmatrix} 2 & 1 & 3 \\ 1 & 2 & 0 \\ 2 & -3 & -5 \end{pmatrix}.$$

We create three new vectors in L by the formulas

$$\boldsymbol{w}_1 = \boldsymbol{v}_1 + \boldsymbol{v}_3, \qquad \boldsymbol{w}_2 = \boldsymbol{v}_1 - \boldsymbol{v}_2 + 2\boldsymbol{v}_3, \qquad \boldsymbol{w}_3 = \boldsymbol{v}_1 + 2\boldsymbol{v}_2.$$

This is equivalent to multiplying the matrix A on the left by the matrix

$$U = \begin{pmatrix} 1 & 0 & 1 \\ 1 & -1 & 2 \\ 1 & 2 & 0 \end{pmatrix},$$

and we find that $\boldsymbol{w}_1, \boldsymbol{w}_2, \boldsymbol{w}_3$ are the rows of the matrix

$$B = UA = \begin{pmatrix} 4 & -2 & -2 \\ 5 & -7 & -7 \\ 4 & 5 & 3 \end{pmatrix}.$$

The matrix U has determinant -1, so the vectors $\boldsymbol{w}_1, \boldsymbol{w}_2, \boldsymbol{w}_3$ are also a basis for L. The inverse of U is

$$U^{-1} = \begin{pmatrix} 4 & -2 & -1 \\ -2 & 1 & 1 \\ -3 & 2 & 1 \end{pmatrix},$$

and the rows of U^{-1} tell us how to express the $\boldsymbol{v}_i$ as linear combinations of the $\boldsymbol{w}_j$,

$$\boldsymbol{v}_1 = 4\boldsymbol{w}_1 - 2\boldsymbol{w}_2 - \boldsymbol{w}_3, \quad \boldsymbol{v}_2 = -2\boldsymbol{w}_1 + \boldsymbol{w}_2 + \boldsymbol{w}_3, \quad \boldsymbol{v}_3 = -3\boldsymbol{w}_1 + 2\boldsymbol{w}_2 + \boldsymbol{w}_3.$$

Remark 7.16. If $L \subset \mathbb{R}^m$ is a lattice of dimension n, then a basis for L may be written as the rows of an n-by-m matrix A, that is, a matrix with n rows and m columns. A new basis for L may be obtained by multiplying the matrix A on the left by an n-by-n matrix U such that U has integer entries and determinant ± 1. The set of such matrices U is called the *general linear group* (over $\mathbb{Z}$) and is denoted by $\mathrm{GL}_n(\mathbb{Z})$; cf. Example 2.11(g). It is the group of matrices with integer entries whose inverses also have integer entries.

There is an alternative, more abstract, way to define lattices that intertwines geometry and algebra.

Definition. A subset L of $\mathbb{R}^m$ is an *additive subgroup* if it is closed under addition and subtraction. It is called a *discrete additive subgroup* if there is a positive constant $\epsilon > 0$ with the following property: for every $\boldsymbol{v} \in L$,

$$L \cap \{\boldsymbol{w} \in \mathbb{R}^m : \|\boldsymbol{v} - \boldsymbol{w}\| < \epsilon\} = \{\boldsymbol{v}\}. \tag{7.8}$$

In other words, if you take any vector $\boldsymbol{v}$ in L and draw a solid ball of radius ϵ around $\boldsymbol{v}$, then there are no other points of L inside the ball.

Theorem 7.17. *A subset of $\mathbb{R}^m$ is a lattice if and only if it is a discrete additive subgroup.*

Proof. We leave the proof for the reader; see Exercise 7.9. □

A lattice is similar to a vector space, except that it is generated by all linear combinations of its basis vectors using integer coefficients, rather than using arbitrary real coefficients. It is often useful to view a lattice as an orderly arrangement of points in $\mathbb{R}^m$, where we put a point at the tip of each vector. An example of a lattice in $\mathbb{R}^2$ is illustrated in Fig. 7.1.

Definition. Let L be a lattice of dimension n and let $\boldsymbol{v}_1, \boldsymbol{v}_2, \ldots, \boldsymbol{v}_n$ be a basis for L. The *fundamental domain* (or *fundamental parallelepiped*) for L corresponding to this basis is the set

$$\mathcal{F}(\boldsymbol{v}_1, \ldots, \boldsymbol{v}_n) = \{t_1\boldsymbol{v}_1 + t_2\boldsymbol{v}_2 + \cdots + t_n\boldsymbol{v}_n \; : \; 0 \le t_i < 1\}. \tag{7.9}$$

The shaded area in Fig. 7.1 illustrates a fundamental domain in dimension 2. The next result indicates one reason why fundamental domains are important in studying lattices.

Proposition 7.18. *Let $L \subset \mathbb{R}^n$ be a lattice of dimension n and let $\mathcal{F}$ be a fundamental domain for L. Then every vector $\boldsymbol{w} \in \mathbb{R}^n$ can be written in the form*

$$\boldsymbol{w} = \boldsymbol{t} + \boldsymbol{v} \quad \textit{for a unique } \boldsymbol{t} \in \mathcal{F} \textit{ and a unique } \boldsymbol{v} \in L.$$

Equivalently, the union of the translated fundamental domains

$$\mathcal{F} + \boldsymbol{v} = \{\boldsymbol{t} + \boldsymbol{v} : \boldsymbol{t} \in \mathcal{F}\}$$

as $\boldsymbol{v}$ ranges over the vectors in the lattice L exactly covers $\mathbb{R}^n$; see Fig. 7.2.

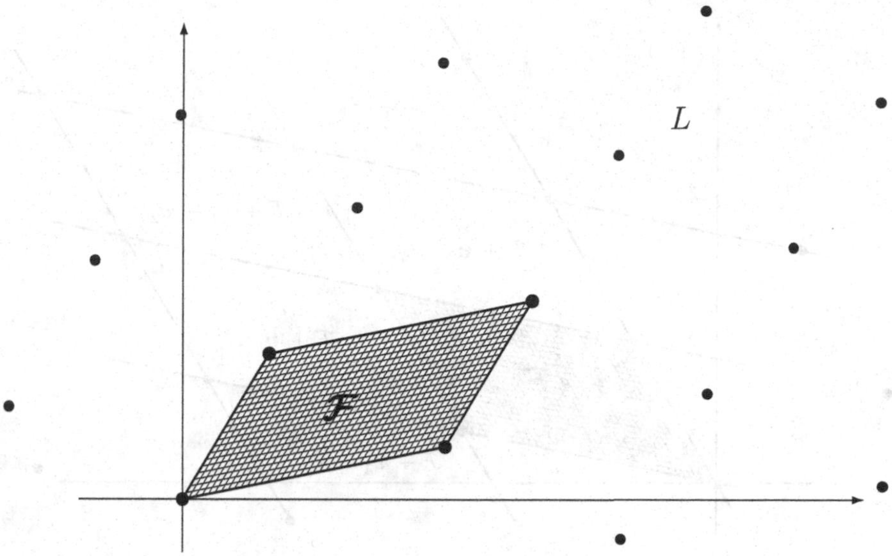

Figure 7.1: A lattice L and a fundamental domain $\mathcal{F}$

Proof. Let $\boldsymbol{v}_1, \ldots, \boldsymbol{v}_n$ be a basis of L that gives the fundamental domain $\mathcal{F}$. Then $\boldsymbol{v}_1, \ldots, \boldsymbol{v}_n$ are linearly independent in $\mathbb{R}^n$, so they are a basis of $\mathbb{R}^n$. This means that any $\boldsymbol{w} \in \mathbb{R}^n$ can be written in the form

$$\boldsymbol{w} = \alpha_1\boldsymbol{v}_1 + \alpha_2\boldsymbol{v}_2 + \cdots + \alpha_n\boldsymbol{v}_n \qquad \text{for some } \alpha_1, \ldots, \alpha_n \in \mathbb{R}.$$

We now write each α_i as

$$\alpha_i = t_i + a_i \qquad \text{with } 0 \le t_i < 1 \text{ and } a_i \in \mathbb{Z}.$$

Then

$$\boldsymbol{w} = \overbrace{t_1\boldsymbol{v}_1 + t_2\boldsymbol{v}_2 + \cdots + t_n\boldsymbol{v}_n}^{\text{this is a vector } \boldsymbol{t} \in \mathcal{F}} + \overbrace{a_1\boldsymbol{v}_1 + a_2\boldsymbol{v}_2 + \cdots + a_n\boldsymbol{v}_n}^{\text{this is a vector } \boldsymbol{v} \in L}.$$

This shows that $\boldsymbol{w}$ can be written in the desired form.

Next suppose that $\boldsymbol{w} = \boldsymbol{t} + \boldsymbol{v} = \boldsymbol{t}' + \boldsymbol{v}'$ has two representations as a sum of a vector in $\mathcal{F}$ and a vector in L. Then

$$(t_1 + a_1)\boldsymbol{v}_1 + (t_2 + a_2)\boldsymbol{v}_2 + \cdots + (t_n + a_n)\boldsymbol{v}_n = (t_1' + a_1')\boldsymbol{v}_1 + (t_2' + a_2')\boldsymbol{v}_2 + \cdots + (t_n' + a_n')\boldsymbol{v}_n.$$

Since $\boldsymbol{v}_1, \ldots, \boldsymbol{v}_n$ are independent, it follows that

$$t_i + a_i = t_i' + a_i' \qquad \text{for all } i = 1, 2, \ldots, n.$$

Hence

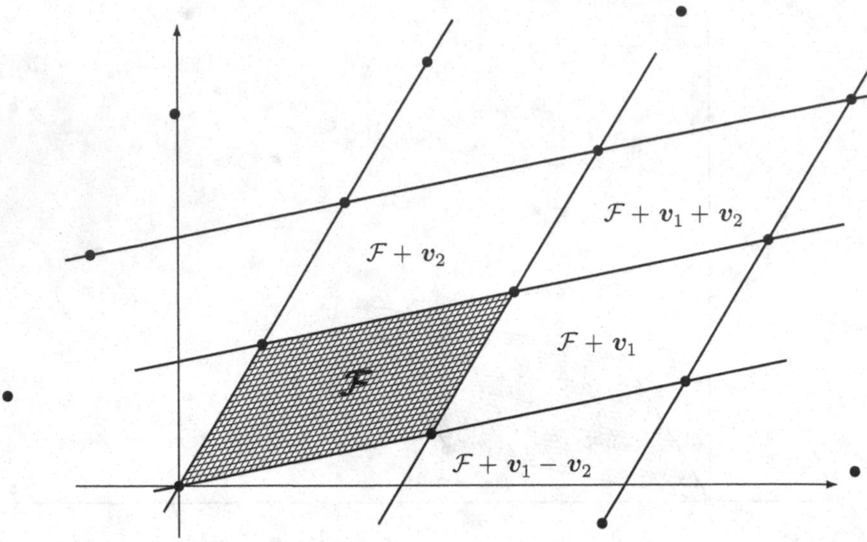

Figure 7.2: Translations of $\mathcal{F}$ by vectors in L exactly covers $\mathbb{R}^n$

$$t_i - t_i' = a_i' - a_i \in \mathbb{Z}$$

is an integer. But we also know that t_i and t_i' are greater than or equal to 0 and strictly smaller than 1, so the only way for $t_i - t_i'$ to be an integer is if $t_i = t_i'$. Therefore $\boldsymbol{t} = \boldsymbol{t}'$, and then also

$$\boldsymbol{v} = \boldsymbol{w} - \boldsymbol{t} = \boldsymbol{w} - \boldsymbol{t}' = \boldsymbol{v}'.$$

This completes the proof that $\boldsymbol{t} \in \mathcal{F}$ and $\boldsymbol{v} \in L$ are uniquely determined by $\boldsymbol{w}$. □

It turns out that all fundamental domains of a lattice L have the same volume. We prove this later (Corollary 7.22) for lattices of dimension n in $\mathbb{R}^n$. The volume of a fundamental domain turns out to be an extremely important invariant of the lattice.

Definition. Let L be a lattice of dimension n and let $\mathcal{F}$ be a fundamental domain for L. Then the n-dimensional volume of $\mathcal{F}$ is called the *determinant of* L (or sometimes the *covolume*[4] of L). It is denoted by $\det(L)$.

If you think of the basis vectors $\boldsymbol{v}_1, \ldots, \boldsymbol{v}_n$ as being vectors of a given length that describe the sides of the parallelepiped $\mathcal{F}$, then for basis vectors

[4]Note that the lattice L itself has no volume, since it is a countable collection of points. If $L \subset \mathbb{R}^n$ has dimension n, then the *covolume of* L is defined to be the volume of the quotient group $\mathbb{R}^n/L$.

of given lengths, the largest volume is obtained when the vectors are pairwise orthogonal to one another. This leads to the following important upper bound for the determinant of a lattice.

Proposition 7.19 (Hadamard's Inequality). *Let L be a lattice, take any basis $\boldsymbol{v}_1, \ldots, \boldsymbol{v}_n$ for L, and let $\mathcal{F}$ be a fundamental domain for L. Then*

$$\det L = \mathrm{Vol}(\mathcal{F}) \le \|\boldsymbol{v}_1\| \, \|\boldsymbol{v}_2\| \cdots \|\boldsymbol{v}_n\|. \tag{7.10}$$

The closer that the basis is to being orthogonal, the closer that Hadamard's inequality (7.10) comes to being an equality.

It is fairly easy to compute the determinant of a lattice L if its dimension is the same as its ambient space, i.e., if L is contained in $\mathbb{R}^n$ and L has dimension n. This formula, which luckily is the case that is of most interest to us, is described in the next proposition. See Exercise 7.14 to learn how to compute the determinant of a lattice in the general case.

Proposition 7.20. *Let $L \subset \mathbb{R}^n$ be a lattice of dimension n, let $\boldsymbol{v}_1, \boldsymbol{v}_2, \ldots, \boldsymbol{v}_n$ be a basis for L, and let $\mathcal{F} = \mathcal{F}(\boldsymbol{v}_1, \ldots, \boldsymbol{v}_n)$ be the associated fundamental domain as defined by (7.9). Write the coordinates of the ith basis vector as*

$$\boldsymbol{v}_i = (r_{i1}, r_{i2}, \ldots, r_{in})$$

and use the coordinates of the $\boldsymbol{v}_i$ as the rows of a matrix,

$$F = F(\boldsymbol{v}_1, \ldots, \boldsymbol{v}_n) = \begin{pmatrix} r_{11} & r_{12} & \cdots & r_{1n} \\ r_{21} & r_{22} & \cdots & r_{2n} \\ \vdots & \vdots & \ddots & \vdots \\ r_{n1} & r_{n2} & \cdots & r_{nn} \end{pmatrix}. \tag{7.11}$$

Then the volume of $\mathcal{F}$ is given by the formula

$$\mathrm{Vol}\big(\mathcal{F}(\boldsymbol{v}_1, \ldots, \boldsymbol{v}_n)\big) = \big|\det\big(F(\boldsymbol{v}_1, \ldots, \boldsymbol{v}_n)\big)\big|.$$

Proof. The proof uses multivariable calculus. We can compute the volume of $\mathcal{F}$ as the integral of the constant function 1 over the region $\mathcal{F}$,

$$\mathrm{Vol}(\mathcal{F}) = \int_{\mathcal{F}} dx_1 \, dx_2 \cdots dx_n.$$

The fundamental domain $\mathcal{F}$ is the set described by (7.9), so we make a change of variables from $\boldsymbol{x} = (x_1, \ldots, x_n)$ to $\boldsymbol{t} = (t_1, \ldots, t_n)$ according to the formula

$$(x_1, x_2, \ldots, x_n) = t_1 \boldsymbol{v}_1 + t_2 \boldsymbol{v}_2 + \cdots + t_n \boldsymbol{v}_n.$$

In terms of the matrix $F = F(\boldsymbol{v}_1, \ldots, \boldsymbol{v}_n)$ defined by (7.11), the change of variables is given by the matrix equation $\boldsymbol{x} = \boldsymbol{t}F$. The Jacobian matrix of this change of variables is F, and the fundamental domain $\mathcal{F}$ is the image under F of the unit cube $C_n = [0,1]^n$, so the change of variables formula for integrals yields

$$\int_{\mathcal{F}} dx_1\, dx_2 \cdots dx_n = \int_{FC_n} dx_1\, dx_2 \cdots dx_n = \int_{C_n} |\det F|\, dt_1\, dt_2 \cdots dt_n$$
$$= |\det F| \operatorname{Vol}(C_n) = |\det F|. \qquad \square$$

Example 7.21. The lattice in Example 7.15 has determinant

$$\det L = |\det A| = \left| \det \begin{pmatrix} 2 & 1 & 3 \\ 1 & 2 & 0 \\ 2 & -3 & -5 \end{pmatrix} \right| = |-36| = 36.$$

Corollary 7.22. *Let $L \subset \mathbb{R}^n$ be a lattice of dimension n. Then every fundamental domain for L has the same volume. Hence* $\det(L)$ *is an invariant of the lattice L, independent of the particular fundamental domain used to compute it.*

Proof. Let $\boldsymbol{v}_1, \ldots, \boldsymbol{v}_n$ and $\boldsymbol{w}_1, \ldots, \boldsymbol{w}_n$ be two fundamental domains for L, and let $F(\boldsymbol{v}_1, \ldots, \boldsymbol{v}_n)$ and $F(\boldsymbol{w}_1, \ldots, \boldsymbol{w}_n)$ be the associated matrices (7.11) obtained by using the coordinates of the vectors as the rows of the matrices. Then Proposition 7.14 tells us that

$$F(\boldsymbol{v}_1, \ldots, \boldsymbol{v}_n) = AF(\boldsymbol{w}_1, \ldots, \boldsymbol{w}_n) \tag{7.12}$$

for some n-by-n matrix with integer entries and $\det(A) = \pm 1$. Now applying Proposition 7.20 twice yields

$$\begin{aligned}
&\operatorname{Vol}\bigl(\mathcal{F}(\boldsymbol{v}_1, \ldots, \boldsymbol{v}_n)\bigr) \\
&\quad = \bigl|\det\bigl(F(\boldsymbol{v}_1, \ldots, \boldsymbol{v}_n)\bigr)\bigr| && \text{from Proposition 7.20,} \\
&\quad = \bigl|\det\bigl(AF(\boldsymbol{w}_1, \ldots, \boldsymbol{w}_n)\bigr)\bigr| && \text{from (7.12),} \\
&\quad = \bigl|\det(A)\bigr|\bigl|\det\bigl(F(\boldsymbol{w}_1, \ldots, \boldsymbol{w}_n)\bigr)\bigr| && \text{since } \det(AB) = \det(A)\det(B), \\
&\quad = \bigl|\det\bigl(F(\boldsymbol{w}_1, \ldots, \boldsymbol{w}_n)\bigr)\bigr| && \text{since } \det(A) = \pm 1, \\
&\quad = \operatorname{Vol}\bigl(\mathcal{F}(\boldsymbol{w}_1, \ldots, \boldsymbol{w}_n)\bigr) && \text{from Proposition 7.20.} \qquad \square
\end{aligned}$$

7.5 Short Vectors in Lattices

The fundamental computational problems associated to a lattice are those of finding a shortest nonzero vector in the lattice and of finding a vector in the lattice that is closest to a given nonlattice vector. In this section we discuss these problems, mainly from a theoretical perspective. Section 7.13 is devoted to a practical method for finding short and close vectors in a lattice.

7.5.1 The Shortest Vector Problem and the Closest Vector Problem

We begin with a description of two fundamental lattice problems.

The Shortest Vector Problem (SVP)**:** Find a shortest nonzero vector in a lattice L, i.e., find a nonzero vector $\boldsymbol{v} \in L$ that minimizes the Euclidean norm $\|\boldsymbol{v}\|$.

The Closest Vector Problem (CVP)**:** Given a vector $\boldsymbol{w} \in \mathbb{R}^m$ that is not in L, find a vector $\boldsymbol{v} \in L$ that is closest to $\boldsymbol{w}$, i.e., find a vector $\boldsymbol{v} \in L$ that minimizes the Euclidean norm $\|\boldsymbol{w} - \boldsymbol{v}\|$.

Remark 7.23. Note that there may be more than one shortest nonzero vector in a lattice. For example, in $\mathbb{Z}^2$, all four of the vectors $(0, \pm 1)$ and $(\pm 1, 0)$ are solutions to SVP. This is why SVP asks for "a" shortest vector and not "the" shortest vector. A similar remark applies to CVP.

We have seen in Sects. 7.1 and 7.2 that a solution to SVP can be used to break various cryptosystems. We will see more examples later in this chapter.

Both SVP and CVP are profound problems, and both become computationally difficult as the dimension n of the lattice grows. On the other hand, even approximate solutions to SVP and CVP turn out to have surprisingly many applications in different fields of pure and applied mathematics. In full generality, CVP is known to be $\mathcal{NP}$-hard and SVP is $\mathcal{NP}$-hard under a certain "randomized reduction hypothesis."[5]

In practice, CVP is considered to be "a little bit harder" than SVP, since CVP can often be reduced to SVP in a slightly higher dimension. For example, the $(n+1)$-dimensional SVP used to solve the knapsack cryptosystem in Sect. 7.2 can be naturally formulated as an n-dimensional CVP. For a proof that SVP is no harder than CVP, see [50], and for a thorough discussion of the complexity of different types of lattice problems, see [86].

Remark 7.24. In full generality, both SVP and CVP are considered to be extremely hard problems, but in practice it is difficult to achieve this idealized "full generality." In real world scenarios, cryptosystems based on $\mathcal{NP}$-hard

[5]This hypothesis means that the class of polynomial-time algorithms is enlarged to include those that are not deterministic, but will, with high probability, terminate in polynomial time with a correct result. See Ajtai [3] for details.

or $\mathcal{NP}$-complete problems tend to rely on a particular subclass of problems, either to achieve efficiency or to allow the creation of a trapdoor. When this is done, there is always the possibility that some special property of the chosen subclass of problems makes them easier to solve than the general case. We have already seen this with the knapsack cryptosystem in Sect. 7.2. The general knapsack problem is $\mathcal{NP}$-complete, but the disguised superincreasing knapsack problem that was suggested for use in cryptography is much easier to solve than the general knapsack problem.

There are many important variants of SVP and CVP that arise both in theory and in practice. We describe a few of them here.

Shortest Basis Problem (SBP) Find a basis $\boldsymbol{v}_1, \ldots, \boldsymbol{v}_n$ for a lattice that is shortest in some sense. For example, we might require that

$$\max_{1 \le i \le n} \|\boldsymbol{v}_i\| \quad \text{or} \quad \sum_{i=1}^{n} \|\boldsymbol{v}_i\|^2$$

be minimized. There are thus many different versions of SBP, depending on how one decides to measure the "size" of a basis.

Approximate Shortest Vector Problem (apprSVP) Let $\psi(n)$ be a function of n. In a lattice L of dimension n, find a nonzero vector that is no more than $\psi(n)$ times longer than a shortest nonzero vector. In other words, if $\boldsymbol{v}_{\text{shortest}}$ is a shortest nonzero vector in L, find a nonzero vector $\boldsymbol{v} \in L$ satisfying

$$\|\boldsymbol{v}\| \le \psi(n) \|\boldsymbol{v}_{\text{shortest}}\|.$$

Each choice of function $\psi(n)$ gives a different apprSVP. As specific examples, one might ask for an algorithm that finds a nonzero $\boldsymbol{v} \in L$ satisfying

$$\|\boldsymbol{v}\| \le 3\sqrt{n} \|\boldsymbol{v}_{\text{shortest}}\| \quad \text{or} \quad \|\boldsymbol{v}\| \le 2^{n/2} \|\boldsymbol{v}_{\text{shortest}}\|.$$

Clearly an algorithm that solves the former is much stronger than one that solves the latter, but even the latter may be useful if the dimension is not too large.

Approximate Closest Vector Problem (apprCVP) This is the same as apprSVP, but now we are looking for a vector that is an approximate solution to CVP, instead of an approximate solution to SVP.

7.5.2 Hermite's Theorem and Minkowski's Theorem

How long is the shortest nonzero vector in a lattice L? The answer depends to some extent on the dimension and the determinant of L. The next result gives an explicit upper bound in terms of $\dim(L)$ and $\det(L)$ for the shortest nonzero vector in L.

Theorem 7.25 (Hermite's Theorem). *Every lattice L of dimension n contains a nonzero vector $\boldsymbol{v} \in L$ satisfying*

$$\|\boldsymbol{v}\| \le \sqrt{n}\det(L)^{1/n}.$$

Remark 7.26. For a given dimension n, *Hermite's constant* γ_n is the smallest value such that every lattice L of dimension n contains a nonzero vector $\boldsymbol{v} \in L$ satisfying

$$\|\boldsymbol{v}\|^2 \le \gamma_n \det(L)^{2/n}.$$

Our version of Hermite's theorem (Theorem 7.25) says that $\gamma_n \le n$. The exact value of γ_n is known only for $1 \le n \le 8$ and for $n = 24$:

γ_2^2	γ_3^3	γ_4^4	γ_5^5	γ_6^6	γ_7^7	γ_8^8	γ_{24}^{24}
$\frac{4}{3}$	2	4	8	$\frac{64}{3}$	64	256	4

For cryptographic purposes, we are mainly interested in the value of γ_n when n is large. For large values of n it is known that Hermite's constant satisfies

$$\frac{n}{2\pi e} \le \gamma_n \le \frac{n}{\pi e}, \tag{7.13}$$

where $\pi = 3.14159\ldots$ and $e = 2.71828\ldots$ are the usual constants.

Remark 7.27. There are versions of Hermite's theorem that deal with more than one vector. For example, one can prove that an n-dimensional lattice L always has a basis $\boldsymbol{v}_1, \ldots, \boldsymbol{v}_n$ satisfying

$$\|\boldsymbol{v}_1\|\,\|\boldsymbol{v}_2\| \cdots \|\boldsymbol{v}_n\| \le n^{n/2}(\det L).$$

This complements Hadamard's inequality (Proposition 7.19), which says that every basis satisfies

$$\|\boldsymbol{v}_1\|\,\|\boldsymbol{v}_2\| \cdots \|\boldsymbol{v}_n\| \ge \det L.$$

We define the *Hadamard ratio of the basis* $\mathcal{B} = \{\boldsymbol{v}_1, \ldots, \boldsymbol{v}_n\}$ to be the quantity

$$\mathcal{H}(\mathcal{B}) = \left(\frac{\det L}{\|\boldsymbol{v}_1\|\,\|\boldsymbol{v}_2\| \cdots \|\boldsymbol{v}_n\|}\right)^{1/n}.$$

Thus $0 < \mathcal{H}(\mathcal{B}) \le 1$, and the closer that the value is to 1, the more orthogonal are the vectors in the basis. (The reciprocal of the Hadamard ratio is sometimes called the *orthogonality defect.* We also note that some authors define the Hadamard ratio without taking the nth root.)

The proof of Hermite's theorem uses a result of Minkowski that is important in its own right. In order to state Minkowski's theorem, we set one piece of useful notation and give some basic definitions.

Definition. For any $\boldsymbol{a} \in \mathbb{R}^n$ and any $R > 0$, the *(closed) ball of radius R centered at $\boldsymbol{a}$* is the set

$$\mathbb{B}_R(\boldsymbol{a}) = \bigl\{\boldsymbol{x} \in \mathbb{R}^n : \|\boldsymbol{x} - \boldsymbol{a}\| \le R\bigr\}.$$

Definition. Let S be a subset of $\mathbb{R}^n$.

(a) S is *bounded* if the lengths of the vectors in S are bounded. Equivalently, S is bounded if there is a radius R such that S is contained within the ball $\mathbb{B}_R(\mathbf{0})$.

(b) S is *symmetric* if for every point $\boldsymbol{a}$ in S, the negation $-\boldsymbol{a}$ is also in S.

(c) S is *convex* if whenever two points $\boldsymbol{a}$ and $\boldsymbol{b}$ are in S, then the entire line segment connecting $\boldsymbol{a}$ to $\boldsymbol{b}$ lies completely in S.

(d) S is *closed* if it has the following property: If $\boldsymbol{a} \in \mathbb{R}^n$ is a point such that every ball $\mathbb{B}_R(\boldsymbol{a})$ contains a point of S, then $\boldsymbol{a}$ is in S.

Theorem 7.28 (Minkowski's Theorem). *Let $L \subset \mathbb{R}^n$ be a lattice of dimension n and let $S \subset \mathbb{R}^n$ be a bounded symmetric convex set whose volume satisfies*

$$\mathrm{Vol}(S) > 2^n \det(L).$$

Then S contains a nonzero lattice vector.

If S is also closed, then it suffices to take $\mathrm{Vol}(S) \geq 2^n \det(L)$.

Proof. Let $\mathcal{F}$ be a fundamental domain for L. Proposition 7.18 tells us that every vector $\boldsymbol{a} \in S$ can be written uniquely in the form

$$\boldsymbol{a} = \boldsymbol{v}_{\boldsymbol{a}} + \boldsymbol{w}_{\boldsymbol{a}} \qquad \text{with } \boldsymbol{v}_{\boldsymbol{a}} \in L \text{ and } \boldsymbol{w}_{\boldsymbol{a}} \in \mathcal{F}.$$

(See Fig. 7.2 for an illustration.) We dilate S by a factor of $\frac{1}{2}$, i.e., shrink S by a factor of 2,

$$\frac{1}{2}S = \left\{\frac{1}{2}\boldsymbol{a} : \boldsymbol{a} \in S\right\},$$

and consider the map

$$\frac{1}{2}S \longrightarrow \mathcal{F}, \qquad \frac{1}{2}\boldsymbol{a} \longmapsto \boldsymbol{w}_{\frac{1}{2}\boldsymbol{a}}. \tag{7.14}$$

Shrinking S by a factor of 2 changes its volume by a factor of 2^n, so

$$\mathrm{Vol}\left(\frac{1}{2}S\right) = \frac{1}{2^n}\mathrm{Vol}(S) > \det(L) = \mathrm{Vol}(\mathcal{F}).$$

(Here is where we are using our assumption that the volume of S is larger than $2^n \det(L)$.)

The map (7.14) is given by a finite collection of translation maps (this is where we are using the assumption that S is bounded), so the map (7.14) is volume preserving. Hence the fact that the domain $\frac{1}{2}S$ has volume strictly larger than the volume of the range $\mathcal{F}$ implies that there exist distinct points $\frac{1}{2}\boldsymbol{a}_1$ and $\frac{1}{2}\boldsymbol{a}_2$ with the same image in $\mathcal{F}$.

We have thus found distinct points in S satisfying

$$\frac{1}{2}\boldsymbol{a}_1 = \boldsymbol{v}_1 + \boldsymbol{w} \quad \text{and} \quad \frac{1}{2}\boldsymbol{a}_2 = \boldsymbol{v}_2 + \boldsymbol{w} \quad \text{with } \boldsymbol{v}_1, \boldsymbol{v}_2 \in L \text{ and } \boldsymbol{w} \in \mathcal{F}.$$

Subtracting them yields a nonzero vector

$$\frac{1}{2}\boldsymbol{a}_1 - \frac{1}{2}\boldsymbol{a}_2 = \boldsymbol{v}_1 - \boldsymbol{v}_2 \in L.$$

We now observe that the vector

$$\underbrace{\frac{1}{2}\boldsymbol{a}_1 + \overbrace{\left(-\frac{1}{2}\boldsymbol{a}_2\right)}^{\substack{S \text{ is symmetric,}\\ \text{so } -\boldsymbol{a}_2 \text{ is in } S}}}_{\substack{\text{this is the midpoint of the line}\\ \text{segment from } \boldsymbol{a}_1 \text{ to } -\boldsymbol{a}_2,\\ \text{so it is in } S \text{ by convexity}}}$$

is in the set S. Therefore

$$\mathbf{0} \neq \boldsymbol{v}_1 - \boldsymbol{v}_2 \in S \cap L,$$

so we have constructed a nonzero lattice point in S.

This completes the proof of Minkowski's theorem assuming that the volume of S is strictly larger than $2^n \det(L)$. We now assume that S is closed and allow $\operatorname{Vol}(S) = 2^n \det(L)$. For every $k \geq 1$, we expand S by a factor of $1 + \frac{1}{k}$ and apply the earlier result to find a nonzero vector

$$\mathbf{0} \neq \boldsymbol{v}_k \in \left(1 + \frac{1}{k}\right) S \cap L.$$

Each of the lattice vectors $\boldsymbol{v}_1, \boldsymbol{v}_2, \ldots$ is in the bounded set $2S$, so the discreteness of L tells us that the sequence contains only finitely many distinct vectors. Thus we can choose some $\boldsymbol{v}$ that appears infinitely often in the sequence, so we have found a nonzero lattice vector $\boldsymbol{v} \in L$ in the intersection

$$\bigcap_{k=1}^{\infty} \left(1 + \frac{1}{k}\right) S. \tag{7.15}$$

The assumption that S is closed implies that the intersection (7.15) is equal to S, so $\mathbf{0} \neq \boldsymbol{v} \in S \cap L$. □

Proof of Hermite's theorem (*Theorem 7.25*). The proof is a simple application of Minkowski's theorem. Let $L \subset \mathbb{R}^n$ be a lattice and let S be the hypercube in $\mathbb{R}^n$, centered at $\mathbf{0}$, whose sides have length $2B$,

$$S = \bigl\{(x_1, \ldots, x_n) \in \mathbb{R}^n : -B \leq x_i \leq B \quad \text{for all } 1 \leq i \leq n\bigr\}.$$

The set S is symmetric, closed, and bounded, and its volume is

$$\operatorname{Vol}(S) = (2B)^n.$$

So if we set $B = \det(L)^{1/n}$, then $\mathrm{Vol}(S) = 2^n \det(L)$ and we can apply Minkowski's theorem to deduce that there is a vector $\mathbf{0} \neq \boldsymbol{a} \in S \cap L$. Writing the coordinates of $\boldsymbol{a}$ as $(a_1, \ldots, a_n)$, by definition of S we have

$$\|\boldsymbol{a}\| = \sqrt{a_1^2 + \cdots + a_n^2} \leq \sqrt{n}\, B = \sqrt{n} \det(L)^{1/n}.$$

This completes the proof of Theorem 7.25. □

7.5.3 The Gaussian Heuristic

It is possible to improve the constant appearing in Hermite's theorem (Theorem 7.25) by applying Minkowski's theorem (Theorem 7.28) to a hypersphere, rather than a hypercube. In order to do this, we need to know the volume of a ball in $\mathbb{R}^n$. The following material is generally covered in advanced calculus classes.

Definition. The *gamma function* $\Gamma(s)$ is defined for $s > 0$ by the integral

$$\Gamma(s) = \int_0^\infty t^s e^{-t} \frac{dt}{t}. \tag{7.16}$$

The gamma function is a very important function that appears in many mathematical formulas. We list a few of its basic properties.

Proposition 7.29. (a) *The integral* (7.16) *defining* $\Gamma(s)$ *is convergent for all* $s > 0$.
(b) $\Gamma(1) = 1$ *and* $\Gamma(s+1) = s\Gamma(s)$. *This allows us to extend* $\Gamma(s)$ *to all* $s \in \mathbb{R}$ *with* $s \neq 0, -1, -2, \ldots$.
(c) *For all integers* $n \geq 1$ *we have* $\Gamma(n+1) = n!$. *Thus* $\Gamma(s)$ *interpolates the values of the factorial function to all real (and even complex) numbers.*
(d) $\Gamma(\frac{1}{2}) = \sqrt{\pi}$.
(e) (Stirling's formula) *For large values of* s *we have*

$$\Gamma(1+s)^{1/s} \approx \frac{s}{e}. \tag{7.17}$$

(*More precisely,* $\ln \Gamma(1+s) = \ln(s/e)^s + \frac{1}{2}\ln(2\pi s) + O(1)$ *as* $s \to \infty$.)

Proof. The properties of the gamma function are described in real and complex analysis textbooks; see for example [2] or [43]. □

The formula for the volume of a ball in n-dimensional space involves the gamma function.

Theorem 7.30. *Let* $\mathbb{B}_R(\boldsymbol{a})$ *be a ball of radius* R *in* $\mathbb{R}^n$. *Then the volume of* $\mathbb{B}_R(\boldsymbol{a})$ *is*

$$\mathrm{Vol}\big(\mathbb{B}_R(\boldsymbol{a})\big) = \frac{\pi^{n/2} R^n}{\Gamma(1 + n/2)}. \tag{7.18}$$

For large values of n, the volume of the ball $\mathbb{B}_R(\boldsymbol{a}) \subset \mathbb{R}^n$ is approximately given by

$$\operatorname{Vol}\big(\mathbb{B}_R(\boldsymbol{a})\big)^{1/n} \approx \sqrt{\frac{2\pi e}{n}}\,R. \tag{7.19}$$

Proof. See [43, §5.9], for example, for a proof of the formula (7.18) giving the volume of a ball.

We can use (7.18) and Stirling's formula (7.17) to prove (7.19). Thus

$$\operatorname{Vol}\big(\mathbb{B}_R(\boldsymbol{a})\big)^{1/n} = \frac{\pi^{1/2}R}{\Gamma(1+n/2)^{1/n}} \approx \frac{\pi^{1/2}R}{(n/2e)^{1/2}} = \sqrt{\frac{2\pi e}{n}}\,R. \qquad \square$$

Remark 7.31. Theorem 7.30 allows us to improve Theorem 7.25 for large values of n. The ball $\mathbb{B}_R(\mathbf{0})$ is bounded, closed, convex, and symmetric, so Minkowski's theorem (Theorem 7.28) says that if we choose R such that

$$\operatorname{Vol}\big(\mathbb{B}_R(\mathbf{0})\big) \ge 2^n \det(L),$$

then the ball $\mathbb{B}_R(\mathbf{0})$ contains a nonzero lattice point. Assuming that n is large, we can use (7.19) to approximate the volume of $\mathbb{B}_R(\mathbf{0})$, so we need to choose R to satisfy

$$\sqrt{\frac{2\pi e}{n}}\,R \gtrapprox 2\det(L)^{1/n}.$$

Hence for large n there exists a nonzero vector $\boldsymbol{v} \in L$ satisfying

$$\|\boldsymbol{v}\| \lessapprox \sqrt{\frac{2n}{\pi e}} \cdot (\det L)^{1/n}.$$

This improves the estimate in Theorem 7.25 by a factor of $\sqrt{2/\pi e} \approx 0.484$.

Although exact bounds for the size of a shortest vector are unknown when the dimension n is large, we can estimate its size by a probabilistic argument that is based on the following principle:

> Let $\mathbb{B}_R(\mathbf{0})$ be a large ball centered at $\mathbf{0}$. Then the number of lattice points in $\mathbb{B}_R(\mathbf{0})$ is approximately equal to the volume of $\mathbb{B}_R(\mathbf{0})$ divided by the volume of a fundamental domain $\mathcal{F}$.

This is reasonable, since $\#\big(\mathbb{B}_R(\mathbf{0}) \cap L\big)$ should be approximately the number of copies of $\mathcal{F}$ that fit into $\mathbb{B}_R(\mathbf{0})$. (See Exercise 7.15 for a more rigorous justification.)

For example, if we let $L = \mathbb{Z}^2$, then this principle says that the area of a circle is approximately the number of integer points inside the circle. The problem of estimating the error term in

$$\#\big\{(x,y) \in \mathbb{Z}^2 : x^2 + y^2 \le R^2\big\} = \pi R^2 + (\text{error term})$$

is a famous classical problem. In higher dimensions, the problem becomes more difficult because, as n increases, the error created by lattice points near the boundary of the ball can be quite large until R becomes very large. Thus the estimate

$$\#\{\boldsymbol{v} \in L : \|\boldsymbol{v}\| \leq R\} \approx \frac{\mathrm{Vol}(\mathbb{B}_R(\mathbf{0}))}{\mathrm{Vol}(\mathcal{F})} \tag{7.20}$$

is somewhat problematic when n is large and R is not too large. Still, one can ask for the value of R that makes the right-hand side (7.20) equal to 1, since in some sense this is the value of R for which we might expect to first find a nonzero lattice point in the ball.

Assuming that n is large, we use the estimate (7.19) from Theorem 7.30. We set

$$\left(\frac{2\pi e}{n}\right)^{n/2} R^n \approx \mathrm{Vol}(\mathbb{B}_R(\mathbf{0})) \qquad \text{equal to} \qquad \mathrm{Vol}(\mathcal{F}) = \det(L),$$

and we solve for

$$R \approx \sqrt{\frac{n}{2\pi e}} (\det L)^{1/n}.$$

This leads to the following heuristic.

Definition. Let L be a lattice of dimension n. The *Gaussian expected shortest length* is

$$\sigma(L) = \sqrt{\frac{n}{2\pi e}} (\det L)^{1/n}. \tag{7.21}$$

The *Gaussian heuristic* says that a shortest nonzero vector in a "randomly chosen lattice" will satisfy

$$\|\boldsymbol{v}_{\text{shortest}}\| \approx \sigma(L).$$

More precisely, if $\epsilon > 0$ is fixed, then for all sufficiently large n, a randomly chosen lattice of dimension n will satisfy

$$(1-\epsilon)\sigma(L) \leq \|\boldsymbol{v}_{\text{shortest}}\| \leq (1+\epsilon)\sigma(L).$$

(See [133] for some mathematical justification of this heuristic principle.)

Remark 7.32. For small values of n, it is better to use the exact formula (7.18) for the volume of $\mathcal{B}_R(\mathbf{0})$, so the Gaussian expected shortest length for small n is

$$\sigma(L) = \big(\Gamma(1+n/2)\det(L)\big)^{1/n}/\sqrt{\pi}. \tag{7.22}$$

For example, when $n = 6$, then (7.21) gives $\sigma(L) = 0.5927(\det L)^{1/6}$, while (7.22) gives $\sigma(L) = 0.7605(\det L)^{1/6}$, which is a significant difference. On the other hand, if $n = 100$, then they give

$$\sigma(L) = 2.420(\det L)^{1/100} \qquad \text{and} \qquad \sigma(L) = 2.490(\det L)^{1/100},$$

respectively, so the difference is much smaller.

Example 7.33. Let $(m_1, \ldots, m_n, S)$ be a knapsack problem. The associated lattice $L_{M,S}$ is generated by the rows of the matrix (7.4) given on page 383. The matrix $L_{M,S}$ has dimension $n+1$ and determinant $\det L_{M,S} = 2^n S$. As explained in Sect. 7.2, the number S satisfies $S = \mathcal{O}(2^{2n})$, so $S^{1/n} \approx 4$. This allows us to approximate the Gaussian shortest length as

$$\begin{aligned}\sigma(L_{M,S}) &= \sqrt{\frac{n+1}{2\pi e}}(\det L_{M,S})^{1/(n+1)} = \sqrt{\frac{n+1}{2\pi e}}(2^n S)^{1/(n+1)} \\ &\approx \sqrt{\frac{n}{2\pi e}} \cdot 2S^{1/n} \approx \sqrt{\frac{n}{2\pi e}} \cdot 8 \approx 1.936\sqrt{n}.\end{aligned}$$

On the other hand, as explained in Sect. 7.2, the lattice $L_{M,S}$ contains a vector $\boldsymbol{t}$ of length $\sqrt{n}$, and knowledge of $\boldsymbol{t}$ reveals the solution to the subset-sum problem. Hence solving SVP for the lattice $L_{M,S}$ is very likely to solve the subset-sum problem. For a further discussion of the use of lattice methods to solve subset-sum problems, see Sect. 7.14.2.

We will find that the Gaussian heuristic is useful in quantifying the difficulty of locating short vectors in lattices. In particular, if the actual shortest vector of a particular lattice L is significantly shorter than $\sigma(L)$, then lattice reduction algorithms such as LLL seem to have a much easier time locating the shortest vector.

A similar argument leads to a Gaussian heuristic for CVP. Thus if $L \subset \mathbb{R}^n$ is a random lattice of dimension n and $\boldsymbol{w} \in \mathbb{R}^n$ is a random point, then we expect that the lattice vector $\boldsymbol{v} \in L$ closest to $\boldsymbol{w}$ satisfies

$$\|\boldsymbol{v} - \boldsymbol{w}\| \approx \sigma(L).$$

And just as for SVP, if L contains a point that is significantly closer than $\sigma(L)$ to $\boldsymbol{w}$, then lattice reduction algorithms have an easier time solving CVP.

7.6 Babai's Algorithm and Using a "Good" Basis to Solve apprCVP

If a lattice $L \subset \mathbb{R}^n$ has a basis $\boldsymbol{v}_1, \ldots, \boldsymbol{v}_n$ consisting of vectors that are pairwise orthogonal, i.e., such that

$$\boldsymbol{v}_i \cdot \boldsymbol{v}_j = 0 \quad \text{for all } i \neq j,$$

then it is easy to solve both SVP and CVP. Thus to solve SVP, we observe that the length of any vector in L is given by the formula

$$\|a_1\boldsymbol{v}_1 + a_2\boldsymbol{v}_2 + \cdots + a_n\boldsymbol{v}_n\|^2 = a_1^2\|\boldsymbol{v}_1\|^2 + a_2^2\|\boldsymbol{v}_2\|^2 + \cdots + a_n^2\|\boldsymbol{v}_n\|^2.$$

Since $a_1, \ldots, a_n \in \mathbb{Z}$, we see that the shortest nonzero vector(s) in L are simply the shortest vector(s) in the set $\{\pm\boldsymbol{v}_1, \ldots, \pm\boldsymbol{v}_n\}$.

Similarly, suppose that we want to find the vector in L that is closest to a given vector $\boldsymbol{w} \in \mathbb{R}^n$. We first write

$$\boldsymbol{w} = t_1\boldsymbol{v}_1 + t_2\boldsymbol{v}_2 + \cdots + t_n\boldsymbol{v}_n \quad \text{with } t_1, \ldots, t_n \in \mathbb{R}.$$

Then for $\boldsymbol{v} = a_1\boldsymbol{v}_1 + \cdots + a_n\boldsymbol{v}_n \in L$, we have

$$\|\boldsymbol{v} - \boldsymbol{w}\|^2 = (a_1 - t_1)^2\|\boldsymbol{v}_1\|^2 + (a_2 - t_2)^2\|\boldsymbol{v}_2\|^2 + \cdots + (a_n - t_n)^2\|\boldsymbol{v}_n\|^2. \quad (7.23)$$

The a_i are required to be integers, so (7.23) is minimized if we take each a_i to be the integer closest to the corresponding t_i.

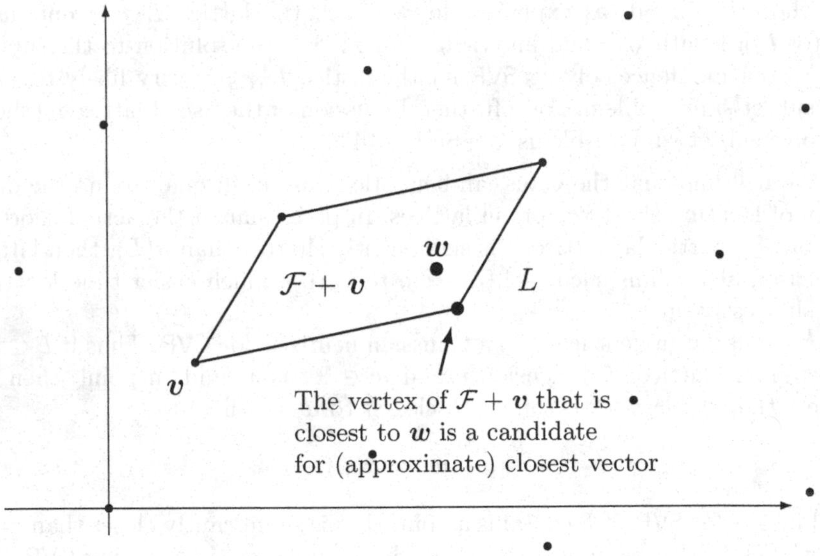

Figure 7.3: Using a given fundamental domain to try to solve CVP

It is tempting to try a similar procedure with an arbitrary basis of L. If the vectors in the basis are reasonably orthogonal to one another, then we are likely to be successful in solving CVP; but if the basis vectors are highly non-orthogonal, then the algorithm does not work well. We briefly discuss the underlying geometry, then describe the general method, and conclude with a 2-dimensional example.

A basis $\{\boldsymbol{v}_1, \ldots, \boldsymbol{v}_n\}$ for L determines a fundamental domain $\mathcal{F}$ in the usual way, see (7.9). Proposition 7.18 says that the translates of $\mathcal{F}$ by the elements of L fill up the entire space $\mathbb{R}^n$, so any $\boldsymbol{w} \in \mathbb{R}^n$ is in a unique translate $\mathcal{F} + \boldsymbol{v}$ of $\mathcal{F}$ by an element $\boldsymbol{v} \in L$. We take the vertex of the parallelepiped $\mathcal{F} + \boldsymbol{v}$ that is closest to $\boldsymbol{w}$ as our hypothetical solution to CVP. This procedure is illustrated in Fig. 7.3. It is easy to find the closest vertex, since

$$\boldsymbol{w} = \boldsymbol{v} + \epsilon_1\boldsymbol{v}_1 + \epsilon_2\boldsymbol{v}_2 + \cdots + \epsilon_n\boldsymbol{v}_n \quad \text{for some } 0 \le \epsilon_1, \epsilon_2, \ldots, \epsilon_n < 1,$$

so we simply replace ϵ_i by 0 if it is less than $\frac{1}{2}$ and replace it by 1 if it is greater than or equal to $\frac{1}{2}$.

Looking at Fig. 7.3 makes it seem that this procedure is bound to work, but that's because the basis vectors in the picture are reasonably orthogonal to one another. Figure 7.4 illustrates two different bases for the same lattice. The first basis is "good" in the sense that the vectors are fairly orthogonal; the second basis is "bad" because the angle between the basis vectors is small.

If we try to solve CVP using a bad basis, we are likely run into problems as illustrated in Fig. 7.5. The nonlattice target point is actually quite close to a lattice point, but the parallelogram is so elongated that the closest vertex

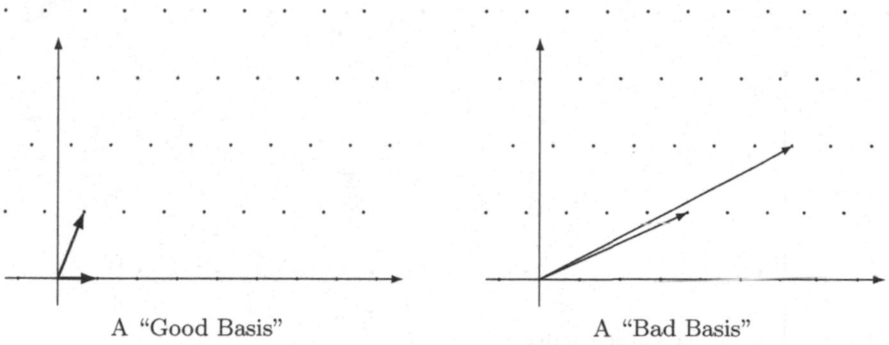

Figure 7.4: Two different bases for the same lattice

to the target point is quite far away. And it is important to note that the difficulties get much worse as the dimension of the lattice increases. Examples visualized in dimension 2 or 3, or even dimension 4 or 5, do not convey the extent to which the following closest vertex algorithm generally fails to solve even apprCVP unless the basis is quite orthogonal.

Theorem 7.34 (Babai's Closest Vertex Algorithm). *Let $L \subset \mathbb{R}^n$ be a lattice with basis $\boldsymbol{v}_1, \ldots, \boldsymbol{v}_n$, and let $\boldsymbol{w} \in \mathbb{R}^n$ be an arbitrary vector. If the vectors in the basis are sufficiently orthogonal to one another, then the following algorithm solves* CVP.

Write $\boldsymbol{w} = t_1\boldsymbol{v}_1 + t_2\boldsymbol{v}_2 + \cdots + t_n\boldsymbol{v}_n$ with $t_1, \ldots, t_n \in \mathbb{R}$.
Set $a_i = \lfloor t_i \rceil$ for $i = 1, 2, \ldots, n$.
Return the vector $\boldsymbol{v} = a_1\boldsymbol{v}_1 + a_2\boldsymbol{v}_2 + \cdots + a_n\boldsymbol{v}_n$.

In general, if the vectors in the basis are reasonably orthogonal to one another, then the algorithm solves some version of apprCVP, *but if the basis vectors are highly nonorthogonal, then the vector returned by the algorithm is generally far from the lattice vector that is closest to $\boldsymbol{w}$.*

Example 7.35. Let $L \subset \mathbb{R}^2$ be the lattice given by the basis

$$\boldsymbol{v}_1 = (137, 312) \quad \text{and} \quad \boldsymbol{v}_2 = (215, -187).$$

We are going to use Babai's algorithm (Theorem 7.34) to find a vector in L that is close to the vector

$$\boldsymbol{w} = (53172, 81743).$$

The first step is to express $\boldsymbol{w}$ as a linear combination of $\boldsymbol{v}_1$ and $\boldsymbol{v}_2$ using real coordinates. We do this using linear algebra. Thus we need to find $t_1, t_2 \in \mathbb{R}$ such that

$$\boldsymbol{w} = t_1\boldsymbol{v}_1 + t_2\boldsymbol{v}_2.$$

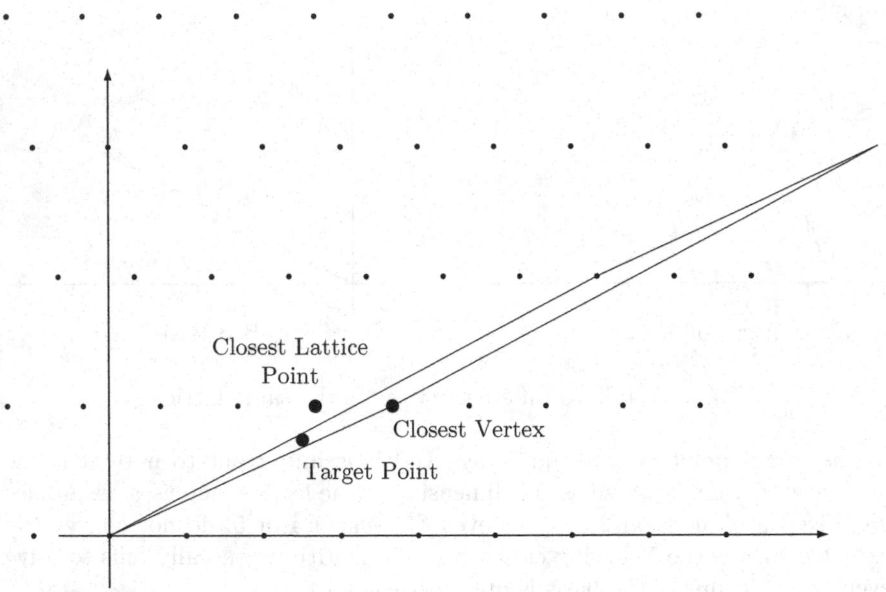

Figure 7.5: Babai's algorithm works poorly if the basis is "bad"

This gives the two linear equations

$$53172 = 137t_1 + 215t_2 \quad \text{and} \quad 81743 = 312t_1 - 187t_2, \tag{7.24}$$

or, for those who prefer matrix notation,

$$(53172, 81743) = (t_1, t_2)\begin{pmatrix} 137 & 312 \\ 215 & -187 \end{pmatrix}. \tag{7.25}$$

It is easy to solve for (t_1, t_2), either by solving the system (7.24) or by inverting the matrix in (7.25). We find that $t_1 \approx 296.85$ and $t_2 \approx 58.15$. Babai's algorithm tells us to round t_1 and t_2 to the nearest integer and then compute

$$\boldsymbol{v} = \lfloor t_1 \rceil \boldsymbol{v}_1 + \lfloor t_2 \rceil \boldsymbol{v}_2 = 297(137, 312) + 58(215, -187) = (53159, 81818).$$

Then $\boldsymbol{v}$ is in L and $\boldsymbol{v}$ should be close to $\boldsymbol{w}$. We find that

$$\|\boldsymbol{v}-\boldsymbol{w}\| \approx 76.12$$

is indeed quite small. This is to be expected, since the vectors in the given basis are fairly orthogonal to one another, as is seen by the fact that the Hadamard ratio

$$\mathcal{H}(\boldsymbol{v}_1,\boldsymbol{v}_2) = \left(\frac{\det(L)}{\|\boldsymbol{v}_1\|\|\boldsymbol{v}_2\|}\right)^{1/2} \approx \left(\frac{92699}{(340.75)(284.95)}\right)^{1/2} \approx 0.977$$

is reasonably close to 1.

We now try to solve the same closest vector problem in the same lattice, but using the new basis

$$\boldsymbol{v}_1' = (1975, 438) = 5\boldsymbol{v}_1 + 6\boldsymbol{v}_2 \quad \text{and} \quad \boldsymbol{v}_2' = (7548, 1627) = 19\boldsymbol{v}_1 + 23\boldsymbol{v}_2.$$

The system of linear equations

$$(53172, 81743) = (t_1, t_2)\begin{pmatrix} 1975 & 438 \\ 7548 & 1627 \end{pmatrix} \tag{7.26}$$

has the solution $(t_1, t_2) \approx (5722.66, -1490.34)$, so we set

$$\boldsymbol{v}' = 5723\boldsymbol{v}_1' - 1490\boldsymbol{v}_2' = (56405, 82444).$$

Then $\boldsymbol{v}' \in L$, but $\boldsymbol{v}'$ is not particularly close to $\boldsymbol{w}$, since

$$\|\boldsymbol{v}' - \boldsymbol{w}\| \approx 3308.12.$$

The nonorthogonality of the basis $\{\boldsymbol{v}_1', \boldsymbol{v}_2'\}$ is shown by the smallness of the Hadamard ratio

$$\mathcal{H}(\boldsymbol{v}_1',\boldsymbol{v}_2') = \left(\frac{\det(L)}{\|\boldsymbol{v}_1\|\|\boldsymbol{v}_2\|}\right)^{1/2} \approx \left(\frac{92699}{(2022.99)(7721.36)}\right)^{1/2} \approx 0.077.$$

7.7 Cryptosystems Based on Hard Lattice Problems

During the mid-1990s, several cryptosystems were introduced whose underlying hard problem was SVP and/or CVP in a lattice L of large dimension n. The most important of these, in alphabetical order, were the Ajtai–Dwork cryptosystem [4], the GGH cryptosystem of Goldreich, Goldwasser, and Halevi [49], and the NTRU cryptosystem proposed by Hoffstein, Pipher, and Silverman [54].

The motivation for the introduction of these cryptosystems was twofold. First, it is certainly of interest to have cryptosystems based on a variety of hard mathematical problems, since then a breakthrough in solving one mathematical problem does not compromise the security of all systems. Second, lattice-based cryptosystems are frequently much faster than factorization or discrete logarithm-based systems such as Elgamal, RSA, and ECC. Roughly speaking, in order to achieve k bits of security, encryption and decryption for Elgamal, RSA, and ECC require $\mathcal{O}(k^3)$ operations, while encryption and decryption for lattice-based systems require only $\mathcal{O}(k^2)$ operations.[6] Further, the simple linear algebra operations used by lattice-based systems are very easy to implement in hardware and software. However, it must be noted that the security analysis of lattice-based cryptosystems is not nearly as well understood as it is for factorization and discrete logarithm-based systems. So although lattice-based systems are the subject of much current research, their real-world implementations are few in comparison with older systems.

The Ajtai–Dwork system is particularly interesting because Ajtai and Dwork showed that their system is provably secure unless a worst-case lattice problem can be solved in polynomial time. Offsetting this important theoretical result is the practical limitation that the key size turns out to be $\mathcal{O}(n^4)$, which leads to enormous keys. Nguyen and Stern [95] subsequently showed that any practical and efficient implementation of the Ajtai–Dwork system is insecure.

The basic GGH cryptosystem, which we explain in more detail in Sect. 7.8, is a straightforward application of the ideas that we have already discussed. Alice's private key is a good basis $\mathcal{B}_{\text{good}}$ for a lattice L and her public key is a bad basis $\mathcal{B}_{\text{bad}}$ for L. Bob's message is a binary vector $\boldsymbol{m}$, which he uses to form a linear combination $\sum m_i \boldsymbol{v}_i^{\text{bad}}$ of the vectors in $\mathcal{B}_{\text{bad}}$. He then perturbs the sum by adding a small random vector $\boldsymbol{r}$. The resulting vector $\boldsymbol{w}$ differs from a lattice vector $\boldsymbol{v}$ by the vector $\boldsymbol{r}$. Since Alice knows a good basis for L, she can use Babai's algorithm to find $\boldsymbol{v}$, and then she expresses $\boldsymbol{v}$ in terms of the bad basis to recover $\boldsymbol{m}$. Eve, on the other hand, knows only the bad basis $\mathcal{B}_{\text{bad}}$, so she is unable to solve CVP in L.

A public key in the GGH cryptosystem is a bad basis for the lattice L, so it consists of n^2 (large) numbers. In the original proposal, the key size was $\mathcal{O}(n^3 \log n)$, but using an idea of Micciancio [85], it is possible to reduce the key size to $\mathcal{O}(n^2 \log n)$ bits.

Goldreich, Goldwasser and Halevi conjectured that for $n > 300$, the CVP underlying GGH would be intractable. However, the effectiveness of LLL-type lattice reduction algorithms on lattices of high dimension had not, at that time, been closely studied. Nguyen [92] showed that a transformation of the original GGH encryption scheme reduced the problem to an easier CVP. This enabled him to solve the proposed GGH challenge problems in dimensions up to 350. For $n > 400$, the public key is approximately 128 kB.

The NTRU public key cryptosystem [54], whose original public presentation took place at the Crypto '96 rump session, is most naturally described in terms of quotients of polynomial rings. However, the hard problem underlying

[6]There are various tricks that one can use to reduce these estimates. For example, using a small encryption exponent reduces RSA encryption to $\mathcal{O}(k^2)$ operations, while using product-form polynomials reduces NTRU encryption to $\mathcal{O}(k \log k)$ operations.

NTRU is easily transformed into an SVP (for key recovery) or a CVP (for plaintext recovery) in a special class of lattices. The NTRU lattices, which are described in Sect. 7.11, are lattices of even dimension $n = 2N$ consisting of all vectors $(\boldsymbol{x}, \boldsymbol{y}) \in \mathbb{Z}^{2N}$ satisfying

$$\boldsymbol{y} \equiv \boldsymbol{x}H \pmod{q}$$

for some fixed positive integer q that is a public parameter. (In practice, $q = \mathcal{O}(n)$.) The matrix H, which is the public key, is an N-by-N circulant matrix. This means that each successive row of H is a rotation of the previous row, so in order to describe H, it suffices to specify its first row. Thus the public key has size $\mathcal{O}(n \log n)$, which is significantly smaller than GGH.

The NTRU private key is a single short vector $(\boldsymbol{f}, \boldsymbol{g}) \in L$. The set consisting of the short vector $(\boldsymbol{f}, \boldsymbol{g})$, together with its partial rotations, gives $N = \frac{1}{2}\dim(L)$ independent short vectors in L. This allows the owner of $(\boldsymbol{f}, \boldsymbol{g})$ to solve certain instances of CVP in L and thereby recover the encrypted plaintext. (For details, see Sect. 7.11 and Exercise 7.36.) Thus the security of the plaintext relies on the difficulty of solving CVP in the NTRU lattice. Further, the vector $(\boldsymbol{f}, \boldsymbol{g})$ and its rotations are almost certainly the shortest nonzero vectors in L, so NTRU is also vulnerable to a solution of SVP.

7.8 The GGH Public Key Cryptosystem

Alice begins by choosing a set of linearly independent vectors

$$\boldsymbol{v}_1, \boldsymbol{v}_2, \ldots, \boldsymbol{v}_n \in \mathbb{Z}^n$$

that are reasonably orthogonal to one another. One way to do this is to fix a parameter d and choose the coordinates of $\boldsymbol{v}_1, \ldots, \boldsymbol{v}_n$ randomly between $-d$ and d. Alice can check that her choice of vectors is good by computing the Hadamard ratio (Remark 7.27) of her basis and verifying that it is not too small. The vectors $\boldsymbol{v}_1, \ldots, \boldsymbol{v}_n$ are Alice's private key. For convenience, we let V be the n-by-n matrix whose rows are the vectors $\boldsymbol{v}_1, \ldots, \boldsymbol{v}_n$, and we let L be the lattice generated by these vectors.

Alice next chooses an n-by-n matrix U with integer coefficients and $\det(U) = \pm 1$. One way to create U is as a product of a large number of randomly chosen elementary matrices. She then computes

$$W = UV.$$

The row vectors $\boldsymbol{w}_1, \ldots, \boldsymbol{w}_n$ of W are a new basis for L. They are Alice's public key.

When Bob wants to send a message to Alice, he selects a small vector $\boldsymbol{m}$ with integer coordinates as his plaintext, e.g., $\boldsymbol{m}$ might be a binary vector. Bob also chooses a small random perturbation vector $\boldsymbol{r}$ that acts as a random element. For example, he might choose the coordinates of $\boldsymbol{r}$ randomly between $-\delta$ and δ, where δ is a fixed public parameter. He then computes the vector

$$\boldsymbol{e} = \boldsymbol{m}W + \boldsymbol{r} = \sum_{i=1}^{n} m_i \boldsymbol{w}_i + \boldsymbol{r},$$

which is his ciphertext. Notice that $\boldsymbol{e}$ is not a lattice point, but it is close to the lattice point $\boldsymbol{m}W$, since $\boldsymbol{r}$ is small.

Alice	Bob
Key creation	
Choose a good basis $\boldsymbol{v}_1, \ldots, \boldsymbol{v}_n$. Choose an integer matrix U satisfying $\det(U) = \pm 1$. Compute a bad basis $\boldsymbol{w}_1, \ldots, \boldsymbol{w}_n$ as the rows of $W = UV$. Publish the public key $\boldsymbol{w}_1, \ldots, \boldsymbol{w}_n$.	
Encryption	
	Choose small plaintext vector $\boldsymbol{m}$. Choose random small vector $\boldsymbol{r}$. Use Alice's public key to compute $\boldsymbol{e} = x_1\boldsymbol{w}_1 + \cdots + x_n\boldsymbol{w}_n + \boldsymbol{r}$. Send the ciphertext $\boldsymbol{e}$ to Alice.
Decryption	
Use Babai's algorithm to compute the vector $\boldsymbol{v} \in L$ closest to $\boldsymbol{e}$. Compute $\boldsymbol{v}W^{-1}$ to recover $\boldsymbol{m}$.	

Table 7.3: The GGH cryptosystem

Decryption is straightforward. Alice uses Babai's algorithm, as described in Theorem 7.34, with the good basis $\boldsymbol{v}_1, \ldots, \boldsymbol{v}_n$ to find a vector in L that is close to $\boldsymbol{e}$. Since she is using a good basis and $\boldsymbol{r}$ is small, the lattice vector that she finds is $\boldsymbol{m}W$. She then multiplies by W^{-1} to recover $\boldsymbol{m}$. The GGH cryptosystem is summarized in Table 7.3.

Example 7.36. We illustrate the GGH cryptosystem with a 3-dimensional example. For Alice's private good basis we take

$$\boldsymbol{v}_1 = (-97, 19, 19), \quad \boldsymbol{v}_2 = (-36, 30, 86), \quad \boldsymbol{v}_3 = (-184, -64, 78).$$

The lattice L spanned by $\boldsymbol{v}_1$, $\boldsymbol{v}_2$, and $\boldsymbol{v}_3$ has determinant $\det(L) = 859516$, and the Hadamard ratio of the basis is

$$\mathcal{H}(\boldsymbol{v}_1, \boldsymbol{v}_2, \boldsymbol{v}_3) = \bigl(\det(L)/\|\boldsymbol{v}_1\|\,\|\boldsymbol{v}_2\|\,\|\boldsymbol{v}_3\|\bigr)^{1/3} \approx 0.74620.$$

Alice multiplies her private basis by the matrix

$$U = \begin{pmatrix} 4327 & -15447 & 23454 \\ 3297 & -11770 & 17871 \\ 5464 & -19506 & 29617 \end{pmatrix},$$

which has determinant $\det(U) = -1$, to create her public basis

$$\begin{aligned} \boldsymbol{w}_1 &= (-4179163, -1882253, 583183), \\ \boldsymbol{w}_2 &= (-3184353, -1434201, 444361), \\ \boldsymbol{w}_3 &= (-5277320, -2376852, 736426). \end{aligned}$$

The Hadamard ratio of the public basis is very small,

$$\mathcal{H}(\boldsymbol{v}_1, \boldsymbol{v}_2, \boldsymbol{v}_3) = \big(\det(L)/\|\boldsymbol{w}_1\|\,\|\boldsymbol{w}_2\|\,\|\boldsymbol{w}_3\|\big)^{1/3} \approx 0.0000208.$$

Bob decides to send Alice the plaintext $\boldsymbol{m} = (86, -35, -32)$ using the random element $\boldsymbol{r} = (-4, -3, 2)$. The corresponding ciphertext is

$$\begin{aligned} \boldsymbol{e} &= (86, -35, -32) \begin{pmatrix} -4179163 & -1882253 & 583183 \\ -3184353 & -1434201 & 444361 \\ -5277320 & -2376852 & 736426 \end{pmatrix} + (-4, -3, 2) \\ &= (-79081427, -35617462, 11035473). \end{aligned}$$

Alice uses Babai's algorithm to decrypt. She first writes $\boldsymbol{e}$ as a linear combination of her private basis with real coefficients,

$$\boldsymbol{e} \approx 81878.97\boldsymbol{v}_1 - 292300.00\boldsymbol{v}_2 + 443815.04\boldsymbol{v}_3.$$

She rounds the coefficients to the nearest integer and computes a lattice vector

$$\boldsymbol{v} = 81879\boldsymbol{v}_1 - 292300\boldsymbol{v}_2 + 443815\boldsymbol{v}_3 = (-79081423, -35617459, 11035471)$$

that is close to $\boldsymbol{e}$. She then recovers $\boldsymbol{m}$ by expressing $\boldsymbol{v}$ as a linear combination of the public basis and reading off the coefficients,

$$\boldsymbol{v} = 86\boldsymbol{w}_1 - 35\boldsymbol{w}_2 - 32\boldsymbol{w}_3.$$

Now suppose that Eve tries to decrypt Bob's message, but she knows only the public basis $\boldsymbol{w}_1, \boldsymbol{w}_2, \boldsymbol{w}_3$. If she applies Babai's algorithm using the public basis, she finds that

$$\boldsymbol{e} \approx 75.76\boldsymbol{w}_1 - 34.52\boldsymbol{w}_2 - 24.18\boldsymbol{w}_3.$$

Rounding, she obtains a lattice vector

$$\boldsymbol{v}' = 76\boldsymbol{w}_1 - 35\boldsymbol{w}_2 - 24\boldsymbol{w}_3 = (-79508353, -35809745, 11095049)$$

that is somewhat close to $\boldsymbol{e}$. However, this lattice vector gives the incorrect plaintext $(76, -35, -24)$, not the correct plaintext $\boldsymbol{m} = (86, -35, -32)$. It is instructive to compare how well Babai's algorithm did for the different bases. We find that

$$\|\boldsymbol{e} - \boldsymbol{v}\| \approx 5.39 \qquad \text{and} \qquad \|\boldsymbol{e} - \boldsymbol{v}'\| \approx 472004.09$$

Of course, the GGH cryptosystem is not secure in dimension 3, since even if we use numbers that are large enough to make an exhaustive search impractical, there are efficient algorithms to find good bases in low dimension. In dimension 2, an algorithm for finding a good basis dates back to Gauss. A powerful generalization to arbitrary dimension, known as the LLL algorithm, is covered in Sect. 7.13.

Remark 7.37. We observe that GGH is an example of a probabilistic cryptosystem (see Sect. 3.10), since a single plaintext leads to many different ciphertexts due to the choice of the random perturbation $\boldsymbol{r}$. This leads to a potential danger if Bob sends the same message twice using different random perturbations, or sends different messages using the same random perturbation. One possible solution is to choose the random perturbation $\boldsymbol{r}$ deterministically by applying a hash function (Sect. 8.1) to the plaintext $\boldsymbol{m}$, but this causes other security issues. See Exercises 7.20 and 7.21 for a further discussion.

Remark 7.38. An alternative version of GGH reverses the roles of $\boldsymbol{m}$ and $\boldsymbol{r}$, so the ciphertext has the form $\boldsymbol{e} = \boldsymbol{r}W + \boldsymbol{m}$. Alice finds $\boldsymbol{r}W$ by computing the lattice vector closest to $\boldsymbol{e}$, and then she recovers the plaintext as $\boldsymbol{m} = \boldsymbol{e} - \boldsymbol{r}W$.

7.9 Convolution Polynomial Rings

In this section we describe the special sort of polynomial quotient rings that are used by the NTRU public key cryptosystem, which is the topic of Sects. 7.10 and 7.11. The reader who is unfamiliar with basic ring theory should read Sect. 2.10 before continuing.

Definition. Fix a positive integer N. The *ring of convolution polynomials* (*of rank* N) is the quotient ring

$$R = \frac{\mathbb{Z}[x]}{(x^N - 1)}.$$

Similarly, the *ring of convolution polynomials* (*modulo* q) is the quotient ring

$$R_q = \frac{(\mathbb{Z}/q\mathbb{Z})[x]}{(x^N - 1)}.$$

Proposition 2.50 tells us that every element of R or R_q has a unique representative of the form

$$a_0 + a_1 x + a_2 x^2 + \cdots + a_{N-1} x^{N-1}$$

with the coefficients in $\mathbb{Z}$ or $\mathbb{Z}/q\mathbb{Z}$, respectively. We observe that it is easier to do computations in the rings R and R_q than it is in more general polynomial quotient rings, because the polynomial $x^N - 1$ has such a simple form. The point is that when we mod out by $x^N - 1$, we are simply requiring x^N to equal 1. So any time x^N appears, we replace it by 1. For example, if we have a term x^k, then we write $k = iN + j$ with $0 \le j < N$ and set

$$x^k = x^{iN+j} = (x^N)^i \cdot x^j = 1^i \cdot x^j = x^j.$$

In brief, the exponents on the powers of x may be reduced modulo N.

It is often convenient to identify a polynomial

$$\boldsymbol{a}(x) = a_0 + a_1 x + a_2 x^2 + \cdots + a_{N-1} x^{N-1} \in R$$

with its vector of coefficients

$$(a_0, a_1, a_2, \ldots, a_{N-1}) \in \mathbb{Z}^N,$$

and similarly with polynomials in R_q. Addition of polynomials corresponds to the usual addition of vectors,

$$\boldsymbol{a}(x) + \boldsymbol{b}(x) \longleftrightarrow (a_0 + b_0, a_1 + b_1, a_2 + b_2, \ldots, a_{N-1} + b_{N-1}).$$

The rule for multiplication in R is a bit more complicated. We write $\star$ for multiplication in R and R_q, to distinguish it from standard multiplication of polynomials.

Proposition 7.39. *The product of two polynomials $\boldsymbol{a}(x), \boldsymbol{b}(x) \in R$ is given by the formula*

$$\boldsymbol{a}(x) \star \boldsymbol{b}(x) = \boldsymbol{c}(x) \quad \textit{with} \quad c_k = \sum_{i+j \equiv k \pmod{N}} a_i b_{k-i}, \tag{7.27}$$

where the sum defining c_k is over all i and j between 0 and $N-1$ satisfying the condition $i + j \equiv k \pmod{N}$. The product of two polynomials $\boldsymbol{a}(x), \boldsymbol{b}(x) \in R_q$ is given by the same formula, except that the value of c_k is reduced modulo q.

Proof. We first compute the usual polynomial product of $\boldsymbol{a}(x)$ and $\boldsymbol{b}(x)$, after which we use the relation $x^N = 1$ to combine the terms. Thus

$$\begin{aligned}
\boldsymbol{a}(x) \star \boldsymbol{b}(x) &= \left(\sum_{i=0}^{N-1} a_i x^i \right) \star \left(\sum_{j=0}^{N-1} b_j x^j \right) \\
&= \sum_{k=0}^{2N-2} \left(\sum_{i+j=k} a_i b_j \right) x^k \\
&= \sum_{k=0}^{N-1} \left(\sum_{i+j=k} a_i b_j \right) x^k + \sum_{k=N}^{2N-2} \left(\sum_{i+j=k} a_i b_j \right) x^{k-N}
\end{aligned}$$

$$= \sum_{k=0}^{N-1}\left(\sum_{i+j=k} a_i b_j\right) x^k + \sum_{k=0}^{N-2}\left(\sum_{i+j=k+N} a_i b_j\right) x^k$$
$$= \sum_{k=0}^{N-1}\left(\sum_{i+j\equiv k \pmod{N}} a_i b_j\right) x^k. \qquad \square$$

Example 7.40. We illustrate multiplication in the convolution rings R and R_q with an example. We take $N = 5$ and let $\boldsymbol{a}(x), \boldsymbol{b}(x) \in R$ be the polynomials

$$\boldsymbol{a}(x) = 1 - 2x + 4x^3 - x^4 \quad \text{and} \quad \boldsymbol{b}(x) = 3 + 4x - 2x^2 + 5x^3 + 2x^4.$$

Then

$$\begin{aligned} \boldsymbol{a}(x) \star \boldsymbol{b}(x) &= 3 - 2x - 10x^2 + 21x^3 + 5x^4 - 16x^5 + 22x^6 + 3x^7 - 2x^8 \\ &= 3 - 2x - 10x^2 + 21x^3 + 5x^4 - 16 + 22x + 3x^2 - 2x^3 \\ &= -13 + 20x - 7x^2 + 19x^3 + 5x^4 \quad \text{in } R = \mathbb{Z}[x]/(x^5 - 1). \end{aligned}$$

If we work instead in the ring R_{11}, then we reduce the coefficients modulo 11 to obtain

$$\boldsymbol{a}(x) \star \boldsymbol{b}(x) = 9 + 9x + 4x^2 + 8x^3 + 5x^4 \quad \text{in } R_{11} = (\mathbb{Z}/11\mathbb{Z})[x]/(x^5 - 1).$$

Remark 7.41. The *convolution product* of two vectors is given by

$$(a_0, a_1, a_2, \ldots, a_{N-1}) \star (b_0, b_1, b_2, \ldots, b_{N-1}) = (c_0, c_1, c_2, \ldots, c_{N-1}),$$

where the c_k are defined by (7.27). We use $\star$ interchangeably to denote convolution multiplication in the rings R and R_q and the convolution product of vectors.

There is a natural map from R to R_q in which we simply reduce the coefficients of a polynomial modulo q. This reduction modulo q map satisfies

$$\big(\boldsymbol{a}(x) + \boldsymbol{b}(x)\big) \bmod q = \big(\boldsymbol{a}(x) \bmod q\big) + \big(\boldsymbol{b}(x) \bmod q\big), \tag{7.28}$$
$$\big(\boldsymbol{a}(x) \star \boldsymbol{b}(x)\big) \bmod q = \big(\boldsymbol{a}(x) \bmod q\big) \star \big(\boldsymbol{b}(x) \bmod q\big). \tag{7.29}$$

(In mathematical terminology, the map $R \to R_q$ is a ring homomorphism.)

It is often convenient to have a consistent way of going in the other direction. Among the many ways of lifting, we choose the following.

Definition. Let $\boldsymbol{a}(x) \in R_q$. The *center-lift of* $\boldsymbol{a}(x)$ *to* R is the unique polynomial $\boldsymbol{a}'(x) \in R$ satisfying

$$\boldsymbol{a}'(x) \bmod q = \boldsymbol{a}(x)$$

whose coefficients are chosen in the interval

$$-\frac{q}{2} < a_i' \le \frac{q}{2}.$$

For example, if $q = 2$, then the center-lift of $\boldsymbol{a}(x)$ is a binary polynomial.

Remark 7.42. It is important to observe that the lifting map does not satisfy the analogs of (7.28) and (7.29). In other words, the sum or product of the lifts need *not* be equal to the lift of the sum or product.

Example 7.43. Let $N = 5$ and $q = 7$, and consider the polynomial

$$\boldsymbol{a}(x) = 5 + 3x - 6x^2 + 2x^3 + 4x^4 \in R_7.$$

The coefficients of the center-lift of $\boldsymbol{a}(x)$ are chosen from $\{-3, -2, \ldots, 2, 3\}$, so

$$\text{Center-lift of } \boldsymbol{a}(x) = -2 + 3x + x^2 + 2x^3 - 3x^4 \in R.$$

Similarly, the lift of $\boldsymbol{b}(x) = 3 + 5x^2 - 6x^3 + 3x^4$ is $3 - 2x^2 + x^3 + 3x^4$. Notice that

$$(\text{Lift of } \boldsymbol{a}) \star (\text{Lift of } \boldsymbol{b}) = 20x + 10x^2 - 11x^3 - 14x^4$$

and

$$(\text{Lift of } \boldsymbol{a} \star \boldsymbol{b}) = -x + 3x^2 + 3x^3$$

are not equal to one another, although they are congruent modulo 7.

Example 7.44. Very few polynomials in R have multiplicative inverses, but the situation is quite different in R_q. For example, let $N = 5$ and $q = 2$. Then the polynomial $1 + x + x^4$ has an inverse in R_2, since in R_2 we have

$$(1 + x + x^4) \star (1 + x^2 + x^3) = 1 + x + x^2 + 2x^3 + 2x^4 + x^6 + x^7 = 1.$$

(Since $N = 5$, we have $x^6 = x$ and $x^7 = x^2$.) When q is a prime, the extended Euclidean algorithm for polynomials (Proposition 2.46) tells us which polynomials are units and how to compute their inverses in R_q.

Proposition 7.45. *Let q be prime. Then $\boldsymbol{a}(x) \in R_q$ has a multiplicative inverse if and only if*

$$\gcd(\boldsymbol{a}(x), x^N - 1) = 1 \quad \text{in } (\mathbb{Z}/q\mathbb{Z})[x]. \tag{7.30}$$

If (7.30) is true, then the inverse $\boldsymbol{a}(x)^{-1} \in R_q$ can be computed using the extended Euclidean algorithm (Proposition 2.46) to find polynomials $\boldsymbol{u}(x), \boldsymbol{v}(x) \in (\mathbb{Z}/q\mathbb{Z})[x]$ satisfying

$$\boldsymbol{a}(x)\boldsymbol{u}(x) + (x^N - 1)\boldsymbol{v}(x) = 1.$$

Then $\boldsymbol{a}(x)^{-1} = \boldsymbol{u}(x)$ in R_q.

Proof. Proposition 2.46 says that we can find polynomials $\boldsymbol{u}(x)$ and $\boldsymbol{v}(x)$ in the polynomial ring $(\mathbb{Z}/q\mathbb{Z})[x]$ satisfying

$$\boldsymbol{a}(x)\boldsymbol{u}(x) + (x^N - 1)\boldsymbol{v}(x) = \gcd(\boldsymbol{a}(x), x^N - 1).$$

If the gcd is equal to 1, then reducing modulo $x^N - 1$ yields $\boldsymbol{a}(x) \star \boldsymbol{u}(x) = 1$ in R_q. Conversely, if $\boldsymbol{a}(x)$ is a unit in R_q, then we can find a polynomial $\boldsymbol{u}(x)$ such that $\boldsymbol{a}(x) \star \boldsymbol{u}(x) = 1$ in R_q. By definition of R_q, this means that

$$\boldsymbol{a}(x)\boldsymbol{u}(x) \equiv 1 \pmod{(x^N - 1)},$$

so by definition of congruences, there is a polynomial $\boldsymbol{v}(x)$ satisfying

$$\boldsymbol{a}(x)\boldsymbol{u}(x) - 1 = (x^N - 1)\boldsymbol{v}(x) \quad \text{in } (\mathbb{Z}/q\mathbb{Z})[x]. \qquad \square$$

Example 7.46. We let $N = 5$ and $q = 2$ and give the full details for computing $(1 + x + x^4)^{-1}$ in R_2. First we use the Euclidean algorithm to compute the greatest common divisor of $1 + x + x^4$ and $1 - x^5$ in $(\mathbb{Z}/2\mathbb{Z})[x]$. (Note that since we are working modulo 2, we have $1 - x^5 = 1 + x^5$.) Thus

$$\begin{aligned} x^5 + 1 &= x \cdot (x^4 + x + 1) + (x^2 + x + 1), \\ x^4 + x + 1 &= (x^2 + x)(x^2 + x + 1) + 1. \end{aligned}$$

So the gcd is equal to 1, and using the usual substitution method yields

$$\begin{aligned} 1 &= (x^4 + x + 1) + (x^2 + x)(x^2 + x + 1) \\ &= (x^4 + x + 1) + (x^5 + 1 + x(x^4 + x + 1)) \\ &= (x^4 + x + 1)(x^3 + x^2 + 1) + (x^5 + 1)(x^2 + x). \end{aligned}$$

Hence

$$(1 + x + x^4)^{-1} = 1 + x^2 + x^3 \quad \text{in } R_2.$$

(See Exercise 1.12 for an efficient computer algorithm and Fig. 1.3 for the "magic box method" to compute $\boldsymbol{a}(x)^{-1}$ in R_q.)

Remark 7.47. The ring R_q makes perfect sense regardless of whether q is prime, and indeed there are situations in which it can be advantageous to take q composite, for example $q = 2^k$. In general, if q is a power of a prime p, then in order to compute the inverse of $\boldsymbol{a}(x)$ in R_q, one first computes the inverse in R_p, then "lifts" this value to an inverse in R_{p^2}, and then lifts to an inverse in R_{p^4}, and so on. (See Exercise 7.27.) Similarly, if $q = q_1 q_2 \cdots q_r$, where each $q_i = p_i^{k_i}$ is a prime power, one first computes inverses in R_{q_i} and then combines the inverses using the Chinese remainder theorem.

7.10 The NTRU Public Key Cryptosystem

Cryptosystems based on the difficulty of integer factorization or the discrete logarithm problem are group-based cryptosystems, because the underlying hard problem involves only one operation. For RSA, Diffie–Hellman, and Elgamal, the group is the group of units modulo m for some modulus m that may be prime or composite, and the group operation is multiplication modulo m. For ECC, the group is the set of points on an elliptic curve modulo p and the group operation is elliptic curve addition.

Rings are algebraic objects that have two operations, addition and multiplication, which are connected via the distributive law. In this section we describe NTRUEncrypt, the NTRU public key cryptosystem. NTRUEncrypt is most naturally described using convolution polynomial rings, but the underlying hard mathematical problem can also be interpreted as SVP or CVP in a lattice. We discuss the connection with lattices in Sect. 7.11.

7.10.1 NTRUEncrypt

In this section we describe NTRUEncrypt, the NTRU (pronounced *en-trū*) public key cryptosystem. We begin by fixing an integer $N \geq 1$ and two moduli p and q, and we let R, R_p, and R_q be the convolution polynomial rings

$$R = \frac{\mathbb{Z}[x]}{(x^N - 1)}, \qquad R_p = \frac{(\mathbb{Z}/p\mathbb{Z})[x]}{(x^N - 1)}, \qquad R_q = \frac{(\mathbb{Z}/q\mathbb{Z})[x]}{(x^N - 1)},$$

described in Sect. 7.9. As usual, we may view a polynomial $\boldsymbol{a}(x) \in R$ as an element of R_p or R_q by reducing its coefficients modulo p or q. In the other direction, we use center-lifts to move elements from R_p or R_q to R. We make various assumptions on the parameters N, p and q, in particular we require that N be prime and that $\gcd(N, q) = \gcd(p, q) = 1$. (The reasons for these assumptions are explained in Exercises 7.32 and 7.37.)

We need one more piece of notation before describing NTRUEncrypt.

Definition. For any positive integers d_1 and d_2, we let

$$\mathcal{T}(d_1, d_2) = \left\{ \boldsymbol{a}(x) \in R : \begin{array}{l} \boldsymbol{a}(x) \text{ has } d_1 \text{ coefficients equal to } 1, \\ \boldsymbol{a}(x) \text{ has } d_2 \text{ coefficients equal to } -1, \\ \boldsymbol{a}(x) \text{ has all other coefficients equal to } 0 \end{array} \right\}.$$

Polynomials in $\mathcal{T}(d_1, d_2)$ are called *ternary* (or *trinary*) *polynomials.* They are analogous to *binary polynomials*, which have only 0's and 1's as coefficients.

We are now ready to describe NTRUEncrypt. Alice (or some trusted authority) chooses public parameters (N, p, q, d) satisfying the guidelines described earlier (or see Table 7.4). Alice's private key consists of two randomly chosen polynomials

$$\boldsymbol{f}(x) \in \mathcal{T}(d+1, d) \quad \text{and} \quad \boldsymbol{g}(x) \in \mathcal{T}(d, d). \tag{7.31}$$

Alice computes the inverses

$$\boldsymbol{F}_q(x) = \boldsymbol{f}(x)^{-1} \text{ in } R_q \quad \text{and} \quad \boldsymbol{F}_p(x) = \boldsymbol{f}(x)^{-1} \text{ in } R_p. \tag{7.32}$$

(If either inverse fails to exist, she discards this $\boldsymbol{f}(x)$ and chooses a new one. We mention that Alice chooses $\boldsymbol{f}(x)$ in $\mathcal{T}(d+1,d)$, rather than in $\mathcal{T}(d,d)$, because elements in $\mathcal{T}(d,d)$ never have inverses in R_q; see Exercise 7.24.)

Alice next computes

$$\boldsymbol{h}(x) = \boldsymbol{F}_q(x) \star \boldsymbol{g}(x) \quad \text{in } R_q. \tag{7.33}$$

The polynomial $\boldsymbol{h}(x)$ is Alice's public key. Her private key, which she'll need to decrypt messages, is the pair $\big(\boldsymbol{f}(x), \boldsymbol{F}_p(x)\big)$. Alternatively, Alice can just store $\boldsymbol{f}(x)$ and recompute $\boldsymbol{F}_p(x)$ when she needs it.

Bob's plaintext is a polynomial $\boldsymbol{m}(x) \in R$ whose coefficients satisfy $-\frac{1}{2}p < m_i \le \frac{1}{2}p$, i.e., the plaintext $\boldsymbol{m}$ is a polynomial in R that is the center-lift of a polynomial in R_p. Bob chooses a random polynomial (a random element) $\boldsymbol{r}(x) \in \mathcal{T}(d,d)$ and computes[7]

$$\boldsymbol{e}(x) \equiv p\boldsymbol{h}(x) \star \boldsymbol{r}(x) + \boldsymbol{m}(x) \pmod{q}. \tag{7.34}$$

Bob's ciphertext $\boldsymbol{e}(x)$ is in the ring R_q.

On receiving Bob's ciphertext, Alice starts the decryption process by computing

$$\boldsymbol{a}(x) \equiv \boldsymbol{f}(x) \star \boldsymbol{e}(x) \pmod{q}. \tag{7.35}$$

She then center lifts $\boldsymbol{a}(x)$ to an element of R and does a mod p computation,

$$\boldsymbol{b}(x) \equiv \boldsymbol{F}_p(x) \star \boldsymbol{a}(x) \pmod{p}. \tag{7.36}$$

Assuming that the parameters have been chosen properly, we now verify that the polynomial $\boldsymbol{b}(x)$ is equal to the plaintext $\boldsymbol{m}(x)$.

NTRUEncrypt, the NTRU public key cryptosystem, is summarized in Table 7.4.

Proposition 7.48. *If the NTRUEncrypt parameters (N,p,q,d) are chosen to satisfy*

$$q > (6d+1)\,p, \tag{7.37}$$

then the polynomial $\boldsymbol{b}(x)$ computed by Alice in (7.36) is equal to Bob's plaintext $\boldsymbol{m}(x)$.

Proof. We first determine more precisely the shape of Alice's preliminary calculation of $\boldsymbol{a}(x)$. Thus

$$\begin{aligned}
\boldsymbol{a}(x) &\equiv \boldsymbol{f}(x) \star \boldsymbol{e}(x) \pmod{q} && \text{from (7.35)},\\
&\equiv \boldsymbol{f}(x) \star \big(p\boldsymbol{h}(x) \star \boldsymbol{r}(x) + \boldsymbol{m}(x)\big) \pmod{q} && \text{from (7.34)},\\
&\equiv p\boldsymbol{f}(x) \star \boldsymbol{F}_q(x) \star \boldsymbol{g}(x) \star \boldsymbol{r}(x) + \boldsymbol{f}(x) \star \boldsymbol{m}(x) \pmod{q} && \text{from (7.33)}.\\
&\equiv p\boldsymbol{g}(x) \star \boldsymbol{r}(x) + \boldsymbol{f}(x) \star \boldsymbol{m}(x) \pmod{q} && \text{from (7.32)}.
\end{aligned}$$

[7]Note that when we write a congruence of polynomials modulo q, we really mean that the computation is being done in R_q.

Public parameter creation	
A trusted party chooses public parameters (N, p, q, d) with N and p prime, $\gcd(p, q) = \gcd(N, q) = 1$, and $q > (6d + 1)p$.	
Alice	**Bob**
Key creation	
Choose private $\boldsymbol{f} \in \mathcal{T}(d+1, d)$ that is invertible in R_q and R_p. Choose private $\boldsymbol{g} \in \mathcal{T}(d, d)$. Compute $\boldsymbol{F}_q$, the inverse of $\boldsymbol{f}$ in R_q. Compute $\boldsymbol{F}_p$, the inverse of $\boldsymbol{f}$ in R_p. Publish the public key $\boldsymbol{h} = \boldsymbol{F}_q \star \boldsymbol{g}$.	
Encryption	
	Choose plaintext $\boldsymbol{m} \in R_p$. Choose a random $\boldsymbol{r} \in \mathcal{T}(d, d)$. Use Alice's public key $\boldsymbol{h}$ to compute $\boldsymbol{e} \equiv p\boldsymbol{r} \star \boldsymbol{h} + \boldsymbol{m} \pmod{q}$. Send ciphertext $\boldsymbol{e}$ to Alice.
Decryption	
Compute $\boldsymbol{f} \star \boldsymbol{e} \equiv p\boldsymbol{g} \star \boldsymbol{r} + \boldsymbol{f} \star \boldsymbol{m} \pmod{q}$. Center-lift to $\boldsymbol{a} \in R$ and compute $\boldsymbol{m} \equiv \boldsymbol{F}_p \star \boldsymbol{a} \pmod{p}$.	

Table 7.4: NTRUEncryt: the NTRU public key cryptosystem

Consider the polynomial

$$p\boldsymbol{g}(x) \star \boldsymbol{r}(x) + \boldsymbol{f}(x) \star \boldsymbol{m}(x), \tag{7.38}$$

computed exactly in R, rather than modulo q. We need to bound its largest possible coefficient. The polynomials $\boldsymbol{g}(x)$ and $\boldsymbol{r}(x)$ are in $\mathcal{T}(d, d)$, so if, in the convolution product $\boldsymbol{g}(x) \star \boldsymbol{r}(x)$, all of their 1's match up and all of their -1's match up, the largest possible coefficient of $\boldsymbol{g}(x) \star \boldsymbol{r}(x)$ is $2d$. Similarly, $\boldsymbol{f}(x) \in \mathcal{T}(d+1, d)$ and the coefficients of $\boldsymbol{m}(x)$ are between $-\frac{1}{2}p$ and $\frac{1}{2}p$, so the largest possible coefficient of $\boldsymbol{f}(x) \star \boldsymbol{m}(x)$ is $(2d+1) \cdot \frac{1}{2}p$. So even if the largest coefficient of $\boldsymbol{g}(x) \star \boldsymbol{r}(x)$ happens to coincide with the largest coefficient of $\boldsymbol{r}(x) \star \boldsymbol{m}(x)$, the largest coefficient of (7.38) has magnitude at most

$$p \cdot 2d + (2d+1) \cdot \frac{1}{2}p = \left(3d + \frac{1}{2}\right)p.$$

Thus our assumption (7.37) ensures that every coefficient of (7.38) has magnitude strictly smaller than $\frac{1}{2}q$. Hence when Alice computes $\boldsymbol{a}(x)$ modulo q

(i.e., in R_q) and then lifts it to R, she recovers the exact value (7.38). In other words,

$$\boldsymbol{a}(x) = p\boldsymbol{g}(x) \star \boldsymbol{r}(x) + \boldsymbol{f}(x) \star \boldsymbol{m}(x) \tag{7.39}$$

exactly in R, and not merely modulo q.

The rest is easy. Alice multiplies $\boldsymbol{a}(x)$ by $\boldsymbol{F}_p(x)$, the inverse of $\boldsymbol{f}(x)$ modulo p, and reduces the result modulo p to obtain

$$\begin{aligned}
\boldsymbol{b}(x) &\equiv \boldsymbol{F}_p(x) \star \boldsymbol{a}(x) \pmod{p} && \text{from (7.36)},\\
&\equiv \boldsymbol{F}_p(x) \star \big(p\boldsymbol{g}(x) \star \boldsymbol{r}(x) + \boldsymbol{f}(x) \star \boldsymbol{m}(x)\big) \pmod{p} && \text{from (7.39)},\\
&\equiv \boldsymbol{F}_p(x) \star \boldsymbol{f}(x) \star \boldsymbol{m}(x) \pmod{p} && \text{reducing mod } p,\\
&\equiv \boldsymbol{m}(x) \pmod{p}. && \text{from (7.32)}.
\end{aligned}$$

Hence $\boldsymbol{b}(x)$ and $\boldsymbol{m}(x)$ are the same modulo p. □

Remark 7.49. The condition $q > (6d + 1)p$ in Proposition 7.48 ensures that decryption never fails. However, an examination of the proof shows that decryption is likely to succeed even for considerably smaller values of q, since it is highly unlikely that the positive and negative coefficients of $\boldsymbol{g}(x)$ and $\boldsymbol{r}(x)$ will exactly line up, and similarly for $\boldsymbol{f}(x)$ and $\boldsymbol{m}(x)$. So for additional efficiency and to reduce the size of the public key, it may be advantageous to choose a smaller value of q. It then becomes a delicate problem to estimate the probability of decryption failure. It is important that the probability of decryption failure be very small (e.g., smaller than 2^{-80}), since decryption failures have the potential to reveal private key information to an attacker.

Remark 7.50. Notice that NTRUEncrypt is an example of a probabilistic cryptosystem (Sect. 3.10), since a single plaintext $\boldsymbol{m}(x)$ has many different encryptions $p\boldsymbol{h}(x) \star \boldsymbol{r}(x) + \boldsymbol{m}(x)$ corresponding to different choices of the random element $\boldsymbol{r}(x)$. As is common for such systems, cf. Remark 7.37 for GGH, it is a bad idea for Bob to send the same message twice using different random elements, just as it is inadvisable for Bob to use the same random element to send two different plaintexts; see Exercise 7.34. Various ways of ameliorating this danger for GGH, which also apply *mutatis mutandis* to NTRUEncrypt, are described in Exercises 7.20 and 7.21.

Remark 7.51. The polynomial $\boldsymbol{f}(x) \in \mathcal{T}(d+1, d)$ has small coefficients, but the coefficients of its inverse $\boldsymbol{F}_q(x) \in R_q$ tend to be randomly and uniformly distributed modulo q. (This is not a theorem, but it is an experimentally observed fact.) For example, let $N = 11$ and $q = 73$ and take a random polynomial

$$\boldsymbol{f}(x) = x^{10} + x^8 - x^3 + x^2 - 1 \in \mathcal{T}(3, 2).$$

Then $\boldsymbol{f}(x)$ is invertible in R_q, and its inverse

$$\boldsymbol{F}_q(x) = 22x^{10} + 33x^9 + 15x^8 + 33x^7 - 10x^6 + 36x^5 - 33x^4 - 30x^3 + 12x^2 - 32x + 28$$

has random-looking coefficients. Similarly, in practice the coefficients of the public key and the ciphertext,

$$\boldsymbol{h}(x) \equiv \boldsymbol{F}_q(x) \star \boldsymbol{g}(x) \pmod{q} \quad \text{and} \quad \boldsymbol{e}(x) \equiv p\boldsymbol{r}(x) \star \boldsymbol{h}(x) + \boldsymbol{m}(x) \pmod{q},$$

also appear to be randomly distributed modulo q.

Remark 7.52. As noted in Sect. 7.7, a motivation for using lattice-based cryptosystems is their high speed compared to discrete logarithm and factorization-based cryptosystems. How fast is NTRUEncrypt? The most time consuming part of encryption and decryption is the convolution product. In general, a convolution product $\boldsymbol{a} \star \boldsymbol{b}$ requires N^2 multiplications, since each coefficient is essentially the dot product of two vectors. However, the convolution products required by NTRUEncrypt have the form $\boldsymbol{r} \star \boldsymbol{h}$, $\boldsymbol{f} \star \boldsymbol{e}$, and $\boldsymbol{F}_p \star \boldsymbol{a}$, where $\boldsymbol{r}$, $\boldsymbol{f}$, and $\boldsymbol{F}_p$ are ternary polynomials. Thus these convolution products can be computed without any multiplications; they each require approximately $\frac{2}{3}N^2$ additions and subtractions. (If d is smaller than $N/3$, the first two require only $\frac{2}{3}dN$ additions and subtractions.) Thus NTRUEncrypt encryption and decryption take $\mathcal{O}(N^2)$ steps, where each step is extremely fast.

Example 7.53. We present a small numerical example of NTRUEncrypt with public parameters

$$(N, p, q, d) = (7, 3, 41, 2).$$

We have

$$41 = q > (6d + 1)p = 39,$$

so Proposition 7.48 ensures that decryption will work. Alice chooses

$$\boldsymbol{f}(x) = x^6 - x^4 + x^3 + x^2 - 1 \in \mathcal{T}(3, 2) \quad \text{and} \quad \boldsymbol{g}(x) = x^6 + x^4 - x^2 - x \in \mathcal{T}(2, 2).$$

She computes the inverses

$$\boldsymbol{F}_q(x) = \boldsymbol{f}(x)^{-1} \bmod q = 8x^6 + 26x^5 + 31x^4 + 21x^3 + 40x^2 + 2x + 37 \in R_q,$$
$$\boldsymbol{F}_p(x) = \boldsymbol{f}(x)^{-1} \bmod p = x^6 + 2x^5 + x^3 + x^2 + x + 1 \in R_p.$$

She stores $(\boldsymbol{f}(x), \boldsymbol{F}_p(x))$ as her private key and computes and publishes her public key

$$\boldsymbol{h}(x) = \boldsymbol{F}_q(x) \star \boldsymbol{g}(x) = 20x^6 + 40x^5 + 2x^4 + 38x^3 + 8x^2 + 26x + 30 \in R_q.$$

Bob decides to send Alice the message

$$\boldsymbol{m}(x) = -x^5 + x^3 + x^2 - x + 1$$

using the random element

$$\boldsymbol{r}(x) = x^6 - x^5 + x - 1.$$

Bob computes and sends to Alice the ciphertext

$$\boldsymbol{e}(x) \equiv p\boldsymbol{r}(x)\star\boldsymbol{h}(x)+\boldsymbol{m}(x) \equiv 31x^6+19x^5+4x^4+2x^3+40x^2+3x+25 \pmod{q}.$$

Alice's decryption of Bob's message proceeds smoothly. First she computes

$$\boldsymbol{f}(x) \star \boldsymbol{e}(x) \equiv x^6 + 10x^5 + 33x^4 + 40x^3 + 40x^2 + x + 40 \pmod{q}. \quad (7.40)$$

She then center-lifts (7.40) modulo q to obtain

$$\boldsymbol{a}(x) = x^6 + 10x^5 - 8x^4 - x^3 - x^2 + x - 1 \in R.$$

Finally, she reduces $\boldsymbol{a}(x)$ modulo p and computes

$$\boldsymbol{F}_p(x) \star \boldsymbol{a}(x) \equiv 2x^5 + x^3 + x^2 + 2x + 1 \pmod{p}. \quad (7.41)$$

Center-lifting (7.41) modulo p retrieves Bob's plaintext $\boldsymbol{m}(x) = -x^5 + x^3 + x^2 - x + 1$.

7.10.2 Mathematical Problems for NTRUEncrypt

As noted in Remark 7.51, the coefficients of the public key $\boldsymbol{h}(x)$ appear to be random integers modulo q, but there is a hidden relationship

$$\boldsymbol{f}(x) \star \boldsymbol{h}(x) \equiv \boldsymbol{g}(x) \pmod{q}, \quad (7.42)$$

where $\boldsymbol{f}(x)$ and $\boldsymbol{g}(x)$ have very small coefficients. Thus breaking NTRUEncrypt by finding the private key comes down to solving the following problem:

The NTRU Key Recovery Problem
Given $\boldsymbol{h}(x)$, find ternary polynomials $\boldsymbol{f}(x)$ and $\boldsymbol{g}(x)$ satisfying $\boldsymbol{f}(x) \star \boldsymbol{h}(x) \equiv \boldsymbol{g}(x) \pmod{q}$.

Remark 7.54. The solution to the NTRU key recovery problem is not unique, because if $\big(\boldsymbol{f}(x), \boldsymbol{g}(x)\big)$ is one solution, then $\big(x^k \star \boldsymbol{f}(x), x^k \star \boldsymbol{g}(x)\big)$ is also a solution for every $0 \le k < N$. The polynomial $x^k \star \boldsymbol{f}(x)$ is called a *rotation of* $\boldsymbol{f}(x)$ because the coefficients have been cyclically rotated k positions. Rotations act as private decryption keys in the sense that decryption with $x^k \star \boldsymbol{f}(x)$ yields the rotated plaintext $x^k \star \boldsymbol{m}(x)$.

More generally, any pair of polynomials $\big(\boldsymbol{f}(x), \boldsymbol{g}(x)\big)$ with sufficiently small coefficients and satisfying (7.42) serves as an NTRU decryption key. For example, if $\boldsymbol{f}(x)$ is the original decryption key and if $\boldsymbol{\theta}(x)$ has tiny coefficients, then $\boldsymbol{\theta}(x) \star \boldsymbol{f}(x)$ may also work as a decryption key.

Remark 7.55. Why would one expect the NTRU key recovery problem to be a hard mathematical problem? A first necessary requirement is that the problem not be practically solvable by a brute-force or collision search. We discuss such searches later in this section. More importantly, in Sect. 7.11.2

we prove that solving the NTRU key recovery problem is (almost certainly) equivalent to solving SVP in a certain class of lattices. This relates the NTRU problem to a well-studied problem, albeit for a special collection of lattices. The use of lattice reduction is currently the best known method to recover an NTRU private key from the public key. Is lattice reduction the best possible method? Just as with integer factorization and the various discrete logarithm problems underlying other cryptosystems, no one knows for certain whether faster algorithms exist. So the only way to judge the difficulty of the NTRU key recovery problem is to note that it has been well studied by the mathematical and cryptographic community. Then a quantitative estimate of the difficulty of solving the problem is obtained by applying the fastest algorithm currently known.

How hard is Eve's task if she tries a brute-force search of all possible private keys? Note that Eve can determine whether she has found the private key $\boldsymbol{f}(x)$ by verifying that $\boldsymbol{f}(x) \star \boldsymbol{h}(x) \pmod q$ is a ternary polynomial. (In all likelihood, the only polynomials with this property are the rotations of $\boldsymbol{f}(x)$, but if Eve happens to find another ternary polynomial with this property, it will serve as a decryption key.)

So we need to compute the size of the set of ternary polynomials. In general, we can specify an element of $\mathcal{T}(d_1, d_2)$ by first choosing d_1 coefficients to be 1 and then choosing d_2 of the remaining $N - d$ coefficients to be -1. Hence

$$\#\mathcal{T}(d_1, d_2) = \binom{N}{d_1}\binom{N-d_1}{d_2} = \frac{N!}{d_1!\,d_2!\,(N-d_1-d_2)!}. \tag{7.43}$$

We remark that this number is maximized if d_1 and d_2 are both approximately $N/3$.

For a brute-force search, Eve must try each polynomial in $\mathcal{T}(d+1, d)$ until she finds a decryption key, but note that all of the rotations of $\boldsymbol{f}(x)$ are decryption keys, so there are N winning choices. Hence it will take Eve approximately $\#\mathcal{T}(d+1, d)/N$ tries to find some rotation of $\boldsymbol{f}(x)$.

Example 7.56. We consider the set of NTRUEncrypt parameters

$$(N, p, q, d) = (251, 3, 257, 83).$$

(This set does not satisfy the $q > (6d+1)p$ requirement, so there may be a rare decryption failure; see Remark 7.49.) Eve expects to check approximately

$$\frac{\mathcal{T}(84, 83)}{251} = \frac{1}{251}\binom{251}{84}\binom{167}{83} \approx 2^{381.6}$$

polynomials before finding a decryption key.

Remark 7.57. Not surprisingly, if Eve has a sufficient amount of storage, she can use a collision algorithm to search for the private key. (This was first observed by Andrew Odlyzko.) We describe the basic idea. Eve searches through pairs of ternary polynomials

$$\boldsymbol{f}_1(x) = \sum_{0\le i<N/2} a_i x^i \qquad \text{and} \qquad \boldsymbol{f}_2(x) = \sum_{N/2\le i<N} a_i x^i$$

having the property that $\boldsymbol{f}_1(x) + \boldsymbol{f}_2(x) \in \mathcal{T}(d+1,d)$. She computes

$$\boldsymbol{f}_1(x) \star \boldsymbol{h}(x) \pmod q \qquad \text{and} \qquad -\boldsymbol{f}_2(x) \star \boldsymbol{h}(x) \pmod q$$

and puts them into bins depending on their coefficients. The bins are set up so that when a polynomial from each list lands in the same bin, the quantity

$$\big(\boldsymbol{f}_1(x) + \boldsymbol{f}_2(x)\big) \star \boldsymbol{h}(x) \pmod q$$

has small coefficients, and hence $\boldsymbol{f}_1(x) + \boldsymbol{f}_2(x)$ is a decryption key. For further details, see [101].

The net effect of the collision algorithm is, as usual, to more or less take the square root of the number of steps required to find a key, so the collision-search security is approximately the square root of (7.43). Returning to Example 7.56, a collision search takes on the order of $\sqrt{2^{381.6}} \approx 2^{190.8}$ steps.

In general, if we maximize the size of $\mathcal{T}(d+1,d)$ by setting $d \approx N/3$, then we can use Stirling's formula (Proposition 7.29) to estimate

$$\#\mathcal{T}(d+1,d) \approx \frac{N!}{((N/3)!)^3} \approx \left(\frac{N}{e}\right)^N \cdot \left(\left(\frac{N}{3e}\right)^{N/3}\right)^{-3} \approx 3^N.$$

So a collision search in this case take $\mathcal{O}(3^{N/2}/\sqrt{N}\,)$ steps.

Remark 7.58. We claimed earlier that $\boldsymbol{f}(x)$ and its rotations are probably the only decryption keys in $\mathcal{T}(d+1,d)$. To see why this is true, we ask for the probability that some random $\boldsymbol{f}(x) \in \mathcal{T}(d+1,d)$ has the property that

$$\boldsymbol{f}(x) \star \boldsymbol{h}(x) \pmod q \quad \text{is a ternary polynomial.} \tag{7.44}$$

Treating the coefficients of (7.44) as independent[8] random variables that are uniformly distributed modulo q, the probability that any particular coefficient is ternary is $3/q$, and hence the probability that every coefficient is ternary is approximately $(3/q)^N$. Hence

$$\begin{pmatrix}\text{Expected number of decryp-}\\ \text{tion keys in } \mathcal{T}(d+1,d)\end{pmatrix} \approx \Pr\begin{pmatrix}\boldsymbol{f}(x) \in \mathcal{T}(d+1,d)\\ \text{is a decryption key}\end{pmatrix} \times \#\mathcal{T}(d+1,d)$$
$$= \left(\frac{3}{q}\right)^N \binom{N}{d+1}\binom{N-d-1}{d}.$$

[8]The coefficients of $\boldsymbol{f}(x) \star \boldsymbol{h}(x) \pmod q$ are not entirely independent, but they are sufficiently independent for this to be a good approximation.

Returning to Example 7.56, we see that the expected number of decryption keys in $\mathcal{T}(84,83)$ for $N = 251$ and $q = 257$ is

$$\left(\frac{3}{257}\right)^{251}\binom{251}{84}\binom{167}{83} \approx 2^{-1222.02}. \tag{7.45}$$

Of course, if $\boldsymbol{h}(x)$ is an NTRUEncrypt public key, then there do exist decryption keys, since we built the decryption key $\boldsymbol{f}(x)$ into the construction of $\boldsymbol{h}(x)$. But the probability calculation (7.45) makes it unlikely that there are any additional decryption keys beyond $\boldsymbol{f}(x)$ and its rotations.

7.11 NTRUEncrypt as a Lattice Cryptosystem

In this section we explain how NTRU key recovery can be formulated as a shortest vector problem in a certain special sort of lattice. Exercise 7.36 sketches a similar description of NTRU plaintext recovery as a closest vector problem.

7.11.1 The NTRU Lattice

Let

$$\boldsymbol{h}(x) = h_0 + h_1 x + \cdots + h_{N-1}x^{N-1}$$

be an NTRUEncrypt public key. The *NTRU lattice* $L_{\boldsymbol{h}}^{\mathrm{NTRU}}$ associated to $\boldsymbol{h}(x)$ is the $2N$-dimensional lattice spanned by the rows of the matrix

$$M_{\boldsymbol{h}}^{\mathrm{NTRU}} = \left(\begin{array}{cccc|cccc} 1 & 0 & \cdots & 0 & h_0 & h_1 & \cdots & h_{N-1} \\ 0 & 1 & \cdots & 0 & h_{N-1} & h_0 & \cdots & h_{N-2} \\ \vdots & \vdots & \ddots & \vdots & \vdots & \vdots & \ddots & \vdots \\ 0 & 0 & \cdots & 1 & h_1 & h_2 & \cdots & h_0 \\ \hline 0 & 0 & \cdots & 0 & q & 0 & \cdots & 0 \\ 0 & 0 & \cdots & 0 & 0 & q & \cdots & 0 \\ \vdots & \vdots & \ddots & \vdots & \vdots & \vdots & \ddots & \vdots \\ 0 & 0 & \cdots & 0 & 0 & 0 & \cdots & q \end{array}\right).$$

Notice that $M_{\boldsymbol{h}}^{\mathrm{NTRU}}$ is composed of four N-by-N blocks:

Upper left block = Identity matrix,
Lower left block = Zero matrix,
Lower right block = q times the identity matrix,
Upper right block = Cyclical permutations of the coefficients of $\boldsymbol{h}(x)$.

It is often convenient to abbreviate the NTRU matrix as

$$M_{\boldsymbol{h}}^{\text{NTRU}} = \begin{pmatrix} I & \boldsymbol{h} \\ 0 & qI \end{pmatrix}, \tag{7.46}$$

where we view (7.46) as a 2-by-2 matrix with coefficients in R.

We are going to identify each pair of polynomials

$$\boldsymbol{a}(x) = a_0 + a_1x + \cdots + a_{N-1}x^{N-1} \quad \text{and} \quad \boldsymbol{b}(x) = b_0 + b_1x + \cdots + b_{N-1}x^{N-1}$$

in R with a $2N$-dimensional vector

$$(\boldsymbol{a}, \boldsymbol{b}) = (a_0, a_1, \ldots, a_{N-1}, b_0, b_1, \ldots, b_{N-1}) \in \mathbb{Z}^{2N}.$$

We now suppose that the NTRUEncrypt public key $\boldsymbol{h}(x)$ was created using the private polynomials $\boldsymbol{f}(x)$ and $\boldsymbol{g}(x)$ and compute what happens when we multiply the NTRU matrix by a carefully chosen vector.

Proposition 7.59. *Assuming that* $\boldsymbol{f}(x) \star \boldsymbol{h}(x) \equiv \boldsymbol{g}(x) \pmod{q}$, *let* $\boldsymbol{u}(x) \in R$ *be the polynomial satisfying*

$$\boldsymbol{f}(x) \star \boldsymbol{h}(x) = \boldsymbol{g}(x) + q\boldsymbol{u}(x). \tag{7.47}$$

Then

$$(\boldsymbol{f}, -\boldsymbol{u})M_{\boldsymbol{h}}^{\text{NTRU}} = (\boldsymbol{f}, \boldsymbol{g}), \tag{7.48}$$

so the vector $(\boldsymbol{f}, \boldsymbol{g})$ *is in the* NTRU *lattice* $L_{\boldsymbol{h}}^{\text{NTRU}}$.

Proof. It is clear that the first N coordinates of the product (7.48) are the vector $\boldsymbol{f}$, since the left-hand side of $M_{\boldsymbol{h}}^{\text{NTRU}}$ is the identity matrix atop the zero matrix. Next consider what happens when we multiply the column of $M_{\boldsymbol{h}}^{\text{NTRU}}$ whose top entry is h_k by the vector $(\boldsymbol{f}, -\boldsymbol{u})$. We get the quantity

$$h_kf_0 + h_{k-1}f_1 + \cdots + h_{k+1}f_{N-1} - qu_k,$$

which is the kth entry of the vector $\boldsymbol{f}(x) \star \boldsymbol{h}(x) - q\boldsymbol{u}(x)$. From (7.47), this is the kth entry of the vector $\boldsymbol{g}$, so the second N coordinates of the product (7.48) form the vector $\boldsymbol{g}$. Finally, (7.48) says that we can get the vector $(\boldsymbol{f}, \boldsymbol{g})$ by taking a certain linear combination of the rows of $M_{\boldsymbol{h}}^{\text{NTRU}}$. Hence $(\boldsymbol{f}, \boldsymbol{g}) \in L_{\boldsymbol{h}}^{\text{NTRU}}$. □

Remark 7.60. Using the abbreviation (7.46) and multiplying 2-by-2 matrices having coefficients in R, the proof of Proposition 7.59 becomes the succinct computation

$$(\boldsymbol{f}, -\boldsymbol{u})\begin{pmatrix} 1 & \boldsymbol{h} \\ 0 & q \end{pmatrix} = (\boldsymbol{f}, \boldsymbol{f} \star \boldsymbol{h} - q\boldsymbol{u}) = (\boldsymbol{f}, \boldsymbol{g}).$$

Proposition 7.61. *Let* (N, p, q, d) *be NTRUEncrypt parameters, where for simplicity we will assume that*

$$p = 3 \qquad \text{and} \qquad d \approx N/3 \qquad \text{and} \qquad q \approx 6pd \approx 2pN.$$

Let $L_{\boldsymbol{h}}^{\text{NTRU}}$ *be an NTRU lattice associated to the private key* $(\boldsymbol{f}, \boldsymbol{g})$.

(a) $\det(L_h^{\text{NTRU}}) = q^N$.
(b) $\|(\boldsymbol{f},\boldsymbol{g})\| \approx \sqrt{4d} \approx \sqrt{4N/3} \approx 1.155\sqrt{N}$.
(c) *The Gaussian heuristic predicts that the shortest nonzero vector in the* NTRU *lattice has length*

$$\sigma(L_h^{\text{NTRU}}) \approx \sqrt{Nq/\pi e} \approx 0.838N.$$

Hence if N is large, then there is a high probability that the shortest nonzero vectors in L_h^{NTRU} are $(\boldsymbol{f},\boldsymbol{g})$ and its rotations. Further,

$$\frac{\|(\boldsymbol{f},\boldsymbol{g})\|}{\sigma(L)} \approx \frac{1.38}{\sqrt{N}},$$

so the vector $(\boldsymbol{f},\boldsymbol{g})$ is a factor of $\mathcal{O}(1/\sqrt{N})$ shorter than predicted by the Gaussian heuristic.

Proof. (a) Proposition 7.20 says that $\det(L_h^{\text{NTRU}})$ is equal to the determinant of the matrix M_h^{NTRU}. The matrix is upper triangular, so its determinant is the product of the diagonal entries, which equals q^N.
(b) Each of $\boldsymbol{f}$ and $\boldsymbol{g}$ has (approximately) d coordinates equal to 1 and d coordinates equal to -1.
(c) Using (a) and keeping in mind that L_h^{NTRU} has dimension $2N$, we estimate the Gaussian expected shortest length using the formula (7.21),

$$\sigma(L_h^{\text{NTRU}}) = \sqrt{\frac{2N}{2\pi e}}(\det L)^{1/2N} = \sqrt{\frac{Nq}{\pi e}} \approx \sqrt{\frac{6}{\pi e}}\,N. \qquad \square$$

7.11.2 Quantifying the Security of an NTRU Lattice

Proposition 7.61 says that Eve can determine Alice's private NTRU key if she can find a shortest vector in the NTRU lattice L_h^{NTRU}. Thus the security of NTRUEncrypt depends at least on the difficulty of solving SVP in L_h^{NTRU}. More generally, if Eve can solve apprSVP in L_h^{NTRU} to within a factor of approximately N^ϵ for some $\epsilon < \frac{1}{2}$, then the short vector that she finds will probably serve as a decryption key.

This leads to the question of how to estimate the difficulty of finding a short, or shortest, vector in an NTRU lattice. The LLL algorithm that we describe in Sect. 7.13.2 runs in polynomial time and solves apprSVP to within a factor of 2^N, but if N is large, LLL does not find very small vectors in L_h^{NTRU}. In Sect. 7.13.4 we describe a generalization of the LLL algorithm, called BKZ-LLL, that is able to find very small vectors. The BKZ-LLL algorithm includes a blocksize parameter β, and it solves apprSVP to within a factor of $\beta^{2N/\beta}$, but its running time is exponential in β.

Unfortunately, the operating characteristics of standard lattice reduction algorithms such as BKZ-LLL are not nearly as well understood as are the operating characteristics of sieves, the index calculus, or Pollard's ρ method. This makes it difficult to predict theoretically how well a lattice reduction algorithm will perform on any given class of lattices. Thus in practice, the security of a lattice-based cryptosystem such as NTRUEncrypt must be determined experimentally.

Roughly, one takes a sequence of parameters (N, q, d) in which N grows and such that certain ratios involving N, q, and d are held approximately constant. For each set of parameters, one runs many experiments using BKZ-LLL with increasing block size β until the algorithm finds a short vector in L_h^{NTRU}. Then one plots the logarithm of the average running time against N, verifies that the points approximately lie on line, and computes the best-fitting line

$$\log(\text{Running Time}) = AN + B. \tag{7.49}$$

After doing this for many values of N up to the point at which the computations become infeasible, one can use the line (7.49) to extrapolate the expected amount of time it would take to find a private key vector in an NTRU lattice L_h^{NTRU} for larger values of N. Such experiments suggest that values of N in the range from 250 to 1000 yield security levels comparable to currently secure implementations of RSA, Elgamal, and ECC. Details of such experiments are described in [102].

Remark 7.62. Proposition 7.61 says that the short target vectors in an NTRU lattice are $\mathcal{O}(\sqrt{N})$ shorter than predicted by the Gaussian heuristic. Theoretically and experimentally, it is true that if a lattice of dimension n has a vector that is extremely small, say $\mathcal{O}(2^n)$ shorter than the Gaussian prediction, then lattice reduction algorithms such as LLL and its variants are very good at finding the tiny vector. It is a natural and extremely interesting question to ask whether vectors that are only $\mathcal{O}(n^\epsilon)$ shorter than the Gaussian prediction might similarly be easier to find. At this time, no one knows the answer to this question.

7.12 Lattice-Based Digital Signature Schemes

We have already seen digital signatures schemes whose security depends on the integer factorization problem (Sect. 4.2) and on the discrete logarithm problem in the multiplicative group (Sect. 4.3) or in an elliptic curve (Sect. 6.4.3). In this section we briefly discuss how digital signature schemes may be constructed from hard lattice problems.

7.12.1 The GGH Digital Signature Scheme

It is easy to convert the CVP idea underlying GGH encryption into a lattice-based digital signature scheme. Samantha knows a good (i.e., short and reasonably orthogonal) private basis $\mathcal{B}$ for a lattice L, so she can use Babai's algorithm (Theorem 7.34) to solve, at least approximately, the closest vector problem in L for a given vector $\boldsymbol{d} \in \mathbb{R}^n$. She expresses her solution $\boldsymbol{s} \in L$ in terms of a bad public basis $\mathcal{B}'$. The vector $\boldsymbol{s}$ is Samantha's signature on the document $\boldsymbol{d}$. Victor can easily check that $\boldsymbol{s}$ is in L and is close to $\boldsymbol{d}$. The GGH digital signature scheme is summarized in Table 7.5.

Samantha	Victor
Key creation	
Choose a good basis $\boldsymbol{v}_1, \ldots, \boldsymbol{v}_n$ and a bad basis $\boldsymbol{w}_1, \ldots, \boldsymbol{w}_n$ for L. Publish the public key $\boldsymbol{w}_1, \ldots, \boldsymbol{w}_n$.	
Signing	
Choose document $\boldsymbol{d} \in \mathbb{Z}^n$ to sign. Use Babai's algorithm with the good basis to compute a vector $\boldsymbol{s} \in L$ that is close to $\boldsymbol{d}$. Write $\boldsymbol{s} = a_1\boldsymbol{w}_1 + \cdots + a_n\boldsymbol{w}_n$. Publish the signature $(a_1, \ldots, a_n)$.	
Verification	
	Compute $\boldsymbol{s} = a_1\boldsymbol{w}_1 + \cdots + a_n\boldsymbol{w}_n$. Verify that $\boldsymbol{s}$ is sufficiently close to $\boldsymbol{d}$.

Table 7.5: The GGH digital signature scheme

Notice the tight fit between the digital signature and the underlying hard problem. The signature $\boldsymbol{s} \in L$ is a solution to apprCVP for the vector $\boldsymbol{d} \in \mathbb{R}^n$, so signing a document is equivalent to solving apprCVP.

Remark 7.63. In a lattice-based digital signature scheme, the digital document to be signed is a vector in $\mathbb{R}^n$. Just as with other signature schemes, in practice Samantha applies a hash function to her actual document in order to create a short document of just a few hundred bits, which is then signed. (See Remark 4.2.) For lattice-based signatures, one uses a hash function whose output is a vector in $\mathbb{Z}^n$ having coordinates in some specified range.

Example 7.64. We illustrate the GGH digital signature scheme using the lattice and the good and bad bases from Example 7.36 on page 410. Samantha decides to sign the document

$$\boldsymbol{d} = (678846, 651685, 160467) \in \mathbb{Z}^3.$$

She uses Babai's algorithm to find a vector

$$\boldsymbol{s} = 2213\boldsymbol{v}_1 + 7028\boldsymbol{v}_2 - 6231\boldsymbol{v}_3 = (678835, 651671, 160437) \in L$$

that is quite close to $\boldsymbol{d}$,

$$\|\boldsymbol{s} - \boldsymbol{d}\| \approx 34.89.$$

Samantha next uses linear algebra to express $\boldsymbol{s}$ in terms of the bad basis,

$$\boldsymbol{s} = 1531010\boldsymbol{w}_1 - 553385\boldsymbol{w}_2 - 878508\boldsymbol{w}_3,$$

where $\boldsymbol{w}_1, \boldsymbol{w}_2, \boldsymbol{w}_3$ are the vectors on page 410. She publishes

$$(1531010, -553385, -878508)$$

as her signature for the document $\boldsymbol{d}$. Victor verifies the signature by using the public basis to compute

$$\boldsymbol{s} = 1531010\boldsymbol{w}_1 - 553385\boldsymbol{w}_2 - 878508\boldsymbol{w}_3 = (678835, 651671, 160437),$$

which is automatically a vector in L, and then verifying that $\|\boldsymbol{s} - \boldsymbol{d}\| \approx 34.89$ is small.

We observe that if Eve attempts to sign $\boldsymbol{d}$ using Babai's algorithm with the bad basis $\{\boldsymbol{w}_1, \boldsymbol{w}_2, \boldsymbol{w}_3\}$, then the signature that she obtains is

$$\boldsymbol{s}' = (2773584, 1595134, -131844) \in L.$$

This vector is not a good solution to apprCVP, since $\|\boldsymbol{s}' - \boldsymbol{d}\| > 10^6$.

Remark 7.65 (Key Size Issues). The GGH signature scheme suffers the same drawback as the GGH cryptosystem, namely security requires lattices of high dimension, which in turn lead to very large public verification keys; cf. Sect. 7.7. It is thus tempting to use an NTRU lattice L^{NTRU} as the public key, but there is an initial difficulty because L^{NTRU} has dimension $2N$, so the known (secret) short vector $(\boldsymbol{f}, \boldsymbol{g})$ and its rotations $(x^i \star \boldsymbol{f}, x^i \star \boldsymbol{g})$ for $0 \le i < N$ give only half a very short basis for L^{NTRU}. Using a technique described in [55], it is possible to extend the half-basis to a full basis that is short enough to make an NTRU signature scheme feasible. However, both GGH and NTRU signature schemes have a more serious shortcoming which we now describe.

7.12.2 Transcript Analysis

In any digital signature scheme, each document/signature pair $(\boldsymbol{d}, \boldsymbol{s})$ reveals some information about the private signing key $\boldsymbol{v}$, since at the very least, it reveals that the document $\boldsymbol{d}$ signed with the private key $\boldsymbol{v}$ yields the signature $\boldsymbol{s}$. Hence a sufficiently long *transcript* of signed documents

$$(\boldsymbol{d}_1, \boldsymbol{s}_1), (\boldsymbol{d}_2, \boldsymbol{s}_2), (\boldsymbol{d}_3, \boldsymbol{s}_3), \ldots, (\boldsymbol{d}_r, \boldsymbol{s}_r) \tag{7.50}$$

may reveal information about either the signing key or how to sign additional documents.

We illustrate with the GGH signature scheme. By construction, the signature $\boldsymbol{s}$ is created using Babai's algorithm to solve apprCVP with the good basis $\boldsymbol{v}_1, \ldots, \boldsymbol{v}_n$ and target vector $\boldsymbol{d}$. It follows that the difference $\boldsymbol{d} - \boldsymbol{s}$ has the form

$$\boldsymbol{d} - \boldsymbol{s} = \sum_{i=1}^{n} \epsilon_i(\boldsymbol{d}, \boldsymbol{s})\boldsymbol{v}_i \qquad \text{with} \qquad \bigl|\epsilon_i(\boldsymbol{d}, \boldsymbol{s})\bigr| \le \frac{1}{2}.$$

As $\boldsymbol{d}$ and $\boldsymbol{s}$ vary, the $\epsilon_i(\boldsymbol{d}, \boldsymbol{s})$ values are more or less randomly distributed between $-\frac{1}{2}$ and $\frac{1}{2}$. Hence the transcript (7.50) reveals to an adversary a large number of points that are randomly scattered in the fundamental domain

$$\mathcal{F} = \bigl\{\epsilon_1\boldsymbol{v}_1 + \epsilon_2\boldsymbol{v}_2 + \cdots + \epsilon_n\boldsymbol{v}_n : -\tfrac{1}{2} < \epsilon_1, \ldots, \epsilon_n \le \tfrac{1}{2}\bigr\}$$

spanned by the good secret basis $\boldsymbol{v}_1, \ldots, \boldsymbol{v}_n$. Using this collection of points, it may be possible to (approximately) recover the basis vectors spanning the fundamental domain $\mathcal{F}$. An algorithm to perform this task was given by Nguyen and Regev [93, 94]. They used their algorithm to break instances of GGH in dimension n with a transcript consisting of roughly n^2 signatures, and they gave similar applications to NTRU signatures. It is possible to blunt these attacks by introducing small biased perturbations into each signature [55, 56], but the process is inefficient and may still be subject to transcript attacks [39].

7.12.3 Rejection Sampling

An alternative method of thwarting transcript attacks was proposed by Lyubashevsky [80, 79, 78]. It is based on an idea from statistics called rejection sampling, in which one generates samples from a desired probability distribution by using samples from another distribution. There are now a number of proposed digital signature schemes that use rejection sampling to achieve transcript security. In this section we discuss rejection sampling as a general technique, after which we apply rejection sampling to an abstract signature scheme (Sect. 7.12.4) and illustrate the method with a specific lattice-based scheme (Sect. 7.12.5).

The notion of rejection sampling was introduced by J. von Neumann in 1951 [146]. His aim was to produce samples from a distribution $F(x)$, which is itself hard to sample, by using another distribution $G(x)$ whose samples are easy to produce.[9] The use of rejection sampling in the context of foiling a transcript attack on a digital signature scheme amounts to a clever reversal of this situation. Imagine that the signature scheme somehow generates samples that can be used to produce a distribution $G(x)$. The signature scheme is vulnerable to a transcript attack if this distribution $G(x)$ possesses features that provide information about the private key, since then sufficiently many samples may reveal the key. In order to foil a transcript attack, one wants to hide the unique identifying features of $G(x)$. Under certain circumstances,

[9] Distribution functions are discussed in Sect. 5.3.4.

there is a Monte Carlo type algorithm which does this.[10] It works by rejecting certain samples so that the resulting collection is disguised as a generic desired distribution $F(x)$.

Let $F(x)$ and $G(x)$ be probability distribution functions having the property that

$$F(x) \leq MG(x) \quad \text{for some constant } M.$$

The goal is to generate samples that are distributed according to $F(x)$ from samples generated from $G(x)$. To do this, let $U(x)$ be the uniform distribution on the unit interval $[0, 1]$. One repeatedly takes samples x from $G(x)$ and samples u from $U(x)$. The pair

$$(x, u) \text{ is } \begin{cases} \text{accepted if } u < F(x)/MG(x), \\ \text{rejected otherwise.} \end{cases}$$

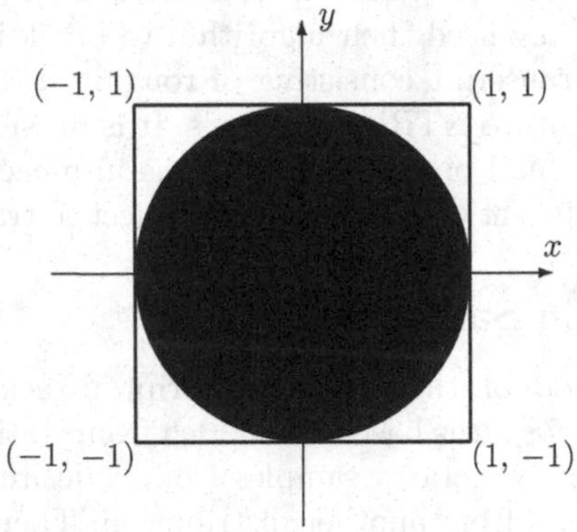

Figure 7.6: Rejection sampling on the circle

Suppose that $(x_1, u_1), (x_2, u_2), \ldots$ is the list of accepted pairs. Then one can show, using Bayes's formula, that the collection of points

$$\big\{(x_i, u_i MG(x_i)) : i = 1, 2, 3, \ldots\big\}$$

is uniformly distributed under the graph of $F(x)$. We do not give the proof, but instead consider the following example where the situation is particularly intuitive.

Suppose that we have a way of uniformly choosing numbers in the square

$$S = \big\{(x, y) : -1 \leq x \leq 1,\ -1 \leq y \leq 1\big\}$$

and that we want to choose points that are uniformly distributed in the circle

$$C = \big\{(x, y) : x^2 + y^2 \leq 1\big\}$$

[10] Monte Carlo algorithms are discussed in Sect. 5.3.3.

as illustrated in Fig. 7.6. So our samples are points (x, y) in the plane, and our uniform distribution functions on the circle and the square are, respectively,[11]

$$F_C(x,y) = \begin{cases} \dfrac{1}{\pi} & \text{if } x^2+y^2 \le 1, \\ 0 & \text{otherwise;} \end{cases}$$

$$G_S(x,y) = \begin{cases} \dfrac{1}{4} & \text{if } -1 \le x \le 1 \text{ and } 1 \le y \le 1, \\ 0 & \text{otherwise.} \end{cases}$$

For all (x, y) we clearly have

$$F_C(x,y) \le MG_S(x,y) \quad \text{with } M = \frac{4}{\pi},$$

since $F_C(x,y)/G_S(x,y) = M$ if (x, y) is in the circle, and equals 0 otherwise. Now the pair $\bigl((x,y),u\bigr)$, where (x, y) is uniformly sampled from the square and u is uniformly sampled from the interval $[0, 1]$, will be accepted if and only if $F_C(x,y)/MG_S(x,y) = 1$, which means it is accepted if and only if (x, y) is in the circle. In brief, rejection sampling amounts to choosing points uniformly in the square and rejecting those points which do not lie in the circle. The result is a collection of points uniformly distributed in the circle.

7.12.4 Rejection Sampling Applied to an Abstract Signature Scheme

In this section we describe, abstractly, how rejection sampling can be used to protect a digital signature scheme from transcript attacks. Note that we are describing the properties that such a scheme should have, without giving any indication of how one might create such a scheme. (Just as Diffie and Hellman described what a public key cryptosystem should do, without providing an example of such a system.)

We consider an abstract digital signature scheme $(K^{\mathrm{Pri}}, K^{\mathrm{Pub}}, \mathsf{Sign}, \mathsf{Verify})$ as described in Sect. 4.1. We assume further that the signing algorithm Sign uses three inputs, the private key K^{Pri}, the document hash D being signed, and a random number R. Rejection sampling introduces a conditional property $\mathcal{P}$. In order to sign D, Samantha chooses a random R and computes the signature $S = \mathsf{Sign}(K^{\mathrm{Pri}}, D, R)$. If S has property $\mathcal{P}$, then she publishes S as the signature on D; but if S does not have property $\mathcal{P}$, then she rejects S, chooses a new value for R, and repeats the process. This means that in any transcript of Samantha's signatures $(D_1, S_1), (D_2, S_2), \ldots$, every S_i has property $\mathcal{P}$.

Now for the tricky part. We want the attacker Eve, using only the public key K^{Pub}, to be able to create a list of pairs $(D_1', S_1'), (D_2', S_2'), \ldots$ satisfying:

[11] Why does F_C have the $1/\pi$ and G_S have the $1/4$? It's because the total probabilities $\int_{-\infty}^{\infty} F_C(x,y)\,dx\,dy$ and $\int_{-\infty}^{\infty} G_S(x,y)\,dx\,dy$ must equal 1.

(i) S_i' is a valid signature on D_i' for the key K^{Pub}, i.e.,

$$\mathsf{Verify}(K^{\text{Pub}}, D_i', S_i') = \text{TRUE} \quad \text{for all } i.$$

(ii) The distribution of Eve's fake transcript $(D_1', S_1'), (D_2', S_2'), \ldots$ is indistinguishable from a transcript that Samantha creates using her private key K^{Pri}.

Property (i) may seem problematic, since it says that Eve can produce an unlimited number of valid document/signature pairs (D, S). However, recall that D is really a *hash* of the actual document being signed. (See Sects. 4.2 and 8.1 for a discussion of hash functions and their uses.) So although we want Eve to be able to easily create valid (D, S) pairs, she will not know what document she has signed, because she is not able to invert the hash function. In other words, although Eve can create valid pairs (D, S), if someone hands her a particular D, she will not be able to find an associated S. Thus security, as always, relies on various (reasonable) assumptions, in this case that we have a sufficiently cryptographically secure hash function.

7.12.5 The NTRU Modular Lattice Signature Scheme

Lyubashevsky gave an example of a transcript-secure signature scheme based on the learning with errors (LWE) problem. We briefly sketch a new rejection-sampling signature scheme called NTRUMLS (NTRU modular lattice signature scheme) that uses NTRU lattices [57].[12] We set one piece of notation. The *sup norm* of a polynomial $\boldsymbol{a}(x) = a_0 + a_1 x + \cdots + a_{N-1} x^{N-1}$ is denoted

$$\|\boldsymbol{a}\|_\infty = \max\{|a_0|, |a_1|, \ldots, |a_{N-1}|\}.$$

The basic set-up for NTRUMLS is similar to the set-up for NTRUEncrypt in Sect. 7.10, with parameters (N, p, q), private key polynomials $\boldsymbol{f}$ and $\boldsymbol{g}$ with small coefficients, and public key polynomial $\boldsymbol{h} \equiv \boldsymbol{f}^{-1} \star \boldsymbol{g} \pmod{q}$.[13] NTRUMLS also uses a public rejection parameter B and a public hash function that takes a digital document μ and a public key $\boldsymbol{h}$ and creates a pair of mod p polynomials:

$$\mathsf{Hash} : \{\text{documents}\} \times \{\text{public keys}\} \longrightarrow \{\text{pairs of mod } p \text{ polynomials}\}$$
$$\mathsf{Hash}(\mu, \boldsymbol{h}) = (\boldsymbol{s}_p, \boldsymbol{t}_p).$$

An NTRUMLS signature on the document μ for the public key $\boldsymbol{h}$ is a pair of polynomials $(\boldsymbol{s}, \boldsymbol{t})$ satisfying the following three conditions:

[12] NTRUMLS was released in 2014, so it is very new. We present it as an illustration of how rejection sampling might work in practice, but as with all new systems, NTRUMLS will require years of scrutiny before it can be deemed secure.

[13] There are some further minor requirements that we omit, since our aim is to illustrate the idea of rejection sampling. See Exercise 7.42.

(a) $\boldsymbol{t} \equiv \boldsymbol{s} \star \boldsymbol{h} \pmod{q}$.
(b) $\boldsymbol{s} \equiv \boldsymbol{s}_p \pmod{p}$ and $\boldsymbol{t} \equiv \boldsymbol{t}_p \pmod{p}$, where $(\boldsymbol{s}_p, \boldsymbol{t}_p) = \mathsf{Hash}(\mu, \boldsymbol{h})$.
(c) $\|\boldsymbol{s}\|_\infty \le \frac{1}{2}q - B$ and $\|\boldsymbol{t}\|_\infty \le \frac{1}{2}q - B$.

Here are some further remarks on the three signing conditions:

(a) This ties the signature to the signing key. It is equivalent to the assertion that $(\boldsymbol{s}, \boldsymbol{t})$ is in the lattice $L_{\boldsymbol{h}}^{\text{NTRU}}$ associated to $\boldsymbol{h}$; cf. Sect. 7.11.1.

(b) This ties the signature $(\boldsymbol{s}, \boldsymbol{t})$ to the document hash $(\boldsymbol{s}_p, \boldsymbol{t}_p)$. It is equivalent to the assertion that the difference $(\boldsymbol{s}, \boldsymbol{t}) - (\boldsymbol{s}_p, \boldsymbol{t}_p)$ is in the lattice $(p\mathbb{Z})^{2N}$.

(c) This is the rejection sampling condition, since it says that we reject the signature $(\boldsymbol{s}, \boldsymbol{t})$ if it is too large. Note the tension inherent in this condition. If B is too large, then it will be difficult to generate signatures, while one can show that if B is not large enough, then transcripts leak private key information.

Using the private key $(\boldsymbol{f}, \boldsymbol{g})$, it is not hard to create a pair $(\boldsymbol{s}, \boldsymbol{t})$ satisfying conditions (a) and (b). Further, for appropriately chosen values of p, q, and B, one can show that it will not take too many tries to find an $(\boldsymbol{s}, \boldsymbol{t})$ that also satisfies condition (c). (See Exercise 7.42 for details.)

The transcript security analysis relies on the following two facts. The proof, which we omit (see [57]), relies on various reasonable randomness assumptions.

- When the signing algorithm is applied to a given document hash $(\boldsymbol{s}_p, \boldsymbol{t}_p)$, each pair $(\boldsymbol{s}, \boldsymbol{t})$ satisfying conditions (a), (b), (c) has an equal probability of being chosen as the signature.

- Suppose that an attacker creates a list of $(\boldsymbol{s}, \boldsymbol{t})$ pairs by randomly choosing $\boldsymbol{s}$'s satisfying $\|\boldsymbol{s}\|_\infty \le \frac{1}{2}q - B$, computing $\boldsymbol{t} \equiv \boldsymbol{s} \star \boldsymbol{h} \pmod{q}$, and keeping the pair $(\boldsymbol{s}, \boldsymbol{t})$ if $\|\boldsymbol{t}\|_\infty \le \frac{1}{2}q - B$. Then the reduction of his list modulo p is uniformly randomly distributed among all pairs of mod p polynomials. (Note that the each of the attacker's $(\boldsymbol{s}, \boldsymbol{t})$ pairs is a valid signature on the document hash $(\boldsymbol{s} \bmod p, \boldsymbol{t} \bmod p)$ for the verification key $\boldsymbol{h}$. He is thus able to create an arbitrarily long transcript of valid signatures, but he not able to specify, a priori, the $\boldsymbol{t}_p$ parts of the document hashes that he is signing.)

These two facts show that an attacker, using only the public key $\boldsymbol{h}$, can create a transcript of signed document hashes that is indistinguishable from a transcript created using the private key $(\boldsymbol{f}, \boldsymbol{g})$. Hence the latter transcript contains no information about the private key. We refer the reader to the references [57, 80, 79, 78] for further details on NTRUMLS and other transcript-secure lattice-based signature schemes.

7.13 Lattice Reduction Algorithms

We have now seen several cryptosystems whose security depends on the difficulty of solving apprSVP and/or apprCVP in various types of lattices. In this section we describe an algorithm called LLL that solves these problems to within a factor of C^n, where C is a small constant and n is the dimension of the lattice. Thus in small dimensions, the LLL algorithm comes close to solving SVP and CVP, but in large dimensions it does not do as well. Ultimately, the security of lattice-based cryptosystems depends on the inability of LLL and other lattice reduction algorithms to efficiently solve apprSVP and apprCVP to within a factor of, say, $\mathcal{O}(\sqrt{n})$. We begin in Sect. 7.13.1 with Gauss's lattice reduction algorithm, which rapidly solves SVP in lattices of dimension 2. Next, in Sect. 7.13.2, we describe and analyze the LLL algorithm. Section 7.13.3 explains how to combine LLL and Babai's algorithm to solve apprCVP, and we conclude in Sect. 7.13.4 by briefly describing some generalizations of LLL.

7.13.1 Gaussian Lattice Reduction in Dimension 2

The algorithm for finding an optimal basis in a lattice of dimension 2 is essentially due to Gauss. The underlying idea is to alternately subtract multiples of one basis vector from the other until further improvement is not possible.

So suppose that $L \subset \mathbb{R}^2$ is a 2-dimensional lattice with basis vectors $\boldsymbol{v}_1$ and $\boldsymbol{v}_2$. Swapping $\boldsymbol{v}_1$ and $\boldsymbol{v}_2$ if necessary, we may assume that $\|\boldsymbol{v}_1\| < \|\boldsymbol{v}_2\|$. We now try to make $\boldsymbol{v}_2$ smaller by subtracting a multiple of $\boldsymbol{v}_1$. If we were allowed to subtract an arbitrary multiple of $\boldsymbol{v}_1$, then we could replace $\boldsymbol{v}_2$ with the vector

$$\boldsymbol{v}_2^* = \boldsymbol{v}_2 - \frac{\boldsymbol{v}_1 \cdot \boldsymbol{v}_2}{\|\boldsymbol{v}_1\|^2}\boldsymbol{v}_1,$$

which is orthogonal to $\boldsymbol{v}_1$. The vector $\boldsymbol{v}_2^*$ is the projection of $\boldsymbol{v}_2$ onto the orthogonal complement of $\boldsymbol{v}_1$. (See Fig. 7.7.)

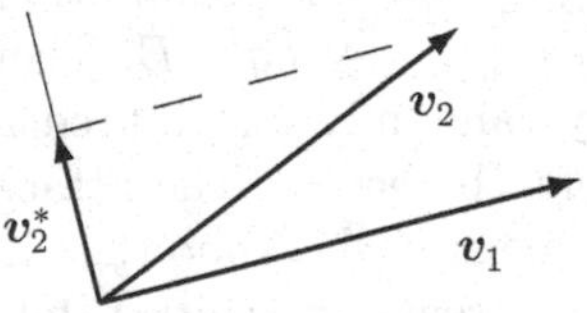

Figure 7.7: $\boldsymbol{v}_2^*$ is the projection of $\boldsymbol{v}_2$ onto the orthogonal complement of $\boldsymbol{v}_1$

Of course, this is cheating, since the vector $\boldsymbol{v}_2^*$ is unlikely to be in L. In reality we are allowed to subtract only integer multiples of $\boldsymbol{v}_1$ from $\boldsymbol{v}_2$. So we do the best that we can and replace $\boldsymbol{v}_2$ with the vector

$$\boldsymbol{v}_2 - m\boldsymbol{v}_1 \qquad \text{with} \quad m = \left\lfloor \frac{\boldsymbol{v}_1 \cdot \boldsymbol{v}_2}{\|\boldsymbol{v}_1\|^2} \right\rceil.$$

If $\boldsymbol{v}_2$ is still longer than $\boldsymbol{v}_1$, then we stop. Otherwise, we swap $\boldsymbol{v}_1$ and $\boldsymbol{v}_2$ and repeat the process. Gauss proved that this process terminates and that the resulting basis for L is extremely good. The next proposition makes this precise.

Proposition 7.66 (Gaussian Lattice Reduction). *Let $L \subset \mathbb{R}^2$ be a 2-dimensional lattice with basis vectors $\boldsymbol{v}_1$ and $\boldsymbol{v}_2$. The following algorithm terminates and yields a good basis for L.*

```
Loop
    If ‖v2‖ < ‖v1‖, swap v1 and v2.
    Compute m = ⌊v1 · v2/‖v1‖²⌉.
    If m = 0, return the basis vectors v1 and v2.
    Replace v2 with v2 − m v1.
Continue Loop
```

More precisely, when the algorithm terminates, the vector $\boldsymbol{v}_1$ is a shortest nonzero vector in L, so the algorithm solves SVP. *Further, the angle θ between $\boldsymbol{v}_1$ and $\boldsymbol{v}_2$ satisfies $|\cos\theta| \le \|\boldsymbol{v}_1\|/2\|\boldsymbol{v}_2\|$, so in particular, $\frac{\pi}{3} \le \theta \le \frac{2\pi}{3}$.*

Proof. We prove that $\boldsymbol{v}_1$ is a smallest nonzero lattice vector and leave the other parts of the proof to the reader. So we suppose that the algorithm has terminated and returned the vectors $\boldsymbol{v}_1$ and $\boldsymbol{v}_2$. This means that $\|\boldsymbol{v}_2\| \ge \|\boldsymbol{v}_1\|$ and that

$$\frac{|\boldsymbol{v}_1 \cdot \boldsymbol{v}_2|}{\|\boldsymbol{v}_1\|^2} \le \frac{1}{2}. \tag{7.51}$$

(Geometrically, condition (7.51) says that we cannot make $\boldsymbol{v}_2$ smaller by subtracting an integral multiple of $\boldsymbol{v}_1$ from $\boldsymbol{v}_2$.) Now suppose that $\boldsymbol{v} \in L$ is any nonzero vector in L. Writing

$$\boldsymbol{v} = a_1\boldsymbol{v}_1 + a_2\boldsymbol{v}_2 \quad \text{with } a_1, a_2 \in \mathbb{Z},$$

we find that

$$\begin{aligned}
\|\boldsymbol{v}\|^2 &= \|a_1\boldsymbol{v}_1 + a_2\boldsymbol{v}_2\|^2 \\
&= a_1^2\|\boldsymbol{v}_1\|^2 + 2a_1a_2(\boldsymbol{v}_1\cdot\boldsymbol{v}_2) + a_2^2\|\boldsymbol{v}_2\|^2 \\
&\ge a_1^2\|\boldsymbol{v}_1\|^2 - 2|a_1a_2|\,|\boldsymbol{v}_1\cdot\boldsymbol{v}_2| + a_2^2\|\boldsymbol{v}_2\|^2 \\
&\ge a_1^2\|\boldsymbol{v}_1\|^2 - |a_1a_2|\|\boldsymbol{v}_1\|^2 + a_2^2\|\boldsymbol{v}_2\|^2 \quad \text{from (7.51)}, \\
&\ge a_1^2\|\boldsymbol{v}_1\|^2 - |a_1a_2|\|\boldsymbol{v}_1\|^2 + a_2^2\|\boldsymbol{v}_1\|^2 \quad \text{since } \|\boldsymbol{v}_2\| \ge \|\boldsymbol{v}_1\|, \\
&= \left(a_1^2 - |a_1|\,|a_2| + a_2^2\right)\|\boldsymbol{v}_1\|^2.
\end{aligned}$$

For any real numbers t_1 and t_2, the quantity

$$t_1^2 - t_2t_2 + t_2^2 = \left(t_1 - \frac{1}{2}t_2\right)^2 + \frac{3}{4}t_2^2 = \frac{3}{4}t_1^2 + \left(\frac{1}{2}t_1 - t_2\right)^2$$

is not zero unless $t_1 = t_2 = 0$. So the fact that a_1 and a_2 are integers and not both 0 tells us that $\|\boldsymbol{v}\|^2 \geq \|\boldsymbol{v}_1\|^2$. This proves that $\boldsymbol{v}_1$ is a smallest nonzero vector in L. □

Example 7.67. We illustrate Gauss's lattice reduction algorithm (Proposition 7.66) with the lattice L having basis

$$\boldsymbol{v}_1 = (66586820, 65354729) \quad \text{and} \quad \boldsymbol{v}_2 = (6513996, 6393464).$$

We first compute $\|\boldsymbol{v}_1\|^2 \approx 8.71 \cdot 10^{15}$ and $\|\boldsymbol{v}_2\|^2 \approx 8.33 \cdot 10^{13}$. Since $\boldsymbol{v}_2$ is shorter than $\boldsymbol{v}_1$, we swap them, so now $\boldsymbol{v}_1 = (6513996, 6393464)$ and $\boldsymbol{v}_2 = (66586820, 65354729)$.

Next we subtract a multiple of $\boldsymbol{v}_1$ from $\boldsymbol{v}_2$. The multiplier is

$$m = \left\lfloor \frac{\boldsymbol{v}_1 \cdot \boldsymbol{v}_2}{\|\boldsymbol{v}_1\|^2} \right\rceil = \lfloor 10.2221 \rceil = 10,$$

so we replace $\boldsymbol{v}_2$ with

$$\boldsymbol{v}_2 - m\boldsymbol{v}_1 = (1446860, 1420089).$$

This new vector has norm $\|\boldsymbol{v}_2\|^2 \approx 4.11 \cdot 10^{12}$, which is smaller than $\|\boldsymbol{v}_1\|^2 \approx 8.33 \cdot 10^{13}$, so again we swap,

$$\boldsymbol{v}_1 = (1446860, 1420089) \quad \text{and} \quad \boldsymbol{v}_2 = (6513996, 6393464).$$

We repeat the process with $m = \lfloor \boldsymbol{v}_1 \cdot \boldsymbol{v}_2 / \|\boldsymbol{v}_1\|^2 \rceil = \lfloor 4.502 \rceil = 5$, which gives the new vector

$$\boldsymbol{v}_2 - m\boldsymbol{v}_1 = (-720304, -706981)$$

having norm $\|\boldsymbol{v}_2\|^2 \approx 1.01 \cdot 10^{12}$, so again we swap $\boldsymbol{v}_1$ and $\boldsymbol{v}_2$. Continuing this process leads to smaller and smaller bases until, finally, the algorithm terminates. The step by step results of the algorithm, including the value of m used at each stage, are listed in the following table:

Step	$\boldsymbol{v}_1$	$\boldsymbol{v}_2$	m
1	(6513996, 6393464)	(66586820, 65354729)	10
2	(1446860, 1420089)	(6513996, 6393464)	5
3	(−720304, −706981)	(1446860, 1420089)	−2
4	(6252, 6127)	(−720304, −706981)	−115
5	(−1324, −2376)	(6252, 6127)	−3
6	(2280, −1001)	(−1324, −2376)	0

The final basis is quite small, and (2280, −1001) is a solution to SVP for the lattice L.

7.13.2 The LLL Lattice Reduction Algorithm

Gauss's lattice reduction algorithm (Proposition 7.66) gives an efficient way to find a shortest nonzero vector in a lattice of dimension 2, but as the dimension increases, the shortest vector problem becomes much harder . A major advance came in 1982 with the publication of the LLL algorithm [77]. In this section we give a full description of the LLL algorithm, and in the next section we briefly describe some of its generalizations.

Suppose that we are given a basis $\{\boldsymbol{v}_1, \boldsymbol{v}_2, \ldots, \boldsymbol{v}_n\}$ for a lattice L. Our object is to transform the given basis into a "better" basis. But what do we mean by a better basis? We would like the vectors in the better basis to be as short as possible, beginning with the shortest vector that we can find, and then with vectors whose lengths increase as slowly as possible until we reach the last vector in the basis. Alternatively, we would like the vectors in the better basis to be as orthogonal as possible to one another, i.e., so that the dot products $\boldsymbol{v}_i \cdot \boldsymbol{v}_j$ are as close to zero as possible.

Recall that Hadamard's inequality (Proposition 7.19) says that

$$\det L = \operatorname{Vol}(\mathcal{F}) \leq \|\boldsymbol{v}_1\| \, \|\boldsymbol{v}_2\| \cdots \|\boldsymbol{v}_n\|, \tag{7.52}$$

where $\operatorname{Vol}(\mathcal{F})$ is the volume of a fundamental domain for L. The closer that the basis comes to being orthogonal, the closer that the inequality (7.52) comes to being an equality.

To assist us in creating an improved basis, we begin by constructing a Gram–Schmidt orthogonal basis as described in Theorem 7.13. Thus we start with $\boldsymbol{v}_1^* = \boldsymbol{v}_1$, and then for $i \geq 2$ we let

$$\boldsymbol{v}_i^* = \boldsymbol{v}_i - \sum_{j=1}^{i-1} \mu_{i,j} \boldsymbol{v}_j^*, \quad \text{where} \quad \mu_{i,j} = \frac{\boldsymbol{v}_i \cdot \boldsymbol{v}_j^*}{\|\boldsymbol{v}_j^*\|^2} \quad \text{for} \quad 1 \leq j \leq i-1. \tag{7.53}$$

The collection of vectors $\mathcal{B}^* = \{\boldsymbol{v}_1^*, \boldsymbol{v}_2^*, \ldots, \boldsymbol{v}_n^*\}$ is an orthogonal basis for the *vector space* spanned by $\mathcal{B} = \{\boldsymbol{v}_1, \boldsymbol{v}_2, \ldots, \boldsymbol{v}_n\}$, but note that $\mathcal{B}^*$ is *not* a basis for the *lattice* L spanned by $\mathcal{B}$, because the Gram–Schmidt process (7.53) involves taking linear combinations with nonintegral coefficients. However, as we now prove, it turns out that the two bases have the same determinant.

Proposition 7.68. *Let* $\mathcal{B} = \{\boldsymbol{v}_1, \boldsymbol{v}_2, \ldots, \boldsymbol{v}_n\}$ *be a basis for a lattice L and let* $\mathcal{B}^* = \{\boldsymbol{v}_1^*, \boldsymbol{v}_2^*, \ldots, \boldsymbol{v}_n^*\}$ *be the associated Gram–Schmidt orthogonal basis as described in Theorem* 7.13. *Then*

$$\det(L) = \prod_{i=1}^{n} \|\boldsymbol{v}_i^*\|.$$

Proof. Let $F = F(\boldsymbol{v}_1, \ldots, \boldsymbol{v}_n)$ be the matrix (7.11) described in Proposition 7.20. This is the matrix whose rows are the coordinates of $\boldsymbol{v}_1, \ldots, \boldsymbol{v}_n$. The proposition tells us that $\det(L) = |\det F|$.

Let $F^* = F(\boldsymbol{v}_1^*, \ldots, \boldsymbol{v}_n^*)$ be the analogous matrix whose rows are the vectors $\boldsymbol{v}_1^*, \ldots, \boldsymbol{v}_n^*$. Then (7.53) tells us that the matrices F and F^* are related by

$$MF^* = F,$$

where M is the change of basis matrix

$$M = \begin{pmatrix} 1 & 0 & 0 & \cdots & 0 & 0 \\ \mu_{2,1} & 1 & 0 & \cdots & 0 & 0 \\ \mu_{3,1} & \mu_{3,2} & 1 & \cdots & 0 & 0 \\ \vdots & \vdots & \vdots & \ddots & & \vdots \\ \mu_{n-1,1} & \mu_{n-1,2} & \mu_{n-1,3} & \cdots & 1 & 0 \\ \mu_{n,1} & \mu_{n,2} & \mu_{n,3} & \cdots & \mu_{n,n-1} & 1 \end{pmatrix}.$$

Note that M is lower diagonal with 1's on the diagonal, so $\det(M) = 1$. Hence

$$\det(L) = |\det F| = |\det(MF^*)| = |(\det M)(\det F^*)| = |\det F^*| = \prod_{i=1}^{n} \|\boldsymbol{v}_i^*\|.$$

(The last equality follows from the fact that the $\boldsymbol{v}_i^*$, which are the rows of F^*, are pairwise orthogonal.) □

Definition. Let V be a vector space, and let $W \subset V$ be a vector subspace of V. The *orthogonal complement of* W *(in* V*)* is

$$W^{\perp} = \{\boldsymbol{v} \in V : \boldsymbol{v} \cdot \boldsymbol{w} = 0 \quad \text{for all } \boldsymbol{w} \in W\}.$$

It is not hard to see that $W^{\perp}$ is also a vector subspace of V and that every vector $\boldsymbol{v} \in V$ can be written as a sum $\boldsymbol{v} = \boldsymbol{w} + \boldsymbol{w}'$ for unique vectors $\boldsymbol{w} \in W$ and $\boldsymbol{w}' \in W^{\perp}$. (See Exercise 7.46.)

Using the notion of orthogonal complement, we can describe the intuition behind the Gram–Schmidt construction as follows:

$$\boldsymbol{v}_i^* = \text{Projection of } \boldsymbol{v}_i \text{ onto } \operatorname{Span}(\boldsymbol{v}_1, \ldots, \boldsymbol{v}_{i-1})^{\perp}.$$

Although $\mathcal{B}^* = \{\boldsymbol{v}_1^*, \boldsymbol{v}_2^*, \ldots, \boldsymbol{v}_n^*\}$ is not a basis for the original lattice L, we use the set $\mathcal{B}^*$ of associated Gram–Schmidt vectors to define a concept that is crucial for the LLL algorithm.

Definition. Let $\mathcal{B} = \{\boldsymbol{v}_1, \boldsymbol{v}_2, \ldots, \boldsymbol{v}_n\}$ be a basis for a lattice L and let $\mathcal{B}^* = \{\boldsymbol{v}_1^*, \boldsymbol{v}_2^*, \ldots, \boldsymbol{v}_n^*\}$ be the associated Gram–Schmidt orthogonal basis as described in Theorem 7.13. The basis $\mathcal{B}$ is said to be *LLL reduced* if it satisfies the following two conditions:

$$\textbf{(Size Condition)} \quad |\mu_{i,j}| = \frac{|\boldsymbol{v}_i \cdot \boldsymbol{v}_j^*|}{\|\boldsymbol{v}_j^*\|^2} \le \frac{1}{2} \qquad \text{for all } 1 \le j < i \le n.$$

$$\textbf{(Lovász Condition)} \quad \|\boldsymbol{v}_i^*\|^2 \ge \left(\frac{3}{4} - \mu_{i,i-1}^2\right) \|\boldsymbol{v}_{i-1}^*\|^2 \quad \text{for all } 1 < i \le n.$$

There are several different ways to state the Lovász condition. For example, it is equivalent to the inequality

$$\|\boldsymbol{v}_i^* + \mu_{i,i-1}\boldsymbol{v}_{i-1}^*\|^2 \geq \frac{3}{4}\|\boldsymbol{v}_{i-1}^*\|^2,$$

and it is also equivalent to the statement that

$$\begin{aligned}&\big\|\text{Projection of } \boldsymbol{v}_i \text{ onto } \operatorname{Span}(\boldsymbol{v}_1,\ldots,\boldsymbol{v}_{i-2})^\perp\big\| \\ &\qquad\qquad \geq \frac{3}{4}\big\|\text{Projection of } \boldsymbol{v}_{i-1} \text{ onto } \operatorname{Span}(\boldsymbol{v}_1,\ldots,\boldsymbol{v}_{i-2})^\perp\big\|.\end{aligned}$$

The fundamental result of Lenstra, Lenstra, and Lovász [77] says that an LLL reduced basis is a good basis and that it is possible to compute an LLL reduced basis in polynomial time. We start by showing that an LLL reduced basis has desirable properties, after which we describe the LLL lattice reduction algorithm.

Theorem 7.69. *Let L be a lattice of dimension n. Any LLL reduced basis $\{\boldsymbol{v}_1, \boldsymbol{v}_2, \ldots, \boldsymbol{v}_n\}$ for L has the following two properties:*

$$\prod_{i=1}^{n} \|\boldsymbol{v}_i\| \leq 2^{n(n-1)/4} \det L, \tag{7.54}$$

$$\|\boldsymbol{v}_j\| \leq 2^{(i-1)/2}\|\boldsymbol{v}_i^*\| \quad \text{for all } 1 \leq j \leq i \leq n. \tag{7.55}$$

Further, the initial vector in an LLL reduced basis satisfies

$$\|\boldsymbol{v}_1\| \leq 2^{(n-1)/4}|\det L|^{1/n} \quad \text{and} \quad \|\boldsymbol{v}_1\| \leq 2^{(n-1)/2} \min_{\mathbf{0}\neq\boldsymbol{v}\in\mathbf{L}} \|\boldsymbol{v}\|. \tag{7.56}$$

Thus an LLL reduced basis solves apprSVP *to within a factor of $2^{(n-1)/2}$.*

Proof. The Lovász condition and the fact that $|\mu_{i,i-1}| \leq \frac{1}{2}$ imply that

$$\|\boldsymbol{v}_i^*\|^2 \geq \left(\frac{3}{4} - \mu_{i,i-1}^2\right)\|\boldsymbol{v}_{i-1}^*\|^2 \geq \frac{1}{2}\|\boldsymbol{v}_{i-1}^*\|^2. \tag{7.57}$$

Applying (7.57) repeatedly yields the useful estimate

$$\|\boldsymbol{v}_j^*\|^2 \leq 2^{i-j}\|\boldsymbol{v}_i^*\|^2. \tag{7.58}$$

We now compute

$$\begin{aligned}\|\boldsymbol{v}_i\|^2 &= \left\|\boldsymbol{v}_i^* + \sum_{j=1}^{i-1}\mu_{i,j}\boldsymbol{v}_j^*\right\|^2 && \text{from (7.53)},\\ &= \|\boldsymbol{v}_i^*\|^2 + \sum_{j=1}^{i-1}\mu_{i,j}^2\|\boldsymbol{v}_j^*\|^2 && \text{since } \boldsymbol{v}_1^*,\ldots,\boldsymbol{v}_n^* \text{ are orthogonal},\end{aligned}$$

$$
\begin{aligned}
&\le \|v_i^*\|^2 + \sum_{j=1}^{i-1} \frac{1}{4}\|v_j^*\|^2 && \text{since } |\mu_{i,j}| \le \frac{1}{2},\\
&\le \|v_i^*\|^2 + \sum_{j=1}^{i-1} 2^{i-j-2}\|v_i^*\|^2 && \text{from (7.58)},\\
&= \frac{1+2^{i-1}}{2}\|v_i^*\|^2\\
&\le 2^{i-1}\|v_i^*\|^2 && \text{since } 1 \le 2^{i-1} \text{ for all } i \ge 1. \qquad (7.59)
\end{aligned}
$$

Multiplying (7.59) by itself for $1 \le i \le n$ yields

$$\prod_{i=1}^{n} \|v_i\|^2 \le \prod_{i=1}^{n} 2^{i-1}\|v_i^*\|^2 = 2^{n(n-1)/2} \prod_{i=1}^{n} \|v_i^*\|^2 = 2^{n(n-1)/2}(\det L)^2,$$

where for the last equality we have used Proposition 7.68. Taking square roots completes the proof of (7.54).

Next, for any $j \le i$, we use (7.59) (with $i = j$) and (7.58) to estimate

$$\|v_j\|^2 \le 2^{j-1}\|v_j^*\|^2 \le 2^{j-1} \cdot 2^{i-j}\|v_i^*\|^2 = 2^{i-1}\|v_i^*\|^2.$$

Taking square roots gives (7.55).

Now we set $j = 1$ in (7.55), multiply over $1 \le i \le n$, and use Proposition 7.68 to obtain

$$\|v_1\|^n \le \prod_{i=1}^{n} 2^{(i-1)/2}\|v_i^*\| = 2^{n(n-1)/4} \prod_{i=1}^{n} \|v_i^*\| = 2^{n(n-1)/4} \det L.$$

Taking nth roots gives the first estimate in (7.56).

To prove the second estimate, let $v \in L$ be a nonzero lattice vector and write

$$v = \sum_{j=1}^{i} a_j v_j = \sum_{j=1}^{i} b_j v_j^*$$

with $a_i \ne 0$. Note that $a_1, \ldots, a_i$ are integers, while $b_i, \ldots, b_i$ are real numbers. In particular, $|a_i| \ge 1$.

By construction, for any k we know that the vectors $v_1^*, \ldots, v_k^*$ are pairwise orthogonal, and we proved (Theorem 7.13) that they span the same space as the vectors $v_1, \ldots, v_k$. Hence

$$v \cdot v_i^* = a_i v_i \cdot v_i^* = b_i v_i^* \cdot v_i^* \quad \text{and} \quad v_i \cdot v_i^* = v_i^* \cdot v_i^*,$$

from which we conclude that $a_i = b_i$. Therefore $|b_i| = |a_i| \ge 1$, and using this and (7.55) (with $j = 1$) gives the estimate

$$\|v\|^2 = \sum_{j=1}^{i} b_j^* \|v_j^*\|^2 \ge b_i^2\|v_i^*\|^2 \ge \|v_i^*\|^2 \ge 2^{-(i-1)}\|v_1\|^2 \ge 2^{-(n-1)}\|v_1\|^2.$$

Taking square roots gives the second estimate in (7.56). □

Remark 7.70. Before describing the technicalities of the LLL algorithm, we make some brief remarks indicating the general underlying idea. Given a basis $\{\boldsymbol{v}_1, \boldsymbol{v}_2, \ldots, \boldsymbol{v}_n\}$, it is easy to form a new basis that satisfies the Size Condition. Roughly speaking, we do this by subtracting from $\boldsymbol{v}_k$ appropriate integer multiples of the previous vectors $\boldsymbol{v}_1, \ldots, \boldsymbol{v}_{k-1}$ so as to make $\boldsymbol{v}_k$ smaller. In the LLL algorithm, we do this in stages, rather than all at once, and we'll see that the size reduction condition depends on the ordering of the vectors. After doing size reduction, we check to see whether the Lovász condition is satisfied. If it is, then we have a (nearly) optimal ordering of the vectors. If not, then we reorder the vectors and do further size reduction.

For simplicity, and because it is the case that we need, we state and analyze the LLL algorithm for lattices in $\mathbb{Z}^n$. See Exercise 7.54 for the general case.

Theorem 7.71 (LLL Algorithm). *Let $\{\boldsymbol{v}_1, \ldots, \boldsymbol{v}_n\}$ be a basis for a lattice L that is contained in $\mathbb{Z}^n$. The algorithm described in Fig. 7.8 terminates in a finite number of steps and returns an LLL reduced basis for L.*

More precisely, let $B = \max \|\boldsymbol{v}_i\|$. Then the algorithm executes the main k loop (Steps [4–14]) no more than $\mathcal{O}(n^2 \log n + n^2 \log B)$ times. In particular, the LLL algorithm is a polynomial-time algorithm.

Remark 7.72. The problem of efficiently implementing the LLL algorithm presents many challenges. First, size reduction and the Lovász condition use the Gram–Schmidt orthogonalized basis $\boldsymbol{v}_1^*, \ldots, \boldsymbol{v}_n^*$ and the associated projection factors $\mu_{i,j} = \boldsymbol{v}_i \cdot \boldsymbol{v}_j^* / \|\boldsymbol{v}_j^*\|^2$. In an efficient implementation of the LLL algorithm, one should compute these quantities as needed and store them for future use, recomputing only when necessary. We have not addressed this issue in Fig. 7.8, since it is not relevant for understanding the LLL algorithm, nor for proving that it returns an LLL reduced basis in polynomial time. See Exercise 7.50 for a more efficient version of the LLL algorithm.

Another major challenge arises from the fact that if one attempts to perform LLL reduction on an integer lattice using exact values, the intermediate calculations involve enormous numbers. Thus in working with lattices of high dimension, it is generally necessary to use floating point approximations, which leads to problems with round-off errors. We do not have space here to discuss this practical difficulty, but the reader should be aware that it exists.

Remark 7.73. Before embarking on the somewhat technical proof of Theorem 7.71, we discuss the intuition behind the swap step (Step [11]). The swap step is executed when the Lovász condition fails for $\boldsymbol{v}_k$, so

$$\begin{aligned} &\|\text{Projection of } \boldsymbol{v}_k \text{ onto } \operatorname{Span}(\boldsymbol{v}_1, \ldots, \boldsymbol{v}_{k-2})^\perp\| \\ &\qquad < \frac{3}{4}\|\text{Projection of } \boldsymbol{v}_{k-1} \text{ onto } \operatorname{Span}(\boldsymbol{v}_1, \ldots, \boldsymbol{v}_{k-2})^\perp\|. \end{aligned} \tag{7.60}$$

The goal of LLL is to produce a list of short vectors in increasing order of length. For each $1 \le \ell \le n$, let L_ℓ denote the lattice spanned by $\boldsymbol{v}_1, \ldots, \boldsymbol{v}_\ell$.

Input a basis $\{\boldsymbol{v}_1, \ldots, \boldsymbol{v}_n\}$ for a lattice L
Set $k = 2$
Set $\boldsymbol{v}_1^* = \boldsymbol{v}_1$
Loop while $k \le n$
  Loop Down $j = k-1, k-2, \ldots, 2, 1$
    Set $\boldsymbol{v}_k = \boldsymbol{v}_k - \lfloor \mu_{k,j} \rceil \boldsymbol{v}_j$ [Size Reduction]
  End j Loop
  If $\|\boldsymbol{v}_k^*\|^2 \ge \left(\frac{3}{4} - \mu_{k,k-1}^2\right) \|\boldsymbol{v}_{k-1}^*\|^2$ [Lovász Condition]
    Set $k = k + 1$
  Else
    Swap $\boldsymbol{v}_{k-1}$ and $\boldsymbol{v}_k$ [Swap Step]
    Set $k = \max(k-1, 2)$
  End If
End k Loop
Return LLL reduced basis $\{\boldsymbol{v}_1, \ldots, \boldsymbol{v}_n\}$

Note: At each step, $\boldsymbol{v}_1^*, \ldots, \boldsymbol{v}_k^*$ is the orthogonal set of vectors obtained by applying Gram–Schmidt (Theorem 7.13) to the current values of $\boldsymbol{v}_1, \ldots, \boldsymbol{v}_k$, and $\mu_{i,j}$ is the associated quantity $(\boldsymbol{v}_i \cdot \boldsymbol{v}_j^*)/\|\boldsymbol{v}_j^*\|^2$.

Figure 7.8: The LLL lattice reduction algorithm

Note that as LLL progresses, the sublattices L_ℓ change due to the swap step; only L_n remains the same, since it is the entire lattice. What LLL attempts to do is to find an ordering of the basis vectors (combined with size reductions whenever possible) that minimizes the determinants $\det(L_\ell)$, i.e., LLL attempts to minimize the volumes of the fundamental domains of the sublattices $L_1, \ldots, L_n$.

If the number 3/4 in (7.60) is replaced by the number 1, then the LLL algorithm does precisely this; it swaps $\boldsymbol{v}_k$ and $\boldsymbol{v}_{k-1}$ whenever doing so reduces the value of $\det L_{k-1}$. Unfortunately, if we use 1 instead of 3/4, then it is an open problem whether the LLL algorithm terminates in polynomial time.

If we use 3/4, or any other constant strictly less than 1, then LLL runs in polynomial time, but we may miss an opportunity to reduce the size of a determinant by passing up a swap. For example, in the very first step, we swap only if $\|\boldsymbol{v}_2\| < \frac{3}{4}\|\boldsymbol{v}_1\|$, while we could reduce the determinant by swapping whenever $\|\boldsymbol{v}_2\| < \|\boldsymbol{v}_1\|$. In practice, one often takes a constant larger than 3/4, but less than 1, in the Lovász condition. (See Exercise 7.51.)

Note that an immediate effect of swapping at stage k is (usually) to make the new value of $\mu_{k,k-1}$ larger. This generally allows us to size reduce the new $\boldsymbol{v}_k$ using the new $\boldsymbol{v}_{k-1}$, so swapping results in additional size reduction among the basis vectors, making them more orthogonal.

Proof (sketch) of Theorem 7.71. For simplicity, and because it is the case that we need, we will assume that $L \subset \mathbb{Z}^n$ is a lattice whose vectors have integral coordinates.

It is clear that if the LLL algorithm terminates, then it terminates with an LLL reduced basis, since the j-loop (Steps [5–7]) ensures that the basis satisfies the size condition, and the fact that $k = n + 1$ on termination means that every vector in the basis has passed the Lovász condition test in Step [8].

However, it is not clear that the algorithm actually terminates, because the k-increment in Step [9] is offset by the k-decrement in Step [12]. What we will do is show that Step [12] is executed only a finite number of times. Since either Step [9] or Step [12] is executed on each iteration of the k-loop, this ensures that k eventually becomes larger than n and the algorithm terminates.

Let $\boldsymbol{v}_1, \ldots, \boldsymbol{v}_n$ be a basis of L and let $\boldsymbol{v}_1^*, \ldots, \boldsymbol{v}_n^*$ be the associated Gram–Schmidt orthogonalized basis from Theorem 7.13. For each $\ell = 1, 2, \ldots, n$, we let

$$L_\ell = \text{lattice spanned by } \boldsymbol{v}_1, \ldots, \boldsymbol{v}_\ell,$$

and we define quantities

$$d_\ell = \prod_{i=1}^{\ell} \|\boldsymbol{v}_i^*\|^2 \quad \text{and} \quad D = \prod_{\ell=1}^{n} d_\ell = \prod_{i=1}^{n} \|\boldsymbol{v}_i^*\|^{2(n+1-i)}.$$

Using an argument similar to the proof of Theorem 7.68, one can show that $\det(L_\ell)^2 = d_\ell$; see Exercise 7.14(b,d).

During the LLL algorithm, the value of D changes only when we execute the swap step (Step [11]). More precisely, when [11] is executed, the only d_ℓ that changes is d_{k-1}, since if $\ell < k-1$, then d_ℓ involves neither $\boldsymbol{v}_{k-1}^*$ nor $\boldsymbol{v}_k^*$, while if $\ell \geq k$, then the product defining d_ℓ includes both $\boldsymbol{v}_{k-1}^*$ and $\boldsymbol{v}_k^*$, so the product doesn't change if we swap them.

We can estimate the change in d_{k-1} by noting that when [11] is executed, the Lovász condition in Step [8] is false, so we have

$$\|\boldsymbol{v}_k^*\|^2 < \left(\frac{3}{4} - \mu_{k,k-1}^2\right) \|\boldsymbol{v}_{k-1}^*\|^2 \leq \frac{3}{4} \|\boldsymbol{v}_{k-1}^*\|^2.$$

Hence the effect of swapping $\boldsymbol{v}_k^*$ and $\boldsymbol{v}_{k-1}^*$ in Step [11] is to change the value of d_{k-1} as follows:

$$\begin{aligned} d_{k-1}^{\text{new}} &= \|\boldsymbol{v}_1^*\|^2 \cdot \|\boldsymbol{v}_2^*\|^2 \cdots \|\boldsymbol{v}_{k-2}^*\|^2 \cdot \|\boldsymbol{v}_k^*\|^2 \\ &= \|\boldsymbol{v}_1^*\|^2 \cdot \|\boldsymbol{v}_2^*\|^2 \cdots \|\boldsymbol{v}_{k-2}^*\|^2 \cdot \|\boldsymbol{v}_{k-1}^*\|^2 \cdot \frac{\|\boldsymbol{v}_k^*\|^2}{\|\boldsymbol{v}_{k-1}^*\|^2} \\ &= d_{k-1}^{\text{old}} \cdot \frac{\|\boldsymbol{v}_k^*\|^2}{\|\boldsymbol{v}_{k-1}^*\|^2} \leq \frac{3}{4} d_{k-1}^{\text{old}}. \end{aligned}$$

Hence if the swap step [11] is executed N times, then the value of D is reduced by a factor of at least $(3/4)^N$, since each swap reduces the value of some d_ℓ by at least a factor of 3/4 and D is the product of all of the d_ℓ's.

Since we have assumed that the lattice L is contained in $\mathbb{Z}^n$, the basis vectors $\boldsymbol{v}_1, \ldots, \boldsymbol{v}_\ell$ of L_ℓ have integer coordinates. It follows from the definition of d_ℓ and Exercise 7.14(d) that

$$d_\ell = \prod_{i=1}^{\ell} \|\boldsymbol{v}_i^*\|^2 = \det\Bigl(\bigl(\boldsymbol{v}_i \cdot \boldsymbol{v}_j\bigr)_{1\le i,j\le \ell}\Bigr),$$

which shows d_ℓ is a positive *integer*. Hence

$$D = \prod_{\ell=1}^{n} d_\ell \ge 1. \tag{7.61}$$

Hence D is bounded away from 0 by a constant depending only on the dimension of the lattice L, so it can be multiplied by 3/4 only a finite number of times. This proves that the LLL algorithm terminates.

In order to give an upper bound on the running time, we do some further estimations. Let D_{init} denote the initial value of D for the original basis, let D_{final} denote the value of D for the basis when the LLL algorithm terminates, and as above, let N denote the number of times that the swap step (Step [11]) is executed. (Note that the k loop is executed at most $2N + n$ times, so it suffices to find a bound for N.) The lower bound for D is valid for every basis produced during the execution of the algorithm, so by our earlier results we know that

$$1 \le D_{\text{final}} \le (3/4)^N D_{\text{init}}.$$

Taking logarithms yields (note that $\log(3/4) < 1$)

$$N = \mathcal{O}(\log D_{\text{init}}).$$

To complete the proof, we need to estimate the size of D_{init}. But this is easy, since by the Gram–Schmidt construction we certainly have $\|\boldsymbol{v}_i^*\| \le \|\boldsymbol{v}_i\|$, so

$$D_{\text{init}} = \prod_{i=1}^{n} \|\boldsymbol{v}_i^*\|^{n+1-i} \le \prod_{i=1}^{n} \|\boldsymbol{v}_i\|^{n+1-i} \le \bigl(\max_{1\le i\le n} \|\boldsymbol{v}_i\|\bigr)^{2(1+2+\cdots+n)} = B^{n^2+n}.$$

Hence $\log D_{\text{init}} = \mathcal{O}(n^2 \log B)$. □

Remark 7.74. Rather than counting the number of times that the main loop is executed, we might instead count the number of basic arithmetic operations required by LLL. This means counting how many times the internal j-loop is executed and also how many times we perform operations on the coordinates of a vector. For example, adding two vectors or multiplying a vector by a constant is n basic operations. Counted in this way, it is proven in [77] that the LLL algorithm (if efficiently implemented) terminates after no more than $\mathcal{O}\bigl(n^6(\log B)^3\bigr)$ basic operations.

Example 7.75. We illustrate the LLL algorithm on the 6-dimensional lattice L with (ordered) basis given by the rows of the matrix

$$M = \begin{pmatrix} 19 & 2 & 32 & 46 & 3 & 33 \\ 15 & 42 & 11 & 0 & 3 & 24 \\ 43 & 15 & 0 & 24 & 4 & 16 \\ 20 & 44 & 44 & 0 & 18 & 15 \\ 0 & 48 & 35 & 16 & 31 & 31 \\ 48 & 33 & 32 & 9 & 1 & 29 \end{pmatrix}.$$

The smallest vector in this basis is $\|\boldsymbol{v}_2\| = 51.913$.

The output from LLL is the basis consisting of the rows of the matrix

$$M^{LLL} = \begin{pmatrix} 7 & -12 & -8 & 4 & 19 & 9 \\ -20 & 4 & -9 & 16 & 13 & 16 \\ 5 & 2 & 33 & 0 & 15 & -9 \\ -6 & -7 & -20 & -21 & 8 & -12 \\ -10 & -24 & 21 & -15 & -6 & -11 \\ 7 & 4 & -9 & -11 & 1 & 31 \end{pmatrix}.$$

We check that both matrices have the same determinant,

$$\det(M) = \det(M^{LLL}) = \pm 777406251.$$

Further, as expected, the LLL reduced matrix has a much better (i.e., larger) Hadamard ratio than the original matrix,

$$\mathcal{H}(M) = 0.46908 \quad \text{and} \quad \mathcal{H}(M^{LLL}) = 0.88824,$$

so the vectors in the LLL basis are more orthogonal. (The Hadamard ratio is defined in Remark 7.27.) The smallest vector in the LLL reduced basis is $\|\boldsymbol{v}_1\| = 26.739$, which is a significant improvement over the original basis. This may be compared with the Gaussian expected shortest length (Remark 7.32) of $\sigma(L) = (3! \det L)^{1/3}/\sqrt{\pi} = 23.062$.

The LLL algorithm executed 19 swap steps (Step [11] in Fig. 7.8). The sequence of k values from start to finish was

$$2, 2, 3, 2, 3, 4, 3, 2, 2, 3, 4, 5, 4, 3, 2, 3, 4, 5, 4, 3, 4, 5, 6, 5,$$
$$4, 3, 4, 5, 6, 5, 4, 3, 2, 2, 3, 2, 3, 4, 5, 6.$$

Notice how the algorithm almost finished twice (it got to $k = 6$) before finally terminating the third time. This illustrates how the value of k moves up and down as the algorithm proceeds.

We next reverse the order of the rows of M and apply LLL. Then LLL executes only 11 swap steps and gives the basis

$$M^{LLL} = \begin{pmatrix} -7 & 12 & 8 & -4 & -19 & -9 \\ 20 & -4 & 9 & -16 & -13 & -16 \\ -28 & 11 & 12 & -9 & 17 & -14 \\ -6 & -7 & -20 & -21 & 8 & -12 \\ -7 & -4 & 9 & 11 & -1 & -31 \\ 10 & 24 & -21 & 15 & 6 & 11 \end{pmatrix}.$$

We find the same smallest vector, but the Hadamard ratio $\mathcal{H}(M^{LLL}) = 0.878973$ is a bit lower, so the basis isn't quite as good. This illustrates the fact that the output from LLL is dependent on the order of the basis vectors.

We also ran LLL with the original matrix, but using 0.99 instead of $\frac{3}{4}$ in the Lovász Step [8]. The algorithm did 22 swap steps, which is more than the 19 swap steps required using $\frac{3}{4}$. This is not surprising, since increasing the constant makes the Lovász condition more stringent, so it is harder for the algorithm to get to the k-increment step. Using 0.99, the LLL algorithm returns the basis

$$M^{LLL} = \begin{pmatrix} -7 & 12 & 8 & -4 & -19 & -9 \\ -20 & 4 & -9 & 16 & 13 & 16 \\ 6 & 7 & 20 & 21 & -8 & 12 \\ -28 & 11 & 12 & -9 & 17 & -14 \\ -7 & -4 & 9 & 11 & -1 & -31 \\ -10 & -24 & 21 & -15 & -6 & -11 \end{pmatrix}.$$

Again we get the same smallest vector, but now the basis has $\mathcal{H}(M^{LLL}) = 0.87897$. This is actually slightly worse than the basis obtained using $\frac{3}{4}$, again illustrating the unpredictable dependence of the LLL algorithm's output on its parameters.

7.13.3 Using LLL to Solve apprCVP

We explained in Sect. 7.6 that if a lattice L has an orthogonal basis, then it is very easy to solve both SVP and CVP. The LLL algorithm does not return an orthogonal basis, but it does produce a basis in which the basis vectors are *quasi-orthogonal*, i.e., they are reasonably orthogonal to one another. Thus we can combine the LLL algorithm (Fig. 7.8) with Babai's algorithm (Theorem 7.34) to form an algorithm that solves apprCVP.

Theorem 7.76 (LLL apprCVP Algorithm). *There is a constant C such that for any lattice L of dimension n given by a basis $\boldsymbol{v}_1, \ldots, \boldsymbol{v}_n$, the following algorithm solves* apprCVP *to within a factor of C^n.*

Apply LLL to $\boldsymbol{v}_1, \ldots, \boldsymbol{v}_n$ to find an LLL reduced basis. Apply Babai's algorithm using the LLL reduced basis.

Proof. We leave the proof for the reader; see Exercise 7.52. □

Remark 7.77. In [8], Babai suggested two ways to use LLL as part of an apprCVP algorithm. The first method uses the closest vertex algorithm that we described in Theorem 7.34. The second method uses the closest plane algorithm. Combining the closest plane method with an LLL reduced basis tends to give a better result than using the closest vertex method. See Exercise 7.53 for further details.

7.13.4 Generalizations of LLL

There have been many improvements to and generalizations of the LLL algorithm. Most of these methods involve trading increased running time for improved output. We briefly describe two of these improvements in order to give the reader some idea of how they work and the trade-offs involved. For further reading, see [71, 115, 116, 117, 118, 119].

The first variant of LLL is called the *deep insertion method.* In standard LLL, the swap step involves switching $\boldsymbol{v}_k$ and $\boldsymbol{v}_{k-1}$, which then usually allows some further size reduction of the new $\boldsymbol{v}_k$. In the deep insertion method, one instead inserts $\boldsymbol{v}_k$ between $\boldsymbol{v}_{i-1}$ and $\boldsymbol{v}_i$, where i is chosen to allow a large amount of size reduction. In the worst case, the resulting algorithm may no longer terminate in polynomial time, but in practice, when run on most lattices, LLL with deep insertions runs quite rapidly and often returns a significantly better basis than basic LLL.

The second variant of LLL is based on the notion of a Korkin–Zolotarev reduced basis. For any list of vectors $\boldsymbol{v}_1, \boldsymbol{v}_2, \ldots$ and any $i \geq 1$, let $\boldsymbol{v}_1^*, \boldsymbol{v}_2^*, \ldots$ denote the associated Gram–Schmidt orthogonalized vectors and define a map

$$\pi : L \longrightarrow \mathbb{R}^n, \qquad \pi_i(\boldsymbol{v}) = \boldsymbol{v} - \sum_{j=1}^{i} \frac{\boldsymbol{v} \cdot \boldsymbol{v}_j^*}{\|\boldsymbol{v}_j^*\|^2} \boldsymbol{v}_j^*.$$

(We also define π_0 to be the identity map, $\pi_0(\boldsymbol{v}) = \boldsymbol{v}$.) Geometrically, we may describe π_i as the projection map

$$\pi_i : L \longrightarrow \operatorname{Span}(\boldsymbol{v}_1, \ldots, \boldsymbol{v}_i)^\perp \subset \mathbb{R}^n$$

from L onto the orthogonal complement of the space spanned by $\boldsymbol{v}_1, \ldots, \boldsymbol{v}_i$.

Definition. Let L be a lattice. A basis $\boldsymbol{v}_1, \ldots, \boldsymbol{v}_n$ for L is called *Korkin–Zolotarev (KZ) reduced* if it satisfies the following three conditions:

1. $\boldsymbol{v}_1$ is a shortest nonzero vector in L.
2. For $i = 2, 3, \ldots, n$, the vector $\boldsymbol{v}_i$ is chosen such that $\pi_{i-1}(\boldsymbol{v}_i)$ is the shortest nonzero vector in $\pi_{i-1}(L)$.
3. For all $1 \leq i < j \leq n$, we have $\left|\pi_{i-1}(\boldsymbol{v}_i) \cdot \pi_{i-1}(\boldsymbol{v}_j)\right| \leq \frac{1}{2}\left\|\pi_{i-1}(\boldsymbol{v}_i)\right\|^2$.

A KZ-reduced basis is generally much better than an LLL-reduced basis. In particular, the first vector in a KZ-reduced basis is always a solution to SVP. Not surprisingly, the fastest known methods to find a KZ-reduced basis take time that is exponential in the dimension.

The *block Korkin–Zolotarev* variant of the LLL algorithm, which is abbreviated BKZ-LLL, replaces the swap step in the standard LLL algorithm by a block reduction step. One way to view the "swap and size reduction" process in LLL is Gaussian lattice reduction on the 2-dimensional lattice spanned by $\boldsymbol{v}_{k-1}$ and $\boldsymbol{v}_k$. In BKZ-LLL, one works instead with a block of vectors of length β, say

$$\boldsymbol{v}_k, \boldsymbol{v}_{k+1}, \ldots, \boldsymbol{v}_{k+\beta-1},$$

and one replaces the vectors in this block with a KZ-reduced basis spanning the same sublattice. If β is large, there is an obvious disadvantage in that it takes a long time to compute a KZ-reduced basis. Compensating for this extra time is the fact that the eventual output of the algorithm is improved, both in theory and in practice.

Theorem 7.78. *If the BKZ-LLL algorithm is run on a lattice L of dimension n using blocks of size β, then the algorithm is guaranteed to terminate in no more than $O(\beta^{c\beta}n^d)$ steps, where c and d are small constants. Further, the smallest vector $\boldsymbol{v}_1$ found by the algorithm is guaranteed to satisfy*

$$\|\boldsymbol{v}_1\| \le \left(\frac{\beta}{\pi e}\right)^{\frac{n-1}{\beta-1}} \min_{\mathbf{0}\ne \boldsymbol{v}\in \mathbf{L}} \|\boldsymbol{v}\|.$$

Remark 7.79. Theorem 7.78 says that BKZ-LLL solves apprSVP to within a factor of approximately $\beta^{n/\beta}$. This may be compared with standard LLL, which solves apprSVP to within a factor of approximately $2^{n/2}$. As β increases, the accuracy of BKZ-LLL increases, at the cost of increased running time. However, if we want to solve apprSVP to within, say, $\mathcal{O}(n^\delta)$ for some fixed exponent δ and large dimension n, then we need to take $\beta \approx n/\delta$, so the running time of BKZ-LLL becomes exponential in n. And although these are just worst-case running time estimates, experimental evidence also leads to the conclusion that using BKZ-LLL to solve apprSVP to within $\mathcal{O}(n^\delta)$ requires a block size that grows linearly with n, and hence has a running time that grows exponentially in n.

7.14 Applications of LLL to Cryptanalysis

The LLL algorithm has many applications to cryptanalysis, ranging from attacks on knapsack public key cryptosystems to more recent analysis of lattice-based cryptosystems such as Ajtai–Dwork, GGH, and NTRU. There are also lattice reduction attacks on RSA in certain situations, see for example [19, 18, 32, 33, 58]. Finally, we want to stress that LLL and its generalizations have a wide variety of applications in pure and applied mathematics outside of their uses in cryptography.

In this section we illustrate the use of LLL in the cryptanalysis of the four cryptosystems (congruential, knapsack, GGH, NTRU) described earlier in this chapter. We note that LLL has no trouble breaking the examples in this section because the dimensions that we use are so small. In practice, secure instances of these cryptosystems require lattices of dimension 500–1000, which, except for NTRUEncrypt, lead to impractical key lengths.

7.14.1 Congruential Cryptosystems

Recall the congruential cipher described in Sect. 7.1. Alice chooses a modulus q and two small secret integers f and g, and her public key is the integer $h \equiv f^{-1}g \pmod{q}$. Eve knows the public values of q and h, and she wants to recover the private key f. One way for Eve to find the private key is to look for small vectors in the lattice L generated by

$$\boldsymbol{v}_1 = (1, h) \quad \text{and} \quad \boldsymbol{v}_2 = (0, q),$$

since as we saw, the vector (f, g) is in L, and given the size constraints on f and g, it is likely to be the shortest nonzero vector in L.

We illustrate by breaking Example 7.1. In that example,

$$q = 122430513841 \quad \text{and} \quad h = 39245579300.$$

We apply Gaussian lattice reduction (Proposition 7.66) to the lattice generated by

$$(1, 39245579300) \quad \text{and} \quad (0, 122430513841).$$

The algorithm takes 11 iterations to find the short basis

$$(-231231, -195698) \quad \text{and} \quad (-368222, 217835).$$

Up to an irrelevant change of sign, this gives Alice's private key $f = 231231$ and $g = 195698$.

7.14.2 Applying LLL to Knapsacks

In Sect. 7.2 we described how to reformulate a knapsack (subset-sum) problem described by $\boldsymbol{M} = (m_1, \ldots, m_n)$ and S as a lattice problem using the lattice $L_{\boldsymbol{M},S}$ with basis given by the rows of the matrix (7.4) on page 383. We further explained in Example 7.33 why the target vector $\boldsymbol{t} \in L_{\boldsymbol{M},S}$, which has length $\|\boldsymbol{t}\| = \sqrt{n}$, is probably about half the size of all other nonzero vectors in $L_{\boldsymbol{M},S}$.

We illustrate the use of the LLL algorithm to solve the knapsack problem

$$\boldsymbol{M} = (89, 243, 212, 150, 245) \quad \text{and} \quad S = 546$$

considered in Example 7.7. We apply LLL to the lattice generated by the rows of the matrix

$$A_{\boldsymbol{M},S} = \begin{pmatrix} 2 & 0 & 0 & 0 & 0 & 89 \\ 0 & 2 & 0 & 0 & 0 & 243 \\ 0 & 0 & 2 & 0 & 0 & 212 \\ 0 & 0 & 0 & 2 & 0 & 150 \\ 0 & 0 & 0 & 0 & 2 & 245 \\ 1 & 1 & 1 & 1 & 1 & 546 \end{pmatrix}.$$

LLL performs 21 swaps and returns the reduced basis

$$\begin{pmatrix} -1 & 1 & -1 & 1 & -1 & 0 \\ 1 & -1 & -1 & 1 & -1 & -1 \\ -1 & -1 & -1 & 1 & 1 & 2 \\ 1 & -1 & -1 & -1 & -1 & 2 \\ -2 & -2 & 4 & 0 & -2 & 0 \\ -6 & -4 & -6 & -6 & 0 & -3 \end{pmatrix}.$$

We write the short vector

$$(-1, 1, -1, 1, -1, 0)$$

in the top row as a linear combination of the original basis vectors given by the rows of the matrix $A_{M,S}$,

$$(-1, 1, -1, 1, -1, 0) = (-1, 0, -1, 0, -1, 1)A_{M,S}.$$

The vector $(-1, 0, -1, 0, -1, 1)$ gives the solution to the knapsack problem,

$$-89 - 212 - 245 + 546 = 0.$$

Remark 7.80. When using LLL to solve subset-sum problems, it is often helpful to multiply $m_1, \ldots, m_n, S$ by a large constant C. This has the effect of multiplying the last column of the matrix (7.4) by C, so the determinant is multiplied by C and the Gaussian expected shortest vector is multiplied by $C^{1/(n+1)}$. The target vector $\boldsymbol{t}$ still has length $\sqrt{n}$, so if C is large, the target vector becomes much smaller than the likely next shortest vector. This tends to make it easier for LLL to find $\boldsymbol{t}$.

7.14.3 Applying LLL to GGH

We apply LLL to Example 7.36, in which the Alice's public lattice L is generated by the rows $\boldsymbol{w}_1, \boldsymbol{w}_2, \boldsymbol{w}_3$ of the matrix

$$\begin{pmatrix} -4179163 & -1882253 & 583183 \\ -3184353 & -1434201 & 444361 \\ -5277320 & -2376852 & 736426 \end{pmatrix}$$

and Bob's encrypted message is

$$\boldsymbol{e} = (-79081427, -35617462, 11035473).$$

Eve wants to find a vector in L that is close to $\boldsymbol{e}$. She first applies LLL (Theorem 7.71) to the lattice L and finds the quasi-orthogonal basis

$$\begin{pmatrix} 36 & -30 & -86 \\ 61 & 11 & 67 \\ -10 & 102 & -40 \end{pmatrix}.$$

This basis has Hadamard ratio $\mathcal{H} = 0.956083$, which is even better than Alice's good basis. Eve next applies Babai's algorithm (Theorem 7.34) to find a lattice vector

$$\boldsymbol{v} = (-79081423, -35617459, 11035471)$$

that is very close to $\boldsymbol{e}$. Finally she writes $\boldsymbol{v}$ in terms of the original lattice vectors,

$$\boldsymbol{v} = 86\boldsymbol{w}_1 - 35\boldsymbol{w}_2 - 32\boldsymbol{w}_3,$$

which retrieves Bob's plaintext $\boldsymbol{m} = (86, -35, -32)$.

7.14.4 Applying LLL to NTRU

We apply LLL to the NTRU cryptosystem described in Example 7.53. Thus $N = 7$, $q = 41$, and the public key is the polynomial

$$\boldsymbol{h}(x) = 30 + 26x + 8x^2 + 38x^3 + 2x^4 + 40x^5 + 20x^6.$$

As explained in Sect. 7.11, the associated NTRU lattice is generated by the rows of the matrix

$$M_{\boldsymbol{h}}^{\mathrm{NTRU}} = \begin{pmatrix} 1&0&0&0&0&0&0&30&26&8&38&2&40&20\\ 0&1&0&0&0&0&0&20&30&26&8&38&2&40\\ 0&0&1&0&0&0&0&40&20&30&26&8&38&2\\ 0&0&0&1&0&0&0&2&40&20&30&26&8&38\\ 0&0&0&0&1&0&0&38&2&40&20&30&26&8\\ 0&0&0&0&0&1&0&8&38&2&40&20&30&26\\ 0&0&0&0&0&0&1&26&8&38&2&40&20&30\\ 0&0&0&0&0&0&0&41&0&0&0&0&0&0\\ 0&0&0&0&0&0&0&0&41&0&0&0&0&0\\ 0&0&0&0&0&0&0&0&0&41&0&0&0&0\\ 0&0&0&0&0&0&0&0&0&0&41&0&0&0\\ 0&0&0&0&0&0&0&0&0&0&0&41&0&0\\ 0&0&0&0&0&0&0&0&0&0&0&0&41&0\\ 0&0&0&0&0&0&0&0&0&0&0&0&0&41 \end{pmatrix}.$$

Eve applies LLL reduction to $M_{\boldsymbol{h}}^{\mathrm{NTRU}}$. The algorithm performs 96 swap steps and returns the LLL reduced matrix

$$M_{\mathrm{red}}^{\mathrm{NTRU}} = \begin{pmatrix} 1&0&-1&1&0&-1&-1&-1&0&-1&0&1&1&0\\ 0&1&1&-1&0&1&-1&-1&-1&0&1&0&1&0\\ -1&1&0&-1&-1&1&0&-1&0&1&1&0&-1&0\\ -1&-1&1&0&-1&1&0&1&0&-1&0&-1&0&1\\ -1&1&0&-1&1&0&-1&0&-1&0&-1&0&1&1\\ -1&-1&-1&-1&-1&-1&-1&0&0&0&0&0&0&0\\ 0&1&0&1&0&-1&1&-1&-1&0&0&2&0&0\\ -8&-1&0&9&0&-1&0&-4&2&6&0&-4&7&-7\\ 8&1&0&0&-8&-1&2&0&-5&8&-7&-3&1&6\\ 0&-9&-2&1&9&-1&0&-6&-3&2&5&0&-5&7\\ 0&8&0&-9&-1&-8&8&2&7&-11&3&-5&2&2\\ 1&0&0&9&2&-1&-9&5&-7&6&3&-2&-5&0\\ -2&1&9&-1&0&0&-9&2&5&0&-5&7&-6&-3\\ 3&2&3&3&-6&2&-6&11&6&8&0&9&5&2 \end{pmatrix}.$$

We can compare the relative quasi-orthogonality of the original and the reduced bases by computing the Hadamard ratios,

$$\mathcal{H}(M_{\boldsymbol{h}}^{\text{NTRU}}) = 0.1184 \quad \text{and} \quad \mathcal{H}(M_{\text{red}}^{\text{NTRU}}) = 0.8574.$$

The smallest vector in the reduced basis is the top row of the reduced matrix,

$$(1, 0, -1, 1, 0, -1, -1, -1, 0, -1, 0, 1, 1, 0).$$

Splitting this vector into two pieces gives polynomials

$$\boldsymbol{f}'(x) = 1 - x^2 + x^3 - x^5 - x^6 \quad \text{and} \quad \boldsymbol{g}'(x) = -1 - x^2 + x^4 + x^5.$$

Note that $\boldsymbol{f}'(x)$ and $\boldsymbol{g}'(x)$ are not the same as Alice's original private key polynomials $\boldsymbol{f}(x)$ and $\boldsymbol{g}(x)$ from Example 7.53. However, they are simple rotations of Alice's key,

$$\boldsymbol{f}'(x) = -x^3 \star \boldsymbol{f}(x) \quad \text{and} \quad \boldsymbol{g}'(x) = -x^3 \star \boldsymbol{g}(x),$$

so Eve can use $\boldsymbol{f}'(x)$ and $\boldsymbol{g}'(x)$ to decrypt messages.

Exercises

Section 7.1. A Congruential Public Key Cryptosystem

7.1. Alice uses the congruential cryptosystem with $q = 918293817$ and private key $(f, g) = (19928, 18643)$.
(a) What is Alice's public key h?
(b) Alice receives the ciphertext $e = 619168806$ from Bob. What is the plaintext?
(c) Bob sends Alice a second message by encrypting the plaintext $m = 10220$ using the random element $r = 19564$. What is the ciphertext that Bob sends to Alice?

Section 7.2. Subset-Sum Problems and Knapsack Cryptosystems

7.2. Use the algorithm described in Proposition 7.5 to solve each of the following subset-sum problems. If the "solution" that you get is not correct, explain what went wrong.
(a) $\boldsymbol{M} = (3, 7, 19, 43, 89, 195), \quad S = 260.$
(b) $\boldsymbol{M} = (5, 11, 25, 61, 125, 261), \quad S = 408.$
(c) $\boldsymbol{M} = (2, 5, 12, 28, 60, 131, 257), \quad S = 334.$
(d) $\boldsymbol{M} = (4, 12, 15, 36, 75, 162), \quad S = 214.$

7.3. Alice's public key for a knapsack cryptosystem is

$$\boldsymbol{M} = (5186, 2779, 5955, 2307, 6599, 6771, 6296, 7306, 4115, 637).$$

Eve intercepts the encrypted message $S = 4398$. She also breaks into Alice's computer and steals Alice's secret multiplier $A = 4392$ and secret modulus $B = 8387$. Use this information to find Alice's superincreasing private sequence $\boldsymbol{r}$ and then decrypt the message.

7.4. Proposition 7.3 gives an algorithm that solves an n-dimensional knapsack problem in $\mathcal{O}(2^{n/2})$ steps, but it requires $\mathcal{O}(2^{n/2})$ storage. Devise an algorithm, similar to Pollard's ρ algorithm (Sect. 5.5), that takes $\mathcal{O}(2^{n/2})$ steps, but requires only $\mathcal{O}(1)$ storage.

Section 7.3. A Brief Review of Vector Spaces

7.5. (a) Let

$$\mathcal{B} = \{(1,3,2),(2,-1,3),(1,0,2)\}, \qquad \mathcal{B}' = \{(-1,0,2),(3,1,-1),(1,0,1)\}.$$

Each of the sets $\mathcal{B}$ and $\mathcal{B}'$ is a basis for $\mathbb{R}^3$. Find the change of basis matrix that transforms $\mathcal{B}'$ into $\mathcal{B}$.

(b) Let $\boldsymbol{v} = (2,3,1)$ and $\boldsymbol{w} = (-1,4,-2)$. Compute the lengths $\|\boldsymbol{v}\|$ and $\|\boldsymbol{w}\|$ and the dot product $\boldsymbol{v} \cdot \boldsymbol{w}$. Compute the angle between $\boldsymbol{v}$ and $\boldsymbol{w}$.

7.6. Use the Gram–Schmidt algorithm (Theorem 7.13) to find an orthogonal basis from the given basis.

(a) $\boldsymbol{v}_1 = (1,3,2)$, $\boldsymbol{v}_2 = (4,1,-2)$, $\boldsymbol{v}_3 = (-2,1,3)$.

(b) $\boldsymbol{v}_1 = (4,1,3,-1)$, $\boldsymbol{v}_2 = (2,1,-3,4)$, $\boldsymbol{v}_3 = (1,0,-2,7)$.

Section 7.4. Lattices: Basic Definitions and Properties

7.7. Let L be the lattice generated by $\{(1,3,-2),(2,1,0),(-1,2,5)\}$. Draw a picture of a fundamental domain for L and find its volume.

7.8. Let $L \subset \mathbb{R}^m$ be an additive subgroup with the property that there is a positive constant $\epsilon > 0$ such that

$$L \cap \{\boldsymbol{w} \in \mathbb{R}^m : \|\boldsymbol{w}\| < \epsilon\} = \{\boldsymbol{0}\}.$$

Prove that L is discrete, and hence is a lattice. (In other words, show that in the definition of discrete subgroup, it suffices to check that (7.8) is true for the single vector $\boldsymbol{v} = \boldsymbol{0}$.)

7.9. Prove that a subset of $\mathbb{R}^m$ is a lattice if and only if it is a discrete additive subgroup.

7.10. This exercise describes a result that you may have seen in your linear algebra course.

Let A be an n-by-n matrix with entries a_{ij}, and for each pair of indices i and j, let A_{ij} denote the $(n-1)$-by-$(n-1)$ matrix obtained by deleting the ith row of A and the jth column of A. Define a new matrix B whose ijth entry b_{ij} is given by the formula

$$b_{ij} = (-1)^{i+j} \det(A_{ji}).$$

(Note that b_{ij} is the determinant of the submatrix A_{ji}, i.e., the indices are reversed.) The matrix B is called the *adjoint of* A.

(a) Prove that

$$AB = BA = \det(A) I_n,$$

where I_n is the n-by-n identity matrix.

(b) Deduce that if $\det(A) \neq 0$, then

$$A^{-1} = \frac{1}{\det(A)} B.$$

(c) Suppose that A has integer entries. Prove that A^{-1} exists and has integer entries if and only if $\det(A) = \pm 1$.
(d) For those who know ring theory from Sect. 2.10 or from some other source, suppose that A has entries in a ring R. Prove that A^{-1} exists and has entries in R if and only if $\det(A)$ is a unit in R.

7.11. Recall from Remark 7.16 that the general linear group $\mathrm{GL}_n(\mathbb{Z})$ is the group of n-by-n matrices with integer coefficients and determinant ± 1. Let A and B be matrices in $\mathrm{GL}_n(\mathbb{Z})$.
(a) Prove that $AB \in \mathrm{GL}_n(\mathbb{Z})$.
(b) Prove that $A^{-1} \in \mathrm{GL}_n(\mathbb{Z})$.
(c) Prove that the n-by-n identity matrix is in $\mathrm{GL}_n(\mathbb{Z})$.
(d) Prove that $\mathrm{GL}_n(\mathbb{Z})$ is a group. (*Hint.* You have already done most of the work in proving (a), (b), and (c). For the associative law, either prove it directly or use the fact that you know that it is true for matrices with real coefficients.)
(e) Is $\mathrm{GL}_n(\mathbb{Z})$ a commutative group?

7.12. Which of the following matrices are in $\mathrm{GL}_n(\mathbb{Z})$? Find the inverses of those matrices that are in $\mathrm{GL}_n(\mathbb{Z})$.

(a) $A_1 = \begin{pmatrix} 3 & 1 \\ 2 & 2 \end{pmatrix}$ (b) $A_2 = \begin{pmatrix} 3 & -2 \\ 2 & -1 \end{pmatrix}$

(c) $A_3 = \begin{pmatrix} 3 & 2 & 2 \\ 2 & 1 & 2 \\ -1 & 3 & 1 \end{pmatrix}$ (d) $A_4 = \begin{pmatrix} -3 & -1 & 2 \\ 1 & -3 & -1 \\ 3 & 0 & -2 \end{pmatrix}$

7.13. Let L be the lattice given by the basis

$$\mathcal{B} = \{(3, 1, -2), (1, -3, 5), (4, 2, 1)\}.$$

Which of the following sets of vectors are also bases for L? For those that are, express the new basis in terms of the basis $\mathcal{B}$, i.e., find the change of basis matrix.
(a) $\mathcal{B}_1 = \{(5, 13, -13), (0, -4, 2), (-7, -13, 18)\}$.
(b) $\mathcal{B}_2 = \{(4, -2, 3), (6, 6, -6), (-2, -4, 7)\}$.

7.14. Let $L \subset \mathbb{R}^m$ be a lattice of dimension n and let $\boldsymbol{v}_1, \ldots, \boldsymbol{v}_n$ be a basis for L. Note that we are allowing n to be smaller than m. The *Gram matrix of* $\boldsymbol{v}_1, \ldots, \boldsymbol{v}_n$ is the matrix

$$\mathrm{Gram}(\boldsymbol{v}_1, \ldots, \boldsymbol{v}_n) = (\boldsymbol{v}_i \cdot \boldsymbol{v}_j)_{1 \le i,j \le n}.$$

(a) Let $F(\boldsymbol{v}_1, \ldots, \boldsymbol{v}_n)$ be the matrix (7.11) described in Proposition (7.20), except that now $F(\boldsymbol{v}_1, \ldots, \boldsymbol{v}_n)$ is an n-by-m matrix, so it need not be square. Prove that

$$\mathrm{Gram}(\boldsymbol{v}_1, \ldots, \boldsymbol{v}_n) = F(\boldsymbol{v}_1, \ldots, \boldsymbol{v}_n) F(\boldsymbol{v}_1, \ldots, \boldsymbol{v}_n)^t,$$

where $F(\boldsymbol{v}_1, \ldots, \boldsymbol{v}_n)^t$ is the transpose matrix, i.e., the matrix with rows and columns interchanged.

(b) Prove that

$$\det\bigl(\mathrm{Gram}(\boldsymbol{v}_1,\ldots,\boldsymbol{v}_n)\bigr) = \det(L)^2, \tag{7.62}$$

where note that $\det(L)$ is the volume of the parallelepiped spanned by any basis for L. (You may find it easier to first do the case $n = m$.)

(c) Let $L \subset \mathbb{R}^4$ be the 3-dimensional lattice with basis

$$\boldsymbol{v}_1 = (1,0,1,-1), \quad \boldsymbol{v}_2 = (1,2,0,4), \quad \boldsymbol{v}_3 = (1,-1,2,1).$$

Compute the Gram matrix of this basis and use it to compute $\det(L)$.

(d) Let $\boldsymbol{v}_1^*,\ldots,\boldsymbol{v}_n^*$ be the Gram–Schmidt orthogonalized vectors (Theorem 7.13) associated to $\boldsymbol{v}_1,\ldots,\boldsymbol{v}_n$. Prove that

$$\det\bigl(\mathrm{Gram}(\boldsymbol{v}_1,\ldots,\boldsymbol{v}_n)\bigr) = \|\boldsymbol{v}_1^*\|^2\|\boldsymbol{v}_2^*\|^2\cdots\|\boldsymbol{v}_n^*\|^2.$$

Section 7.5. The Shortest and Closest Vector Problems

7.15. Let L be a lattice and let $\mathcal{F}$ be a fundamental domain for L. This exercise sketches a proof that

$$\lim_{R\to\infty} \frac{\#\bigl(\mathbb{B}_R(\mathbf{0})\cap L\bigr)}{\mathrm{Vol}\bigl(\mathbb{B}_R(\mathbf{0})\bigr)} = \frac{1}{\mathrm{Vol}(\mathcal{F})}. \tag{7.63}$$

(a) Consider the translations of $\mathcal{F}$ that are entirely contained within $\mathbb{B}_R(\mathbf{0})$, and also those that have nontrivial intersection with $\mathbb{B}_R(\mathbf{0})$. Prove the inclusion of sets

$$\bigcup_{\substack{\boldsymbol{v}\in L\\ \mathcal{F}+\boldsymbol{v}\subset\mathbb{B}_R(\mathbf{0})}} (\mathcal{F}+\boldsymbol{v}) \subset \mathbb{B}_R(\mathbf{0}) \subset \bigcup_{\substack{\boldsymbol{v}\in L\\ (\mathcal{F}+\boldsymbol{v})\cap\mathbb{B}_R(\mathbf{0})\neq\emptyset}} (\mathcal{F}+\boldsymbol{v}).$$

(b) Take volumes in (a) and prove that

$$\begin{aligned}\#\{\boldsymbol{v}\in L : \mathcal{F}+\boldsymbol{v}\subset\mathbb{B}_R(\mathbf{0})\}\cdot\mathrm{Vol}(\mathcal{F})\\ \le \mathrm{Vol}\bigl(\mathbb{B}_R(\mathbf{0})\bigr) \le \#\{\boldsymbol{v}\in L : (\mathcal{F}+\boldsymbol{v})\cap\mathbb{B}_R(\mathbf{0})\neq\emptyset\}\cdot\mathrm{Vol}(\mathcal{F}).\end{aligned}$$

(*Hint.* Proposition 7.18 says that the different translates of $\mathcal{F}$ are disjoint.)

(c) Prove that the number of translates $\mathcal{F}+\boldsymbol{v}$ that intersect $\mathbb{B}_R(\mathbf{0})$ without being entirely contained within $\mathbb{B}_R(\mathbf{0})$ is comparatively small compared to the number of translates $\mathcal{F}_{\boldsymbol{v}}$ that are entirely contained within $\mathbb{B}_R(\mathbf{0})$. (This is the hardest part of the proof.)

(d) Use (b) and (c) to prove that

$$\mathrm{Vol}\bigl(\mathbb{B}_R(\mathbf{0})\bigr) = \#\bigl(\mathbb{B}_R(\mathbf{0})\cap L\bigr)\cdot\mathrm{Vol}(\mathcal{F}) + (\text{smaller term}).$$

Divide by $\mathrm{Vol}\bigl(\mathbb{B}_R(\mathbf{0})\bigr)$ and let $R\to\infty$ to complete the proof of (7.63).

7.16. A lattice L of dimension $n = 251$ has determinant $\det(L) \approx 2^{2251.58}$. With no further information, approximately how large would you expect the shortest nonzero vector to be?

Section 7.6. Babai's Algorithm and Solving CVP with a "Good" Basis

7.17. Let $L \subset \mathbb{R}^2$ be the lattice given by the basis $\boldsymbol{v}_1 = (213,-437)$ and $\boldsymbol{v}_2 = (312,105)$, and let $\boldsymbol{w} = (43127,11349)$.

(a) Use Babai's algorithm to find a vector $\boldsymbol{v} \in L$ that is close to $\boldsymbol{w}$. Compute the distance $\|\boldsymbol{v} - \boldsymbol{w}\|$.
(b) What is the value of the Hadamard ratio $\left(\det(L)/\|\boldsymbol{v}_1\|\|\boldsymbol{v}_2\|\right)^{1/2}$? Is the basis $\{\boldsymbol{v}_1, \boldsymbol{v}_2\}$ a "good" basis?
(c) Show that the vectors $\boldsymbol{v}_1' = (2937, -1555)$ and $\boldsymbol{v}_2' = (11223, -5888)$ are also a basis for L by expressing them as linear combinations of $\boldsymbol{v}_1$ and $\boldsymbol{v}_2$ and checking that the change-of-basis matrix has integer coefficients and determinant ± 1.
(d) Use Babai's algorithm with the basis $\{\boldsymbol{v}_1', \boldsymbol{v}_2'\}$ to find a vector $\boldsymbol{v}' \in L$. Compute the distance $\|\boldsymbol{v}' - \boldsymbol{w}\|$ and compare it to your answer from (a).
(e) Compute the Hadamard ratio using $\boldsymbol{v}_1'$ and $\boldsymbol{v}_2'$. Is $\{\boldsymbol{v}_1', \boldsymbol{v}_2'\}$ a good basis?

Section 7.8. The GGH Public Key Cryptosystem

7.18. Alice uses the GGH cryptosystem with private basis

$$\boldsymbol{v}_1 = (4, 13), \quad \boldsymbol{v}_2 = (-57, -45),$$

and public basis

$$\boldsymbol{w}_1 = (25453, 9091), \quad \boldsymbol{w}_2 = (-16096, -5749).$$

(a) Compute the determinant of Alice's lattice and the Hadamard ratio of the private and public bases.
(b) Bob sends Alice the encrypted message $\boldsymbol{e} = (155340, 55483)$. Use Alice's private basis to decrypt the message and recover the plaintext. Also determine Bob's random perturbation $\boldsymbol{r}$.
(c) Try to decrypt Bob's message using Babai's algorithm with the public basis $\{\boldsymbol{w}_1, \boldsymbol{w}_2\}$. Is the output equal to the plaintext?

7.19. Alice uses the GGH cryptosystem with private basis

$$\boldsymbol{v}_1 = (58, 53, -68), \quad \boldsymbol{v}_2 = (-110, -112, 35), \quad \boldsymbol{v}_3 = (-10, -119, 123)$$

and public basis

$$\begin{aligned} \boldsymbol{w}_1 &= (324850, -1625176, 2734951), \\ \boldsymbol{w}_2 &= (165782, -829409, 1395775), \\ \boldsymbol{w}_3 &= (485054, -2426708, 4083804). \end{aligned}$$

(a) Compute the determinant of Alice's lattice and the Hadamard ratio of the private and public bases.
(b) Bob sends Alice the encrypted message $\boldsymbol{e} = (8930810, -44681748, 75192665)$. Use Alice's private basis to decrypt the message and recover the plaintext. Also determine Bob's random perturbation $\boldsymbol{r}$.
(c) Try to decrypt Bob's message using Babai's algorithm with the public basis $\{\boldsymbol{w}_1, \boldsymbol{w}_2, \boldsymbol{w}_3\}$. Is the output equal to the plaintext?

7.20. Bob uses the GGH cryptosystem to send some messages to Alice.
(a) Suppose that Bob sends the same message $\boldsymbol{m}$ twice, using different random elements $\boldsymbol{r}$ and $\boldsymbol{r}'$. Explain what sort of information Eve can deduce from the ciphertexts $\boldsymbol{e} = \boldsymbol{m}W + \boldsymbol{r}$ and $\boldsymbol{e}' = \boldsymbol{m}W + \boldsymbol{r}'$.

(b) For example, suppose that $n = 5$ and that random permutations are chosen with coordinates in the set $\{-2, -1, 0, 1, 2\}$. This means that there are $5^5 = 3125$ possibilities for $\boldsymbol{r}$. Suppose further that Eve intercepts two ciphertexts

$$\boldsymbol{e} = (-9, -29, -48, 18, 48) \quad \text{and} \quad \boldsymbol{e}' = (-6, -26, -51, 20, 47)$$

having the same plaintext. With this information, how many possibilities are there for $\boldsymbol{r}$?

(c) Suppose that Bob is lazy and uses the same perturbation to send two different messages. Explain what sort of information Eve can deduce from the ciphertexts $\boldsymbol{e} = \boldsymbol{m}W + \boldsymbol{r}$ and $\boldsymbol{e}' = \boldsymbol{m}'W + \boldsymbol{r}$.

7.21. The previous exercise shows the danger of using GGH to send a single message $\boldsymbol{m}$ twice using different values of $\boldsymbol{r}$.

(a) In order to guard against this danger, suppose that Bob generates $\boldsymbol{r}$ by applying a publicly available hash function Hash to $\boldsymbol{m}$, i.e., Bob's encrypted message is

$$\boldsymbol{e} = \boldsymbol{m}W + \mathsf{Hash}(\boldsymbol{m}).$$

(See Sect. 8.1 for a discussion of hash functions.) If Eve guesses that Bob's message might be $\boldsymbol{m}'$, explain why she can check whether her guess is correct.

(b) Explain why the following algorithm eliminates both the problem with repeated messages and the problem described in (a), while still allowing Alice to decrypt Bob's message. Bob chooses an message $\boldsymbol{m}_0$ and a random string $\boldsymbol{r}_0$. He then computes

$$\boldsymbol{m} = (\boldsymbol{m}_0 \ \mathsf{xor}\ \boldsymbol{r}_0) \parallel \boldsymbol{r}_0, \qquad \boldsymbol{r} = \mathsf{Hash}(\boldsymbol{m}), \qquad \boldsymbol{e} = \boldsymbol{m}W + \boldsymbol{r}.$$

(c) In (b), the advantage of constructing $\boldsymbol{m}$ from $\boldsymbol{m}_0 \,\mathsf{xor}\, \boldsymbol{r}_0$ is that none of the bits of the actual plaintext $\boldsymbol{m}_0$ appear unaltered in $\boldsymbol{m}$. In practice, people replace $(\boldsymbol{m}_0 \ \mathsf{xor}\ \boldsymbol{r}_0) \parallel \boldsymbol{r}_0$ with more complicated mixing functions $M(\boldsymbol{m}_0, \boldsymbol{r}_0)$ having the following two properties: (1) M is easily invertible. (2) If even one bit of either $\boldsymbol{m}_0$ or $\boldsymbol{r}_0$ changes, then the value of every bit of $M(\boldsymbol{m}_0, \boldsymbol{r}_0)$ changes in an unpredictable manner. Try to construct a mixing function M having these properties.

Section 7.9. Convolution Polynomial Rings

7.22. Compute (by hand!) the polynomial convolution product $\boldsymbol{c} = \boldsymbol{a} \star \boldsymbol{b}$ using the given value of N.

(a) $N = 3$, $\quad \boldsymbol{a}(x) = -1 + 4x + 5x^2$, $\quad \boldsymbol{b}(x) = -1 - 3x - 2x^2$;

(b) $N = 5$, $\quad \boldsymbol{a}(x) = 2 - x + 3x^3 - 3x^4$, $\quad \boldsymbol{b}(x) = 1 - 3x^2 - 3x^3 - x^4$;

(c) $N = 6$, $\quad \boldsymbol{a}(x) = x + x^2 + x^3$, $\quad \boldsymbol{b}(x) = 1 + x + x^5$;

(d) $N = 10$, $\quad \boldsymbol{a}(x) = x + x^2 + x^3 + x^4 + x^6 + x^7 + x^9$,
$\quad \boldsymbol{b}(x) = x^2 + x^3 + x^6 + x^8$.

7.23. Compute the polynomial convolution product $\boldsymbol{c} = \boldsymbol{a} \star \boldsymbol{b}$ modulo q using the given values of q and N.

(a) $N = 3, \quad q = 7, \quad \boldsymbol{a}(x) = 1 + x, \quad \boldsymbol{b}(x) = -5 + 4x + 2x^2;$

(b) $N = 5, \quad q = 4, \quad \boldsymbol{a}(x) = 2 + 2x - 2x^2 + x^3 - 2x^4,$
$\boldsymbol{b}(x) = -1 + 3x - 3x^2 - 3x^3 - 3x^4;$

(c) $N = 7, \quad q = 3, \quad \boldsymbol{a}(x) = x + x^3, \quad \boldsymbol{b}(x) = x + x^2 + x^4 + x^6;$

(d) $N = 10, \quad q = 2, \quad \boldsymbol{a}(x) = x^2 + x^5 + x^7 + x^8 + x^9,$
$\boldsymbol{b}(x) = 1 + x + x^3 + x^4 + x^5 + x^7 + x^8 + x^9.$

7.24. Let $\boldsymbol{a}(x) \in (\mathbb{Z}/q\mathbb{Z})[x]$, where q is a prime.
(a) Prove that

$$\boldsymbol{a}(1) \equiv 0 \pmod{q} \quad \text{if and only if} \quad (x-1) \mid \boldsymbol{a}(x) \quad \text{in } (\mathbb{Z}/q\mathbb{Z})[x].$$

(b) Suppose that $\boldsymbol{a}(1) \equiv 0 \pmod{q}$. Prove that $\boldsymbol{a}(x)$ is not invertible in R_q.

7.25. Let $N = 5$ and $q = 3$ and consider the two polynomials

$$\boldsymbol{a}(x) = 1 + x^2 + x^3 \in R_3 \quad \text{and} \quad \boldsymbol{b}(x) = 1 + x^2 - x^3 \in R_3.$$

One of these polynomials has an inverse in R_3 and the other does not. Compute the inverse that exists, and explain why the other doesn't exist.

7.26. For each of the following values of N, q, and $\boldsymbol{a}(x)$, either find $\boldsymbol{a}(x)^{-1}$ in R_q or show that the inverse does not exist.
(a) $N = 5$, $q = 11$, and $\boldsymbol{a}(x) = x^4 + 8x + 3$;
(b) $N = 5$, $q = 13$, and $\boldsymbol{a}(x) = x^3 + 2x - 3$.
(c) $N = 7$, $q = 23$, and $\boldsymbol{a}(x) = 20x^6 + 8x^5 + 4x^4 + 15x^3 + 19x^2 + x + 8$.

7.27. This exercise illustrates how to find inverses in

$$R_m = \frac{(\mathbb{Z}/m\mathbb{Z})[x]}{(x^N - 1)}$$

when m is a prime power p^e.
(a) Let $f(x) \in \mathbb{Z}[x]/(X^N - 1)$ be a polynomial, and suppose that we have already found a polynomial $F(x)$ such that

$$f(x) \star F(x) \equiv 1 \pmod{p^i}$$

for some $i \geq 1$. Prove that the polynomial

$$G(x) = F(x) \star (2 - f(x) \star F(x))$$

satisfies

$$f(x) \star G(x) \equiv 1 \pmod{p^{2i}}.$$

(b) Suppose that we know an inverse of $f(x)$ modulo p. Using (a) repeatedly, how many convolution multiplications does it take to compute the inverse of $f(x)$ modulo p^e?

(c) Use the method in (a) to compute the following inverses modulo $m = p^e$, where to ease your task, we have given you the inverse modulo p.

$$\begin{aligned}
&\text{(i)} \quad N = 5, \quad m = 2^4, \quad && f(x) = 7 + 3x + x^2, \\
& && f(x)^{-1} \equiv 1 + x^2 + x^3 \pmod{2}. \\
&\text{(ii)} \quad N = 5, \quad m = 2^7, \quad && f(x) = 22 + 11x + 5x^2 + 7x^3, \\
& && f(x)^{-1} \equiv 1 + x^2 + x^3 \pmod{2}. \\
&\text{(iii)} \quad N = 7, \quad m = 5^5, \quad && f(x) = 112 + 34x + 239x^2 + 234x^3 + 105x^4 \\
& && \qquad + 180x^5 + 137x^6, \\
& && f(x)^{-1} \equiv 1 + 3x^2 + 2x^4 \pmod{5}.
\end{aligned}$$

7.28. Let $\boldsymbol{a} \in \mathbb{R}^N$ be a fixed vector.

(a) Suppose that $\boldsymbol{b}$ is an N-dimensional vector whose coefficients are chosen randomly from the set $\{-1, 0, 1\}$. Prove that the expected values of $\|\boldsymbol{b}\|^2$ and $\|\boldsymbol{a} \star \boldsymbol{b}\|^2$ are given by

$$E(\|\boldsymbol{b}\|^2) = \frac{2}{3}N \quad \text{and} \quad E(\|\boldsymbol{a} \star \boldsymbol{b}\|^2) = \|\boldsymbol{a}\|^2 E(\|\boldsymbol{b}\|^2).$$

(b) More generally, suppose that the coefficients of $\boldsymbol{b}$ are chosen at random from the set of integers $\{-T, -T+1, \ldots, T-1, T\}$. Compute the expected values of $\|\boldsymbol{b}\|^2$ and $\|\boldsymbol{a} \star \boldsymbol{b}\|^2$ as in (a).

(c) Suppose now that the coefficients of $\boldsymbol{b}$ are real numbers that are chosen uniformly and independently in the interval from $-R$ to R. Prove that

$$E(\|\boldsymbol{b}\|^2) = \frac{R^2 N}{3} \quad \text{and} \quad E(\|\boldsymbol{a} \star \boldsymbol{b}\|^2) = \|\boldsymbol{a}\|^2 E(\|\boldsymbol{b}\|^2).$$

(*Hint.* The most direct way to do (c) is to use continuous probability theory. As an alternative, let the coefficients of $\boldsymbol{b}$ be chosen uniformly and independently from the set $\{jR/T : -T \le j \le T\}$, redo the computation from (b), and then let $T \to \infty$.)

(d) For each of the scenarios described in (a), (b), and (c), prove that

$$E(\|\boldsymbol{a} + \boldsymbol{b}\|^2) = \|\boldsymbol{a}\|^2 + E(\|\boldsymbol{b}\|^2).$$

Section 7.10. The NTRU Public Key Cryptosystem

7.29. Alice and Bob agree to communicate using NTRUEncrypt with

$$(N, p, q) = (7, 3, 37).$$

Alice's private key is

$$\begin{aligned}
\boldsymbol{f}(x) &= -1 + X - X^3 + X^4 + X^5, \\
\boldsymbol{F}_3(x) &= 1 + X - X^2 + X^4 + X^5 + X^6.
\end{aligned}$$

(You can check that $\boldsymbol{f} \star \boldsymbol{F}_3 \equiv 1 \pmod{3}$.) Alice receives the ciphertext

$$\boldsymbol{e}(x) = 2 + 8X^2 - 16X^3 - 9X^4 - 18X^5 - 3X^6.$$

from Bob. Decipher the message and find the plaintext.

7.30. Alice and Bob decide to communicate using NTRUEncrypt with parameters $(N, p, q) = (7, 3, 29)$. Alice's public key is

$$\boldsymbol{h}(x) = 3 + 14X - 4X^2 + 13X^3 - 6X^4 + 2X^5 + 7X^6.$$

Bob sends Alice the plaintext message $\boldsymbol{m}(x) = 1 + X - X^2 - X^3 - X^6$ using the random element $\boldsymbol{r}(x) = -1 + X^2 - X^5 + X^6$.

(a) What ciphertext does Bob send to Alice?

(b) Alice's private key is $\boldsymbol{f}(x) = -1 + X - X^2 + X^4 + X^6$ and $\boldsymbol{F}_3(x) = 1 + X + X^2 + X^4 + X^5 - X^6$. Check your answer in (a) by using $\boldsymbol{f}$ and $\boldsymbol{F}_3$ to decrypt the message.

7.31. What is the message expansion of NTRUEncrypt in terms of N, p, and q?

7.32. The guidelines for choosing NTRUEncrypt public parameters (N, p, q, d) require that $\gcd(p, q) = 1$. Prove that if $p \mid q$, then it is very easy for Eve to decrypt the message without knowing the private key. (*Hint.* First do the case that $p = q$.)

7.33. The guidelines for choosing NTRUEncrypt public parameters (N, p, q, d) include the assumption that $\gcd(N, q) = 1$. Suppose instead that Alice takes $q = N$, where as always, N is an odd prime.

(a) Make a change of variables $x = y + 1$ in the ring $\mathbb{Z}[x]/(x^N - 1)$, and show that the NTRU lattice takes a simpler form.

(b) Can you find an efficient way to break NTRU in the case that $q = N$ that does involve lattice reduction? (This appears to be an open problem.)

7.34. Alice uses NTRUEncrypt with $p = 3$ to send messages to Bob.

(a) Suppose that Alice uses the same random element $\boldsymbol{r}(x)$ to encrypt two different plaintexts $\boldsymbol{m}_1(x)$ and $\boldsymbol{m}_2(x)$. Explain how Eve can use the two ciphertexts $\boldsymbol{e}_1(x)$ and $\boldsymbol{e}_2(x)$ to determine approximately $\frac{2}{9}$ of the coefficients of $\boldsymbol{m}_1(x)$. (See Exercise 7.38 for a way to exploit this information.)

(b) For example, suppose that $N = 8$, so there are 3^8 possibilities for $\boldsymbol{m}_1(x)$. Suppose that Eve intercepts two ciphertexts

$$\boldsymbol{e}_1(x) = 32 + 21x - 9x^2 - 20x^3 - 29x^4 - 29x^5 - 19x^6 + 38x^7,$$
$$\boldsymbol{e}_2(x) = 33 + 21x - 7x^2 - 19x^3 - 31x^4 - 27x^5 - 19x^6 + 38x^7,$$

that were encrypted using the same random element $\boldsymbol{r}(x)$. How many coefficients of $\boldsymbol{m}_1(x)$ can she determine exactly? How many possibilities are there for $\boldsymbol{m}_1(x)$?

(c) Formulate a similar attack if Alice uses two different random elements $\boldsymbol{r}_1(x)$ and $\boldsymbol{r}_2(x)$ to encrypt the same plaintext $\boldsymbol{m}(x)$. (*Hint.* Do it first assuming that $\boldsymbol{h}(x)$ has an inverse in R_q. The problem is harder without this assumption.)

7.35. This exercise describes a variant of NTRUEncrypt that eliminates a step in the decryption algorithm at the cost of requiring slightly larger parameters. Suppose that the NTRUEncrypt private key polynomials $\boldsymbol{f}(x)$ and $\boldsymbol{g}(x)$ are chosen to satisfy

$$\boldsymbol{f}(x) = 1 + p\boldsymbol{f}_0(x) \equiv 1 \pmod{p} \quad \text{and} \quad \boldsymbol{g}(x) = p\boldsymbol{g}_0(x) \equiv 0 \pmod{p},$$

and that NTRU encryption is changed to

$$\boldsymbol{e}(x) \equiv \boldsymbol{h}(x) \star \boldsymbol{r}(x) + \boldsymbol{m}(x) \pmod{q}.$$

(The change is the omission of p before $\boldsymbol{h}(x)$.)

(a) Prove that if q is sufficiently large, then the following algorithm correctly decrypts the message:
- Compute $\boldsymbol{a}(x) \equiv \boldsymbol{f}(x) \star \boldsymbol{e}(x) \pmod{q}$ and center-lift to an element of R.
- Compute $\boldsymbol{a}(x) \pmod{p}$. The result is $\boldsymbol{m}(x)$.

Note that this eliminates the necessity to multiply $\boldsymbol{a}(x)$ by $\boldsymbol{f}(x)^{-1} \pmod{p}$.

(a) Suppose that we choose $\boldsymbol{f}_0, \boldsymbol{g}_0 \in \mathcal{T}(d,d)$, and that we also assume that $\boldsymbol{m}$ is ternary. Prove that decryption works provided $q > 8dp + 2$. (*Hint.* Mimic the proof of Proposition 7.48.)

Section 7.11. NTRU as a Lattice Cryptosystem

7.36. This exercise explains how to formulate NTRU message recovery as a closest vector problem. Let $\boldsymbol{h}(x)$ be an NTRU public key and let

$$\boldsymbol{e}(x) \equiv \boldsymbol{pr}(x) \star \boldsymbol{h}(x) + \boldsymbol{m}(x) \pmod{q}$$

be a message encrypted using $\boldsymbol{h}(x)$.

(a) Prove that the vector $(\boldsymbol{pr}, \boldsymbol{e} - \boldsymbol{m})$ is in $L_{\boldsymbol{h}}^{\text{NTRU}}$.

(b) Prove that the lattice vector in (a) is almost certainly the closest lattice vector to the known vector $(0, \boldsymbol{e})$. Hence solving CVP reveals the plaintext $\boldsymbol{m}$. (For simplicity, you may assume that $d \approx N/3$ and $q \approx 2N$, as we did in Proposition 7.61.)

(c) Show how one can reduce the lattice-to-target distance, without affecting the determinant, by using instead a modified NTRU lattice of the form

$$\begin{pmatrix} 1 & p\boldsymbol{h} \\ 0 & q \end{pmatrix}.$$

7.37. The guidelines for choosing NTRUEncrypt public parameters (N, p, q, d) include the requirement that N be prime. To see why, suppose (say) that N is even. Explain how Eve can recover the private key by solving a lattice problem in dimension N, rather than in dimension $2N$. *Hint.* Use the natural map

$$\mathbb{Z}[x]/(x^N - 1) \to \mathbb{Z}[x]/(x^{N/2} - 1).$$

7.38. Suppose that Bob and Alice are using NTRUEncrypt to exchange messages and that Eve intercepts a ciphertext $\boldsymbol{e}(x)$ for which she already knows part of the plaintext $\boldsymbol{m}(x)$. (This is not a ludicrous assumption; see Exercise 7.34, for example.) More precisely, suppose that Eve knows t of the coefficients of $\boldsymbol{m}(x)$. Explain how to set up a CVP to find $\boldsymbol{m}(x)$ using a lattice of dimension $2N - 2t$.

Section 7.12. Lattice-Based Digital Signature Schemes

7.39. Samantha uses the GGH digital signature scheme with private and public bases

$$\begin{aligned} \boldsymbol{v}_1 &= (-20, -8, 1), & \boldsymbol{w}_1 &= (-248100, 220074, 332172), \\ \boldsymbol{v}_2 &= (14, 11, 23), & \boldsymbol{w}_2 &= (-112192, 99518, 150209), \\ \boldsymbol{v}_3 &= (-18, 1, -12), & \boldsymbol{w}_3 &= (-216150, 191737, 289401). \end{aligned}$$

What is her signature on the document

$$\boldsymbol{d} = (834928, 123894, 7812738)?$$

7.40. Samantha uses the GGH digital signature scheme with public basis

$$\begin{aligned} \boldsymbol{w}_1 &= (3712318934, -14591032252, 11433651072),\\ \boldsymbol{w}_2 &= (-1586446650, 6235427140, -4886131219),\\ \boldsymbol{w}_3 &= (305711854, -1201580900, 941568527). \end{aligned}$$

She publishes the signature

$$(6987814629, 14496863295, -9625064603)$$

on the document

$$\boldsymbol{d} = (5269775, 7294466, 1875937).$$

If the maximum allowed distance from the signature to the document is 60, verify that Samantha's signature is valid.

7.41. Samantha uses the GGH digital signature scheme with public basis

$$\begin{aligned} \boldsymbol{w}_1 &= (-1612927239, 1853012542, 1451467045),\\ \boldsymbol{w}_2 &= (-2137446623, 2455606985, 1923480029),\\ \boldsymbol{w}_3 &= (2762180674, -3173333120, -2485675809). \end{aligned}$$

Use LLL or some other lattice reduction algorithm to find a good basis for Samantha's lattice, and then use the good basis to help Eve forge a signature on the document

$$\boldsymbol{d} = (87398273893, 763829184, 118237397273).$$

What is the distance from your forged signature lattice vector to the target vector? (You should be able to get a distance smaller than 100.)

7.42. This exercise gives further details of the NTRUMLS signature scheme. We fix parameters (N, p, q) and set

$$B = \left\lceil \frac{p^2 N}{4} \right\rceil \quad \text{and} \quad A = \left\lfloor \frac{q}{2p} - \frac{1}{2} \right\rfloor.$$

We choose private key polynomials $\boldsymbol{f}$ and $\boldsymbol{g}$ as follows. For $\boldsymbol{f}$ we first choose a polynomial $\boldsymbol{F}$ whose coefficients are randomly selected from the set $\{-1, 0, 1\}$ and then let $\boldsymbol{f} = p\boldsymbol{F}$. For $\boldsymbol{g}$ we choose a polynomial whose coefficients are randomly selected to lie between $-p/2$ and $p/2$. We further assume that both $\boldsymbol{F}$ and $\boldsymbol{g}$ are invertible modulo p and that $\boldsymbol{f}$ is invertible modulo q, otherwise we discard them and choose new polynomials.

(a) If $\boldsymbol{a}$ and $\boldsymbol{b}$ are polynomials whose coefficients lie between $-p/2$ and $p/2$, prove that $\|\boldsymbol{a} \star \boldsymbol{b}\|_\infty \le B$.

(b) Prove that the following algorithm outputs a pair of polynomials $(\boldsymbol{s}, \boldsymbol{t})$ satisfying

$$\boldsymbol{t} \equiv \boldsymbol{h} \star \boldsymbol{s} \pmod{q} \quad \text{and} \quad \boldsymbol{s} \equiv \boldsymbol{s}_p \pmod{p} \quad \text{and} \quad \boldsymbol{t} \equiv \boldsymbol{t}_p \pmod{p}.$$

0: Input polynomials $\boldsymbol{s}_p$ and $\boldsymbol{t}_p$ with coefficients between $-\frac{1}{2}p$ and $\frac{1}{2}p$.

1: Choose a random polynomial $\boldsymbol{r}$ with coefficients between $-A$ and A.

2: Set $\boldsymbol{s}_0 = \boldsymbol{s}_p + p\boldsymbol{r}$.

3: Set $\boldsymbol{t}_0 \equiv \boldsymbol{h} \star \boldsymbol{s}_0 \pmod{q}$ with $\|\boldsymbol{t}_0\|_\infty \le \frac{1}{2}q$.

4: Set $\boldsymbol{a} \equiv \boldsymbol{g}^{-1} \star (\boldsymbol{t}_p - \boldsymbol{t}_0) \pmod{p}$ with $\|\boldsymbol{a}\|_\infty \le \frac{1}{2}p$.
5: Set $\boldsymbol{s} = \boldsymbol{s}_0 + \boldsymbol{a} \star \boldsymbol{f}$ and $\boldsymbol{t} = \boldsymbol{t}_0 + \boldsymbol{a} \star \boldsymbol{g}$.

(c) Prove that the output from the algorithm in (b) satisfies

$$\|\boldsymbol{s}\|_\infty \le \frac{q}{2} + B \quad \text{and} \quad \|\boldsymbol{t}\|_\infty \le \frac{q}{2} + B.$$

(d) Make the simplifying assumption that the output produces polynomials whose coefficients are uniformly and independently distributed between $-\frac{1}{2}q - B$ and $\frac{1}{2}q + B$. Assume further that $k := q/NB$ is not too large, say $2 \le k \le 50$. Prove that the probability that the algorithm in (b) produces a valid signature is approximately $e^{-8/k}$. (Note that according to (b), the output $(\boldsymbol{s}, \boldsymbol{t})$ will be a valid signature if it satisfies the size criteria $\|\boldsymbol{s}\|_\infty \le \frac{1}{2}q - B$ and $\|\boldsymbol{t}\|_\infty \le \frac{1}{2}q - B$.)

Section 7.13. Lattice Reduction Algorithms

7.43. Let $\boldsymbol{b}_1$ and $\boldsymbol{b}_2$ be vectors, and set

$$t = \boldsymbol{b}_1 \cdot \boldsymbol{b}_2 / \|\boldsymbol{b}_1\|^2 \quad \text{and} \quad \boldsymbol{b}_2^* = \boldsymbol{b}_2 - t\boldsymbol{b}_1.$$

Prove that $\boldsymbol{b}_2^* \cdot \boldsymbol{b}_1 = 0$ and that $\boldsymbol{b}_2^*$ is the projection of $\boldsymbol{b}_2$ onto the orthogonal complement of $\boldsymbol{b}_1$.

7.44. Let $\boldsymbol{a}$ and $\boldsymbol{b}$ be nonzero vectors in $\mathbb{R}^n$.

(a) What value of $t \in \mathbb{R}$ minimizes the distance $\|\boldsymbol{a} - t\boldsymbol{b}\|$? (*Hint.* It's easier to minimize the value of $\|\boldsymbol{a} - t\boldsymbol{b}\|^2$.)

(b) What is the minimum distance in (a)?

(c) If t is chosen as in (a), show that $\boldsymbol{a} - t\boldsymbol{b}$ is the projection of $\boldsymbol{a}$ onto the orthogonal complement of $\boldsymbol{b}$.

(d) If the angle between $\boldsymbol{a}$ and $\boldsymbol{b}$ is θ, use your answer in (b) to show that the minimum distance is $\|a\| \sin\theta$. Draw a picture illustrating this result.

7.45. Apply Gauss's lattice reduction algorithm (Proposition 7.66) to solve SVP for the following two dimensional lattices having the indicated basis vectors. How many steps does the algorithm take?

(a) $\boldsymbol{v}_1 = (120670, 110521)$ and $\boldsymbol{v}_2 = (323572, 296358)$.

(b) $\boldsymbol{v}_1 = (174748650, 45604569)$ and $\boldsymbol{v}_2 = (35462559, 9254748)$.

(c) $\boldsymbol{v}_1 = (725734520, 613807887)$ and $\boldsymbol{v}_2 = (3433061338, 2903596381)$.

7.46. Let V be a vector space, let $W \subset V$ be a vector subspace of V, and let $W^\perp$ be the orthogonal complement of W in V.

(a) Prove that $W^\perp$ is also a vector subspace of V.

(b) Prove that every vector $\boldsymbol{v} \in V$ can be written as a sum $\boldsymbol{v} = \boldsymbol{w} + \boldsymbol{w}'$ for unique vectors $\boldsymbol{w} \in W$ and $\boldsymbol{w}' \in W^\perp$. (One says that V is the *direct sum* of the subspaces W and $W^\perp$.)

(c) Let $\boldsymbol{w} \in W$ and $\boldsymbol{w}' \in W^\perp$ and let $\boldsymbol{v} = a\boldsymbol{w} + b\boldsymbol{w}'$. Prove that

$$\|\boldsymbol{v}\|^2 = a^2\|\boldsymbol{w}\|^2 + b^2\|\boldsymbol{w}'\|^2.$$

7.47. Let L be a lattice with basis vectors $\boldsymbol{v}_1 = (161, 120)$ and $\boldsymbol{v}_2 = (104, 77)$.
(a) Is $(0, 1)$ in the lattice?
(b) Find an LLL reduced basis.
(c) Use the reduced basis to find the closest lattice vector to $\left(-\frac{9}{2}, 11\right)$.

7.48. Use the LLL algorithm to reduce the lattice with basis

$$\boldsymbol{v}_1 = (20, 16, 3), \quad \boldsymbol{v}_2 = (15, 0, 10), \quad \boldsymbol{v}_3 = (0, 18, 9).$$

You should do this exercise by hand, writing out each step.

7.49. Let L be the lattice generated by the rows of the matrix

$$M = \begin{pmatrix} 20 & 51 & 35 & 59 & 73 & 73 \\ 14 & 48 & 33 & 61 & 47 & 83 \\ 95 & 41 & 48 & 84 & 30 & 45 \\ 0 & 42 & 74 & 79 & 20 & 21 \\ 6 & 41 & 49 & 11 & 70 & 67 \\ 23 & 36 & 6 & 1 & 46 & 4 \end{pmatrix}.$$

Implement the LLL algorithm (Fig. 7.8) on a computer and use your program to answer the following questions.
(a) Compute $\det(L)$ and $\mathcal{H}(M)$. What is the shortest basis vector?
(b) Apply LLL to M. How many swaps (Step [11]) are required? What is the value of $\mathcal{H}(M^{LLL})$? What is the shortest basis vector in the LLL reduced basis? How does it compare with the Gaussian expected shortest length?
(c) Reverse the order of the rows of M and apply LLL to the new matrix. How many swaps are required? What is the value of $\mathcal{H}(M^{LLL})$ and what is the shortest basis vector?
(d) Apply LLL to the original matrix M, but in the Lovász condition (Step [8]), use 0.99 instead of $\frac{3}{4}$. How many swaps are required? What is the value of $\mathcal{H}(M^{LLL})$ and what is the shortest basis vector?

7.50. A more efficient way to implement the LLL algorithm is described in Fig. 7.9, with Reduce and Swap subroutines given in Fig. 7.10. (This implementation of LLL follows [28, Algorithm 2.6.3]. We thank Henri Cohen for his permission to include it here.)
(a) Prove that the algorithm described in Figs. 7.9 and 7.10 returns an LLL reduced basis.
(b) For any given N and q, let $L_{N,q}$ be the N-dimensional lattice with basis $\boldsymbol{v}_1, \ldots, \boldsymbol{v}_N$ described by the formulas

$$\boldsymbol{v}_i = (r_{i1}, r_{i2}, \ldots, r_{iN}), \qquad r_{ij} \equiv (i + N)^j \pmod{q}, \qquad 0 \le r_{ij} < q.$$

Implement the LLL algorithm and use it to LLL reduce $L_{N,q}$ for each of the following values of N and q:

(i) $(N, q) = (10, 541)$ (ii) $(N, q) = (20, 863)$
(iii) $(N, q) = (30, 1223)$ (iv) $(N, q) = (40, 3571)$

In each case, compare the Hadamard ratio of the original basis to the Hadamard ratio of the LLL reduced basis, and compare the length of the shortest vector found by LLL to the Gaussian expected shortest length.

```
[1]  Input a basis {v_1, ..., v_n} for a lattice L
[2]  Set k = 2, k_max = 1, v_1^* = v_1, and B_1 = ||v_1||^2
[3]  If k ≤ k_max go to Step [9]
[4]  Set k_max = k and v_k^* = v_k
[5]  Loop j = 1, 2, ..., k − 1
[6]      Set μ_{k,j} = v_k · v_j^*/B_j and v_k^* = v_k^* − μ_{k,j} v_j^*
[7]  End j Loop
[8]  Set B_k = ||v_k^*||^2
[9]  Execute Subroutine RED(k, k − 1)
[10] If B_k < (3/4 − μ_{k,k−1}^2) B_{k−1}
[11]     Execute Subroutine SWAP(k)
[12]     Set k = max(2, k − 1) and go to Step [9]
[13] Else
[14]     Loop ℓ = k − 2, k − 3, ..., 2, 1
[15]         Execute Subroutine RED(k, ℓ)
[16]     End ℓ Loop
[17]     Set k = k + 1
[18] End If
[19] If k ≤ n go to Step [3]
[20] Return LLL reduced basis {v_1, ..., v_n}
```

Figure 7.9: The LLL algorithm—main routine

7.51. Let $\frac{1}{4} < \alpha < 1$ and suppose that we replace the Lovász condition with the condition

$$\|\boldsymbol{v}_i^*\|^2 \geq \left(\alpha - \mu_{i,i-1}^2\right) \|\boldsymbol{v}_{i-1}^*\|^2 \quad \text{for all } 1 < i \leq n. \tag{7.64}$$

(a) Prove a version of Theorem 7.69 assuming the alternative Lovász condition (7.64). What quantity, depending on α, replaces the 2 that appears in the estimates (7.54)–(7.56)?

(b) Prove a version of Theorem 7.71 assuming the alternative Lovász condition (7.64). In particular, how does the upper bound for the number of swap steps depend on α? What happens as $\alpha \to 1$?

7.52. Let $\boldsymbol{v}_1, \ldots, \boldsymbol{v}_n$ be an LLL reduced basis for a lattice L.

(a) Prove that there are constants $C_1 > 1 > C_2 > 0$ such that for all $y_1, \ldots, y_n \in \mathbb{R}$ we have

$$C_1^n \sum_{i=1}^{n} y_i^2 \|\boldsymbol{v}_i\|^2 \geq \left\| \sum_{i=1}^{n} y_i \boldsymbol{v}_i \right\|^2 \geq C_2^n \sum_{i=1}^{n} y_i^2 \|\boldsymbol{v}_i\|^2. \tag{7.65}$$

(This is a hard exercise.) We observe that the inequality (7.65) is another way of saying that the basis $\boldsymbol{v}_1, \ldots, \boldsymbol{v}_n$ is quasi-orthogonal, since if it were truly orthogonal, then we would have an equality $\|\sum y_i \boldsymbol{v}_i\|^2 = \sum y_i^2 \|\boldsymbol{v}_i\|^2$.

—— Subroutine RED(k, ℓ) ——

[1] If $|\mu_{k,\ell}| \le \frac{1}{2}$, return to Main Routine
[2] Set $m = \lfloor \mu_{k,\ell} \rceil$
[3] Set $\boldsymbol{v}_k = \boldsymbol{v}_k - m\boldsymbol{v}_\ell$ and $\mu_{k,\ell} = \mu_{k,\ell} - m$
[4] Loop $i = 1, 2, \ldots, \ell - 1$
[5] Set $\mu_{k,i} = \mu_{k,i} - m\mu_{\ell,i}$
[6] End i Loop
[7] Return to Main Routine

—— Subroutine SWAP(k) ——

[1] Exchange $\boldsymbol{v}_{k-1}$ and $\boldsymbol{v}_k$
[2] Loop $j = 1, 2, \ldots, k - 2$
[3] Exchange $\mu_{k-1,j}$ and $\mu_{k,j}$
[4] End j Loop
[5] Set $\mu = \mu_{k,k-1}$ and $B = B_k + \mu^2 B_{k-1}$
[6] Set $\mu_{k,k-1} = \mu B_{k-1}/B$ and $B_k = B_{k-1}B_k/B$ and $B_{k-1} = B$
[7] Loop $i = k+1, k+2, \ldots, k_{\max}$
[8] Set $m = \mu_{i,k}$ and $\mu_{i,k} = \mu_{i,k-1} - \mu m$ and $\mu_{i,k-1} = m + \mu_{k,k-1}\mu_{i,k}$
[9] End i Loop
[10] Return to Main Routine

Figure 7.10: The LLL algorithm—RED and SWAP subroutines

(b) Prove that there is a constant C such that for any target vector $\boldsymbol{w} \in \mathbb{R}^n$, Babai's algorithm (Theorem 7.34) finds a lattice vector $\boldsymbol{v} \in L$ satisfying

$$\|\boldsymbol{w} - \boldsymbol{v}\| \le C^n \min_{\boldsymbol{u} \in L} \|\boldsymbol{w} - \boldsymbol{u}\|.$$

Thus Babai's algorithm applied with an LLL reduced basis solves apprCVP to within a factor of C^n. This is Theorem 7.76.

(c) Find explicit values for the constants C_1, C_2, and C in (a) and (b).

7.53. Babai's *Closest Plane Algorithm*, which is described in Fig. 7.11, is an alternative rounding method that uses a given basis to solve apprCVP. As usual, the more orthogonal the basis, the better the solution, so generally people first use LLL to create a quasi-orthogonal basis and then apply one of Babai's methods. In both theory and practice, Babai's closest plane algorithm seems to yield better results than Babai's closest vertex algorithm.

Implement both of Babai's algorithms (Theorem 7.34 and Fig. 7.11) and use them to solve apprCVP for each of the following lattices and target vectors. Which one gives the better result?

(a) L is the lattice generated by the rows of the matrix

$$M_L = \begin{pmatrix} -5 & 16 & 25 & 25 & 13 & 8 \\ 26 & -3 & -11 & 14 & 5 & -26 \\ 15 & -28 & 16 & -7 & -21 & -4 \\ 32 & -3 & 7 & -30 & -6 & 26 \\ 15 & -32 & -17 & 32 & -3 & 11 \\ 5 & 24 & 0 & -13 & -46 & 15 \end{pmatrix}$$

```
Input a basis v_1, ..., v_n of a lattice L.
Input a target vector t.
Compute Gram–Schmidt orthogonalized vectors v_1^*, ..., v_n^* (Theorem 7.13).
Set w = t.
Loop i = n, n−1, ..., 2, 1
    Set w = w − ⌊w · v_i^* / ||v_i^*||^2⌉ v_i.
End i Loop
Return the lattice vector t − w.
```

Figure 7.11: Babai's closest plane algorithm

and the target vector is $t = (-178, 117, -407, 419, -4, 252)$. (Notice that the matrix M_L is LLL reduced.)

(b) L is the lattice generated by the rows of the matrix

$$M_L = \begin{pmatrix} -33 & -15 & 22 & -34 & -32 & 41 \\ 10 & 9 & 45 & 10 & -6 & -3 \\ -32 & -17 & 43 & 37 & 29 & -30 \\ 26 & 13 & -35 & -41 & 42 & -15 \\ -50 & 32 & 18 & 35 & 48 & 45 \\ 2 & -5 & -2 & -38 & 38 & 41 \end{pmatrix}$$

and the target vector is $t = (-126, -377, -196, 455, -200, -234)$. (Notice that the matrix M_L is not LLL reduced.)

(c) Apply LLL reduction to the basis in (b), and then use both of Babai's methods to solve apprCVP. Do you get better solutions?

7.54. We proved that the LLL algorithm terminates and has polynomial running time under the assumption that $L \subset \mathbb{Z}^n$; see Theorem 7.71. Show that this assumption is not necessary by proving that LLL terminates in polynomial time for any lattice $L \subset \mathbb{R}^n$. You may assume that your computer can do exact computations in $\mathbb{R}$, although in practice one does need to worry about round-off errors. (*Hint.* Use Hermite's theorem to derive a lower bound, depending on the length of the shortest vector in L, for the quantity D that appears in the proof of Theorem 7.71.)

Section 7.14. Applications of LLL to Cryptanalysis

7.55. You have been spying on George for some time and overhear him receiving a ciphertext $e = 83493429501$ that has been encrypted using the congruential cryptosystem described in Sect. 7.1. You also know that George's public key is $h = 24201896593$ and the public modulus is $q = 148059109201$. Use Gaussian lattice reduction to recover George's private key (f, g) and the message m.

7.56. Let

$$M = (81946, 80956, 58407, 51650, 38136, 17032, 39658, 67468, 49203, 9546)$$

and let $S = 168296$. Use the LLL algorithm to solve the subset-sum problem for M and S, i.e., find a subset of the elements of M whose sum is S.

7.57. Alice and Bob communicate using the GGH cryptosystem. Alice's public key is the lattice generated by the rows of the matrix

$$\begin{pmatrix} 10305608 & -597165 & 45361210 & 39600006 & 12036060 \\ -71672908 & 4156981 & -315467761 & -275401230 & -83709146 \\ -46304904 & 2685749 & -203811282 & -177925680 & -54081387 \\ -68449642 & 3969419 & -301282167 & -263017213 & -79944525 \\ -46169690 & 2677840 & -203215644 & -177405867 & -53923216 \end{pmatrix}.$$

Bob sends her the encrypted message

$$\boldsymbol{e} = (388120266, -22516188, 1708295783, 1491331246, 453299858).$$

Use LLL to find a reduced basis for Alice's lattice, and then use Babai's algorithm to decrypt Bob's message.

7.58. Alice and Bob communicate using NTRUEncrypt with public parameters $(N, p, q, d) = (11, 3, 67, 3)$. Alice's public key is

$$\boldsymbol{h} = 39 + 9x + 33x^2 + 52x^3 + 58x^4 + 11x^5 + 38x^6 + 6x^7 + x^8 + 48x^9 + 41x^{10}.$$

Apply the LLL algorithm to the associated NTRU lattice to find an NTRU private key $(\boldsymbol{f}, \boldsymbol{g})$ for $\boldsymbol{h}$. Check your answer by verifying that $\boldsymbol{g} \equiv \boldsymbol{f} \star \boldsymbol{h} \pmod{q}$. Use the private key to decrypt the ciphertext

$$\boldsymbol{e} = 52 + 50x + 50x^2 + 61x^3 + 61x^4 + 7x^5 + 53x^6 + 46x^7 + 24x^8 + 17x^9 + 50x^{10}.$$

7.59. (a) Suppose that k is a 10 digit integer, and suppose that when $\sqrt{k}$ is computed, the first 15 digits *after* the decimal place are 418400286617716. Find the number k. (*Hint.* Reformulate it as a lattice problem.)

(b) More generally, suppose that you know the first d-digits after the decimal place of $\sqrt{K}$. Explain how to set up a lattice problem to find K.

See Exercise 1.49 for a cryptosystem associated to this problem.

Chapter 8

Additional Topics in Cryptography

The emphasis of this book has been on the mathematical underpinnings of public key cryptography. We have developed most of the mathematics from scratch and in sufficient depth to enable the reader to understand both the underlying mathematical principles and how they are applied in cryptographic constructions. Unfortunately, in achieving this laudable goal, we have now reached the end of a hefty textbook with many important cryptographic topics left untouched.

This final chapter contains a few brief words about some of these additional topics. The reader should keep in mind that each of these areas is important and that the brevity of our coverage reflects only a lack of space, not a lack of interest. We hope that you will view this chapter as a challenge to go out and learn more about mathematical cryptography. In particular, each section in this chapter provides a good starting point for a term paper or class project.

We also note that we have made no attempt to provide a full history of the topics covered, nor have we tried to give credit to all of the researchers working in these areas. For the convenience of the reader and the instructor, here is a list of the topics introduced in this chapter:

J. Hoffstein et al., *An Introduction to Mathematical Cryptography*, Undergraduate Texts in Mathematics, DOI 10.1007/978-1-4939-1711-2_8

8.1 Hash Functions

There are many cryptographic constructions for which one needs a function that is easy to compute, but hard to invert. We have seen a number of examples, including digital signatures (Remark 4.2), randomization of plaintexts in probabilistic cryptosystems (Exercise 7.21), and ID based cryptography (Sect. 6.10.2).

Definition. A *hash function* takes as input an arbitrarily long document D and returns a short bit string H. The primary properties that a hash function Hash should possess are as follows:

- Computation of $\mathsf{Hash}(D)$ should be fast and easy, roughly linear time.
- Inversion of Hash should be difficult, meaning exponential time. More precisely, given a hash value H, it should be difficult to find any document D such that $\mathsf{Hash}(D) = H$.
- For many applications it is also important that Hash be *collision resistant.* This means that it should be hard to find two different documents D_1 and D_2 whose hash values $\mathsf{Hash}(D_1)$ and $\mathsf{Hash}(D_2)$ are the same.

Remark 8.1. Why do we want our hash function to be collision resistant? Suppose that Eve can find two documents D_1 and D_2 that have the same hash value $\mathsf{Hash}(D_1) = \mathsf{Hash}(D_2)$, and suppose that D_1 says "Pay the bearer \$5" and that D_2 says "Pay the bearer \$500." Eve can give Alice \$5 and ask her to sign D_1. Since Alice has actually signed $\mathsf{Hash}(D_1)$, she has also signed $\mathsf{Hash}(D_2)$, so Eve can go to the bank, present the signature as being on D_2, and get paid \$500.

In practice, most hash functions use a mixing algorithm $\mathcal{M}$ that transforms a bit string of length n into another bit string of length n. Then Hash works by breaking a long document into blocks and successively using $\mathcal{M}$ to combine each block with the previously processed material.

Thus to compute $\mathcal{H}(D)$, we first append extra 0 bits to D so that the length of D is an even multiple of n bits. This allows us to write D as a concatenation

$$D = D_1 \parallel D_2 \parallel D_3 \parallel D_4 \parallel \cdots \parallel D_k$$

of bit strings of length n. (See Exercise 3.43 for a discussion of concatenation.)

Having broken D into pieces, we start the computation of $\mathsf{Hash}(D)$ with an initial bit string H_0, which is the always the same. We then compute $\mathcal{M}(D_1)$ and set $H_1 = H_0 \operatorname{xor} \mathcal{M}(D_1)$. We repeat this process k times to obtain $H_2, H_3, \ldots, H_k$, where[1]

$$H_i = H_{i-1} \operatorname{xor} \mathcal{M}(D_i) \qquad \text{for } 1 \le i \le k.$$

Then $\mathsf{Hash}(D)$ is equal to the final output value H_k.

For practical applications, it is very important that a hash function be extremely fast. For example, when digitally signing a document such as a computer program or a video file, the entire document needs to be run through the hash function, so one needs to be able to compute $\mathsf{Hash}(D)$ on megabyte, or even gigabyte, length files.

Since speed is of fundamental importance for hash functions, in the real world one tends to use hash functions constructed using ad hoc mixing operations, rather than basing them on classical hard mathematical problems such as factoring or discrete logarithms. The hash functions in most widespread use today go by the name of SHA (Secure Hash Algorithm). There are several versions of SHA, released at various times, that achieve various levels of security.

How do SHA and other similar hash algorithms work? We briefly illustrate by describing the structure of SHA-1, omitting the specifics of the mixing operations. (See [99] for the official government description of SHA.)

Break document D (with extra bits appended) into 512-bit chunks.
Start with five specific initial values $h_0, \ldots, h_4$.
LOOP over the 512-bit chunks.
    Break a 512 bit chunk into sixteen 32-bit words.
    Create a total of eighty 32-bit words $w_0, \ldots, w_{79}$ by
        rotating the initial words.
    LOOP $i = 0, 1, 2, \ldots, 79$
        Set $a = h_0$, $b = h_1$, $c = h_2$, $d = h_3$, $e = h_4$.
        Compute f using XOR and AND operations on a, b, c, d, e.
        Mix a, b, c, d, e, by rotating some of their bits, permuting
            them, and add f and w_i to a.
    END i LOOP
    Set $h_0 = h_0 + a$, $h_1 = h_1 + b$, $\ldots$, $h_4 = h_4 + e$.
END LOOP over chunks
Output $h_0 \parallel h_1 \parallel h_2 \parallel h_3 \parallel h_4$.

The SHA-1 Hash Algorithm

[1]In practice, the H_{i-1} and $\mathcal{M}(D_i)$ values would be combined in a slightly more complicated fashion to form H_i.

Remark 8.2. Notice that SHA-1 has an inner loop that is repeated 80 times. We say that SHA-1 has 80 *rounds*. Each round involves various mixing operations that use the results from the previous round together with a small amount of new data. For SHA-1, the new data used in the ith round is the 32-bit word w_i. This idea of repeating a simple mixing operation is typical of modern hash functions, pseudorandom number generators (Sect. 8.2), and symmetric ciphers such as DES and AES (Sect. 8.12). In principle, one could make SHA-1 faster by doing fewer rounds, but if one uses too few rounds, then there are known methods to break the system. It is an area of ongoing research to understand how many rounds are necessary to make SHA-1 and similar round-based systems secure.

The original SHA, which was later amended as SHA-1 to fix a minor security flaw, is a hash function whose output has length 160 bits. Starting around 2005 and continuing to the present day, various attacks have been developed for SHA-1, originally for fewer than 80 rounds, but eventually for the full 80-round SHA-1 implementation. To give a rough idea of progress, as of 2012 researchers have theoretical methods that they claim will find a SHA-1 collision in around 2^{61} operations, much less than the ostensible 2^{80} security level.

This prompted the development and publication in 2001 of a suite of hash functions known collectively as SHA-2. The individual hash functions in SHA-2 are called SHA-n, where $n \in \{224, 256, 384, 512\}$ indicates the block size.[2] As with SHA-1, these hash functions were developed internally by the NSA. They are similar in some ways to SHA-1, but have so far resisted any serious attacks when the full number of rounds are used. However, given the attacks on SHA-1, it was felt desirable to create an alternative hash function based on different principles, so in 2007 the United States government opened a 5-year competition to design a new general purpose hash function to be called SHA-3. The winner of the competition, announced in 2012, was created by G. Bertoni, J. Daemen, G. Van Assche, and M. Peeters. As of 2014, a draft standard for SHA-3 is available, but a final standardization document is not yet ready.

8.2 Random Numbers and Pseudorandom Number Generators

We have seen that many cryptographic constructions require the use of random numbers. For example:

[2]This means that a direct search takes approximately 2^n steps to invert SHA-n, and a naive collision algorithm takes approximately $2^{n/2}$ steps to find a collision for SHA-n. Of course, none of this proves that SHA is secure, and indeed the difficulty of inverting or finding collisions for SHA is not related, as far as is known, to any standard mathematical problem.

- The key creation phase of virtually all cryptosystems requires the user to choose one or more random (prime) numbers. The same is true for creating keys in digital signature schemes.
- The Elgamal public key cryptosystem (Sect. 2.4) uses a random element (random number) during the encryption process, and ElGamal-type digital signature schemes such as DSA and ECDSA (Sect. 4.3) use a random element for signing.
- The NTRU public key cryptosystem (Sect. 7.10) also uses a random element during the encryption process.
- The entire premise of probabilistic encryption schemes (Sect. 3.10) is to incorporate randomness into the encryption process.
- Even completely deterministic cryptosystems such as RSA gain important security features when some randomness is incorporated into the plaintext; see Sect. 8.6.

Ideally, we would like a device that generates a completely random list of 0s and 1s. Such devices exist, at least if one believes that quantum theory is correct. They are based on measuring the radioactive decay of atoms. According to quantum theory, given an atom of some radioactive substance, there is a number T such that the atom has a 50 % chance of decaying in the next T seconds, but there is no way of predicting in advance whether the atom will decay. So the device can wait T seconds and then output a 1 if the atom decays and a 0 if it does not decay. The device then chooses another (undecayed) atom and repeats the process. In principle, this gives a completely random unpredictable bit string.[3] Unfortunately, as a practical matter, it is expensive to build a Geiger counter into every computer!

Modern cryptosystems avoid this problem by starting with a random seed and feeding it and other data into a function to produce a long random-looking bit string. A function of this sort is called a *pseudorandom number generator* (PRNG). Notice the contradiction in the terminology. A pseudorandom number generator is a function, so the output that it produces is not random at all, the output is completely determined by the input. However, one hopes that it should be difficult to distinguish output of a PRNG from the output of a true random number generator.

One model of a PRNG is as a function of two variables $F(X, Y)$. In order to get started, Alice chooses a truly random seed value S. (Or if not truly random, then as random as she can make it.) She then computes the numbers

$$R_0 = F(0, S), \qquad R_1 = F(1, S), \qquad R_2 = F(2, S), \quad \dots .$$

The list of bits $R_0 \parallel R_1 \parallel R_2 \parallel \cdots$ is Alice's (pseudo)random bit string.

[3] In practice, more sophisticated measurements of radioactivity are used, but the underlying principle is that quantum theory gives precise probabilities that certain measurable events will occur over a given time period.

In order to be useful for cryptography, a PRNG should have the following two properties:

1. If Eve knows the first k bits of Alice's random bit string, she should have no better than a 50 % chance of predicting whether the next bit will be a 0 or a 1. More precisely, there should not be a fast (e.g., polynomial-time) algorithm that can predict the next bit with better than 50 % chance of success.

2. Suppose that Eve somehow learns part of Alice's random bit string, for example, suppose that she finds out the values of $R_t, R_{t+1}, R_{t+2}, \ldots$. This should not help Eve to determine the earlier part $R_0, R_1, \ldots, R_{t-1}$ of Alice's string.

A PRNG with these properties is said to be *cryptographically secure.*

Example 8.3. One can build a PRNG out of a hash function Hash by choosing an initial random value S and setting

$$R_i = \mathsf{Hash}(i \parallel S).$$

(See Sect. 8.1 for a discussion of hash functions.) Of course, not every hash function yields a cryptographically secure PRNG.

Example 8.4. One can also build a PRNG from a symmetric (i.e., private key) cryptosystem, for example DES or AES (see Sect. 8.12). Here is one way to build a PRNG that has been accepted as a public standard. Start with a random seed S and a key K for the cryptosystem, and let E_K be the associated encryption function. Each time a random number is required, use some system parameters, e.g., the date and time as returned by the computer's CPU, to form a number D and encrypt D using the key K, say

$$C = E_K(D).$$

Then output the "random" number

$$R = E_K(C \mathbin{\mathsf{xor}} S)$$

and replace S with $E_K(R \mathbin{\mathsf{xor}} C)$.

Remark 8.5. Alternatively, a PRNG can be used as a symmetric cipher. The seed value S is Alice and Bob's private key. In order to send a message M, Alice breaks M into pieces $M = M_0 \parallel M_1 \parallel M_2 \parallel \cdots$. She then encrypts the ith piece of the message as

$$C_i = M_i \mathbin{\mathsf{xor}} R_i.$$

Since Bob knows the seed value S, he can compute the same pseudorandom string $R_0 \| R_1 \| R_2 \| \cdots$ that Alice used to encrypt, so he can recover the message as $M_i = C_i \mathbin{\mathsf{xor}} R_i$. (Notice that if the R_i were truly random, then Alice and Bob would be using a one-time pad. However, in that case, they would need to have exchanged all of the R_i's before sending encrypted messages.)

Remark 8.6. PRNGs that are based on hash functions such as SHA or symmetric ciphers such as DES or AES are fast and, as far as is known, cryptographically secure, but the security is not based on reduction to a well-known mathematical problem. There are PRNGs whose security can be reduced to the difficulty of solving a hard mathematical problem such as factoring, but such PRNGs are much slower and thus not used in practice.

8.3 Zero-Knowledge Proofs

In this section we introduce you to two new characters: Peggy, the prover, and Victor, the verifier. Informally, a *zero-knowledge proof* is a procedure that allows Peggy to convince Victor that a certain fact is true without giving Victor any information that would let Victor convince other people that the fact is true. As with many cryptographic constructions, this seems at first glance to be impossible. For example, how could Peggy (in New York) convince Victor (in California) that her house is red without sending Victor a picture of the house? And if she sends Victor a picture, then Victor can show the picture to other people as proof of Peggy's house color.[4]

In practice, an (interactive) zero-knowledge proof generally involves a number of challenge–response communication rounds between Peggy and Victor. In a typical round, Victor sends Peggy a challenge, Peggy sends back a response, and then Victor evaluates the response and decides whether to accept or reject it. After a certain number of rounds, a good zero-knowledge proof showing that a quantity y has some property $\mathcal{P}$ should satisfy the following two conditions:

Completeness If y does have property $\mathcal{P}$, then Victor should always accept Peggy's responses as being valid.

Soundness If y does not have property $\mathcal{P}$, then there should be only a very small probability that Victor accepts all of Peggy's responses as being valid.

In addition to being both sound and complete, a zero-knowledge proof should not convey useful information to Victor, whence the name. Before attempting to describe the somewhat subtle idea contained in the phrase "zero-knowledge," we pause to present a concrete example of a zero-knowledge proof.

Example 8.7. Peggy chooses two large primes p and q and publishes their product N. Peggy's task is to prove to Victor that a certain number y is a square modulo N without revealing to Victor any information that would help him to prove to other people that y is a square modulo N. We note that since Peggy knows how to factor N, if y is a square modulo N, then she can find a square root for y, say x, satisfying

[4]This house color scenario is just an informal analogy. Since Victor and Peggy undoubtedly both own Photoshop, a picture of Peggy's house doesn't actually prove anything!

$$x^2 \equiv y \pmod{N}.$$

In each round, Peggy and Victor perform the following steps:

1. Peggy chooses a random number r modulo N. She computes and sends to Victor the number
$$s \equiv r^2 \pmod{N}.$$
2. Victor randomly chooses a value $\beta \in \{0,1\}$ and sends β to Peggy.
3. Peggy computes and sends to Victor the number
$$z \equiv \begin{cases} r \pmod{N} & \text{if } \beta = 0, \\ xr \pmod{N} & \text{if } \beta = 1. \end{cases}$$
4. Victor computes $z^2 \pmod{N}$ and checks that
$$z^2 \equiv \begin{cases} s \pmod{N} & \text{if } \beta = 0, \\ ys \pmod{N} & \text{if } \beta = 1. \end{cases}$$
If this is true, Victor accepts Peggy's response; otherwise, he rejects it.

Peggy and Victor repeat this procedure n times, where n is reasonably large, say $n = 80$. If all of Peggy's responses are acceptable, then Victor accepts Peggy's proof that y is a square modulo N; otherwise, he rejects her proof.

It is easy to check completeness and soundness. For completeness, note that if y is a square modulo N, then the z that Peggy sends to Victor satisfies $z \equiv x^\beta r \pmod{N}$, so

$$z^2 \equiv x^{2\beta} r^2 \equiv y^\beta s \pmod{N}.$$

Thus Victor always accepts Peggy's response.

Conversely, suppose that y is not a square modulo N. Then regardless of how Peggy chooses s, only one of the two values s and ys is a square modulo N. Hence there is a 50 % chance that Peggy will not be able to answer Victor's challenge, since half the time Victor will require Peggy to prove that s is a square and half the time he will require her to prove that ys is a square. Thus if y is not a square, then the probability that Peggy can provide valid responses to n different challenges is 2^{-n}. So if Peggy is able to send 80 valid responses, Victor should be convinced that y is indeed a square modulo N.

We now consider in what sense Peggy's zero-knowlege proof that y has property $\mathcal{P}$ should not help Victor to subsequently prove to anyone else that y has property $\mathcal{P}$. Informally, the idea is that Victor should be able to generate lists of bogus responses that are indistinguishable from lists of genuine responses created by Peggy. The conclusion is that Peggy's responses do not give Victor any information, because if they did, he could get the same sort of information using his self-generated bogus lists of responses. Rather than giving the precise mathematical formulation of this idea, which involves the statement that two probability distributions are identical, we are content to present an example.

Example 8.8. Continuing with the zero-knowledge proof described in Example 8.7, suppose that Victor has finished talking to Peggy, and now he wants to convince some other verifier, say Valerie, that y is a square modulo N. At the end of his communications with Peggy, Victor has amassed a list of triples

$$(s_1, \beta_1, z_1), (s_2, \beta_2, z_2), (s_3, \beta_3, z_3), \ldots,$$

where each triple satisfies

$$z_i^2 \equiv y^{\beta_i} s_i \pmod{N}.$$

Thus if $\beta_i = 0$, then Victor knows a square root of s_i modulo N, and if $\beta_i = 1$, then Victor knows a square root of ys_i modulo N, but unless the list is extremely long, it is unlikely that there will be any values of s for which Victor knows a square root modulo N of both s and ys. And if Victor knows only one of these two square roots, then there is a 50 % chance that he will be unable to answer each of Valerie's challenges. Hence Peggy's responses are of minimal help to Victor if he wants to prove to Valerie that y is a square modulo N.

Even more is true. Without talking to Peggy at all, Victor can create lists of triples $(s_1, \beta_1, z_1), (s_2, \beta_2, z_2), \ldots$ that are indistinguishable from valid lists generated by Peggy and Victor together. For example, if, when actually talking to Peggy, Victor chooses $\beta = 0$ and $\beta = 1$ randomly in Step 2, then he can generate a similar list of triples (s, β, z) without talking to Peggy by randomly choosing z mod N and $\beta \in \{0, 1\}$ and setting

$$s \equiv z^2 (y^{\beta})^{-1} \pmod{N}.$$

This informal argument shows why the data that Peggy sends to Victor during their interaction does not help Victor prove to anyone else that y is a square modulo N. If it did, then Victor could use his self-generated list of triples for the same purpose.

Remark 8.9. There are various levels of zero knowledge. For example, there is *perfect zero-knowledge*, in which Victor's bogus list of responses is statistically identical to Peggy's actual list; there is *statistical zero-knowledge*, in which the bogus list is statistically extremely close to the actual list; and there is *computational zero-knowledge*, which means that there is no efficient algorithm that can distinguish a bogus list from an actual list. The proof that "y is a square modulo N" described in Example 8.7 is an example of a perfect zero-knowledge proof.

8.4 Secret Sharing Schemes

A *secret sharing scheme* does what its name suggests: it provides a way of sharing a secret among several people. For example, the combination to a vault in a bank might be shared among the president and two vice-presidents by giving each of them one third of the combination. It then requires all three of them to open the vault. Alternatively, we might give half to the president and the other half to each vice-president. Then the vault can be opened by the president and either of his vice-presidents.

However, this example does not meet the requirements of a true secret sharing scheme, since knowledge of any part of the vault's combination makes it easier to guess the full combination. In a true secret sharing scheme among a group of n people, no subgroup of $n-1$ people should be able to gain an advantage in discovering the secret.

It is not hard to construct such a scheme. For example, to share a secret number S mod m among n people, select $n-1$ random numbers

$$D_1, D_2, \ldots, D_{n-1} \quad \text{modulo } m,$$

and set

$$D_n \equiv S - D_1 - D_2 - \cdots - D_{n-1} \pmod{m}.$$

The ith participant receives the value of D_i, and it requires all n values to recover the secret

$$S \equiv D_1 + D_2 + \cdots + D_n \pmod{m}.$$

More generally, suppose that we want to share a secret among n people in such a way that any t of them can recover the secret, but no $t-1$ of them can do so. These are called (t, n) *threshold sharing schemes*, where n is the number of participants and t is the *threshold* of the scheme. Threshold secret sharing schemes with $t < n$ are more difficult to construct; the first ones were invented independently by Adi Shamir [123] and George Blakley [15] in 1979.

We briefly describe *Shamir's secret sharing scheme* for n participants and threshold t. The underlying idea is that it takes $k+1$ values to determine a polynomial of degree k. Thus a linear polynomial $ax+b$ is determined by two values (a line is determined by two points), a quadratic polynomial ax^2+bx+c by three values, etc.

Suppose that we want to share a secret number S among n people so that any t of them can recover S, but fewer than t cannot. We set $a_0 = S$, choose random numbers $a_1, a_2, \ldots, a_{t-1}$, and form the polynomial

$$f(x) = a_0 + a_1 x + a_2 x^2 + \cdots + a_{t-1} x^{t-1}.$$

Next we choose n random values for x, say $x_1, x_2, \ldots, x_n$, and compute

$$y_i = f(x_i) \quad \text{for } 1 \le i \le n.$$

(In practice, one might simply take $x_i = i$.) The ith participant is given the value y_i.

Suppose now that t of the participants want to recover the secret, where to ease notation, we will assume that they are participants 1 through t. After sharing their y_i values, these t participants can form the following system of equations:

$$\begin{aligned} y_1 &= f(x_1) = a_0 + a_1x_1 + a_2x_1^2 + \cdots + a_{t-1}x_1^{t-1}, \\ y_2 &= f(x_2) = a_0 + a_1x_2 + a_2x_2^2 + \cdots + a_{t-1}x_2^{t-1}, \\ &\vdots \qquad\qquad \vdots \\ y_t &= f(x_t) = a_0 + a_1x_t + a_2x_t^2 + \cdots + a_{t-1}x_t^{t-1}. \end{aligned}$$

The participants know all of the x_i and y_i values, so they know this system of t linear equations for the t unknown values $a_0, a_1, \ldots, a_{t-1}$. They can now solve the system, e.g., using Gaussian elimination, to find the a_j values and thus recover the secret value $a_0 = S$. (In practice, a more efficient way to reconstruct $f(x)$ is to use what are known as Lagrange interpolation polynomials.)

8.5 Identification Schemes

An *identification scheme* is an algorithm that permits Alice to prove to Bob that she is really Alice. If Alice is meeting Bob face to face, then Alice might use her driver's license or passport for this purpose. But the problem becomes more difficult when Bob and Alice are communicating over an insecure network. An important feature of a secure identification scheme is that if Eve listens to Bob and Alice's exchange of information, she should not be able to impersonate Alice. Indeed, even Bob should not be able to impersonate Alice.

Identification schemes typically operate by performing a *challenge and response*. This means that Bob starts by sending some sort of challenge to Alice. Alice's response to the challenge demonstrates her identity by showing that she has knowledge that only Alice possesses. Sometimes there is more than one round of challenges and responses.

In practice, the first step is for some trusted authority (TA) to issue private and public identification keys to Alice. Before doing this, the TA actually verifies Alice's identity, say by meeting her and looking at her passport. Then, when Bob issues Alice the challenge, she uses her private key to create the response, and Bob verifies the response using Alice's public key. However, this is too simplistic. How can Bob be sure that Alice's purported public key was created for the real Alice? The answer is that when the TA issues Alice's identification keys, he also gives Alice a digital signature on a hash of her identity and her public key. Part of Bob's verification routine then includes using the TA's public verification key to check the signature on Alice's public information.

There are many identification schemes based on the usual underlying hard problems. For example, there are identification schemes due to Schnorr and to Okamoto that use the discrete logarithm problem, there is an RSA-style scheme due to Guillou and Quisquater that relies on exponentiation modulo a composite modulus, and there is an identification scheme due to Feige, Fiat, and Shamir whose security is based on the difficulty of taking square roots modulo a composite modulus.

Identification schemes are closely related to digital signatures (Chap. 4) and to zero-knowledge proofs (Sect. 8.3). The latter connection is clear, since Alice identifies herself by demonstrating that she has a piece of information. Ideally, Alice should prove that she has this knowledge without giving Bob or Eve any useful information about the proof.

For the relation with digital signatures, it is a standard observation that any challenge–response identification scheme can be turned into a digital signature scheme. The trick is to use a hash of the document being signed as the challenge. Alice's response serves as the signature. Since a secure hash function prevents Alice from having any a priori knowledge of $\mathsf{Hash}(D)$, the hash value truly acts as a random challenge. Bob can then easily verify Alice's signature on D by computing $\mathsf{Hash}(D)$ and checking that Alice's signature is the correct response for the challenge $\mathsf{Hash}(D)$. Conversely, all of the digital signature schemes described in Chap. 4 can be used as identification schemes, with the hash of the document being replaced by Bob's challenge and Alice's signature serving as the response.

8.6 Padding Schemes, the Random Oracle Model, and Provable Security

Alice asks Bob to send her a bit string x consisting of 1024 bits. Bob is supposed to use Alice's 1024-bit RSA public key (N, e) and apply the following algorithm: Bob first computes $y \equiv x^e \pmod{N}$, then computes $z = N \,\mathsf{xor}\, x$, and finally transmits the concatenation $y \,\|\, z$ to Alice.

Of course, this is completely silly. Eve can simply ignore y and recover x immediately after peeling off z. The question is, why is this cryptosystem *particularly* silly? What if another system that we actually use is equally silly for some reason that would be obvious if only we were a bit smarter. For example, Exercise 3.11 describes a somewhat complicated cryptosystem that appears to require knowledge of the factorization of an integer into two primes to break, but in fact, the public key already gives away the factorization.

Our problem boils down to the following. We have convinced ourselves that a certain process, for example RSA encryption, is hard to reverse unless an adversary possesses a key piece of information. We have a cryptosystem that uses this process to encrypt a message. But how do we know for sure that the only way to decrypt the message is to invert the process? It's a little like protecting a treasure in your house by building an incredibly strong lock on the front door, but then walking off and leaving a side door open. What's missing is a guarantee that the only way to steal the treasure is by opening the front door, and by that we mean opening it by picking the lock, not by cutting a hole in the door or knocking it off its hinges.

In order to solve this problem, cryptographers try to do a precise analysis and to give a proof of security for a given cryptosystem. The ultimate goal is to nail down *exactly* the underlying hard problem, and then to construct a proof showing that anyone who can break the cryptosystem can also solve the hard problem. Even more challenging, the argument has to be correct! Such analyses and arguments make up a significant part of modern-day academic cryptography.

Consider the RSA cryptosystem, for example. RSA is based on the hardness of the problem of factorization, but it is not clear at all that breaking RSA is equivalent to factorization. That is, the ability to quickly factor large numbers would enable one to break RSA, but there might be another way to solve the RSA problem without directly solving the problem of factorization. It is tempting to circumvent this difficulty by defining the hard problem that RSA is based on to be precisely the hard problem on which RSA is based. This gains you theoretical security, since your proof of security is now a tautology, and it might even gain you a little bit of practical security, since the passing of years lends credence to the belief that the RSA problem itself is fundamentally hard.

The other cryptosystems that we have studied have similar difficulties. For example, Diffie–Hellman key exchange and the Elgamal cryptosystem are not known to be equivalent to the discrete logarithm problem, and discrete logarithm digital signature schemes such as DSA (Sect. 4.3) rely on the difficulty of solving a strange equation in which the unknown quantity appears as both a base and an exponent. Similarly, the security of NTRUEncrypt is (probabilistically) equivalent to the problem of solving the shortest or closest vector problems in a certain class of lattices. Thus if SVP or CVP were solved for general lattices, then NTRUEncrypt would certainly be broken, but it is also possible that there is an easier way to solve SVP or CVP in the NTRU lattices due to their special form.

In 1979 Rabin [108] introduced a method of public key encryption based on taking square roots modulo a composite modulus $N = pq$. The novelty of Rabin's cryptosystem was that he could *prove* that an adversary capable of decrypting arbitrary ciphertexts could also, with high probability, factor the modulus. This is, on the face of it, quite encouraging. On the other hand, it also means that Rabin's cryptosystem is susceptible to a chosen ciphertext attack, since Rabin's proof essentially says that a decryption oracle allows one to factor the modulus.

At first glance, the whole notion of chosen ciphertext attacks seems counterintuitive and artificial. After all, why would Alice blindly return the decryption of any ciphertext given to her. The answer is that these days Alice is a computer program, and computer programs will do anything that they are programmed to do. In particular, they might be programmed to interchange various types of information, including possibly decrypted messages as a means of identification.

As cryptography developed into a modern science, cryptographers realized that a potential way around this type of problem is to pad encrypted messages using a padding that mixes random data with the message. The object is to somehow create a situation in which Bob can verify that the ciphertext that he is decrypting was actually created by a person (Alice) who had knowledge of the original plaintext. Further, Bob should be confident that when Alice created the ciphertext, she had no significant control over the padding. (An early padding scheme for RSA that lacked this randomness feature was broken by Bleichenbacher [16] by simply sending a large number of messages and seeing which ones were accepted as valid plaintexts, without even being told their decryptions!) The standard way to introduce random-like qualities is to create the padding by applying a hash function to the plaintext. This makes the padding essentially "random" and hence removes it from Alice's direct control, while still leaving it predetermined because it is obtained by evaluating a (hash) function. This crucial assumption, i.e., that hash functions are somehow simultaneously random and deterministic, was introduced by Bellare and Rogaway [12] in 1993. They called security proofs based on this assumption the *random oracle model.*

It was hoped that it would be possible, with precise definitions and careful assumptions, to prove that certain padding schemes really were secure against chosen ciphertext attacks. An early proposal called the Optimal Asymmetric Encryption Padding (OAEP) scheme was proposed by Bellare and Rogaway in 1994 [13]. In this article they proved that OAEP provides security against chosen ciphertext attacks, an assertion that was accepted by the cryptographic community for 7 years, during which time OAEP was written into industry standards.

We illustrate padding schemes by describing OAEP. Bob uses an encryption function E and two hash functions G and H. He chooses a plaintext m and a random bit string r and computes

$$b = G(r) \operatorname{\mathsf{xor}} (m \,\|\, 00\cdots 0),$$
$$c = E\bigl(b \,\|\, \bigl(H(b) \operatorname{\mathsf{xor}} r\bigr)\bigr).$$

Bob sends the ciphertext c to Alice. She decrypts c and breaks it apart to recover b and $H(b) \operatorname{\mathsf{xor}} r$. She uses this to compute first $H(b)$ and then

$$r = H(b) \operatorname{\mathsf{xor}} \bigl(H(b) \operatorname{\mathsf{xor}} r\bigr).$$

Finally, she computes $G(r) \operatorname{\mathsf{xor}} b$ to recover $m \,\|\, 00\cdots 0$. If this string ends with an appropriate number of 0's, Alice accepts m as a valid plaintext; otherwise, she rejects it. Notice how OAEP uses the hash functions to make every bit of c depend on every bit of m and r, in the sense that changing any one bit of either m or r causes every bit of c to have a 50 % chance of changing.

Unfortunately, it was shown by Shoup in 2001 [131] that one of the assumptions in the security proof of OAEP was unreasonable, in the sense that it assumed that no amount of probing of a certain piece of information could produce useful information. This embarrassing incident underlined one of the fundamental limitations of security proofs. Just as it is possible for a complicated cryptographic system to be insecure because the cryptographer has protected it only against the lines of attack that he knows, a "proof of security" is only as secure as the validity of its assumptions (not to mention the correctness of its logic). Shoup proposed a variant called OAEP+ and gave a correct proof of its security in the random oracle model.

One might think that for the purposes of analyzing the security of a cryptosystem, it is very reasonable to assume that hash functions behave exactly as they are supposed to behave, with no hidden flaws, biases, or weaknesses. However, there have been fierce arguments regarding the use of the random oracle model as the basis for the security of cryptosystems, with the result that an alternative (stronger) assumption called the *standard model* has been developed to provide what has become known as *provable security*. For some hint of the controversy that this has engendered, see for example [69].

For an overview of this subject we recommend the highly readable survey articles of Koblitz and Menezes [70] and Bellare [11]. Koblitz and Menezes remark that "The first books on cryptography that the two of us wrote in our naive youth suffer from this defect: the sections on security deal only with the problem of inverting the one-way function." (Also included is a footnote clarifying the meaning of the word "youth.") This quotation highlights the tendency of those trained in pure mathematics, when introduced to the field of public key cryptography, to concentrate primarily on the concept of (trapdoor) one-way functions, to the exclusion of the many practical issues that arise in real-world implementations. Indeed, the authors of the present book must admit that even with full knowledge of this pitfall, it is the quest to construct, understand, and apply new one-way functions to cryptographic systems that draws them to the subject.[5]

8.7 Building Protocols from Cryptographic Primitives

The use of public key cryptography and digital signature schemes in the real world involves far more than simply implementing one or two basic algorithms. Applications almost always involve a number of different cryptographic primitives. For example, a public key might itself be digitally signed, and the public key cryptosystem, whose plaintexts are padded using a hash function, might be used to send the key for a symmetric cipher. So this single application involves

[5]And unfortunately, the authors of this book cannot even offer up youth, by any definition, as an excuse for their behavior.

choosing a public key cryptosystem and digital signature scheme (e.g., RSA), a hash function (e.g., SHA-1), a padding scheme (e.g., OAEP+), and a symmetric cipher (e.g., AES).

However, even this simple description is far from sufficient. For example, are Bob and Alice using RSA with 1024-bit keys or with 2048-bit keys? How long are their AES keys? Exactly how do they use their hash function or symmetric cipher to generate pseudorandom numbers? And even if they specify all of the obvious parameters and decide how to use all of the cryptographic primitives, they're still not ready to communicate. They need to agree on formatting. This may seem pedantic, but it is very important. Before Bob and Alice exchange messages, they need to specify the exact meaning of each byte of the data, e.g., which bytes are the ciphertext, which bytes are the signature, etc. Even something as seemingly trivial as the order in which data is stored in memory and transmitted between computers can cause total system failure if not specifically addressed.[6]

A *cryptographic protocol* is a complete description of everything that is needed in order to implement a cryptographic procedure. The term is not entirely precise, but it generally refers to the way in which one or more cryptographic algorithms are to be implemented and coordinated with one another.

The theory of cryptographic protocols, especially their creation and proofs of security, is a subject in its own right, with numerous articles and books devoted to the topic. We note that even if one assumes that the underlying cryptographic primitives such as RSA or ECC are secure, it is extremely easy to use such primitives to create a seemingly secure protocol that is, in fact, vulnerable to attack. This is especially true if one designs the protocol primarily with a view toward efficiency and flexibility; it is vital that security considerations be given top priority. Further, given the complexity of any protocol that is formed by fitting together several cryptographic primitives, it can be difficult to give even a convincing heuristic argument that the protocol has no security weaknesses. In brief, the construction and analysis of cryptographic protocols is not for the faint of heart, but it is of fundamental importance if modern cryptography is to be of any use in the real world.

In order for computers in far-flung parts of the world to communicate securely (or at all), someone needs to sit down and specify precise cryptographic protocols. This is normally done by standard-setting bodies that are formed either by the government or by representatives from the relevant industries. Even restricting to the field of secure communications, there are many such bodies in existence, each of which consumes countless man-hours of effort and innumerable reams of paper[7] as it spends years issuing draft versions

---

[6]Data stored on a computer as least-significant-byte first is said to be in *little-endian* format, while the reverse order is called *big-endian*. These amusing names come from *Gulliver's Travels*, in which the inhabitants of one kingdom are required to crack their soft-boiled eggs at the "little end," while those in a rival kingdom crack their eggs at the "big end."

[7]We ask the reader to excuse our hyperbole. In particular, the aforementioned reams of paper are figurative, having largely been replaced by megabytes of disk space.

of the eventual final standard. Among the many organizations involved in this process are the Internet Engineering Task Force (IETF), the Institute of Electrical and Electronics Engineers (IEEE), and the American National Standards Institute (ANSI). The IETF supervises Request for Comment (RFC) documents, which are sometimes later released as official standards. The IEEE sponsors the important P1363 standardization project on public key cryptography. There are many reasons why the setting of standards for cryptographic protocols is such an arduous process, including legitimate differences of opinion as to the security of different protocols and the financially serious issue of the extent to which patented algorithms should be incorporated into publicly approved standards. A successful member of a standards-setting board needs not only a solid technical background, but also must have excellent political skills.

8.8 Blind Digital Signatures, Digital Cash, and Bitcoin

Digital cash, and the blind digital signatures on which it is based, were both introduced by David Chaum in 1982 [26]. The idea of a blind digital signature is that the document to be signed is first blinded (concealed) and then signed, yet the signature can still be verified against the original unblinded document. One situation where blind signatures are used is in voting protocols, where a voter might want an election official to validate (sign) her ballot without revealing to the official how she voted.

Example 8.10. Here is a simple example of a blind signature scheme using RSA. Alice has a document hash D, and she wants Samantha to create a blind RSA signature on D using Samantha's signing triple (N, e, d) as described in Table 4.1.

- Alice picks a random number $R \bmod N$ and uses Samantha's public verification key e to compute $D' \equiv R^e D \pmod{N}$.
- Samantha uses her private signing key d to compute the signature
$$S \equiv (D')^d \equiv (R^v D)^d \equiv R^{ed} D^d \equiv RD^d \pmod{N}.$$
(In this computation, we have used the fact that $R^{ed} \equiv R \pmod{N}$.)
- Since Alice knows R, she can compute $R^{-1}S \equiv D^d \pmod{N}$, which is the signature on her document D.

Notice that Alice does not know Samantha's private signing key d, while Samantha does not know the document D that was signed.

A second application of blind digital signatures is to digital cash. The idea is that when Alice withdraws digital cash from her bank account and spends it, her bank is unable to determine who received the money. Thus digital cash should preserve the anonymity of physical cash transactions, since when Alice withdraws actual currency from her account, her bank does not know who she gives it to. (Of course, her bank can record the serial numbers of the bills that she receives and then track those bills as they move through the banking system.) Cryptographic methods adding various enhancements, such as off-line transactions and safeguards against double spending of digital cash units, were described in later work of Chaum, his co-authors, and others; see for example [27].

A potential weakness of these digital cash systems is the necessity of having a trusted central authority, be it a bank, an e-cash company, or a government, that is responsible for issuing cash and safe-guarding the security of the system. In principle, the fiat currencies issued by any government have this weakness, since at any time the issuing body might decide to (say) halve the value of everyone's wealth by doubling the amount of available currency. In the short term, this would make the government very rich, since they would suddenly have a lot of extra cash to spend. In practice, it would lead to runaway inflation.[8]

The article "Bitcoin: A Peer-to-Peer Electronic Cash System" [91] proposed a practical way to create a digital cash system without requiring a trusted central authority. The idea is to digitally sign and cryptographically join every transaction into a massive linked chain. To prevent a small group of users from modifying the chain and, say, repudiating or canceling a payment, the system is designed so that it takes a significant amount of computing power to add a transaction to the chain. People who do these time-consuming *mining* computations that keep the system running are rewarded with a small fraction of a bitcoin. The complete transaction chain is stored on myriad computers and is always available for public scrutiny.

There is much hype and much controversy surrounding Bitcoin, including for example the question of whether a currency that is not backed by anything tangible can have value.[9] This being a mathematical text, we will also not pursue the fascinating economic and sociological issues raised by Bitcoin, nor will we discuss how Bitcoin has been used as a (highly) speculative investment tool and as a way of concealing illegal activities. Instead, we turn to the cryptographic protocols that underlie Bitcoin.

Bitcoin uses the elliptic curve digital signature algorithm (ECDSA, Sect. 6.4.3) to sign transactions. It uses the specific elliptic curve, prime, and point (E, p, P) denoted by `secp256k1` in the Standards for Efficient Cryptography document [142, page 21]. Based on our current knowledge of the difficulty of solving the ECDHP, this provides approximately 128 bit security. Since it's fun to see the actual public parameters used to secure the

---

[8]Many economists feel that a small amount of inflation is a good thing, since it encourages investment while allowing governments to slowly outgrow their accumulated debts.

[9]In fairness, it is not entirely clear in what sense the "full faith and credit of the United States government" that backs US currency is tangible, although one can argue that this faith and credit represents the taxing authority of the government on the US economy, whose productivity and output are eminently tangible.

Bitcoin chain, we list them here. The Bitcoin curve is

$$E: y^2 = x^3 + 7,$$

and the prime

$$p = 2^{256} - 2^{32} - 2^9 - 2^8 - 2^7 - 2^6 - 2^4 - 1$$

was chosen to have low Hamming weight, which speeds up computations in the field $\mathbb{F}_p$. The prime p and curve E have the further property that

$$N = \#E(\mathbb{F}_p) = p + 1 - 432420386565659656852420866390673177327$$

is prime, which means that every non-zero point in $E(\mathbb{F}_p)$ has order N. The public point $P = (x, y) \in E(\mathbb{F}_p)$ specified by the `secp256k1` standard is

$$x = 55066263022277343669578718895168534326250603453777594175500187360389116729240,$$
$$y = 32670510020758816978083085130507043184471273380659243275938904335757337482424.$$

As mentioned earlier, Bitcoin miners perform time-consuming calculations in order to add bitcoin transactions to the chain. They do this by finding an input to the hash function SHA-256[10] (see Sect. 8.1) that is tied to the transaction(s) and has output containing a specified number of leading (or trailing) zeros. Over time, as more CPU power is devoted to mining, the SHA-256 requirements are constantly being strengthened so that it always takes 5–10 min on average for a transaction to be added to the chain. Again, the purpose of this requirement is to prevent any small group from going back and altering the existing chain.

We close this section with a few interesting Bitcoin tidbits:

- Unlike ordinary currency, it is possible to create Bitcoins that can only be spent using two signatures, or two out of three signatures, etc.
- The total number of bitcoins is designed to stabilize at 21,000,000. This will, in principle, eliminate bitcoin inflation (which many would argue is a bug, not a feature, of a currency).
- As with all protocols that rely on factorization or discrete logarithm based cryptography, the construction of quantum computers (Sect. 8.11) will require switching to a quantum secure cryptosystem. The logistics of such a switch are likely to be problematic without considerable preparation and forethought.

[10]It is interesting that the hash function SHA-256 used by Bitcoin provides 256 bit security, while the ECDSA signature scheme `secp256k1` provides only 128 bit security.

8.9 Homomorphic Encryption

The basic RSA PKC has an interesting multiplicativity property. Suppose that Bob sends Alice two ciphertexts c_1 and c_2 associated to the plaintexts m_1 and m_2. Then the ciphertext associated to the product m_1m_2 is the product of the ciphtertexts c_1c_2. This is true because

$$c_1c_2 \equiv m_1^e m_2^e \equiv (m_1m_2)^e \pmod{N}.$$

On the other hand, the ciphertext associated to $m_1 + m_2$ is almost certainly not the sum of the ciphtertexts $c_1 + c_2$, since in general

$$(m_1 + m_2)^e \not\equiv m_1^e + m_2^e \pmod{N}.$$

A fully homomorphic cryptosystem is one that respects both addition and multiplication. Since homomorphic encryption involves two operations, this suggests that the correct setting for homomorphic encryption is ring theory; see Sect. 2.10. In particular, we recall from Exercise 2.31 that if R and S are rings, then a ring homomorphism $\phi : R \to S$ is a function satisfying

$$\phi(a+b) = \phi(a) + \phi(b) \quad \text{and} \quad \phi(a \star b) = \phi(a) \star \phi(b) \qquad \text{for all } a, b, \in R.$$

Definition. A *fully homomorphic encryption scheme* (FHE scheme) is an encryption scheme $(\mathcal{K}, \mathcal{M}, \mathcal{C}, e, d)$ with the property that the plaintext space $\mathcal{M}$ and the ciphertext space $\mathcal{C}$ are rings and such that for every key $k \in \mathcal{K}$, the encryption function

$$e_k : \mathcal{M} \longrightarrow \mathcal{C}$$

is a ring homomorphism.[11]

Example 8.11. Here is an example of an FHE scheme. Let $\mathcal{M} = \mathcal{C} = \mathbb{F}_{p^d}$ be a finite field with p^d elements (see Sect. 2.10.4), and let $\mathcal{K} = \{1, 2, \ldots, d-1\}$. Then for each $k \in \mathcal{K}$, the encryption function

$$e_k(m) = m^{p^k}$$

is fully homomorphic, with decryption function $d_k(c) = c^{p^{d-k}}$. (We are using the fact that $a^{p^d} = a$ and $(a+b)^p = a^p + b^p$ for all elements $a, b \in \mathbb{F}_{p^d}$.) Of course, this cryptosystem is either very insecure, if d is small, or completely impractical, if d is large.

[11]We like this definition of FHE because it stresses the homomorphism in the name, but we note that the standard definition allows more general constructions. Thus a cryptosystem is said to be *fully homomorphic* if it is possible to compute any function or circuit on encrypted data without knowing the decryption key. Thus one only needs to do addition and multiplication on encryptions of $\mathbb{F}_2$, and it is not strictly necessary for the ciphteretxt space to be a ring.

The search for a secure and practical FHE scheme started shortly after the invention of RSA in the 1970s, but it remained an elusive goal until Craig Gentry devised the first such system in 2009 [47, 48]. Although neither Gentry's original system, nor subsequent improvements, are fast enough to be used in practical applications, research continues on this important topic.

Why, one might ask, is homomorphic encryption interesting or useful? The answer is that using the basic operations of addition and multiplication (even just moduo 2), one can perform any computation that can be done by any computer. Consider the following scenario. Bob has a large amount of data to analyze, but the analysis requires a tremendous amount of computing power, far more than Bob possesses. Alice, on the other hand, owns a lot of computers and is happy to rent computing time to Bob.[12] Unfortunately, the data that Bob needs to analyze is confidential; for example, it might include medical or financial records. So what Bob would like to do is the following:

- Encrypt the data, and also encrypt the computer program that will analyze the data.
- Send the encrypted data and the encrypted program to Alice.
- Have Alice run the encrypted program using the encrypted data *without decrypting either.*
- Receive the encrypted output from Alice, which he then decrypts.

The miracle of an FHE scheme is that it allows Alice to run an encrypted computer program on encrypted data and produce encrypted output, without ever seeing any of Bob's unencrypted material.

Gentry's original FHE scheme has two components. First he constructs a system that is "somewhat homomorphic," and then he uses an ingenious bootstrapping procedure to make it fully homomorphic. The following definition will be used to quantify the notion of "somewhat homomorphic."

Definition. A *monic monomial in n variables* is a polynomial of the form

$$M(X_1,\ldots,X_n) = X_1^{e_1}X_2^{e_2}\cdots X_n^{e_n} \quad \text{with } e_1,\ldots,e_n \ge 0.$$

The degree of M is $\deg(M) = e_1 + \cdots + e_n$. More generally, let

$$P(X_1,\ldots,X_n) = M_1(X_1,\ldots,X_n) + \cdots + M_r(X_1,\ldots,X_n)$$

be a sum of monic monomials, and let $d = \max\deg M_i$. Then we say that P has *additive level* r and *multiplicative level* d. We write $\text{Level}_a(P)$ and $\text{Level}_m(P)$, respectively, for the additive and multiplicative levels of P

Every computation in a ring R may be viewed as evaluating some polynomial $P(X_1,\ldots,X_n)$ at some point $(a_1,\ldots,a_n) \in R^n$, although this may first require simplification using the distributive law. For example,

[12]This situation is quite realistic. There are many companies, large and small, from which one can rent computing power.

$$\begin{aligned}(((a_1 + a_2 + a_3) \star a_4) + a_5) \star a_6 &= a_1a_4a_6 + a_2a_4a_6 + a_3a_4a_6 + a_5a_6 \\ &= P(a_1, \ldots, a_6)\end{aligned}$$

with

$$P(X_1, \ldots, X_6) = X_1X_4X_6 + X_2X_4X_6 + X_3X_4X_6 + X_5X_6.$$

We note that in order to compute $P(a_1, \ldots, a_n)$, we need to do at most $\mathrm{Level}_a(P)$ additions of quantities, each of which is a product of at most $\mathrm{Level}_m(P)$ elements of R.

Definition. Let $(\mathcal{K}, \mathcal{M}, \mathcal{C}, e, d)$ be a cryptosystem such that $\mathcal{M}$ and $\mathcal{C}$ are rings. We say that it is a *leveled homomorphic cryptosystem of level* (L_a, L_m) if for every key $k \in \mathcal{K}$, every polynomial $P(X_1, \ldots, X_n)$ satisfying

$$\mathrm{Level}_a(P) \le L_a \quad \text{and} \quad \mathrm{Level}_m(P) \le L_m,$$

and every $m_1, \ldots, m_n \in \mathcal{M}$, we have

$$d_k\bigl(P\bigl(e_k(m_1), \ldots, e_k(m_n)\bigr)\bigr) = P(m_1, \ldots, m_n),$$

i.e., $P\bigl(e_k(m_1), \ldots, e_k(m_n)\bigr)$ is a valid encryption of $P(m_1, \ldots, m_n)$ for the key k.

Informally, a cryptosystem is leveled homomorphic if encryption e_k is a ring homomorphism when it is applied to expressions that involve a limited number of additions and multiplications in the ring $\mathcal{M}$. Leveled homomorphic cryptosystems are interesting for two reasons. First, Gentry's bootstrapping method sometimes allows one to turn a leveled system into a fully homomorphic system, albeit at a significant loss of efficiency. Second, if the levels L_a and L_m are sufficiently large, then even a leveled system may be useful in practice.

A number of leveled homomorphic cryptosystems have been proposed, and the construction of secure and efficient systems is an active area of research. We illustrate by briefly explaining how the NTRU cryptosystem described in Sect. 7.10 can be used as a leveled homomorphic system.

Example 8.12 (Leveled Homomorphic NTRUEncrypt). To simplify the exposition, we use a variant of NTRUEncrypt in which we assume that the private key polynomials $\boldsymbol{f}(x)$ and $\boldsymbol{g}(x)$ are chosen to satisfy

$$\boldsymbol{f}(x) \equiv 1 \pmod{p} \quad \text{and} \quad \boldsymbol{g}(x) \equiv 0 \pmod{p}; \tag{8.1}$$

see Exercise 7.35. The NTRU plaintext and ciphertext spaces are rings $\mathcal{M} = R_p$ and $\mathcal{C} = R_q$, and encryption of a plaintext $\boldsymbol{m}(x)$ using the (public) key $\boldsymbol{h}(x) \equiv \boldsymbol{f}(x)^{-1} \star \boldsymbol{g}(x) \pmod{q}$ and random element $\boldsymbol{r}(x)$ is given by the formula

$$\boldsymbol{e}(x) \equiv \boldsymbol{h}(x) \star \boldsymbol{r}(x) + \boldsymbol{m}(x) \pmod{q}.$$

Decryption is done by first computing the center-lift of $\boldsymbol{f}(x)\star\boldsymbol{e}(x) \pmod q$, and then reducing the result modulo p. Just as in the proof of Proposition 7.48, see also Exercise 7.35, decryption works provided that every coefficient of the polynomial

$$\boldsymbol{g}(x)\star\boldsymbol{r}(x)+\boldsymbol{f}(x)\star\boldsymbol{m}(x)$$

(without any reduction) has magnitude smaller than $\frac{1}{2}q$.

We now consider two ciphtertexts $\boldsymbol{e}_1$ and $\boldsymbol{e}_2$. We claim that if q is large enough, then the decryption of $\boldsymbol{e}_1+\boldsymbol{e}_2$ using the private key $\boldsymbol{f}$ yields the plaintext $\boldsymbol{m}_1+\boldsymbol{m}_2$, and similarly, the decryption of $\boldsymbol{e}_1\star\boldsymbol{e}_2$ using the private key $\boldsymbol{f}^2$ yields the plaintext $\boldsymbol{m}_1\star\boldsymbol{m}_2$. To see why, we compute

$$\begin{aligned}\boldsymbol{f}\star(\boldsymbol{e}_1+\boldsymbol{e}_2) &\equiv (\boldsymbol{g}\star\boldsymbol{r}_1+\boldsymbol{f}\star\boldsymbol{m}_1)+(p\boldsymbol{g}\star\boldsymbol{r}_2+\boldsymbol{f}\star\boldsymbol{m}_2) \pmod q\\ &\equiv \boldsymbol{g}\star(\boldsymbol{r}_1+\boldsymbol{r}_2)+\boldsymbol{f}\star(\boldsymbol{m}_1+\boldsymbol{m}_2) \pmod q,\end{aligned}$$

so if q is large enough, then the center-lift is exactly the polynomial

$$\boldsymbol{g}\star(\boldsymbol{r}_1+\boldsymbol{r}_2)+\boldsymbol{f}\star(\boldsymbol{m}_1+\boldsymbol{m}_2).$$

Using (8.1), we see that when we reduce modulo p, we get $\boldsymbol{m}_1+\boldsymbol{m}_2 \pmod p$.

Similarly,

$$\begin{aligned}\boldsymbol{f}^2\star(\boldsymbol{e}_1\star\boldsymbol{e}_2) &\equiv (\boldsymbol{g}\star\boldsymbol{r}_1+\boldsymbol{f}\star\boldsymbol{m}_1)\star(\boldsymbol{g}\star\boldsymbol{r}_2+\boldsymbol{f}\star\boldsymbol{m}_2) \pmod q\\ &\equiv \boldsymbol{g}\star(\boldsymbol{g}\star\boldsymbol{r}_1\star\boldsymbol{r}_2+\boldsymbol{f}\star\boldsymbol{r}_1\star\boldsymbol{m}_2+\boldsymbol{f}\star\boldsymbol{r}_2\star\boldsymbol{m}_1)\\ &\qquad\qquad +\boldsymbol{f}^2\star(\boldsymbol{m}_1\star\boldsymbol{m}_2) \pmod q.\end{aligned}$$

So again, if q is large enough, then the center-lift gives us exactly the polynomial

$$\boldsymbol{g}\star(\boldsymbol{g}\star\boldsymbol{r}_1\star\boldsymbol{r}_2+\boldsymbol{f}\star\boldsymbol{r}_1\star\boldsymbol{m}_2+\boldsymbol{f}\star\boldsymbol{r}_2\star\boldsymbol{m}_1)+\boldsymbol{f}^2\star(\boldsymbol{m}_1\star\boldsymbol{m}_2),$$

and reducing modulo p gives $\boldsymbol{m}_1\star\boldsymbol{m}_2 \pmod p$.

We have just shown that if q is large enough, then $\boldsymbol{e}_1+\boldsymbol{e}_2$ is an NTRU encryption of $\boldsymbol{m}_1+\boldsymbol{m}_2$ and $\boldsymbol{e}_1\star\boldsymbol{e}_2$ is an NTRU encryption of $\boldsymbol{m}_1\star\boldsymbol{m}_2$, although for the latter the decryption key is $\boldsymbol{f}^2$. In leveled homomorphic terminology, if q is large enough, then the addition property says that NTRUEncrypt is leveled homomorphic with level $(2,1)$, and the multiplication property says that it is leveled homomorphic with level $(1,2)$. Further, it is clear that by increasing q, we can create leveled homomorphic versions of NTRUEncrypt of any level.

In order to derive a rough relationship between the value of q and the levels achieved, we make the simplifying assumption that $\boldsymbol{f}$, $\boldsymbol{g}$, $\boldsymbol{m}$, and $\boldsymbol{r}$ have coefficients between $-p$ and p, and we put no restriction on the number of nonzero coefficients.

Let $P(X_1,\dots,X_n)$ be a sum of monic monomials, and to ease notation, let

$$\alpha = \alpha(P) = \text{Level}_a(P) \quad \text{and} \quad \mu = \mu(P) = \text{Level}_m(P).$$

It is also convenient to use the homogenized version of P defined by

$$P^*(X_0, \ldots, X_n) = X_0^\mu P(X_1/X_0, \ldots, X_n/X_0),$$

so P^* is a sum of monic monomials, each of which has degree exactly μ. Let $\boldsymbol{m}_1, \ldots, \boldsymbol{m}_n$ be plaintexts, and let $\boldsymbol{e}_1, \ldots, \boldsymbol{e}_n$ be associated ciphertexts. Then $P(\boldsymbol{e}_1, \ldots, \boldsymbol{e}_n)$ decrypts correctly to $P(\boldsymbol{m}_1, \ldots, \boldsymbol{m}_n)$ using the decryption key $\boldsymbol{f}^\mu$ if the largest coefficient of

$$P(\boldsymbol{f}, \boldsymbol{g} \star \boldsymbol{r}_1 + \boldsymbol{f} \star \boldsymbol{m}_1, \ldots, \boldsymbol{g} \star \boldsymbol{r}_n + \boldsymbol{f} \star \boldsymbol{m}_n) \tag{8.2}$$

has magnitude smaller than $\frac{1}{2}q$.

We observe that if $\boldsymbol{a}_1, \ldots, \boldsymbol{a}_t \in R$ have coefficients between $-C$ and C, then the largest coefficient of the product $\boldsymbol{a}_1 \star \cdots \star \boldsymbol{a}_t$ is at most $C^t N^{t-1}$, obtained for example if every coefficient of every $\boldsymbol{a}_i$ is C. Hence the largest coefficient of

$$\boldsymbol{g} \star \boldsymbol{r}_i + \boldsymbol{f} \star \boldsymbol{m}_i$$

is at most $2p^2N$. The monomials appearing in (8.2) are products of μ expressions of the form $\boldsymbol{g} \star \boldsymbol{r}_i + \boldsymbol{f} \star \boldsymbol{m}_i$ with $0 \le i \le n$, where for convenience we set $\boldsymbol{r}_0 = 0$ and $\boldsymbol{m}_0 = 1$. So the monomials in (8.2) have coefficients that are at most $(2p^2N)^\mu N^{\mu-1}$, which is smaller than $(2pN)^{2\mu}$. Finally, we note that (8.2) is a sum of at most α monomials, and hence the largest coefficient in (8.2) is at most $\alpha(2pN)^{2\mu}$. This proves that $P(\boldsymbol{e}_1, \ldots, \boldsymbol{e}_n)$ decrypts correctly to $P(\boldsymbol{m}_1, \ldots, \boldsymbol{m}_n)$ using the decryption key $\boldsymbol{f}^\mu$ if q satisfies

$$q > 2\alpha(2pN)^{2\mu}.$$

Thus the size of q increases linearly with the number of additions, but exponentially with the number of multiplications

8.10 Hyperelliptic Curve Cryptography

A *hyperelliptic curve of genus g* is the set of solutions to an equation of the form[13]

$$C : Y^2 = X^{2g+1} + A_1X^{2g} + \cdots + A_{2g}X + A_{2g+1},$$

with the added requirement that the polynomial $F(X) = X^{2g+1} + \cdots + A_{2g+1}$ have distinct roots.[14] And just as for elliptic curves, we throw in one extra point $\mathcal{O}$ that lives "at infinity." Thus an elliptic curve is a curve of genus 1.

[13] When working over a field $\mathbb{F}_{2^k}$, one uses the more general form $Y^2 + Y = F(X)$.

[14] When one is working over $\mathbb{C}$, the distinct roots condition means complex roots. Over a finite field $\mathbb{F}_p$, the condition may be formulated by requiring that the discriminant of $F(X)$ not vanish, or equivalently that $\gcd(F(X), F'(X)) = 1$, where $F'(X)$ is obtained by formally differentiating $F(X)$.

In general there is no addition law for the individual points on a hyperelliptic curve, but it is possible to define an addition law for collections of points. Roughly speaking, we can take two collections of g points, say

$$\{P_1, P_2, \ldots, P_g\} \quad \text{and} \quad \{Q_1, Q_2, \ldots, Q_g\},$$

and "add" them to obtain a new collection of g points $\{R_1, R_2, \ldots, R_g\}$. This generalizes the addition law on an elliptic curve, but a precise formulation is somewhat more complicated.

To describe the addition law exactly, we define a *divisor on C* to be a formal sum of points

$$n_1[P_1] + n_2[P_2] + \cdots + n_r[P_r] \qquad \text{with } P_1, \ldots, P_r \in C \text{ and } n_1, \ldots, n_r \in \mathbb{Z}.$$

Note that a divisor is simply a convenient shorthand for a finite set of points, each of which has an attached multiplicity. In particular, if $f(X,Y)$ is a rational function on C, then we can attach a divisor to $f(X,Y)$ by listing the points where f vanishes and the points where f has poles, with their appropriate multiplicities. The *degree of a divisor* is the sum of its multiplicities,

$$\deg(D) = \deg\bigl(n_1[P_1] + n_2[P_2] + \cdots + n_r[P_r]\bigr) = n_1 + n_2 + \cdots + n_r.$$

(See Sect. 6.8.2 for a discussion of rational functions and divisors on elliptic curves.)

We next define the *divisor group of C*, denoted by $\mathrm{Div}(C)$, to be the set of divisors on C. Note that we can add and subtract divisors by adding and subtracting the multiplicities of each point. We also let $\mathrm{Div}_0(C)$ be the set of divisors of degree 0. One can prove that the divisor of a function always has degree 0. Two divisors D_1 and D_2 are said to be *linearly equivalent* if

$$D_1 - D_2 = \text{divisor of a function}.$$

We write $\mathrm{Jac}_0(C)$ for the set of divisors of degree 0, with the understanding that linearly equivalent divisors are considered to be identical. The set $\mathrm{Jac}_0(C)$, with the addition law obtained by adding the multiplicities of points, is called the *Jacobian variety of C*. It is the higher-genus analogue of elliptic curves and their addition laws.

A crucial property of $\mathrm{Jac}_0(C)$ is that it can be described as the set of solutions to a system of polynomial equations, and the addition law may also be described using polynomials. So if we take solutions with coordinates in $\mathbb{F}_p$, then we obtain a group (i.e., a set with an addition law) that is completely analogous to the group $E(\mathbb{F}_p)$ of points on an elliptic curve. For notational convenience, let $J = \mathrm{Jac}_0(C)$ and $J(\mathbb{F}_p)$ be the points on $\mathrm{Jac}_0(C)$ with coordinates in $\mathbb{F}_p$. The *Hyperelliptic Curve Discrete Logarithm Problem* (HCDLP) is as follows:

Given P and Q in $J(\mathbb{F}_p)$, find an integer m such that $Q \doteq mP$.

It is clear how one can use hyperelliptic curves for public key cryptography by mimicking the constructions for the multiplicative group (Sects. 2.3 and 2.4) and for elliptic curves (Sect. 6.4). This leads to hyperelliptic Diffie–Hellman key exchange and the hyperelliptic Elgamal public key cryptosystem.

The primary, and from cryptographic purposes fundamental, difference between elliptic curves and hyperelliptic curves is that the latter have larger groups of points. More precisely, there is an analogue of Hasse's theorem (Theorem 6.11), due to André Weil, which says that

$$\#J(\mathbb{F}_p) = p^g + \mathcal{O}\left(p^{g-\frac{1}{2}}\right).$$

For example, a hyperelliptic curve of genus 2 has approximately p^2 points.

As with elliptic curves, one hopes that the best algorithms to solve the HCDLP are collision algorithms such as Pollard's ρ algorithm. (But see Remark 8.13.) Since the group $J(\mathbb{F}_p)$ has approximately p^g elements, this means that it takes $\mathcal{O}(p^{g/2})$ steps to solve HCDLP. Thus using curves with $g > 1$ allows us to achieve security levels equivalent to those on elliptic curves while using a smaller prime p.

However, as g gets large, the computational complexity of the addition law becomes formidable (and there are also security issues), so for concreteness, we consider the case $g = 2$. Then $J(\mathbb{F}_p)$ has approximately p^2 elements and it takes $\mathcal{O}(p)$ steps to solve the HCDLP. This may be compared with an elliptic curve, for which $\#E(\mathbb{F}_p) \approx p$ and ECDLP takes $\mathcal{O}(\sqrt{p})$ steps to solve. Thus $J(\mathbb{F}_p)$ allows us to use primes with approximately half as many digits. This does not lead to a significant speed advantage, because the addition law on J is significantly more complicated than the addition law on E. However, it does mean that ciphertexts, and even more importantly, digital signatures, are half as large using J as they are using E. This becomes a large advantage on highly constrained devices such as radio frequency identification (RFID) tags.

Remark 8.13. It is not actually true that the best known methods to solve the HCDLP are collision algorithms. If the genus g of the curve is moderately large compared to the prime p, then Adleman, DeMarrais, and Huang found an index calculus algorithm that solves the HCDLP in subexponential time. They show that if $2g+1 \geq (\ln p)^{1+\epsilon}$ for some $\epsilon > 0$, then the HCDLP can be solved in $L(p^{2g+1})^c$ steps for some small constant c. In the opposite direction, if p is large, say $p > g!$, then Gaudry found an algorithm to solve the HCDLP in $\mathcal{O}(g^3 p^{2+\epsilon})$ steps. This is not helpful if $g = 1$ or 2, but it is significant if $g \geq 3$ and p is large. This is one of the reasons that only hyperelliptic curves of genus 2 and 3 are being seriously considered for use in cryptography.

Finally, just as for elliptic curves, there are various attacks on the HCDLP using versions of the Weil and Tate pairing (see Sect. 6.9.1), but it is easy to avoid such attacks by appropriate choices of parameters.

8.11 Quantum Computing

The value of each bit in a classical computer is either 0 or 1. In a quantum computer, each so-called quantum bit (qubit) may simultaneously takes on every value between 0 and 1 with varying probabilities. This added flexibility allows many computations, including integer factorization and discrete logarithm problems, to be done very quickly. Thus a working quantum computer with sufficiently many qubits would break RSA and both the classical and the elliptic curve versions of Elgamal. (However, there is at present no polynomial-time quantum algorithm that solve the shortest or closest lattice vector problems.)

Tempting though it is, we will not use this opportunity to give a serious introduction to quantum mechanics. The aim of this section is fairly modest. We sketch the basic ideas behind one remarkable application of quantum mechanics to cryptography: Shor's polynomial-time quantum algorithm [128] for factoring integers and for finding discrete logarithms. The following presentation owes a great deal to Shor's accessible and beautifully written exposition [129], which would serve as a nice start for the interested reader familiar with the concept of a Hilbert space. For those with a less robust background in mathematics and quantum theory, see for example [64].

The fundamental unit of information in classical computers is the binary digit (bit), represented as a 0 or 1. Bits are manipulated according to the principles of Boolean logic, in which connectives such as AND and OR operate on pairs of bits in the usual way, and NOT reverses 0 and 1. Sequences of bits are manipulated by Boolean logic gates, using these Boolean rules, and a succession of gates yields an end state, or computation. A quantum computer manipulates *quantum bits* (*qubits*) via quantum logic gates, which are supposed to simulate the laws of quantum mechanics, especially properties such as superposition and entanglement, which give the field of quantum mechanics its distinctive nonclassical characteristics.

A qubit with two states is typically represented using *ket* notation,[15] in which $|0\rangle$ denotes the 0-state and $|1\rangle$ the 1-state. Then the (pure) states of the system have the form

$$\alpha\,|0\rangle + \beta\,|1\rangle\,,$$

where α and β are complex numbers satisfying $|\alpha|^2 + |\beta|^2 = 1$.

In an n-component system, the 2^n basis elements are represented by sequences such as $|s_i\rangle = |0110\ldots0\rangle$ consisting of a list of n zeros and ones, and a state of the system is

$$\sum_{i=0}^{2^n-1} \alpha_i\,|s_i\rangle\,, \qquad \text{where} \sum |\alpha_i|^2 = 1. \tag{8.3}$$

[15]The rather strange word *ket* is the latter half of the word *bracket*. In quantum mechanics there is also *bra* notation $\langle x|$ and a "bracket pairing" $\langle x|y\rangle$.

A sum (8.3) is called a *superposition of states.* There are other quantum states known as "mixed" states that we do not discuss here, so we omit the word pure in the rest of this discussion. Thus a quantum state is represented by a vector of complex numbers of length 2^n such that sum of the squares of their moduli is equal to one. These are called *complex unit vectors.*

Just as for classical computers, manipulating qubits via quantum logic gates requires the notion of a change of state. A quantum change of state is the result of applying a unitary linear transformation[16] to one of the complex unit vectors representing a state. Actually, there are additional restrictions on which unitary transformations are permitted for changes of state. One of these restrictions is the requirement of *locality*: the unitary matrices should operate on only a fixed finite number of bits. It turns out that 2-bit transformations form the building blocks of the allowable transformations.

The quantum-mechanical interpretation of the α_i's is that $|\alpha_i|^2$ represents the probability that a measurement of the system yields state $|s_i\rangle$. It is the probabilistic interpretation of the complex coefficients of these vectors that encodes the physical realities observed in experiment and predicted by physical theory.

In [129], Shor describes a quantum polynomial-time algorithm to find (with high probability) the order r of a number $x \mod n$. (Recall that the order of x is the smallest integer $r \geq 1$ such that $x^r \equiv 1 \mod n$.) Factorization can be reduced to the problem of finding the order of an integer, because if x is chosen randomly and has even order r, then $\gcd(x^{r/2} - 1, n)$ is likely to be a nontrivial odd factor of n. (See [87].) Shor also gives a polynomial-time quantum algorithm to solve the discrete logarithm problem in $\mathbb{F}_p^*$, and such algorithms also exist for the elliptic curve discrete logarithm problem [107]. Interestingly, there are still no polynomial-time (or even subexponential-time) quantum algorithms to solve the shortest or closest vector problems, so lattice-based cryptosystems are currently secure even against the construction of a quantum computer.[17]

The basic building block of Shor's algorithm is a quantum version of the Fast Fourier Transform. In order to find the order r of a number a modulo n, we choose q to be a power of 2 in the interval between n^2 and $2n^2$. Then for any $0 < a < q$, the state $|a\rangle$ is obtained from the binary representation of the number a. The Fourier transform of $|a\rangle$ is the state

$$\frac{1}{q^{1/2}} \sum_{c=0}^{q-1} |c\rangle \exp(2\pi i a c/q).$$

[16]A *unitary linear transformation* is given by a matrix with determinant one whose conjugate transpose is equal to its inverse.

[17]This is not strictly true because there is a general quantum search algorithm that essentially cuts searches by a square root. So if a quantum computer were built, the key size of lattice-based cryptosystems might need to double. But this would be a small price compared to the devastation that a working quantum computer would cause to factorization and discrete logarithm-based systems.

It turns out that this transformation can be achieved in polynomial time. Shor then applies the quantum Fourier transform to a certain superposition of states and measures the resulting system. The key computation shows that the probability of seeing state $|c\rangle$ is relatively large if there exists a rational number $\frac{d}{r} \in \mathbb{Q}$ satisfying

$$\left|\frac{c}{q} - \frac{d}{r}\right| < \frac{1}{2q}.$$

(Recall that r is the order of a.) Using the continued fraction expansion of the known rational number $\frac{c}{q}$, it is not hard to determine the fraction $\frac{d}{r}$ in lowest terms, since $q > n^2$.

There remains only the "minor" challenge of building a functioning quantum computer. Research in this field has focused on the issue of decoherence, which involves controlling the errors in quantum computation introduced by the interaction of the computer with its environment. There is already a vast literature on quantum computing and quantum computers, reflecting to some extent the large amount of government funding that has been allocated to the subject. One place to start gathering resources about quantum computers is the website for NIST's Quantum Information Program at `qubit.nist.gov`.

Finally, we would be remiss if we did not mention the theory of *quantum cryptography*. The idea is to use quantum-mechanical principles such as the Heisenberg uncertainty principle or the entanglement of quantum states to perform a completely secure key exchange. In particular, if Eve attempts to read either Bob's or Alice's transmission, then quantum theory says that she must alter the data, so Bob and Alice will know that their communication has been compromised.

8.12 Modern Symmetric Cryptosystems: DES and AES

In Sect. 1.7.1 we gave an abstract formulation of symmetric ciphers, and in Sect. 1.7.4 we described several elementary examples. Not surprisingly, none of the examples in Sect. 1.7.4 is secure. Modern symmetric ciphers such as the *Data Encryption Standard* (DES) and the *Advanced Encryption Standard* (AES) are based on ad hoc mixing operations, rather than on intractable mathematical problems used by asymmetric ciphers. The reason that DES and AES and other symmetric ciphers are used in practice is that they are much faster than asymmetric ciphers. Thus if Alice wants to send Bob a long message, she first uses an asymmetric cipher such as RSA to send Bob a key for a symmetric cipher, and then she uses a symmetric cipher such as DES or AES to send the actual data.

DES was created by a team of cryptographers at IBM in the early 1970s, and with some modifications suggested by the United States National Security Agency (NSA), it was officially adopted in 1977 as a government standard suitable for use in commercial applications. (See [97].) DES uses a 56-bit private key and encrypts blocks of 64 bits at a time. Most of DES's mixing operations are linear, with the only nonlinear component being the use of eight *S-boxes* (substitution boxes). Each S-box is a look-up table in which six input bits are replaced by four output bits. Figure 8.1 illustrates one of the S-boxes used by DES.

	0	**1**	**2**	**3**	**4**	**5**	**6**	**7**	**8**	**9**	**10**	**11**	**12**	**13**	**14**	**15**
0	14	4	13	1	2	15	11	8	3	10	6	12	5	9	0	7
1	0	15	7	4	14	2	13	1	10	6	12	11	9	5	3	8
2	4	1	14	8	13	6	2	11	15	12	9	7	3	10	5	0
3	15	12	8	2	4	9	1	7	5	11	3	14	10	0	6	13

Figure 8.1: The first of eight S-boxes used by DES

Here is how an S-box is used. The input is a list of six bits, say

$$\text{Input} = \beta_1\beta_2\beta_3\beta_4\beta_5\beta_6.$$

First use the 2-bit binary number $\beta_1\beta_6$, which is a number between 0 and 3, to choose the row of the S-box, and then use the 4-bit binary number $\beta_2\beta_3\beta_4\beta_5$, which is a number between 0 and 15, to choose the column of the S-box. The output is the entry of the S-box for the chosen row and column. This entry, which is between 0 and 15, is converted into a 4-bit number.

For example, suppose that the input string is '110010'. Binary '10' is 2, so we use row 2, and binary '1001' is 9, so we use column 9. The entry of the S-box in Fig. 8.1 for row 2 and column 9 is 12, which we convert to binary '1100'.

The S-boxes were designed to prevent various sorts of attacks, including especially an attack called differential cryptanalysis, which was known to IBM and the NSA in the 1970s, but published only after its rediscovery by Biham and Shamir in the 1980s. Differential cryptanalysis and other non-brute-force attacks are somewhat impractical because they require knowledge of a large number ($>2^{40}$) of plaintext/ciphertext pairs.

A more serious flaw of DES is its comparatively short 56-bit key. As computer hardware became increasingly fast and inexpensive and computing power more distributed in the 1990s, it became feasible to break DES by a brute-force search of all possible keys, either using many machines over the Internet or building a dedicated DES cracking machine. To demonstrate this vulnerability, in 1999 a group broke a 56-bit DES key in less than 24 h.

One solution to this problem, which has been widely adopted, is to use DES multiple times. There are a number of different versions of *Triple DES*, the simplest of which is to simply encrypt the plaintext three times using three different keys. Thus if we write $\text{DES}(k, m)$ for the DES encryption of the message m using the key k, then one version of triple DES is

$$\text{TDES}(k_1, k_2, k_3, m) = \text{DES}(k_3, \text{DES}(k_2, \text{DES}(k_1, m))).$$

A variation replaces the middle DES encryption by a DES decryption; this has the effect that setting $k_1 = k_2 = k_3 = k$ yields ordinary DES encryption. Another variation, used by the electronics payment industry, takes $k_1 = k_3$, which reduces key size at the cost of some security reduction. Finally, since three DES encryptions triple the encryption time, another version called DES-X uses a single DES encryption combined with initial and final XOR operations with two 64-bit keys. Thus DES-X looks like

$$\mathrm{DESX}(k_1, k_2, k_3, m) = k_3 \ \mathsf{xor}\ \mathrm{DES}(k_2, m \ \mathsf{xor}\ k_1).$$

Although DES and its variants were widely deployed, it suffers from short and inflexible key and block sizes. Further, although DES is fast when implemented in specialized hardware, it is comparatively slow in software. So in 1997 the United States National Institute of Standards (NIST) organized an open competition to choose a replacement for DES. There were many submissions, and after several years of analysis and several international conferences devoted to the selection process, NIST announced in 2000 that the Rijndael cipher, invented by the Belgian cryptographers J. Daemen and V. Rijmen, had been chosen as AES. Since that time AES has been widely adopted, although variants of DES are still in use.[18]

AES is a block cipher in which the plaintext–ciphertext blocks are 128 bits in length and the key size may be 128, 192, or 256 bits. AES is similar to DES in that it encrypts and decrypts by repeating a basic operation several times. In the case of AES, there are 10, 12, or 14 rounds depending on the size of the key. AES is also similar to DES in that it uses an S-box to provide the all-important nonlinearity to the encryption process. However, AES's S-box is constructed using the operation of taking multiplicative inverses in the field $\mathbb{F}_{2^8}$ with 2^8 elements. (See Sect. 2.10.4 for a discussion of finite fields with a prime power number of elements.) Many of AES's basic operations use 128 bit blocks, which are broken up into 16 bytes. Each byte consists of 8 bits and is treated as an element of the field $\mathbb{F}_{2^8}$. Then various operations, including inversion, are performed in $\mathbb{F}_{2^8}$. The details of AES are too complicated to give here, but they are designed to be very fast when implemented in either software or hardware. The interested reader will find a full description in the official NIST publication FIPS PUB 197 [96] and in many other sources.

[18] All of the AES finalists (MARS, RC6, Rijndael, Serpent, Twofish) were believed to be secure, and none was clearly superior in all aspects. So the choice of Rijndael was based on its balance of flexibility, ease of implementation, and speed in both hardware and software.

List of Notation

$\mathbb{Z}$	the integers $\{\ldots,-4,-3,-2,-1,0,1,2,3,4,\ldots\}$, 10
$b \mid a$	b divides a (integers), 10
$b \nmid a$	b does not divide a (integers), 10
gcd	greatest common divisor, 11
$a \equiv b \pmod{m}$	a and b are congruent modulo m, 19
$\mathbb{Z}/m\mathbb{Z}$	the ring of integers modulo m, 21
$(\mathbb{Z}/m\mathbb{Z})^*$	the group of units in $\mathbb{Z}/m\mathbb{Z}$, 22
$\mathrm{ord}_p(a)$	order (or exponent) of p in a, 28
$a^{-1} \bmod p$	the multiplicative inverse of a modulo p, 28
$\mathbb{R}$	the field of real numbers, 29
$\mathbb{Q}$	the field of rational numbers, 29
$\mathbb{C}$	the field of complex numbers, 29
$\mathbb{F}_p$	the finite field $\mathbb{Z}/p\mathbb{Z}$, 29
$\mathcal{K}$	space of keys, 37
$\mathcal{M}$	space of messages (plaintexts), 37
$\mathcal{C}$	space of ciphertexts, 37
e or e_k	encryption function, 37
d or d_k	decryption function, 37
$\oplus$	exclusive or (XOR), 43
$\lfloor x \rfloor$	the greatest integer in x, 53
$\log_g(h)$	the discrete logarithm of h to the base g, 65
$\star$	composition operation in a group, 74
$\lvert G \rvert$	the order of the group G, 74
$\#G$	the order of the group G, 74
GL_n	the general linear group, 75
g^x	exponentiation of g in a group G, 75
$\mathcal{O}(g(x))$	big-$\mathcal{O}$ notation, 78
$\star$	multiplication in a ring, 95
$\mathbb{Z}[x]$	ring of polynomials with integer coefficients, 96
$b \mid a$	b divides a (in a ring), 96
$b \nmid a$	b does not divide a (in a ring), 96
$a \equiv b \pmod{m}$	a and b are congruent modulo m (in a ring), 97
R/mR	quotient ring of R by m, 98
$R/(m)$	quotient ring of R by m, 98
$R[x]$	ring of polynomials with coefficients in R, 98

J. Hoffstein et al., *An Introduction to Mathematical Cryptography*,
Undergraduate Texts in Mathematics, DOI 10.1007/978-1-4939-1711-2

deg	degree of a polynomial, 98
$\mathbb{F}_{p^d}$	a finite field with p^d elements, 106
$\mathrm{GF}(p^d)$	a field with p^d elements, 106
$\pi(X)$	number of primes between 2 and X, 133
$\zeta(s)$	Riemann zeta function, 135
$\psi(X,B)$	Number of B-smooth integers between 2 an X, 150
$o(g(x))$	little-o notation, 151
$L(X)$	the function $e^{\sqrt{(\ln X)(\ln\ln X)}}$, 151
$\Omega(g(x))$	big-Ω notation, 152
$\Theta(g(x))$	big-Θ notation, 152
$f(X) \ll g(X)$	alternative for $f(X) = \mathcal{O}(g(X))$, 152
$f(X) \gg g(X)$	alternative for $f(X) = \Omega(g(X))$, 152
$f(X) \gg\ll g(X)$	alternative for $f(X) = \Theta(g(X))$, 152
$\mathbb{Z}[\beta]$	the ring generated by the complex number β, 162
$L_\epsilon(X)$	the function $e^{(\ln X)^\epsilon (\ln\ln X)^{1-\epsilon}}$, 165
$\left(\frac{a}{p}\right)$	the Legendre symbol of a modulo p, 171
$\mathrm{Li}(X)$	the logarithmic integral function, 186
K^{Pri}	private signing key, 194
K^{Pub}	public verification key, 194
Sign	signing algorithm, 194
Verify	verification algorithm, 194
$\binom{n}{r}$	combinatorial symbol n choose r, 212
$\mathrm{IndCo}(\boldsymbol{s},\boldsymbol{t})$	index of coincidence of $\boldsymbol{s}$, 219
$\mathrm{MutIndCo}(\boldsymbol{s},\boldsymbol{t})$	mutual index of coincidence of $\boldsymbol{s}$ and $\boldsymbol{t}$, 221
Pr	a probability function, 229
$\Pr(F \mid E)$	conditional probability of F on E, 234
$f_X(x)$	probability density function of X, 239
$F_X(x)$	probability distribution function of X, 239
$f_{X,Y}(x,y)$	joint density function of X and Y, 241
$f_{X,Y}(x \mid y)$	conditional density function of X and Y, 241
$O_f^+(x)$	orbit of x under iteration of f, 254
$H(X)$	the entropy of the random variable X, 270
$\oplus$	addition on an elliptic curve, 302
$\mathcal{O}$	point at infinity on elliptic curve, 305
Δ_E	discriminant of the elliptic curve E, 306
$E(\mathbb{F}_p)$	points of elliptic curve with coordinates in $\mathbb{F}_p$, 308
$\log_P(Q)$	the elliptic discrete logarithm of Q with respect to P, 313
τ	the p-power Frobenius map $\mathbb{F}_{p^k} \to \mathbb{F}_{p^k}$, 334
τ	the p-power Frobenius map on an elliptic curve $E(\mathbb{F}_{p^k})$, 334
$E[m]$	points of order m on an elliptic curve E, 339
$\deg(D)$	degree of the divisor D, 341
$\mathrm{Sum}(D)$	sum of points in the divisor D, 341
e_m	the Weil pairing on an elliptic curve, 342
$\tau(P,Q)$	Tate pairing on an elliptic curve, 348
$\hat{\tau}(P,Q)$	modified Tate pairing on an elliptic curve, 348
$\hat{e}_\ell$	modified Weil pairing on an elliptic curve, 354
$\boldsymbol{a} \cdot \boldsymbol{b}$	dot product of $\boldsymbol{a}$ and $\boldsymbol{b}$, 385
$\lVert\boldsymbol{a}\rVert$	Euclidean norm of $\boldsymbol{a}$, 385

$\mathrm{GL}_n(\mathbb{Z})$	the special linear group (over $\mathbb{Z}$), 390
$\det(L)$	the determinant (covolume) of the lattice L, 392
γ_n	Hermite constant, 397
$\mathcal{H}(\mathcal{B})$	the Hadamard ratio of the basis $\mathcal{B}$, 397
$\mathbb{B}_R(\boldsymbol{a})$	closed ball of radius R centered at $\boldsymbol{a}$, 397
$\Gamma(s)$	the gamma function, 400
$\sigma(L)$	Gaussian expected shortest length of a vector in L, 402
R	the convolution polynomial ring $\mathbb{Z}[x]/(x^N - 1)$, 412
R_q	the convolution polynomial ring $(\mathbb{Z}/q\mathbb{Z})[x]/(x^N - 1)$, 412
$\boldsymbol{a} \star \boldsymbol{b}$	multiplication in convolution polynomial ring, 413
$\star$	convolution product of vectors, 414
$\mathcal{T}(d_1, d_2)$	ternary polynomial, 417
$L_{\boldsymbol{h}}^{\mathrm{NTRU}}$	the NTRU lattice associated to $\boldsymbol{h}(x)$, 425
$M_{\boldsymbol{h}}^{\mathrm{NTRU}}$	matrix for the NTRU lattice associated to $\boldsymbol{h}(x)$, 425
$\\|\boldsymbol{a}\\|_\infty$	sup norm of $\boldsymbol{a}$, 434
$\mathcal{B}^*$	Gram–Schmidt orthogonal basis associated to $\mathcal{B}$, 439
$W^\perp$	the orthogonal complement of W, 440
Hash	a hash function, 472
$\mathrm{Div}(C)$	group of divisors on a curve, 495
$\mathrm{Div}_0(C)$	group of divisors of degree 0 on a curve, 495
$\mathrm{Jac}_0(C)$	the Jacobian variety of the curve C, 495
$J(\mathbb{F}_p)$	the group of points modulo p on the Jacobian $\mathrm{Jac}_0(C)$, 495
$\|0\rangle$	ket notation in quantum mechanics, 497

References

[1] M. Agrawal, N. Kayal, N. Saxena, PRIMES is in P. Ann. Math. (2) **160**(2), 781–793 (2004)

[2] L.V. Ahlfors, *Complex Analysis: An Introduction to the Theory of Analytic Functions of One Complex Variable*. International Series in Pure and Applied Mathematics, 3rd edn. (McGraw-Hill, New York, 1978)

[3] M. Ajtai, The shortest vector problem in L2 is NP-hard for randomized reductions (extended abstract), in *STOC '98: Proceedings of the Thirtieth Annual ACM Symposium on Theory of Computing*, Dallas (ACM, New York, 1998), pp. 10–19

[4] M. Ajtai, C. Dwork, A public-key cryptosystem with worst-case/average-case equivalence, in *STOC '97*, El Paso (ACM, New York, 1999), pp. 284–293 (electronic)

[5] W.R. Alford, A. Granville, C. Pomerance, There are infinitely many Carmichael numbers. Ann. Math. (2) **139**(3), 703–722 (1994)

[6] ANSI-ECDSA, Public key cryptography for the financial services industry: the elliptic curve digital signature algorithm (ECDSA). ANSI Report X9.62, American National Standards Institute, 1998

[7] T.M. Apostol, *Introduction to Analytic Number Theory*. Undergraduate Texts in Mathematics (Springer, New York, 1976)

[8] L. Babai, On Lovász' lattice reduction and the nearest lattice point problem. Combinatorica **6**(1), 1–13 (1986)

[9] E. Bach, Explicit bounds for primality testing and related problems. Math. Comput. **55**(191), 355–380 (1990)

[10] E. Bach, J. Shallit, *Algorithmic Number Theory: Efficient Algorithms*. Foundations of Computing Series, vol. 1 (MIT, Cambridge, 1996).

[11] M. Bellare, Practice oriented provable-security, in *Proceedings of the First International Workshop on Information Security—ISW '97*, Tatsunokuchi. Volume of 1396 Lecture Notes in Computer Science (Springer, Berlin, 1998)

[12] M. Bellare, P. Rogaway, Random oracles are practical: a paradigm for designing efficient protocols, in *Proceedings of the First Annual Conference on Computer and Communications Security*, Fairfax, 1993, pp. 62–73

J. Hoffstein et al., *An Introduction to Mathematical Cryptography*, Undergraduate Texts in Mathematics, DOI 10.1007/978-1-4939-1711-2

[13] M. Bellare, P. Rogaway, Optimal asymmetric encryption, in *Advances in Cryptology—EUROCRYPT '94*, Perugia. Volume 950 of Lecture Notes in Computer Science (Springer, Berlin, 1995), pp. 92–111

[14] I.F. Blake, G. Seroussi, N.P. Smart, Elliptic Curves in Cryptography. Volume 265 of London Mathematical Society Lecture Note Series (Cambridge University Press, Cambridge, 2000)

[15] G. Blakley, Safeguarding cryptographic keys, in *Proceedings of AFIPS National Computer Conference*, Zurich, vol. 48, 1979, pp. 313–317

[16] D. Bleichenbacher, Chosen ciphertext attacks against protocols based on RSA encryption standard PKCS #1, in *Advances in Cryptology—CRYPTO 1998*, Santa Barbara. Volume 1462 of Lecture Notes in Computer Science (Springer, Berlin, 1998), pp. 1–12

[17] J. Blömer, A. May, Low secret exponent RSA revisited, in *Cryptography and Lattices*, Providence, 2001. Volume 2146 of Lecture Notes in Computer Science (Springer, Berlin, 2001), pp. 4–19

[18] D. Boneh, G. Durfee, Cryptanalysis of RSA with private key d less than $N^{0.292}$, in *Advances in Cryptology—EUROCRYPT '99*, Prague. Volume 1592 of Lecture Notes in Computer Science (Springer, Berlin, 1999), pp. 1–11

[19] D. Boneh, G. Durfee, Cryptanalysis of RSA with private key d less than $N^{0.292}$. IEEE Trans. Inf. Theory **46**(4), 1339–1349 (2000)

[20] D. Boneh, M. Franklin, Identity-based encryption from the Weil pairing, in *Advances in Cryptology—CRYPTO 2001*, Santa Barbara. Volume 2139 of Lecture Notes in Computer Science (Springer, Berlin, 2001), pp. 213–229

[21] D. Boneh, M. Franklin, Identity-based encryption from the Weil pairing. SIAM J. Comput. **32**(3), 586–615 (electronic) (2003)

[22] D. Boneh, R. Venkatesan, Breaking RSA may not be equivalent to factoring (extended abstract), in *Advances in Cryptology—EUROCRYPT '98*, Espoo. Volume 1403 of Lecture Notes in Computer Science (Springer, Berlin, 1998), pp. 59–71

[23] R.P. Brent, An improved Monte Carlo factorization algorithm. BIT **20**(2), 176–184 (1980)

[24] E.R. Canfield, P. Erdős, C. Pomerance, On a problem of Oppenheim concerning "factorisatio numerorum". J. Number Theory **17**(1), 1–28 (1983)

[25] J.W.S. Cassels, *Lectures on Elliptic Curves*. Volume 24 of London Mathematical Society Student Texts (Cambridge University Press, Cambridge, 1991)

[26] D. Chaum, Blind signatures for untraceable payments, in *Advances in Cryptology—CRYPTO '82*, Santa Barbara. Lecture Notes in Computer Science (Plenum Press, New York/London, 1983), pp. 199–203

[27] D. Chaum, A. Fiat, M. Naor, Untraceable electronic cash, in *Advances in Cryptology—CRYPTO 1988*, Santa Barbara. Volume 403 of Lecture Notes in Computer Science (Springer, 1988), pp. 319–327

[28] H. Cohen, *A Course in Computational Algebraic Number Theory*. Volume 138 of Graduate Texts in Mathematics (Springer, Berlin, 1993)

[29] H. Cohen, G. Frey, R. Avanzi, C. Doche, T. Lange, K. Nguyen, F. Vercauteren (eds.), *Handbook of Elliptic and Hyperelliptic Curve Cryptography.* Discrete Mathematics and Its Applications (Boca Raton) (Chapman & Hall/CRC, Boca Raton, 2006)

[30] S.A. Cook, The complexity of theorem-proving procedures, in *STOC '71: Proceedings of the Third Annual ACM Symposium on Theory of Computing*, Shaker Heights (ACM, New York, 1971), pp. 151–158

[31] D. Coppersmith, Solving homogeneous linear equations over GF(2) via block Wiedemann algorithm. Math. Comput. **62**(205), 333–350 (1994)

[32] D. Coppersmith, Small solutions to polynomial equations, and low exponent RSA vulnerabilities. J. Cryptol. **10**(4), 233–260 (1997)

[33] D. Coppersmith, Finding small solutions to small degree polynomials, in *Cryptography and Lattices*, Providence, 2001. Volume 2146 of Lecture Notes in Computer Science (Springer, Berlin, 2001), pp. 20–31

[34] R. Crandall, C. Pomerance, *Prime Numbers* (Springer, New York, 2001)

[35] H. Davenport, *The Higher Arithmetic* (Cambridge University Press, Cambridge, 1999)

[36] M. Dietzfelbinger, *Primality Testing in Polynomial Time: From Randomized Algorithms to "PRIMES is in P"*. Volume 3000 of Lecture Notes in Computer Science (Springer, Berlin, 2004)

[37] W. Diffie, The first ten years of public key cryptology, G.J. Simmons (ed.), in *Contemporary Cryptology* (IEEE, New York, 1992), pp. 135–175

[38] W. Diffie, M.E. Hellman, New directions in cryptography. IEEE Trans. Inf. Theory IT-**22**(6), 644–654 (1976)

[39] L. Ducas, P.Q. Nguyen, Learning a zonotope and more: cryptanalysis of NTRUSign countermeasures, in *Advances in Cryptology—ASIACRYPT 2012*, Beijing. Volume 7658 of Lecture Notes in Computer Science (Springer, Berlin, 2012), pp. 433–450

[40] D.S. Dummit, R.M. Foote, *Abstract Algebra*, 3rd edn. (Wiley, Hoboken, 2004)

[41] T. ElGamal, A public key cryptosystem and a signature scheme based on discrete logarithms. IEEE Trans. Inf. Theory **31**(4), 469–472 (1985)

[42] J. Ellis, The story of non-secret encryption, 1987 (released by CSEG in 1997). `https://cryptocellar.web.cern.ch/cryptocellar/cesg/ellis.pdf`

[43] W. Fleming, *Functions of Several Variables.* Undergraduate Texts in Mathematics, 2nd edn. (Springer, New York, 1977)

[44] M. Fouquet, P. Gaudry, R. Harley, An extension of Satoh's algorithm and its implementation. J. Ramanujan Math. Soc. **15**(4), 281–318 (2000)

[45] J. Fraleigh, *A First Course in Abstract Algebra*, 7th edn. (Addison Welsley, Boston/London, 2002)

[46] M.R. Garey, D.S. Johnson, *Computers and Intractability: A Guide to the Theory of NP-Completeness.* A Series of Books in the Mathematical Sciences (W. H. Freeman, San Francisco, 1979)

[47] C. Gentry, *A Fully Homomorphic Encryption Scheme*, PhD thesis, Stanford University, 2009. `crypto.stanford.edu/craig`

[48] C. Gentry, Fully homomorphic encryption using ideal lattices, in *STOC'09—Proceedings of the 2009 ACM International Symposium on Theory of Computing*, Bethesda (ACM, New York, 2009), pp. 169–178

[49] O. Goldreich, S. Goldwasser, S. Halevi, Public-key cryptosystems from lattice reduction problems, in *Advances in Cryptology—CRYPTO '97*, Santa Barbara, 1997. Volume 1294 of Lecture Notes in Computer Science (Springer, Berlin, 1997), pp. 112–131

[50] O. Goldreich, D. Micciancio, S. Safra, J.-P. Seifert, Approximating shortest lattice vectors is not harder than approximating closest lattice vectors. Inf. Process. Lett. **71**(2), 55–61 (1999)

[51] G.R. Grimmett, D.R. Stirzaker, *Probability and Random Processes*, 3rd edn. (Oxford University Press, New York, 2001)

[52] G.H. Hardy, E.M. Wright, *An Introduction to the Theory of Numbers*, 5th edn. (The Clarendon Press/Oxford University Press, New York, 1979)

[53] I.N. Herstein, *Topics in Algebra*, 2nd edn. (Xerox College Publishing, Lexington, 1975)

[54] J. Hoffstein, J. Pipher, J.H. Silverman, NTRU: a ring-based public key cryptosystem, in *Algorithmic Number Theory*, Portland, 1998. Volume 1423 of Lecture Notes in Computer Science (Springer, Berlin, 1998), pp. 267–288

[55] J. Hoffstein, N. Howgrave-Graham, J. Pipher, J.H. Silverman, W. Whyte, NTRUSign: digital signatures using the NTRU lattice, in *Topics in Cryptology—CT-RSA 2003*. Volume 2612 of Lecture Notes in Computer Science (Springer, Berlin, 2003), pp. 122–140. `https://www.securityinnovation.com/uploads/Crypto/NTRUSign-preV2.pdf`

[56] J. Hoffstein, N. Howgrave-Graham, J. Pipher, J.H. Silverman, W. Whyte, Performance improvements and a baseline parameter generation algorithm for NTRUSign. Presented at Workshop on Mathematical Problems and Techniques in Cryptology, Barcelona, 2005. `https://www.securityinnovation.com/uploads/Crypto/NTRUSignParams-2005-08.pdf`

[57] J. Hoffstein, J. Pipher, J. Schanck, J. Silverman, W. Whyte, Transcript secure signatures based on modular lattices. Cryptology ePrint archive, report 2014/457, 2014. `http://eprint.iacr.org/2014/457`

[58] N. Howgrave-Graham, Approximate integer common divisors, in *Cryptography and Lattices*, Providence, 2001. Volume 2146 of Lecture Notes in Computer Science (Springer, Berlin, 2001), pp. 51–66

[59] K. Ireland, M. Rosen, *A Classical Introduction to Modern Number Theory*. Volume 84 of Graduate Texts in Mathematics (Springer, New York, 1990)

[60] E.T. Jaynes, Information theory and statistical mechanics. Phys. Rev. (2) **106**, 620–630 (1957)

[61] A. Joux, A one round protocol for tripartite Diffie-Hellman, in *Algorithmic Number Theory*, Leiden, 2000. Volume 1838 of Lecture Notes in Computer Science (Springer, Berlin, 2000), pp. 385–393

[62] A. Joux, A one round protocol for tripartite Diffie-Hellman. J. Cryptol. **17**(4), 263–276 (2004)

[63] D. Kahn, *The Codebreakers: The Story of Secret Writing* (Scribner Book, New York, 1996)

[64] P. Kaye, R. Laflamme, M. Mosca, *An Introduction to Quantum Computing* (Oxford University Press, Oxford, 2007)

[65] A.W. Knapp, *Elliptic Curves.* Volume 40 of Mathematical Notes (Princeton University Press, Princeton, 1992)

[66] D. Knuth, *The Art of Computer Programming, Vol. 2: Seminumerical Algorithms*, 2nd edn. (Addison-Wesley, Reading, 1981)

[67] N. Koblitz, Elliptic curve cryptosystems. Math. Comput. **48**(177), 203–209 (1987)

[68] N. Koblitz, *Algebraic Aspects of Cryptography.* Volume 3 of Algorithms and Computation in Mathematics (Springer, Berlin, 1998)

[69] N. Koblitz, The uneasy relationship between mathematics and cryptography. Not. Am. Math. Soc. **54**, 972–979 (2007)

[70] N. Koblitz, A.J. Menezes, Another look at "provable security". J. Cryptol. **20**(1), 3–37 (2007)

[71] J.C. Lagarias, H.W. Lenstra Jr., C.-P. Schnorr, Korkin–Zolotarev bases and successive minima of a lattice and its reciprocal lattice. Combinatorica **10**(4), 333–348 (1990)

[72] B.A. LaMacchia, A.M. Odlyzko, Solving large sparse linear systems over finite fields, in *Advances in Cryptology—CRYPTO '90*, Santa Barbara, 1990. Lecture Notes in Computer Science (Springer, Berlin, 1990)

[73] S. Lang, *Elliptic Curves: Diophantine Analysis.* Volume 231 of Grundlehren der Mathematischen Wissenschaften (Fundamental Principles of Mathematical Sciences) (Springer, Berlin, 1978)

[74] S. Lang, *Elliptic Functions.* Volume 112 of Graduate Texts in Mathematics, 2nd edn. (Springer, New York, 1987). With an appendix by J. Tate

[75] H.W. Lenstra Jr., Factoring integers with elliptic curves. Ann. Math. (2) **126**(3), 649–673 (1987)

[76] H.W. Lenstra jr., C. Pomerance, Primality testing with Gaussian periods (2011). `https://www.math.dartmouth.edu/~carlp/PDF/complexity12.pdf`

[77] A.K. Lenstra, H.W. Lenstra Jr., L. Lovász, Factoring polynomials with rational coefficients. Math. Ann. **261**(4), 515–534 (1982)

[78] V. Lyubashevsky, Lattice-based identification schemes secure under active attacks, in *Public Key Cryptography—PKC 2008*, Barcelona. Volume 4939 of Lecture Notes in Computer Science (Springer, Berlin, 2008), pp. 162–179

[79] V. Lyubashevsky, Fiat-Shamir with aborts: applications to lattice and factoring-based signatures, in *Advances in Cryptology—ASIACRYPT 2009*, Tokyo. Volume 5912 of Lecture Notes in Computer Science (Springer, Berlin, 2009), pp. 598–616

[80] V. Lyubashevsky, Lattice signatures without trapdoors, in *Advances in Cryptology—EUROCRYPT 2012*, Cambridge. Volume 7237 of Lecture Notes in Computer Science (Springer, Heidelberg, 2012), pp. 738–755

[81] A. Menezes, *Elliptic Curve Public Key Cryptosystems*. The Kluwer International Series in Engineering and Computer Science, 234 (Kluwer Academic, Boston, 1993)

[82] A.J. Menezes, T. Okamoto, S.A. Vanstone, Reducing elliptic curve logarithms to logarithms in a finite field. IEEE Trans. Inf. Theory **39**(5), 1639–1646 (1993)

[83] R.C. Merkle, Secure communications over insecure channels, in *Secure Communications and Asymmetric Cryptosystems*, ed. by G.J. Simmons. Volume 69 of AAAS Selected Symposium Series (Westview, Boulder, 1982), pp. 181–196

[84] R.C. Merkle, M.E. Hellman, Hiding information and signatures in trapdoor knapsacks, in *Secure Communications and Asymmetric Cryptosystems*, ed. by G.J. Simmons. Volume 69 of AAAS Selected Symposium Series (Westview, Boulder, 1982), pp. 197–215

[85] D. Micciancio, Improving lattice based cryptosystems using the Hermite normal form, in *Cryptography and Lattices*, Providence, 2001. Volume 2146 of Lecture Notes in Computer Science (Springer, Berlin, 2001), pp. 126–145

[86] D. Micciancio, S. Goldwasser, *Complexity of Lattice Problems: A Cryptographic Perspective*. The Kluwer International Series in Engineering and Computer Science, 671 (Kluwer Academic, Boston, 2002)

[87] G.L. Miller, Riemann's hypothesis and tests for primality. J. Comput. Syst. Sci. **13**(3), 300–317 (1976). Working papers presented at the ACM-SIGACT Symposium on the Theory of Computing, Albuquerque, 1975

[88] V.S. Miller, Use of elliptic curves in cryptography, in *Advances in Cryptology—CRYPTO '85*, Santa Barbara, 1985. Volume 218 of Lecture Notes in Computer Science (Springer, Berlin, 1986), pp. 417–426

[89] V.S. Miller, The Weil pairing, and its efficient calculation. J. Cryptol. **17**(4), 235–261 (2004). Updated and expanded version of unpublished manuscript *Short programs for functions on curves*, 1986

[90] P.L. Montgomery, Speeding the Pollard and elliptic curve methods of factorization. Math. Comput. **48**(177), 243–264 (1987)

[91] S.p. Nakamoto, Bitcoin: a peer-to-peer electronic cash system (2009). `https://bitcoin.org/bitcoin.pdf`

[92] P. Nguyen, Cryptanalysis of the Goldreich–Goldwasser–Halevi cryptosystem from crypto'97, in *Advances in Cryptology—CRYPTO '99*, Santa Barbara, 1999. Volume 1666 of Lecture Notes in Computer Science (Springer, Berlin, 1999), pp. 288–304

[93] P. Nguyen, O. Regev, Learning a parallelepiped: cryptanalysis of GGH and NTRU signatures, in *Advances in Cryptology—EUROCRYPT '06*, St. Petersburg. Volume 4004 of Lecture Notes in Computer Science (Springer, Berlin, 2006)

[94] P.Q. Nguyen, O. Regev, Learning a parallelepiped: cryptanalysis of GGH and NTRU signatures. J. Cryptol. **22**(2), 139–160 (2009)

[95] P. Nguyen, J. Stern, Cryptanalysis of the Ajtai-Dwork cryptosystem, in *Advances in Cryptology—CRYPTO '98*, Santa Barbara, 1998. Volume 1462 of Lecture Notes in Computer Science (Springer, Berlin, 1998), pp. 223–242

[96] NIST–AES, Advanced Encryption Standard (AES). FIPS Publication 197, National Institue of Standards and Technology, 2001. `http://csrc.nist.gov/publications/fips/fips197/fips-197.pdf`

[97] NIST–DES, Data Encryption Standard (DES). FIPS Publication 46-3, National Institue of Standards and Technology, 1999. `http://csrc.nist.gov/publications/fips/fips46-3/fips46-3.pdf`

[98] NIST–DSS, Digital Signature Standard (DSS). FIPS Publication 186-2, National Institue of Standards and Technology, 2004. `http://csrc.nist.gov/publications/fips/fips180-2/fips180-2withchangenotice.pdf`

[99] NIST–SHS, Secure Hash Standard (SHS). FIPS Publication 180-2, National Institue of Standards and Technology, 2003. `http://csrc.nist.gov/publications/fips/fips180-2/fips180-2.pdf`

[100] I. Niven, H.S. Zuckerman, H.L. Montgomery, *An Introduction to the Theory of Numbers* (Wiley, New York, 1991)

[101] NTRU Cryptosystems, A meet-in-the-middle attack on an NTRU private key. Technical report, 1997, updated 2003. Tech. Note 004, `https://www.securityinnovation.com/uploads/Crypto/NTRUTech004v2.pdf`

[102] NTRU Cryptosystems, Estimated breaking times for NTRU lattices. Technical report, 1999, updated 2003. Tech. Note 012, `https://www.securityinnovation.com/uploads/Crypto/NTRUTech012v2.pdf`

[103] A.M. Odlyzko, The rise and fall of knapsack cryptosystems, in *Cryptology and Computational Number Theory*, Boulder, 1989. Volume 42 of Proceedings of Symposia in Applied Mathematics (American Mathematical Society, Providence, 1990), pp. 75–88

[104] J.M. Pollard, Monte Carlo methods for index computation (mod p). Math. Comput. **32**(143), 918–924 (1978)

[105] C. Pomerance, A tale of two sieves. Not. Am. Math. Soc. **43**(12), 1473–1485 (1996)

[106] E.L. Post, A variant of a recursively unsolvable problem. Bull. Am. Math. Soc. **52**, 264–268 (1946)

[107] J. Proos, C. Zalka, Shor's discrete logarithm quantum algorithm for elliptic curves. Quantum Inf. Comput. **3**(4), 317–344 (2003)

[108] M.O. Rabin, Digitized signatures and public-key functions as intractible as factorization. Technical report, MIT Laboratory for Computer Science, 1979. Technical Report LCS/TR-212

[109] H. Riesel, *Prime Numbers and Computer Methods for Factorization*. Volume 126 of Progress in Mathematics (Birkhäuser, Boston, 1994)

[110] R.L. Rivest, A. Shamir, L. Adleman, A method for obtaining digital signatures and public-key cryptosystems. Commun. ACM **21**(2), 120–126 (1978)

[111] K.H. Rosen, *Elementary Number Theory and Its Applications*, 4th edn. (Addison-Wesley, Reading, 2000)

[112] S. Ross, *A First Course in Probability*, 9th edn. (Pearson, England, 2001)

[113] T. Satoh, The canonical lift of an ordinary elliptic curve over a finite field and its point counting. J. Ramanujan Math. Soc. **15**(4), 247–270 (2000)

[114] T. Satoh, K. Araki, Fermat quotients and the polynomial time discrete log algorithm for anomalous elliptic curves. Comment. Math. Univ. St. Paul. **47**(1), 81–92 (1998)

[115] C.-P. Schnorr, A hierarchy of polynomial time lattice basis reduction algorithms. Theor. Comput. Sci. **53**(2–3), 201–224 (1987)

[116] C.P. Schnorr, Fast LLL-type lattice reduction. Inf. Comput. **204**(1), 1–25 (2006)

[117] C.-P. Schnorr, M. Euchner, Lattice basis reduction: improved practical algorithms and solving subset sum problems, in *Fundamentals of Computation Theory*, Gosen, 1991. Volume 529 of Lecture Notes in Computer Science (Springer, Berlin, 1991), pp. 68–85

[118] C.-P. Schnorr, M. Euchner, Lattice basis reduction: improved practical algorithms and solving subset sum problems. Math. Program. **66**(2, Ser. A), 181–199 (1994)

[119] C.P. Schnorr, H.H. Hörner, Attacking the Chor–Rivest cryptosystem by improved lattice reduction, in *Advances in Cryptology—EUROCRYPT '95*, Saint-Malo, 1995. Volume 921 of Lecture Notes in Computer Science (Springer, Berlin, 1995), pp. 1–12

[120] R. Schoof, Elliptic curves over finite fields and the computation of square roots mod p. Math. Comput. **44**(170), 483–494 (1985)

[121] R. Schoof, Counting points on elliptic curves over finite fields. J. Théor. Nombres Bordx. **7**(1), 219–254 (1995). Les Dix-huitièmes Journées Arithmétiques, Bordeaux, 1993

[122] I.A. Semaev, Evaluation of discrete logarithms in a group of p-torsion points of an elliptic curve in characteristic p. Math. Comput. **67**(221), 353–356 (1998)

[123] A. Shamir, How to share a secret. Commun. ACM **22**(11), 612–613 (1979)

[124] A. Shamir, A polynomial-time algorithm for breaking the basic Merkle-Hellman cryptosystem. IEEE Trans. Inf. Theory **30**(5), 699–704 (1984)

[125] A. Shamir, Identity-based cryptosystems and signature schemes, in *Advances in Cryptology*, Santa Barbara, 1984. Volume 196 of Lecture Notes in Computer Science (Springer, Berlin, 1985), pp. 47–53

[126] C.E. Shannon, A mathematical theory of communication. Bell Syst. Tech. J. **27**, 379–423, 623–656 (1948)

[127] C.E. Shannon, Communication theory of secrecy systems. Bell Syst. Tech. J. **28**, 656–715 (1949)

[128] P.W. Shor, Algorithms for quantum computation: discrete logarithms and factoring, in *35th Annual Symposium on Foundations of Computer Science*, Santa Fe, 1994 (IEEE Computer Society, Los Alamitos, 1994), pp. 124–134

[129] P.W. Shor, Polynomial-time algorithms for prime factorization and discrete logarithms on a quantum computer. SIAM J. Comput. **26**(5), 1484–1509 (1997)

[130] V. Shoup, Lower bounds for discrete logarithms and related problems, in *Advances in Cryptology—EUROCRYPT '97*, Konstanz. Volume 1233 of Lecture Notes in Computer Science (Springer, Berlin, 1997), pp. 256–266

[131] V. Shoup, OAEP reconsidered, in *Advances in Cryptology—CRYPTO 2001*, Santa Barbara. Volume 2139 of Lecture Notes in Computer Science (Springer, Berlin, 2001), pp. 239–259

[132] V. Shoup, *A Computational Introduction to Number Theory and Algebra* (Cambridge University Press, 2005). `http://shoup.net/ntb/ntb-b5.pdf`

[133] C.L. Siegel, A mean value theorem in geometry of numbers. Ann. Math. (2) **46**, 340–347 (1945)

[134] J.H. Silverman, *Advanced Topics in the Arithmetic of Elliptic Curves.* Volume 151 of Graduate Texts in Mathematics (Springer, New York, 1994)

[135] J.H. Silverman, Elliptic curves and cryptography, in *Public-Key Cryptography*, Les Diablerets. Volume 62 of Proceedings of Symposia in Applied Mathematics (American Mathematical Society, Providence, 2005), pp. 91–112

[136] J.H. Silverman, *The Arithmetic of Elliptic Curves.* Volume 106 of Graduate Texts in Mathematics, 2nd edn. (Springer, Dordrecht, 2009)

[137] J.H. Silverman, *A Friendly Introduction to Number Theory*, 4th edn. (Pearson, Upper Saddle River, 2013)

[138] J.H. Silverman, J. Tate, *Rational Points on Elliptic Curves.* Undergraduate Texts in Mathematics (Springer, New York, 1992)

[139] S. Singh, *The Code Book: The Science of Secrecy from Ancient Egypt to Quantum Cryptography* Reprint edn. (Anchor, New York, 2000)

[140] B. Skjernaa, Satoh's algorithm in characteristic 2. Math. Comput. **72**(241), 477–487 (electronic) (2003)

[141] N.P. Smart, The discrete logarithm problem on elliptic curves of trace one. J. Cryptol. **12**(3), 193–196 (1999)

[142] Standards for Efficient Cryptography, SEC 2: recommended elliptic curve domain parameters (Version 1), 20 Sept 2000. `http://www.secg.org/collateral/sec2_final.pdf`

[143] J. Talbot, D. Welsh, *Complexity and Cryptography: An Introduction* (Cambridge University Press, Cambridge, 2006)

[144] E. Teske, Speeding up Pollard's rho method for computing discrete logarithms, in *Algorithmic Number Theory*, Portland, 1998. Volume 1423 of Lecture Notes in Computer Science (Springer, Berlin, 1998), pp. 541–554

[145] E. Teske, Square-root algorithms for the discrete logarithm problem (a survey), in *Public-Key Cryptography and Computational Number Theory*, Warsaw, 2000 (de Gruyter, Berlin, 2001), pp. 283–301

[146] J. Von Neumann, Various techniques used in connection with random digits. Natl. Bur. Stand. Appl. Math. Ser. **12**(36–38), 1 (1951). Reprinted in von Neumann's Collected Works, 5 (1963), Pergamon Press, pp. 768–770. `https://dornsifecms.usc.edu/assets/sites/520/docs/VonNeumann-ams12p36-38.pdf`

[147] L.C. Washington, *Elliptic Curves: Number Theory and Cryptography.* Discrete Mathematics and Its Applications (Chapman & Hall/CRC, Boca Raton, 2003)

[148] A.E. Western, J.C.P. Miller, *Tables of Indices and Primitive Roots.* Royal Society Mathematical Tables, vol. 9 (Published for the Royal Society at the Cambridge University Press, London, 1968)

[149] M.J. Wiener, Cryptanalysis of short RSA secret exponents. IEEE Trans. Inf. Theory **36**(3), 553–558 (1990)

[150] S.Y. Yan, *Primality Testing and Integer Factorization in Public-Key Cryptography.* Volume 11 of Advances in Information Security (Kluwer Academic, Boston, 2004)

Index

J. Hoffstein et al., *An Introduction to Mathematical Cryptography*, Undergraduate Texts in Mathematics, DOI 10.1007/978-1-4939-1711-2

Printed in the United States
By Bookmasters